Nutrition Biophysics

Thomas A. Vilgis

Nutrition Biophysics
An Introduction for Students, Professionals and Career Changers

Thomas A. Vilgis
Max-Planck-Institut für Polymerforschung
Mainz, Rheinland-Pfalz, Germany

Foreword by
Hans Konrad Biesalski
Fg. Ernährungswissenschaft
Universität Hohenheim
Stuttgart, Baden-Württemberg, Germany

ISBN 978-3-662-67596-0   ISBN 978-3-662-67597-7 (eBook)
https://doi.org/10.1007/978-3-662-67597-7

Translation from the German language edition: "Biophysik der Ernährung" by Thomas A. Vilgis and Hans Konrad Biesalski, © Springer-Verlag GmbH Deutschland, ein Teil von Springer Nature 2022. Published by Springer Berlin Heidelberg. All Rights Reserved.

This book is a translation of the original German edition "Biophysik der Ernährung" by Vilgis, Thomas A., published by Springer-Verlag GmbH, DE in 2022. The translation was done with the help of an artificial intelligence machine translation tool. A subsequent human revision was done primarily in terms of content, so that the book will read stylistically differently from a conventional translation. Springer Nature works continuously to further the development of tools for the production of books and on the related technologies to support the authors.

© The Editor(s) (if applicable) and The Author(s), under exclusive license to Springer-Verlag GmbH, DE, part of Springer Nature 2023

This work is subject to copyright. All rights are solely and exclusively licensed by the Publisher, whether the whole or part of the material is concerned, specifically the rights of translation, reprinting, reuse of illustrations, recitation, broadcasting, reproduction on microfilms or in any other physical way, and transmission or information storage and retrieval, electronic adaptation, computer software, or by similar or dissimilar methodology now known or hereafter developed.
The use of general descriptive names, registered names, trademarks, service marks, etc. in this publication does not imply, even in the absence of a specific statement, that such names are exempt from the relevant protective laws and regulations and therefore free for general use.
The publisher, the authors, and the editors are safe to assume that the advice and information in this book are believed to be true and accurate at the date of publication. Neither the publisher nor the authors or the editors give a warranty, expressed or implied, with respect to the material contained herein or for any errors or omissions that may have been made. The publisher remains neutral with regard to jurisdictional claims in published maps and institutional affiliations.

This Springer imprint is published by the registered company Springer-Verlag GmbH, DE, part of Springer Nature.
The registered company address is: Heidelberger Platz 3, 14197 Berlin, Germany

Foreword

Biophysics of nutrition - what should I imagine under this? The internet provides little information on these two terms, only a paper from 1990 by A. Grünert, a clinician who mainly dealt with issues of parenteral nutrition. Grünert published an article titled "Biophysical Basics of Nutrient Supply" in the *Journal of Clinical Anesthesia and Intensive Care*. He points out that knowledge of the biophysics of nutrition is particularly important in clinical situations when it comes to assessing the nutrition of those affected. However, that was already it, what the search for biophysics and nutrition has yielded. The questions arise: Is the field as such simply uninteresting or is there no need or do we not recognize the importance of the biophysics of nutrition because we know too little about it? Thomas Vilgis addresses exactly these aspects with his book: He shows how exciting this field is, how much we need this knowledge for research, but also everyday life, and we are amazed at how little we knew about it so far.

Biophysics, according to the Lexicon of Biology, "also referred to as biological physics, is a relatively young scientific field at the border between physics, chemistry, and biology. The subject of research is physical and physico-chemical phenomena in biological systems (life). The goal is to elucidate fundamental processes that form the basis of life using physical methods and within the framework of physical ideas."

To achieve this, it is necessary, first and foremost, to have a deep understanding of physics, as well as a differentiated view of food and nutrition. Thomas Vilgis combines both of these aspects in an outstanding way. As a physicist at the Max Planck Institute for Polymer Research in Mainz, he deals with fundamental questions of the physics and chemistry of soft matter. In doing so, he has increasingly come into contact with food production and its biophysical interactions, and has thus begun to examine food in both simple and complex forms for its biophysical properties. The results of this investigation have been documented in a multitude of popular science books and non-fiction books, and there is hardly any area that he has not examined through the lens of physical-chemical analysis and presented to the reader in an understandable form. This ranges from the physics and chemistry of fine taste, a cookbook of molecular cuisine, experiments for young chefs, to cooking techniques, seasoning, and fermenting, to name just a few of the main focuses. Now, a *Biophysics of Nutrition* serves as a foundation and a meaningful extension for understanding biophysical processes that not only concern our food but also, and especially, the utilization of this food in the organism. Why can a *Biophysics of Nutrition* contribute to a better understanding? For example, a major problem in predominantly poor countries is that the main component of their diet is grain, which contains important micronutrients such as iron and zinc, but these can only be absorbed from grain with great difficulty. Knowledge of biophysical processes that can improve this absorption could significantly contribute to reducing the global deficiency of iron and zinc. The fact that we can hardly absorb the yellow pigment beta-carotene from raw carrots is due to it being packaged in cellulose, which we cannot digest. Only juicing or heating releases the beta-carotene from the cellulose and allows for absorption. This is nothing more than a biophysical process. The situation is similar with fat-soluble compounds, which include fat-soluble vitamins as well as beta-carotene, when we consume them simultaneously with high amounts of dietary fiber. These fibers adsorb, also a biophysical process, the fat-soluble compounds, which are then absorbed in smaller quantities and essentially excreted. Connections like these have contributed to the development of food preparation methods such as cooking, fermenting, or frying playing an important role in human evolution, as they ensured that essential micronutrients were absorbed in sufficient quantities. Vilgis discusses this using various examples, thus also demonstrating the great importance of food processing long before industrial production for humans. *Nutrition Biophysics* is by no means a basic book on food production, but rather a book that expands our understanding of the complex interactions of food and the associated availability of macro- and micronutrients in an area that has been far too little considered so far. The author does not limit himself to a dry description, but, and this is one of his essential qualities, he tells complex content in an equally stimulating way as he critically analyzes and humorously dissects nutritional myths. Again and again, with a biophysical approach, especially when dealing with the development of alternative dietary forms and the newly emerging foods, their benefits, and risks associated with them. It would not be a book by Thomas Vilgis if the question of enjoyment were not also addressed. If you follow the many entries by Vilgis on the Internet, you will repeatedly come across recipes that focus on the actual aspect of a healthy diet, enjoyment. Here, not only is the food considered as the basis of enjoyment, but also "pleasure-hostile" recommendations up to the future of our diet with the concluding chapter on the pleasure, burden, and responsibility of enjoyment.

With *Nutrition Biophysics*, Vilgis presents a book that fills a gap that has not been perceived in this breadth before. It is aimed at nutritionists, technologists, food chemists, and users from gourmet cuisine to community catering. It is also aimed at consumers who simply want to know more about their food and what they should consider when it comes to ensuring a balanced, healthy diet.

I would like to wish this book the dissemination it deserves, in the hope that biophysics of nutrition, which has so far only been mentioned marginally in nutritional sciences, will increasingly find its way into this course of study and thus become part of our nutritional knowledge.

April 2020   Prof. (em.) Dr. H. K. Biesalski

Preface

Why does a physicist, who admittedly enjoys eating and cooking, write a book that primarily deals with the topic of nutrition? Aren't there already too many books on this subject? True, but a book like this one does not yet exist. A book that approaches the topic in a completely different way, that dedicates itself to the subject from the perspective of the basic subjects of physics, chemistry, and biology. The question of whether these subjects are relevant to the topic of nutrition at all is justified when considering the many and often incomprehensible discussions on nutritional issues. Upon closer inspection, it quickly becomes apparent how weak the foundation is for some nutritional rules or recommendations.

The basic realization that every food is nothing more than a complex collection of structured large and small molecules makes the entry easier. If it is then recognized that these food molecules encounter biological systems in humans and must also submit to the elementary laws and interactions of physics, chemistry, and biology, a biophysical-biochemical approach to the topic of nutrition practically emerges by itself. This is also a good thing, because unfortunately many discussions are ideologically driven, and the style of nutrition seems to increasingly serve as a substitute religion. In the thicket of many nutritional promises, in the jungle of opinions, it is certainly appropriate to gain a little more grounding.

In fact, the evolutionary history of life and humans is the best path to knowledge. People 100, 1,000, 10,000, 100,000, or even millions of years ago knew nothing about cholesterol, gluten, saturated or even essential omega-3 fatty acids. However, they learned what they could and had to eat in order to survive at all. This simple principle worked well until after World War II. Even the boundless gluttony in various cultures and periods of the Middle Ages, the human species coped well. As well as great famines, and the consumption of raw, fermented, and heavily roasted food over fire. The *Homo sapiens* developed splendidly under the rules of evolution and the associated selection. Thoughts were free from acrylamide or dioxin, although the Seveso poison had existed on Earth since time immemorial. Today, "free from gluten", "free from egg", "free from animal", "free from lactose" is proclaimed as healthy and preventive, even if the reasons for this are obscure and defy logical conclusions.

Where does this book lead? In essence, it shows ways to relatively carefree enjoyment. Furthermore, it is an attempt to sober-mindedly and objectively, but also clearly refute or classify some opinions. Sometimes this is very easy, sometimes a bit more difficult. For most phenomena, we have to go back to the root. As with incorrectly buttoned shirts. This fundamental error cannot be corrected by simply "rebuttoning" two buttons; the entire shirt must first be unbuttoned and then – hopefully correctly – buttoned up again.

Whenever one does not know what to do, it is advisable to look at the molecular scale. There are plenty of reasons for this: Years of nutrition education through media and guides have led us to talk about bad LDL and good HDL cholesterol today, without questioning what really lies behind "High/Low Density Lipoprotein". The same applies to the value of free triglycerides or triacylglycerols, which appears on every laboratory report during a routine check-up at the family doctor. Many people talk about gluten without really knowing what it is or even what tasks it has in nature. At the same time, many wonder why gluten-free baking is difficult and poses a certain challenge. A look at the gluten molecules answers both questions, and at the same time, one gets an idea of what exactly the nutritional value of the molecule is. The molecular, biophysical view thus solves a whole series of questions in one fell swoop.

For a fundamental understanding, rigorous "downscaling" is necessary. Scaling down a food to ever smaller scales until we "see" its microstructure, its proteins, fats, carbohydrates, the phenols and polyphenols, the gluten and the saturated fatty acids that are being talked about. Once this is achieved, we must strive to recognize their structures and biological functions. When focusing on the molecular structure, a perspective suddenly opens up that becomes more objective, describable with the basic laws of hard sciences - physics, chemistry, and biology. Food, cooking, eating, and even digestion suddenly become understandable, and "universal connections" emerge that primarily have to do with the physics, chemistry, and biology (in this order, which has accompanied us since the origin of this universe with the Big Bang) of soft matter and have nothing to do with opinion. Only after molecular biology does the science of nutrition physiology come into play. The insights gained through the molecular perspective are reassuring. Therefore, this book opens up an idea of "nutrition physics".

This book is aimed at curious readers who no longer want to be confused when it comes to food. And also at those readers who are willing to embark on an often hilly, scientifically driven journey into the molecular world of food. But whoever has managed to do this will never have to be fooled again, but can decide for themselves what they want to know. Or what they can believe.

It is also aimed at students of nutritional sciences and ecotrophology, as well as nutrition physicians who want to understand the basic scientific and nutrition-related principles. It can also be a guide for pleasure-oriented students of physics, chemistry, and biology to look beyond the horizon of their own field of study. Therefore, the book is also interdisciplinary in many places: Whenever appropriate, connections to other scientific disciplines are established. Not only within engineering sciences, such as food technology, but also to cultural studies, sociology, and psychology. Such interdisciplinary excursions are only possible with the topic of food and nutrition.

Preface

The conclusion of the book is basically very simple and clear: If nutrition on the molecular scale gets out of control for a long time, things will eventually go wrong for the whole person.

I would like to thank Professor Hans Konrad Biesalski for many discussions on questions related to nutrition problems. His unbiased approach has shaped many of my views. His willingness to provide a foreword for this book honors me in a very special way. I would like to thank Professor Nicolai Worm for his invaluable, helpful information service and his clear view of many nutrition misconceptions. I would like to thank Springer-Verlag for their trust and courage to print this book at all, especially Ms. Stephanie Preuss. I would like to thank Ms. Carola Lerch for her help and support during the editing process. The production team around Ms. Roopashree Polepalli deserves great thanks for the excellent implementation in the final typesetting.

For the second, revised edition, I would like to thank Dr. Meike Barth and Ken Kissinger for their support, their constant encouragement, and above all for their patience. Last but not least, I would like to thank Dr. Christian Schneider for his valuable work on the videos, for which he sacrificed many hours of his free time.

March, 2022   Thomas A. Vilgis

Contents

1	**Biological Foundations of Our Nutrition**	1
	1.1 Why We Eat	1
	1.2 Oil and Water: More Than Just Essential	4
	1.3 Fatty Acids—A Look into Fat Molecules	5
	1.3.1 Physics and Chemistry Determine Physiology	8
	1.3.2 Plants Show the Way	10
	1.3.3 It's All About the Biological Function	11
	1.3.4 Animal or Plant-Based?	12
	1.4 Macronutrients—Function and Structure	13
	1.4.1 Carbohydrates	13
	1.4.2 Proteins	15
	1.5 Micronutrients—Small Atoms and Molecules, Big Impact	19
	1.5.1 Minerals and Trace Elements	19
	1.5.2 Micronutrients: Vitamins	19
	1.6 Fat and Water—Solvents for Flavors and Taste	23
	1.7 Taste as a Driving Force of Evolution	24
	1.8 Alcohol Dehydrogenases—The Step Towards More Food in the Evolution of Early Humans	25
	1.9 The Control of Fire—The Beginning of Better Nutrition for Modern Humans	27
	1.10 Glutamate Taste—Fire and Protein Hydrolysis	30
	1.11 What Is Really in Plants—The Fundamental Problem of Raw Food	32
	1.12 Decay, Fermentation, and Digestion: Intestines and Microbiome	35
	1.13 How Do We Digest Food? What Are Valuable Ingredients?	36
	1.14 The Evolutionary Advantage of Animal Foods	37
	1.15 The Advantage of Cooking for Nutrition and Bioavailability	38
	1.16 Digestion is Life, is Physical Chemistry	39
	1.16.1 The Advantages of Cooking are Measurable	39
	1.16.2 Physicochemical *in-vitro* Intestinal Models	42
	1.17 Apes Would Choose Cooked Food	44
	1.18 What the Body Wants—and How It Tells Us	45
	References	46

2 Recognizing Food, Learning to Eat: A Look into Evolution ... 51
- 2.1 Hunters, Gatherers, Energy Gainers ... 51
  - 2.1.1 Scavenging and Hunting—Nutrition before the Utilization of Fire ... 52
  - 2.1.2 Effort and Yield of Hunting and Gathering—The Energy Balance ... 52
- 2.2 Animal Foods—High Energy Gain ... 56
- 2.3 Fat, Brain, Bone Marrow—Sources of Essential Omega-3 Fatty Acids ... 58
- 2.4 Fatty Acids: Function, Structure, and Physical Properties ... 60
- 2.5 The Omega-6/Omega-3 Ratio of Foods ... 63
- 2.6 Offal—Forgotten and Valuable Foods ... 67
  - 2.6.1 From Early Hominids to *Homo Sapiens* ... 68
- 2.7 Sedentism—Fundamental Change in Food and Nutrition ... 68
  - 2.7.1 Grains and Starch: Additional Food Sources ... 69
  - 2.7.2 The Fundamental Dietary Change ... 70
  - 2.7.3 Consequences of Sedentism—The Third Revolution in Evolution ... 71
- 2.8 Lactose Tolerance—Point Mutation in DNA: Lactose Tolerance as a Result of Selection Pressure ... 74
- 2.9 Fermenting and Fermentation—in the History of Food Culture ... 79
  - 2.9.1 Fermentation of Vegetables ... 82
  - 2.9.2 Asia—Advanced Culture of Fermentation ... 83
  - 2.9.3 Fermented Beverages, Hot Drinks ... 86
  - 2.9.4 Miso, Fish Sauce, Soy Sauce & Co—Brilliant Examples of (Complete) Fermentation and Flavor ... 87
  - 2.9.5 Safety and Advantage of Fermented Products ... 91
  - 2.9.6 Kokumi— Perpetual Umami Companion for Millennia ... 92
- 2.10 Germination as a Universal Cultural Technique for Food Enrichment ... 93
- 2.11 The Origin of Human Chemical Sensing ... 94
- 2.12 The Cultural Imprint of Universal Molecule Classes—Beloved Flavors ... 94
- 2.13 Culinary Triangle and Structuralism: The Universal Basis of Human Nutrition ... 95
  - 2.13.1 The Molecular-Evolutionary Variant of the Culinary Triangle ... 96
  - 2.13.2 Cooking Cultures in the Light of the Culinary Triangle ... 97
- 2.14 Conclusion ... 100
- References ... 101

3 Consequences of Early Industrialization on the Molecular Composition of Food ... 107
- 3.1 Neolithic—The Modernization of Food ... 107
  - 3.1.1 Natural Fats, Industrial Fats, *trans* Fats ... 109
  - 3.1.2 Cholesterol and Cell Membranes ... 111
  - 3.1.3 Phytosterols ... 114
  - 3.1.4 Cholesterol, LDL, and HDL—What Is Good, What Is Bad? ... 115
  - 3.1.5 Margarine, Fat Mixtures & Co. ... 123
- 3.2 The Early Economic Models Using the Example of Agriculture ... 124
- 3.3 Industrial Animal Production and BSE (Bovine spongiform encephalopathy) ... 127
  - 3.3.1 Contemporary Industrialized Agriculture ... 127
  - 3.3.2 The Early Beginnings of Industrialization ... 127
  - 3.3.3 Animal Food for Ruminants? ... 128
  - 3.3.4 Prion Hypothesis—Physical Infections ... 128
  - 3.3.5 The Well-Intentioned Attempt to Impose Omega-3 Fat on Sheep ... 132
  - 3.3.6 Intensive Fattening, US-Beef—Other *trans* Fats ... 134
- 3.4 Staple Food Bread: Physics, Chemistry, Nutrition ... 137
  - 3.4.1 Grains ... 137
  - 3.4.2 What can be learned from the germination of grain ... 139
  - 3.4.3 Wheat Bread and Aids—Industrial and Natural Methods? ... 140
  - 3.4.4 Dough Properties, Dough Processing, and Gluten ... 142
  - 3.4.5 Yeast and Sourdough—Dough Leavening, Fermentation ... 144
  - 3.4.6 Heating of Gluten: Vulcanization of the Network ... 146
  - 3.4.7 Starch and Water Management in Baked Bread ... 148
  - 3.4.8 The Taste of Bread ... 149
  - 3.4.9 The Early Regional Bread Culture ... 150
  - 3.4.10 Nixtamalization—Ancient Grain Technology of Gluten-Free Cereals ... 151
  - 3.4.11 A Look Into the Modern World of Additives Using Bread as an Example ... 153
- 3.5 Food Spoilage and Preservation Methods ... 162
  - 3.5.1 The Desire for Preservation ... 162
  - 3.5.2 Drying, Canning, and Sterilizing ... 163
  - 3.5.3 Hurdle Concept and Barrier Theory ... 165
- 3.6 Misunderstood Preservation Methods ... 166
  - 3.6.1 Example of Misunderstood Industrialization—Liquid Smoke ... 167
  - 3.6.2 Example of a Misunderstood Water Binding with Polyols—Sorbitol ... 167

		3.6.3	Example of a Misunderstood Preservation—	
			Rosemary Extract.	168
	3.7	Consumer Precariousness		169
	References.			171
4	**Molecules Determine Our Food**.			177
	4.1	Where We Come From.		177
		4.1.1	Genesis 1.0—In the Beginning was the Singularity	178
		4.1.2	Genesis 2.0—Elementary Life is Based on Self-organized Interfaces and Molecular Copy & Paste	179
	4.2	The Beginning of Ancestor Worship		181
		4.2.1	From Ancestor Worship to Religion.	181
		4.2.2	Non-structuralist (Food) Cultures	182
	4.3	Basic Food Meat: Physics, Chemistry, Taste.		183
		4.3.1	Proteins Everywhere	183
		4.3.2	The Modern Western Individual and the Inclination Towards White Meat	185
		4.3.3	Many Studies, Little Clear Insight.	186
		4.3.4	Of Heme Iron and Cancer	187
		4.3.5	Once Guilty, Always Guilty.	189
	4.4	The Chinese Restaurant Syndrome and the Chemistry of Umami Taste		189
		4.4.1	Mother's Milk Sets the Example: Umami and Sweet.	190
		4.4.2	Comparison of Mother's Milk, Dashi, and Chicken Broth	193
		4.4.3	The Umami Taste as a Result of Purine Metabolism.	195
		4.4.4	The Synergy Effect between Glutamic Acid and Nucleotides.	196
		4.4.5	Consequences for the Philosophy of Taste.	198
	4.5	Glutamate and Nucleotides as Flavor Enhancers are the Cause of Global Cooking Cultures		199
		4.5.1	How Taste Chemistry Determines Cooking Culture.	199
		4.5.2	From Hidden Glutamate in Yeast Extract.	201
		4.5.3	Artificial and Natural Glutamate?	202
	4.6	Umami Never Comes Alone		204
		4.6.1	The Chinese Restaurant Syndrome—Biogenic Amines and Secondary Products	205
	4.7	Glutamic Acid and Its Function Beyond Taste		209
	4.8	Physics of Sugar: Taste, Water Binding, Preservation.		210
		4.8.1	Sugar and Natural Sugar Alternatives	210
		4.8.2	The Sweet Receptor.	212

	4.8.3	Metabolism and Sugar	214
	4.8.4	Glycemic Index	216
	4.8.5	Glucose versus Fructose	217
	4.8.6	Honey, Rice, Maple, and Agave Syrup	220
	4.8.7	Fruit Remains Valuable	222
	4.8.8	Sugar is much more than just Sweet: OH loves H_2O	224
	4.8.9	The Appeal of Sucrose, and Why Stevia is not Always an Alternative	225
	4.8.10	What is the Purpose of Additives in Jam?	227
	4.8.11	Conclusion: Is Sugar Poison or Not?	228
4.9	Lipid Digestion: Colloid Physics During the Gastrointestinal Passage		228
	4.9.1	Stomach and Small Intestine—Mainly Colloid Physics	229
	4.9.2	The Path of Fat from the Mouth to the Intestine	230
	4.9.3	Bile Acids, the Other Side of Cholesterol	233
	4.9.4	Long-chain Fats (n > 12)	235
	4.9.5	Medium-chain Fats (12 > n > 6)	235
	4.9.6	Plant Fats, Nuts, and Oilseeds in the Digestive Tract	235
	4.9.7	Oleosins—Very Special Proteins	237
4.10	Basic Food Milk: Physics, Chemistry, Nutrition		238
	4.10.1	Raw Milk Proteins—Macro, Micro, Nano	239
	4.10.2	Structure and Composition of Milk Fats	243
	4.10.3	Milk—Raw and Pasteurized	245
	4.10.4	Whey Protein—A Glutathione Supplier	249
	4.10.5	Milk—Raw vs. Homogenized	250
	4.10.6	Raw Versus Homogenized Milk in Digestion	253
	4.10.7	Homogenized Milk and Atherosclerosis	255
	4.10.8	a1- and a2-β-Casein	256
	4.10.9	Micro-RNA in Milk	260
	4.10.10	Cow's Milk Exosomes as Information and Drug Transporters	262
	4.10.11	Milk Makes the Difference	265
References			266

5 Physical Chemistry of Nutrition and Dietary Forms 273
5.1 Healthy? Harmful? Where are the Dividing Lines 273
5.2 From Observational Studies and Popular Interpretations 276
5.3 The Inevitability of Acrylamide 277
    5.3.1 Acrylamide—Brand New and Yet Ancient 277
    5.3.2 Acrylamide and Non-Enzymatic Browning Reaction 280
    5.3.3 Acrylamide—Amino Acids and Sugar 281
    5.3.4 Acrylamide—Amino Acids and Fats 283

	5.3.5	Glycidamide—Fatty Partner of Acrylamide	284
	5.3.6	How relevant is glycidamide?	285
	5.3.7	The Influence of pH Value, Water Activity, and Fermentation on Acrylamide Formation	286
	5.3.8	Asparaginase—A New Enzymatic Tool for Acrylamide Reduction?	287
	5.3.9	The Other, Good Side of the Maillard Reaction	289
5.4	Biological Value and Food Proteins	290
	5.4.1	Biological Value	290
	5.4.2	Liebig's Minimum Theory	292
	5.4.3	Classic Definitions of Biological Valency	293
	5.4.4	Collagen Drinks—Collagen Against ellulite?	294
	5.4.5	Biological Valency and Bioavailability are Different	296
5.5	Intolerances and Gluten-free Baked Goods	299
	5.5.1	Gluten-Free	299
	5.5.2	Wheat Germ Lectins	301
	5.5.3	Amylase-Trypsin Inhibitors	302
	5.5.4	Why is Wheat Intolerance Increasing?	304
	5.5.5	Is Gluten-Free Healthy?	305
	5.5.6	Gluten-Free—Often Nutrient-Poor	306
	5.5.7	Super Grain Teff?	309
5.6	Carbohydrates: Structure and Digestion	311
	5.6.1	Complex Carbohydrates	311
	5.6.2	Whole Grain Flour during Intestinal Passage	312
	5.6.3	Dietary Fiber	313
5.7	Nitrate and Nitrite	314
	5.7.1	Nitrate, Nitrite, and Stoked Fear	314
	5.7.2	Nitrate, Nitrite—A Search for Traces	316
	5.7.3	Nitrosamines	317
5.8	Raw Food Diet: Physical-Chemical Consequences	319
	5.8.1	Only Raw Food is Healthy? Stories of the *Homo Non Sapiens*	319
	5.8.2	Structure and Mouthfeel—Macronutrients and Dietary Fiber	320
	5.8.3	Micro-nutrients	322
	5.8.4	Temperature and Nutrients	322
	5.8.5	Interfaces Between Enzymes and Vitamins	324
	5.8.6	Blanching, Enzyme Inactivation, and Vitamin C	325
	5.8.7	Pseudo-Raw Enjoyment—A Matter of Food, Time, and Temperature	327
5.9	Paleo-Diet	330
	5.9.1	Back to the Stone Age	330

		5.9.2	The Paleo Hypothesis—Adaptation, Maladaptation ...	332
		5.9.3	For or Against Paleo? What Does Science Say?	333
	5.10	Vegan—Exclusive Exclusion and Missing Links		334
		5.10.1	The Radical Food Elite	334
		5.10.2	Health Benefits?................................	336
		5.10.3	Vitamin D......................................	338
		5.10.4	Essential Fatty Acids: EPA and DHA	340
		5.10.5	Supplementation Necessary......................	340
		5.10.6	Fermentation and Germination as Systematic Methods in Plant-Based Nutrition	341
		5.10.7	Example Nattō...................................	342
		5.10.8	Free From Animal—Vegan Substitute Products	344
		5.10.9	Industrial Processes for Surrogate Products..........	346
		5.10.10	Structuring of Proteins...........................	348
		5.10.11	Leghemoglobin as a Hemoglobin Substitute	350
		5.10.12	The Modified Culinary Triangle of Modern Industrial Culture	351
	5.11	Clean Meat—Cultured Meat from the Petri Dish		352
		5.11.1	Meat without Animals	352
		5.11.2	Technical Problems and Solutions..................	354
	5.12	Insects		354
	5.13	Mushroom Proteins—New Research Findings		357
	5.14	Fast Food, Highly Processed Foods—Curse or Blessing?.......		358
	5.15	Superfoods..		364
	5.16	Secondary Plant Compounds................................		366
		5.16.1	Polyphenols	366
		5.16.2	Carotenoids: Delocalized π-Electron Systems	369
		5.16.3	Why Plants and Seeds Cannot Be Superfoods........	371
	5.17	The Value of Natural Science in Nutrition....................		373
	5.18	What Does "Healthy" Actually Mean?		374
	References...			375
6	**Pleasure and Nutrition**			385
	6.1	Hygiene and Pleasure...................................		385
	6.2	A Dilemma of Food Production...........................		387
	6.3	Detoxifying Food and a Few Contradictions.................		388
	6.4	Understanding of Research is Dwindling		390
		6.4.1	Recognizing Connections	390
		6.4.2	Why Mechanically Separated (or Deboned) Meat is Good in Essence................................	391
	6.5	Tradition and Enjoyment: Meat and Sausage from Home Slaughtering...		392
		6.5.1	What We Know About Meat and Sausage	392

		6.5.2	Warm Meat Processing: Fundamental Advantages on a Molecular Scale	395
		6.5.3	Cold Meat Processing: Systematic Physical Deficits in Microstructure	397
	6.6	Home Cooking is Worth Its Weight in Gold		400
		6.6.1	Control It, Do It Yourself	400
		6.6.2	Forgotten Vegetables, Secondary Vegetable Cuts, Root-to-Leaf	401
		6.6.3	Pure Pleasure	402
		6.6.4	Intramuscular Fat: Fancy Flavour Enhancer	403
	6.7	Nose-to-Tail, Taken Seriously		406
	6.8	Game (Venison), Organic Meat from the Forest		408
		6.8.1	Meat Doesn't Get More Natural Than This	408
		6.8.2	Red and White Muscle Fibers	408
	6.9	The Salt Issue		411
		6.9.1	Salt in the Kitchen	411
		6.9.2	Salt and Osmosis	414
		6.9.3	Salt and Humans	415
		6.9.4	Salt and Interactions at the Atomic and Molecular Level	416
		6.9.5	Salt is Not Equal to Salt	418
		6.9.6	Salt is Not a Poison	419
	6.10	Lot Makes You Full, Complex Makes You Satisfied!		419
		6.10.1	More is Not Always Better	419
		6.10.2	Variety and Combinatorics – Complexity on the Plates	420
		6.10.3	Excitement and Variety	422
	6.11	Hunger—A Western Luxury		425
		6.11.1	The Forgotten Hunger	425
		6.11.2	Autophagy	426
	6.12	What We Will Eat in the Future		429
		6.12.1	We Eat What We Used to Eat	429
		6.12.2	Insects, But Not Only	430
		6.12.3	New Foods to Discover: Duckweeds	431
		6.12.4	Spirulina: Hype or Opportunity?	432
	6.13	*Ikejime*–Gentle, Sustainable, and Umami-Promoting Cultural Technique		433
		6.13.1	Taste-Driven Cultural Technique	433
		6.13.2	The Molecular Aspects of *Ikejime* Slaughtering	436
	References			439
7	Conclusion—Or: What Remains?			445
	7.1	Reading and Better Understanding Critical Studies		445
	7.2	It's Not Just About the Amino Acid Balance		447

7.3	Bioactive Peptides: Small but Essential Features of Protein Origin..	450
7.4	Should We Eat Everything?..............................	452
References...		453

Biological Foundations of Our Nutrition

Abstract

In this introductory chapter, the foundations for the molecular understanding of food are laid. It is shown how chemical structures and molecular interactions can already reveal elementary connections in order to prevent misunderstandings in nutritional issues from arising in the first place. It is about properties and functions of macro- and micronutrients, their effects, but also about elementary aspects of the taste of food. Taste must never be neglected in nutritional issues, as it had a fundamental function in human evolution.

1.1 Why We Eat

In the past, everything was better. At least when it comes to food, it seems, according to many statements in the press, books, and television reports. This may seem subjective, but these statements are not proven or even factual. Never before have food products been so safe, so closely monitored, and so easy to consume. Never in the history of the (Western) world have so many food items been available year-round and in such variety and abundance. As we walk through markets or market halls, we can safely put any piece of native fruit, any vegetable from the field, any raw sausage, any cheese, any fish, any seafood, and even any piece of raw meat into our mouths, chew and swallow it without having to fear food poisoning. This was not always the case. Eating was and is essential for life. Only when we eat can we live. For a long time in human history, the availability of food alone was a necessary prerequisite for survival. This is no longer the case, as the question of survival is no longer tied to the availability of food. People eat today for all sorts of reasons. Mainly because food is simply available everywhere, without having to hunt, produce, or cook it. In fact, our current food is completely controlled and non-toxic, otherwise, it would not be on the market and available. Nevertheless,

© The Author(s), under exclusive license to Springer-Verlag GmbH, DE, part of Springer Nature 2023
T. A. Vilgis, *Nutrition Biophysics*, https://doi.org/10.1007/978-3-662-67597-7_1

power bars with high sugar content, soups and terrines with flavor enhancers, pizzas with high salt and fat content, and beverages highly enriched with sugar and citric acid cause great fear. The "artificial" flavors contained in them also cause concern. However, it would be very easy to avoid these foods by simply not consuming them, as they are not essential for life. Instead, sugar, salt, acids, flavor enhancers, and fats are singled out as culprits, even though they have long accompanied humanity as flavor providers, as is still the case today when food is prepared in one's own kitchen.

However, those who passionately speak of these flavor enhancers as poisons in food are playing a game with fear and forgetting the true dangers of food poisons that have accompanied people for thousands of years in sickness and death and have not lost their horror to this day. These include, for example, *E. coli* bacteria or highly dangerous microbiological germs such as *Campylobacter* or even antibiotic-resistant germs that in most cases lead to severe illnesses and even death. In comparison, even the proven amounts of pesticides, nitrate, or even dioxin are less severe. Anyone who demonizes poison in food should not consume beer or wine, let alone a digestif, and rightly so, cigars, pipes, or cigarettes are stigmatized. Many people die from these every year. No one dies from the amounts of dioxin or fipronil in organic eggs that are just below the detection limit. Even less so from grilled meat, burnt vegetables, and dark bread crust or dark brown toasted bread, although acrylamide can be detected in them. Despite many warnings from politics [1], consumer organizations [2] and nutritional science [3], it would indeed be a small miracle if a direct and causal link between death and acrylamide were proven in an autopsy. We and our ancestors have been ingesting the substance acrylamide for as long as food has been held over fire and exposed to temperatures above 140°C. It just couldn't be detected until a few decades ago [4]. Throughout human history and successful development, people could not afford to throw away food, even if it was darkly grilled, fried, or charred. Food was too precious for that. Anyone who is afraid of acrylamide must live with the consequences: only cooked, only raw, but nothing fried, neither bread nor cocoa nor coffee, not even grain coffee. That alone sounds aromatically boring, tasteless, but it is the price for an acrylamide-free life. Yet, when heating, we only focus on acrylamide. The other toxins, such as heterocyclic aromatic amines (HAA) and polycyclic aromatic hydrocarbons (PAH), are not even mentioned. However, not only acrylamide is formed during roasting, but also "good molecules" that prevent cancer or have a high antioxidant potential, as will be shown in later chapters.

It should also not be forgotten under what adverse circumstances humanity survived despite long phases of wars and natural disasters that forced poor nutrition. Even the people of the war generations are still getting old today, almost 100 years and sometimes beyond, although they were malnourished for years of their lives and did not come close to the concentrations of macro and micronutrients recommended by nutrition societies. The current fears about food are apparently to be critically questioned. It is therefore worthwhile to take a new look at many problems, to subject them to a different perspective, in order to recognize completely new connections that often spread unreflectively in the media and social networks.

Let us return to the nerve poison alcohol for illustration in a thought experiment. If the molecule ethanol did not exist until today and someone were to invent it in a food laboratory to enrich food, preserve drinks, and make them germ-free (as was done thousands of years ago), there would initially be hundreds of tests and clinical surveys to overcome the hurdles of approval. It would quickly be discovered that it is a nerve poison and that animals and humans can die from it. No regulatory authority would grant approval for it. However, the truly dangerous nerve poison alcohol is considered an untouchable cultural asset, whether in beer, wine, or fine spirits. If politics were to ban all alcoholic beverages, there would be great incomprehension, and it would hardly work, as the laws of prohibition show. To better understand human nutrition, a look at the history of evolution is essential, as it opens our eyes to the elementary connections of our food culture, our taste preferences, and the fundamental preparation techniques. At the same time, the fundamental basis of human nutrition is revealed, which has not changed to this day.

Today, it seems we have forgotten the role food really plays in our body, and we have little idea of how our digestive system works. There are plenty of examples. A common rule during almost every doctor's visit is: Saturated fatty acids are unhealthy and increase the "bad" LDL cholesterol. This is said to be responsible for cardiovascular diseases. Further recommendations concern meat, especially red meat, as it causes cancer. Completely unclear are claims such as gluten makes you sick, glutamate causes Alzheimer's, and sugar is poison and addictive.

Should we simply believe these statements without questioning their core messages? A clear "no" to this question would already be a good start. It often begins with ignorance of what cholesterol really is and what functions this vital molecule has for biophysics, biochemistry, and physiology. Likewise, the biophysical fundamental differences between the "bad" Low Density Lipoprotein (LDL) and the "good" High Density Lipoprotein (HDL) are hardly known to anyone. A closer look at biophysics reveals, in essence, two differently structured nanoparticles of natural origin with well-defined physiological tasks, such as transporting fats, cholesterol, and phospholipids safely to cells and back. This fact alone determines both the size and the respective molecular packaging mechanisms. Up to this point, LDL is neither good nor bad, but fulfills the tasks assigned to the nanoparticle based on purely physical-chemical laws. From a biophysical perspective, this difference is strongly relativized, as shown in Chap. 3 when the physical necessities of *low density* and *high density* are discussed.

In this book, we embark on a physical-chemical journey of discovery that leads us into the world of food, its preparation, appreciation, and its physical-chemical properties at the molecular level. Only this molecular perspective shows us what certain molecules can do. Only a closer look into the food, into the structure and function of the existing molecules, opens up the recognition of scientific connections and the expansion of knowledge. Then we can decide for ourselves what we need to know and what we must not believe.

1.2 Oil and Water: More Than Just Essential

The great contrasts in physics, chemistry, and biology are water and oil, that is, liquid fat. It is well known from everyday life that the two do not mix. They are thermodynamically completely incompatible. When they are poured together, stirred, and even shaken with high forces, they separate again after a short time [5]. Oil floats on top, water remains below. What initially sounds banal becomes a fundamental principle for biological systems, food, and also for the effect of food during the gastrointestinal passage.

But why is this the case? This has exclusively molecular causes: water is a polar molecule, a dipole. Water, H_2O, is slightly negatively charged at its oxygen and positively charged at its two hydrogens. Fats and oils, more precisely triacylglycerols, are all non-polar.

The tiny difference in polarity at the molecular level proves to be so strong that both molecules cannot come together and therefore cannot mix. The specific weight, the density of fat, is also lower at about 0.8 g/ml than that of water, whose density is 1 g/ml. Therefore, "oil floats on water." In Fig. 1.1, a water molecule and a typical fat molecule, (triacylglycerol), are schematically represented. The differences are obvious.

However, it is also important to ask the simple question of why the specific weight of fat is lower than that of water. This can only be due to the molecules and their space requirements. The completely different size and structure of fat molecules compared to water molecules already show that water molecules can pack much more densely than the long, bulky, and three-tailed triacylglycerols. Water molecules, on the other hand, can easily attract each other when they come close. The positive side of the hydrogens is more inclined to the negatively charged oxygen. Water can thus pack much more closely due to its simple molecular shape and dipole interaction. The fact that oil has a lower density than water

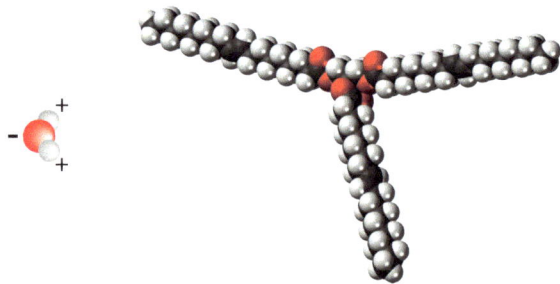

Fig. 1.1 Water, H_2O, (left) with oxygen (red) and the two hydrogen atoms (gray) is polar and has a positively and a negatively charged side. Fat (right) here with three equal-length fatty acids of 18 carbon atoms (black), which are esterified at glycerol (center of the molecule), is non-polar

at all temperatures can also be traced back to molecular properties, as water molecules can arrange themselves within pico- and femtoseconds in such a way that they attract each other via the negative and positive charged sides and pack more densely, even when they move rapidly. The large, bulky triacylglycerols, on the other hand, are much slower.

What do we learn from these simple book-knowledge? The molecular properties, which are evident in the nanoscale range of a few nanometers (1 nm = 0.000000001 m), determine everything, what we can see and experience in the macroscopic range. Therefore, it is advisable to first objectively consider the molecules, their properties, and interactions.

Taking into account elementary physical laws does not allow for misinterpretation, for example, deriving static network structures and thus a memory of water from the polarity of water. Consequently, invigorating or energizing water structures, which are to be achieved through appropriate devices, are physically impossible. Likewise, homeopathic medicines, which after high dilutions have a "memory" of previously added substances, are completely ineffective. The speed of the molecules is much too fast to form long-term structures. An example of hypotheses and the mixing of facts and misinterpretation can be found in the discussions about Exclusion Water [6]. Time and again, water molecules are found on special surfaces, which are, for example, electrically charged. Of course, water molecules must arrange themselves according to the given charges of the surface due to their dipole property, and of course, these structures are more long-lasting, but this does not mean that this water is particularly "energized" or can even develop sanatory powers.

1.3 Fatty Acids—A Look into Fat Molecules

Edible fats are more complex: Unlike water, there is not a single fat molecule. Fats and oils have different compositions and consist of a mixture of various molecules. Although their chemical structure is always the same—three fatty acids are esterified to a glycerol molecule—they can be saturated, unsaturated, polyunsaturated, long-chain, medium-chain, or short-chain. These terms are familiar to us, and for some time now, saturated fatty acids have been referred to as "disease-causing" and unsaturated ones as "healthy," without the origin of this classification being readily assignable to chemical and structural properties. The polyunsaturated, long-chain animal fatty acids are even considered "very healthy" because they are essential, but why do they have an even higher status than the essential plant fatty acids? This is often not clear and understandable, yet the idea that saturated and animal fatty acids are supposed to cause disease has become entrenched for decades without a deeper molecular basis. How questionable these assumptions are was already presented in detail by Taubes in 2001 [7]. Therefore, these questions must be examined more closely in the following chapters, taking into account the physicochemical and structure-forming properties of the molecules. Only then can more valid statements be made. It will turn out that saturated fatty acids are not *per se* harmful, as is often assumed.

First, however, to the definitions. Saturated fatty acids are, at first glance, relatively simple, unspectacular molecules, but there are a few peculiarities. In food fats, with a few exceptions, the total number of carbon atoms in the fatty acid is mathematically an even number, which can be divided by 2. Stearic acid consists of 18 carbon atoms and is chemically an all-*trans* chain with all double bonds in *trans*-position.

Chemically, the carbon atom is tetravalent, meaning that all carbon bonds are saturated in the molecular chain, i.e., each line in the formula in Fig. 1.2 represents a -CH_2-CH_2 sequence. Stearic acid is characterized by C 18:0, where 18 stands for the number of carbon atoms and the zero for full saturation, because the molecule does not carry a double bond. It is also important in this context that each carbon-carbon single bond is freely rotatable around its C-C axis, provided the temperature is high enough. A monounsaturated fatty acid, for example, could form an unsaturated carbon-carbon bond and would then be referred to as C 18:1 (Fig. 1.2). Oleic acid, its trivial name, is found in olive oil, for example, but is also strongly represented in many other fats and oils. The *cis* double bond causes the kink in the structure, the molecule is no longer linear, but has significantly more space requirement. Furthermore, double bonds are no longer freely rotatable around the C=C axis. This is precisely what influences the structure and thus the biophysical properties.

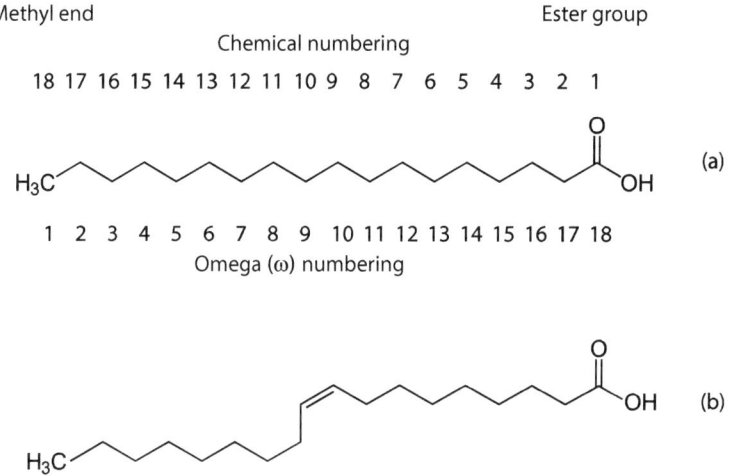

Fig. 1.2 Stearic acid (**a**) consists of 18 carbon atoms and is a so-called all-*trans* chain, whose structure can be described by a linear arrangement. Oleic acid (**b**) also consists of 18 carbon atoms, but has a *cis* double bond at the 9th position, which creates a "kink" and is therefore unsaturated

1.3 Fatty Acids—A Look into Fat Molecules

To understand the structure of the health-promoting Omega-3-fatty acids, one must mentally number the carbon atoms of the molecular chains. There are two ways to do this: chemists like to count from the ester group (COOH), while in nutritional medicine, counting is often done from the side of the methyl group (CH$_3$), i.e., in the opposite direction. An Omega-3 fatty acid then has the first double bond at the third carbon atom after the methyl group. An example of an Omega-3 fatty acid is α-linolenic acid, a triply unsaturated fatty acid, whose first double bond (in Omega numbering) appears at the third carbon atom (Figs. 1.3 and 1.4).

Omega-3 fatty acids are considered essential, as they must be obtained through the diet since they cannot be synthesized by the body. Various fats, therefore, offer a wide range of fatty acids, as shown in Table 1.1 (see, for example, [8]).

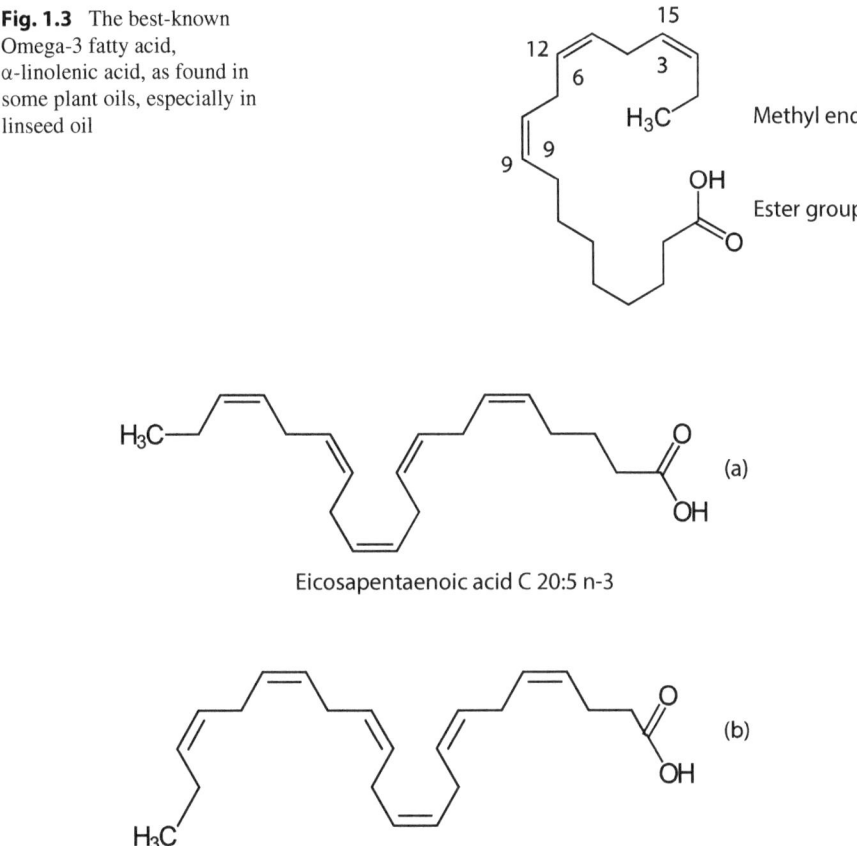

Fig. 1.3 The best-known Omega-3 fatty acid, α-linolenic acid, as found in some plant oils, especially in linseed oil

Eicosapentaenoic acid C 20:5 n-3

Docosahexaenoic acid C 22:6 n-3

Fig. 1.4 Polyunsaturated fatty acids from fish oils, eicosapentaenoic acid (**a**) and docosahexaenoic acid (**b**)

Table 1.1 Typical fatty acid composition in various dietary fats. The dominant fatty acid is bolded. The 0 stands for values less than 1%, shorter fats than C 10 are not listed. Therefore, the numbers in the rows do not add up to 100%

Fat/Oil	C 10:0	C 12:0	C 14:0	C 16:0	C 18:0	C 18:1	C 18:2	C 18:3
Fats of animal origin								
Butterfat	3	3	11	27	12	**29**	2	1
Beef tallow	0	0	3	24	19	**43**	3	1
Pork lard	0	0	2	26	14	**44**	10	0
Goose fat	0	0	0	3	8	**55**	10	0
Fats of vegetable origin								
Peanut oil	0	0	0	11	2	**48**	32	0
Cocoa butter	0	0	0	26	34	**35**	5	0
Coconut oil	12	**48**	16	9	3	6	2	0
Linseed oil	0	0	0	3	7	21	16	**53**
Palm oil	0	0	1	45	4	**40**	10	0
Palm kernel oil	4	**48**	16	8	3	15	2	0
Olive oil	0	0	0	13	3	**71**	10	1
Rapeseed oil	0	0	0	4	2	**62**	22	10
Sunflower oil	0	0	0	7	5	19	**68**	5
Soybean oil	0	0	0	11	4	24	**54**	7
Walnut oil	0	0	0	11	5	28	**51**	5

1.3.1 Physics and Chemistry Determine Physiology

This of various fats and oils have different melting points according to their fatty acid composition. Saturated fats can be arranged much more easily and regularly in crystals than unsaturated ones. The kink and the non-free rotation of the double bond are responsible for this, as they stand in the way of high order. Fats with a higher proportion of saturated and longer fatty acids therefore have high melting points, while fats with short and unsaturated fatty acids have low ones. However, the length of the fatty acids also plays a major role. Short-chain fatty acids have significantly lower melting points compared to long-chain ones, which is why coconut fat becomes liquid at room temperature, even though it consists mostly of saturated fatty acids.

Thus, beef tallow remains solid up to 40°C, containing a high proportion of long-chain, saturated fatty acids C 16:0 and C 18:0. Nevertheless, the beef tallow, classified as "unhealthy," is relatively rich in fatty acid C 18:1, a monounsaturated fatty acid, also found in "healthy" olive oil. Lard has significantly fewer fatty acids of type C 18:0 and therefore melts at lower temperatures. Goose fat begins to melt at room temperature. Fish oil is rich in eicosapentaenoic acid C 20:5 n-3 (EPA) and docosahexaenoic acid C 22:6 n-3 (DHA), which are shown in Fig. 1.3. These

1.3 Fatty Acids—A Look into Fat Molecules

omega-3 fatty acids have an additional structural description (n-3), indicating that the first double bond appears at the third carbon atom after the methyl group, as already hinted at in Fig. 1.3. The two essential fatty acids DHA and EPA are found exclusively in animal fats or microalgae.

However, these polyunsaturated fatty acids can hardly be arranged, i.e., put into a crystal lattice [8]. Fish oils are therefore still liquid even at sub-zero temperatures. Why is this so? The answer can be found in the living conditions and climatic conditions of the animals and plants, as their physiology must function under these circumstances. The living conditions of all living beings and organisms define the composition of the fat. Consequently, the living conditions are crucial: What is the body temperature of the animal, is it warm-blooded or cold-blooded, does it live on land or in water? It quickly becomes clear why fish oils must have a very high number of unsaturated bonds. The animals live at very cold temperatures in the water and sometimes under high pressure in the depths of the sea. It would be fatal if the fats were to crystallize under these conditions. Furthermore, polyunsaturated fatty acids help to reduce the viscosity of the oil. This is vital for fish in the cold sea. Thus, nature has helped itself over the course of evolution with a chemical trick, namely using precisely the exact concentration of polyunsaturated fatty acids at which all necessary physiological functions are ensured through physicochemical properties. Functions are guaranteed.

This variability is not necessary for land animals. Like humans, they compensate for temperature (we always have a temperature of about 37°C in our organs), so a high concentration of polyunsaturated fats would be counterproductive. The fat would be too viscous at our body temperature, and our cell membranes (we will come back to this in detail) would be much too flexible and less resilient. Therefore, all living beings—all insects, all marine animals, and all plants—have a wisely balanced, life-condition-adapted fat composition. That is why goose and duck fat, with a slightly lower melting point (on average), is much more fluid than lard. Waterfowl also live on water, so the thick layer of subcutaneous fat on their breasts must remain supple at water temperatures just above 0°C. However, this is not to say that there are no differences in fat composition among pigs (for example, consider Swabian- Haellian-, Mangaliza, or Pietrain pigs).

From this it immediately becomes clear: For the evaluation of the benefits for human nutrition, several factors must be considered [9], because apart from the melting point and viscosity, the chemically unsaturated double bonds are much more unstable and reactive than the saturated ones. They can easily oxidize, i.e., they break apart spontaneously. In the process, electrons are released, which are highly reactive, as schematically shown in Fig. 1.5. The free electron wants to bond and grabs everything that comes its way. The polyunsaturated fatty acids or PUFAs (*polyunsaturated fatty acids*) form many free radicals, especially at high temperatures, such as our body temperature of 37°C. The double bonds oxidize and release highly reactive electrons.

So, if there is an excess of Omega-3-fatty acids present, they have more negative effects than positive ones. In high pharmacological concentrations, as can occur in some dietary supplements (fish oil capsules) at high doses, they may be

Fig. 1.5 The highly unsaturated essential fatty acid C 22:6 (n-3) also has a downside. The potential for radical formation during peroxidation is very high. The released electrons (red) from the double bond and fragments of the fatty acids become free radicals

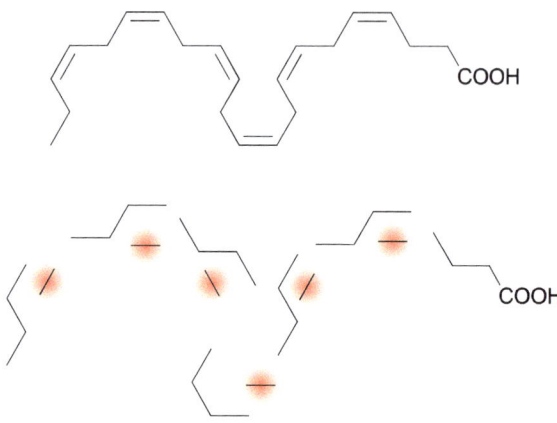

counterproductive. No wonder, in our human body, they are misplaced in the long run and in overdose for purely physicochemical reasons.

This also reveals a significant advantage of the monounsaturated oleic acid, which is always a main representative in many vegetable oils and animal fats: It can oxidize exactly once with a low probability and thus always intercept its own radical. This is just one reason why olive oil and goose/duck fat prove to be unbeatable in many respects in terms of kitchen technology. As we will see in Chap. 2, however, oleic acid (linoleic acid) also has disadvantages when consumed in excess.

1.3.2 Plants Show the Way

The composition of fatty acids in plant fats from fruits (olives, cocoa beans) and seeds such as peanuts, hazelnuts, coconuts, etc. is also adapted to climatic conditions. The viscosity and degree of saturation of the fatty acids play the most significant role in the fatty acid composition. Olive oil serves as a prime example. The Mediterranean oleaginous fruit olive must withstand temperatures ranging from an average of 0°C to 40–50°C, being exposed to the sun and the Mistral wind, as well as sometimes significant temperature fluctuations during the seasons. The dominance of oleic acid with its associated melting point around −5°C is thus the perfect prerequisite. Also, the merely monounsaturated oleic acid is hardly exposed to oxidation processes. In contrast, coconut oil. Coconut palms grow in tropical regions. The constantly humid-hot climate affects even monounsaturated fatty acids like oleic acid. Therefore, the fat consists mostly of saturated fatty acids, which cannot oxidize even under the local temperatures. The smoothness of the oil and the low melting point required for cell function are regulated by the length of the fatty acids instead of the degree of saturation. No wonder that moderately long, medium- and short-chain fatty acids dominate in coconut oil (see Table 1.1). Why these are so particularly healthy [10] is not explained solely by structural considerations but must be examined more closely in the context of fat digestion in later chapters. From the perspective of connoisseurs, however, it is clear that native coconut oil can achieve wonderful aromas, such as creamy-sweet-smelling lactones. The cuisine and dining culture in southern India's Kerala or various regions in Thailand demonstrate this. The odorants referred to as lactones are formed exclusively from medium-chain fatty acids of coconut oil, as also found in milk fat. It is therefore not surprising that coconut-like smells are clearly recognizable when sniffing coffee cream.

The most impressive evidence of the climatic adaptation of fat composition can be seen in cocoa butter. The closer the beans grow to the equator, the less oleic acid and unsaturated fatty acids are incorporated, which would otherwise oxidize. The further away from the equator, the more are found therein. The melting point of different types of cocoa butter is also reflected in the origin of chocolate. A blessing for patisserie and chocolaterie, as melting points for culinary applications can be adjusted to the exact degree. These few examples alone show that it is worthwhile to examine the basic biological functions of food molecules and learn from them.

1.3.3 It's All About the Biological Function

These initial simple considerations and facts show a fundamental principle so far: All nutrients—be they micro- or macronutrients as well as all secondary substances—are simply molecules with very specific functions and properties, not primarily made for humans, but primarily for the metabolism of plants or animals. The better this function fits human physiology, the more available they are for our nutrition. Not because we can directly use the protein, fat, or polyphenols, that would be far too simplistic, but because the essential nutrients contained therein—be they fatty acids, amino acids or secondary nutrients—occur in the necessary and physiologically meaningful concentrations in the food. For example, soy proteins primarily serve as an energy store and amino acid supplier to support the bean during germination. Soy protein is largely broken down during germination and used as an amino acid supplier for plant-relevant proteins and enzymes, while animal muscle proteins perform similar tasks as in humans, enabling movement and being constantly renewed. Consequently, the amino acid composition, shape, and function of plant and animal proteins are inevitably different and adapted to their biological function. These points, which are also important for human nutrition, will be discussed in more detail in the following chapters. Healthy or pleasure?

However, this does not only apply to proteins but also to fats, as already mentioned. The fats occurring in plants and animals, as well as their structures and fatty acid composition, are precisely adapted to the respective relevant physiology.

Attempts to change the fat of animals in the interest of health are therefore doomed to fail, for example, by feeding sheep and lambs mainly flaxseed, fish oil, and algae oil in addition to grass [11]. The hope of introducing plant-based omega-3 fatty acids, i.e. α-linolenic acid from linseed oil, the fat from macroalgae, eicosapentaenoic acid and docosahexaenoic acid from fish oil, into lamb fat did not materialize, and for two reasons: The meat of the slaughtered lambs smelled rancid because rancid-smelling aroma compounds formed from the fragments of the oxidizing omega-3 fatty acids in the intramuscular fat, as illustrated in Fig. 1.5. Furthermore, the fat was of a soft and waxy consistency. The reasons are clear: The physiologically appropriate lamb fat, which is adapted to the living conditions of the lamb, has a composition predetermined by the physiology of the lambs. The fat, like that of all ruminants, is somewhat tallowy and therefore very stable. Although fish oils can be fed, the result for the sensory properties is moderate.

Nevertheless, the positive properties of the long-chain polyunsaturated fatty acids are useful in many respects, provided they are ingested in physiologically appropriate doses through fish meals and not supplemented in excess, as can be seen in this animal experiment. [12].

1.3.4 Animal or Plant-Based?

The blanket view that animal fatty acids are generally harmful, while fatty acids from plant fats are beneficial, is misleading. Practically all fatty acids C n: s, where n stands for the number of carbon atoms and s for their saturation for $n \leq 18$ and $s \leq 3$ occur in both plants and animals. Even triacylglycerols, such as those found in olive oil, rapeseed oil, or cocoa butter, are present in identical composition in pork and goose fat, and even in beef tallow, albeit less frequently in the latter. This is visualized in Fig. 1.6.

The fact that plant and animal fats are not fundamentally distinguishable at the molecular level has trivial reasons. Nature is not so lavishly equipped that it forms different fats and fatty acids for plants and animals. Therefore, there are no markers for animal or plant-based on individual molecules. For human physiology, it is

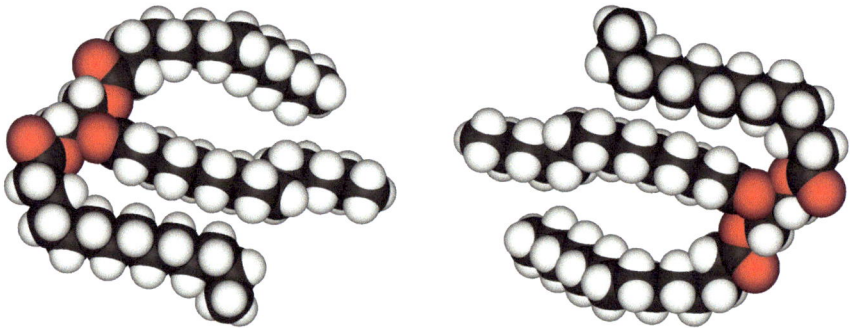

Fig. 1.6 Two different fats? No, identical fats! The molecule is merely mirrored. The two molecules are congruent

completely irrelevant where the individual fatty acids come from. Whether a fatty acid C 18:1 from olive oil or goose fat is incorporated into a membrane is irrelevant for the function and incorporation into cell membranes. Only when the distribution of fatty acids in fats is considered can differences be recognized based on frequencies.

These facts alone are an indication for more serenity and more enjoyment. No cook, no chef in southwestern France, where geese and ducks traditionally make up a large part of the diet, would think of confiting duck legs in olive oil instead of the already accruing duck fat, just because it is plant-based. Nevertheless, the local population has one of the highest life expectancies in the Western world. Despite the traditionally high consumption of duck and goose fat, foie gras, and wine. One part of an explanation for the *"French paradox"* [13] could actually be the biological irrelevance of the origin of fatty acids.

Fats and oils also have an undeserved bad reputation because they contribute many calories. Fat has the highest energy density of macronutrients and boasts a proud 9 kcal/g. But that is precisely the advantage of fat. With 200 g of olive oil, a substantial 1800 kcal can be quickly ingested, which is already very close to the energy of an average meal.

1.4 Macronutrients—Function and Structure

Macronutrients are, in simple terms, "nutrients with calories." They contribute directly to energy supply, and without macronutrients, we cannot survive. The physical energy content of a food is measured in kilocalories (kcal) or kilojoules (kJ). However, this physical value does not always correspond to the energy that the body can derive from it. The actual energy content is therefore referred to as physiological calorific value. For carbohydrates and proteins, it is 4.1 kcal/g or 17 kJ/g, while for fats it is more than double: 9.3 kcal/g or 39 kJ/g.

Apart from fats, which have already been extensively discussed in the previous sections, carbohydrates and proteins are also macronutrients.

1.4.1 Carbohydrates

Carbohydrates essentially consist of sugar, usually glucose or various derivatives in different molecular forms. The best-known representative is table sugar, a disaccharide made of glucose and fructose. Only pure glucose can provide energy without further metabolism, so complex carbohydrates such as Amylose or the highly branched macromolecule Amylopectin are gradually broken down into individual glucose molecules by enzymes called amylases. Carbohydrates, for example starch, are found in legumes, in the seeds of cereals, pseudocereals, and legumes, as well as in root vegetables such as potatoes, sweet potatoes, and taro. This, of course, has its biological purpose, as seeds are meant to germinate and grow. This happens underground, so enough fuel (energy) must be stored in the seeds for this growth process to take place at all. We know that glucose is the number one fuel in the cell and thus also in the seed for germination and growth. However, if pure glucose were stored, it would mean the immediate death of the seed, as osmotic forces would burst the seed upon contact with moisture, making germination impossible. Therefore, during its formation, the plant polymerizes many glucose molecules into long chains, amylose, and branched polymers, amylopectin. The larger and longer the molecules, the weaker the osmotic effect. With the high molecular weight starch molecules in the starch granules, this is practically zero. The starch in seeds and grains thus has only one natural goal: to pack as much glucose as possible without osmotic effects in the smallest space. This is only possible if certain hierarchy principles are adhered to, as shown in Fig. 1.7.

A closer look down to the smallest molecular scales reveals the complete hierarchy of strength. All strengths consist of highly branched amylopectin and linear amylose, as indicated in Fig. 1.8. The linear components of amylopectin as well as amylose consist of polymerized glucose, which is shaped into helices.

Starch is therefore nothing more than "poly sugar". No surprise that a piece of bread alone, in terms of its glucose content, corresponds to several pieces of cube sugar. Bread, potatoes, corn, and other starchy grains are therefore "sugar bombs", even if the molecular structure suggests the term "complex carbohydrates". However, this is often irrelevant for energy input. The distinction between simple and complex carbohydrates is an oversimplification from a physical point of view, as far as energy intake is concerned. Therefore, it is initially unclear whether complex carbohydrates are "healthier" or not. These doubts, made *ad hoc* here, will be deepened in Chap. 5.

What we can already recognize from this example, however, is the actual purpose of the macronutrient carbohydrates: It serves plants, seeds, roots as an energy store, which is enzymatically split as needed during germination. When we eat carbohydrates, we get this rapidly available energy in the form of calories. If we use this in endurance performance, such as physical work or sports, it is not a problem, but it is for "couch potatoes". Excess glucose not needed for immediate energy production is packed as glycogen, also a branched starch, into the muscles

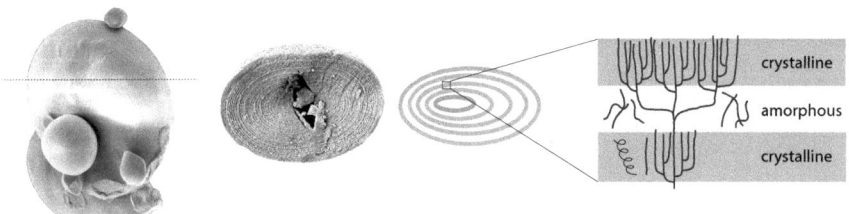

Fig. 1.7 In a hierarchically layered starch grain, polymerized glucose in the form of highly branched amylopectin and linear amylose is tightly packed, as the cut (dashed line) of a grain shows

1.4 Macronutrients—Function and Structure

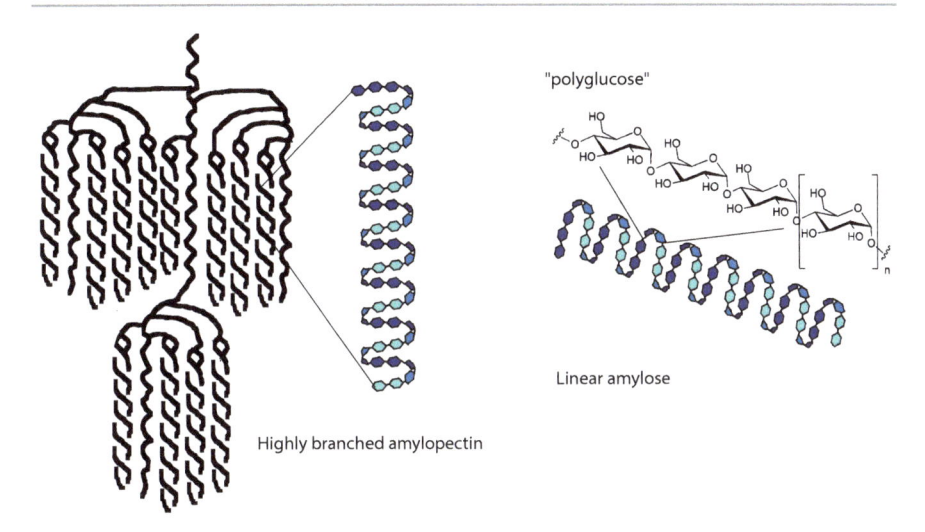

Fig. 1.8 Starch—especially the hyperbranched amylopectin contained therein – is one of the densest packings of glucose produced by nature

to be available as a reserve. In the case of low physical activity, metabolism cannot easily break it down. If the glycogen stores are overfilled, fat is formed from glucose [14] and stored in the depot (see Chap. 4). This inevitably leads initially to padding, later to obesity.

Of course, it is true that the highly branched carbohydrates release their glucose more slowly, as the giant molecules have to be gradually split by enzymes, the amylases of saliva and the pancreas. But per amylose molecule, depending on its length, between 100 and 1500 glucose molecules are released. No less for amylopectin. Starch is therefore pure sugar and is reflected in the energy balance when consuming chips, rolls, and starch-rich snacks, unless one moves accordingly.

1.4.2 Proteins

Proteins also contribute to energy intake. Proteins consist of amino acids and are thus essential components of nutrition. There are 20 different proteinogenic amino acids, which are mainly responsible for the shape and function of proteins. They are shown in Fig. 1.9. In addition to the proteinogenic amino acids, there are many more amino acids in nature, but they are not used to build natural proteins. Shape and function are indeed the central terms, as nature produces proteins in plants and animals according to the blueprint of the organism. Each type of protein has very specific tasks to perform at the molecular level, which are defined by the composition and the specific shape of the folded protein. Therefore, every organism is forced to produce a variety of proteins. As in humans, proteins also have corresponding functions there. 20 different amino acids line up in very specific

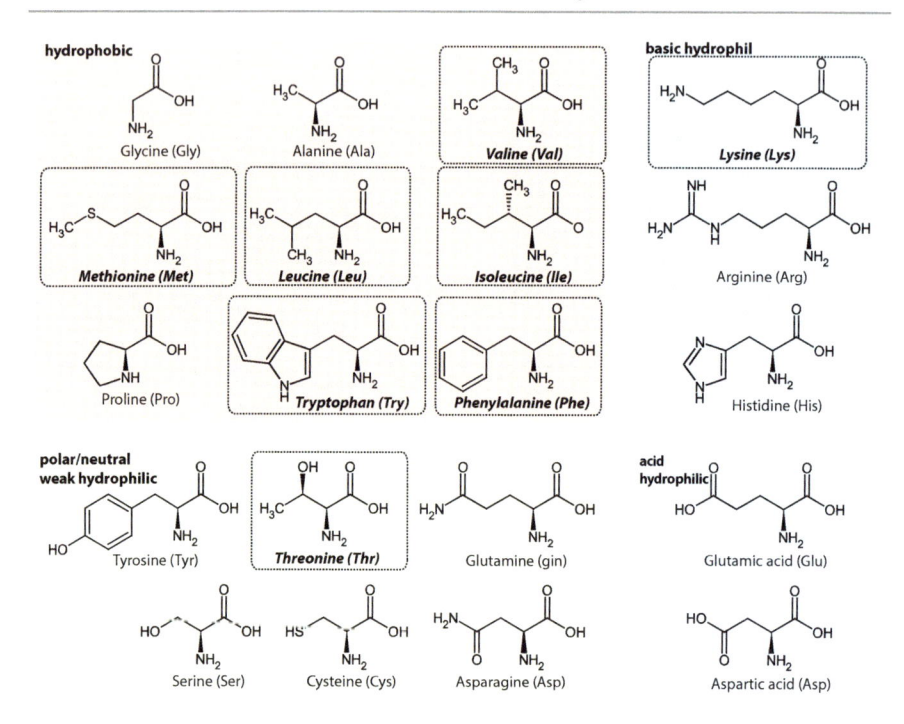

Fig. 1.9 Representation of the 20 proteinogenic amino acids in food. The framed and italicized amino acids are essential for humans

sequences in longer chains. Depending on their function, these protein chains contain between about 50 and several 1000 amino acids. It becomes clear that in this way countless possibilities arise to arrange the amino acids in different sequences, which immediately explains why there are so many different proteins with different properties: muscle proteins in meat, connective tissue proteins, globular muscle colorant proteins, spherical egg white proteins, rubbery elastic proteins from wheat grains, etc. All human proteins, be it muscle proteins, heart muscles, kidney or liver tissue, even the hard, resistant bones, consist of these amino acids.

The physicochemical properties of amino acids play the most important role in the function of proteins. In Fig. 1.9, the chemical structures of amino acids are shown. They can be roughly divided into three groups, which are color-coded in the illustration. Some amino acids are downright water-shy, hydrophobic (in Fig. 1.9 highlighted in red), while some are weakly water-soluble (hydrophilic) (in Fig. 1.9 highlighted in yellow). Furthermore, there are four highly water-soluble, charged amino acids, which are strongly hydrophilic (in Fig. 1.9 highlighted in blue). With these amino acids, (almost) all essential proteins in flora and fauna can be represented. Since they can be strung together in (almost) any order to form chains of different lengths, there is theoretically a huge number of proteins, of which only those with biological function are relevant in organisms. The shape,

1.4 Macronutrients—Function and Structure

and thus also the function, of proteins is determined by the detailed sequence of amino acids (Fig. 1.10).

We do not need to eat proteins because of the proteins themselves, but because of their amino acids. Some amino acids cannot be physiologically synthesized by

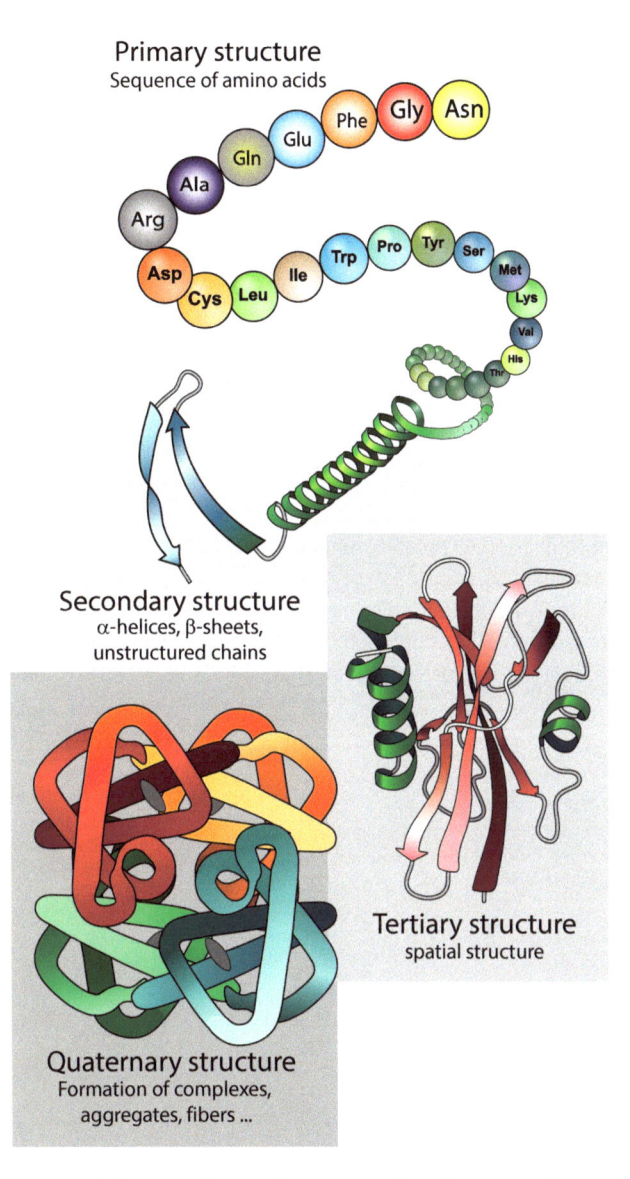

Fig. 1.10 Structure and hierarchy in protein structure. Proteins are made up of 20 different amino acids. The shape in the secondary, tertiary, and quaternary structure defines the function of proteins

the body itself and must be ingested through food. These are also referred to as "essential." In Fig. 1.9, the essential amino acids are framed and printed in bold italics. Upon closer inspection, it can be seen that amino acids, with two exceptions, are "water-shy," i.e., hydrophobic. This purely physical property becomes important at several points.

1.4.2.1 The Protein Lever

As already mentioned, we humans and all animals, besides water, consist largely of proteins. It is no wonder that we need to eat proteins to build brain mass, organs, muscle mass, etc., and maintain them throughout our lives. Protein-rich food is crucial for the function of our Physiology. The fact is that animal protein is the most easily biologically available, as shown in detail in Chap. 5. It is also a fact that the protein composition of animals is closest to that of human proteins. The amino acid patterns of animal proteins are therefore very close to those of humans. It is not surprising how compatible and biologically valuable animal proteins are for human nutrition. The muscles of animals correspond in function and structure to human muscles and therefore provide exactly the right mix of Amino acids. Unlike the macronutrients fat and carbohydrates, amino acids in proteins can be used directly, in simplified terms, without major chemical detours, to build muscle mass.

Proteins satiate faster and more sustainably, as found in a comparison of energy intake and satiety [15]. In a comparative study, the Diet and nutritional data were compared and converted to protein content and ingested energy (megacalories). As Fig. 1.11 illustrates, the total calorie intake decreases significantly with increasing protein content in the food. The authors of the article [15] had systematically evaluated data from the literature. This results in a critical protein content of about 20%, from which the energy intake is almost halved. For this study, many data

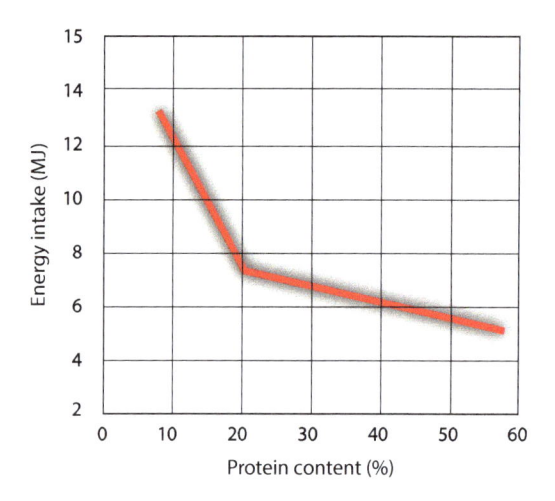

Fig. 1.11 The protein lever shows that with a higher protein content in the food, energy intake decreases

from the specialist literature were collected, and it clearly shows the statistically higher importance of proteins for nutrition compared to carbohydrates, as they satiate more and longer with a similar energy input.

1.5 Micronutrients—Small Atoms and Molecules, Big Impact

1.5.1 Minerals and Trace Elements

Micronutrients do not provide energy and do not contribute to calorie intake, but they enable biological and biofunctional processes in the body. These include vitamins, minerals, trace elements and, strictly speaking, water, the fundamental compound that makes life possible. Minerals such as sodium, potassium, calcium, magnesium, phosphorus, iodine, as well as chlorine, fluorine, and sulfur are present in ionic form. They carry electrical charges and are used as molecular electrical switches, as can be seen in more detail in many places below. Therefore, electrically differently charged ions (+1: sodium, potassium; −1: iodine, chloride, fluorine; +2: calcium, magnesium; −2: sulfur; +3: aluminum; −3: phosphate) are necessary to maintain cell function. These ions are able to interact directly and exclusively with the simply negatively and simply positively charged amino acids in various ways. Proteins can thus be influenced very locally in their structure without disturbing the overall function.

Minerals are found in all foods and salts and must be regularly ingested.

Trace elements such as chromium (Cr), cobalt (Co), iron (Fe), copper (Cu), manganese (Mn), molybdenum (Mo), selenium (Se), zinc (Zn), or silicon (Si) appear in the formation of metal-binding proteins due to their very low concentration. In enzymes, they act as cofactors and thus help enzymes to unfold their effect and work optimally. An adequate intake of micronutrients is therefore of great importance, as many molecular processes are coupled to these atoms, ions, or molecules and have played a prominent role in the development of humanity since the emergence of the first cells, in evolution, and up to the present day [16]. Without these ions and metals, no cellular functioning is possible.

At this point, the basically unclear classification of iron is noticeable. In many cases, iron performs similar tasks as magnesium, which ensures the functioning of the pigment chlorophyll as a doubly positively charged ion, just as the doubly positively charged iron ion in the center of the pigment heme is important for the function. From this functional point of view, iron could also be in the group of minerals.

1.5.2 Micronutrients: Vitamins

Among the micronutrients are also all vitamins; they can be divided into water-soluble and fat-soluble vitamins.

1.5.2.1 Fat-Soluble Vitamins

Vitamin A, Retinol, is an important vitamin for vision, and it is found abundantly in offal and fatty fish. Provitamin A (β-carotene) consists of two chemically linked retinol molecules and is found in colorful vegetables. It is best known in carrots and defines their yellow color. Especially with predominantly plant-based diets, the different bioavailability of provitamin A must be taken into account. It is more readily available from fruits (tomatoes) or leafy vegetables (spinach) than from root vegetables (carrots). Therefore, long cooking or steaming in fat and subsequent pureeing do not harm colorful root vegetables. This is discussed in detail in Chaps. 5 and 6.

Vitamin D, Cholecalciferol, has gained significant importance in recent years. Especially in advanced age, an adequate supply of vitamin D is important. Only through this can calcium be incorporated into bones and thus bone density be regulated. The vitamin is therefore one of the possible molecular keys for the prevention (and treatment) of osteoporosis. Vitamin D in sufficient concentration is hardly obtainable through food. It is found in fatty sea fish (such as mackerel). However, the main supply for humans occurs only under sunlight and the exposure of ultraviolet radiation on the skin. Vitamin D is therefore supplemented in various forms, such as 25-OH-vitamin D3. These molecules are stored, unlike vitamin D, and then converted to cholecalciferol. We will discuss its origin, function, and molecular classification in more detail. To what extent supplementation is really helpful is controversially discussed in observational and meta-studies [17]. A closer look at the molecular processes will also be helpful in this topic, as discussed in Chap. 5.

Vitamin E, Tocopherol, is an important antioxidant vitamin. It is produced only by plants and cyanobacteria [18]. The fat-soluble vitamin is an integral part of the lipid phase of animal cell membranes. It can be ingested with food and is abundant in milk, eggs, as well as in many nuts, seeds, and cereals. Therefore, vitamin E is also found in nut oils, plant, and germ oils.

Vitamin K_1, Phylloquinone, is involved in bone metabolism and blood clotting. As the chemical term suggests, this molecule is found in abundance in green vegetables, such as kale, Brussels sprouts, spinach, or the tubers and leaves of kohlrabi and herbs like chives. It is found in the photosynthesis apparatus of leaves and fruits, such as strawberries, but also in highly pigmented animal products like liver or eggs.

Vitamin K_2, menaquinone, is very similar to phylloquinone but exists as a mixture of substances due to the different lengths of the carbon chain. However, these molecules are also synthesized by the body itself. The occurrence of vitamin K_2 is similar to that of K_1.

From the perspective of their molecular structure, many fat-soluble vitamins strongly resemble fatty acids . Their molecular structure always consists of combined fatty acid and dye derivatives. Therefore, the rule "eat colorful" is right. Fat-soluble vitamins are relatively heat-stable. Compared to water-soluble vitamins, they oxidize much less during cooking and food processing.

1.5.2.2 Water-Soluble Vitamins

Vitamin B_1, Thiamine, is essential for carbohydrate metabolism, as well for the central nervous system. Since Vitamin B_1 has a storage capacity of only about 14 days, it must be constantly supplied. Thiamine can be found, for example, in wheat germ, but also in soybeans and their germs. However, it is also present in higher concentrations in pork, which is not accessible to all ethnic groups. Other good sources are sunflower seeds, macadamia nuts, and sesame seeds. Fresh yeast is also rich in thiamine. Since thiamine oxidizes relatively quickly when heated, short cooking times are essential.

Vitamin B_2, Riboflavin, is present in many foods, so supplementation is only necessary in exceptional cases. The molecule has a heterocyclic part and a sugar alcohol residue (no alcoholic effect), which makes it more heat-stable. Riboflavin is also found in green cabbage and leafy vegetables, as well as in cereals and dairy products. It is also used experimentally to prevent migraine-like headaches.

Vitamin B_3, Niacin, is a heteroaromatic and is stored in the liver. The vitamin is a building block for coenzymes in the entire metabolism. It is found in both all animal and many plant foods, especially in mushrooms (cultivated champignons), nuts, dates, and apricots. Vitamin B_3 is also abundant in all types of legumes. Niacin deficiency is hardly known. The molecule can be produced by the body itself from the essential aromatic amino acid tryptophan.

Vitamin B_5, Pantothenic acid, is also involved in the entire metabolism. It is mainly found in offal and eggs; also in milk and whole grain products, but also in fruits and many vegetables and nuts, especially pine nuts. Avocados are also worth mentioning. Especially in the catering of seniors and in cases of deficiency, this vegetable can play a special role. It is high in calories and provides a whole range of essential fatty acids. Furthermore, ripe avocados with their soft, fiber-free consistency are a welcome food for some forms of swallowing difficulties.

Vitamin B_6, Pyridoxine, is a collective term for three heterocyclic aromatics that are involved in metabolism. The cofactor pyridoxal phosphate cannot be produced by the body itself, but a large number of foods are equipped with pyridoxine, so that deficiency states are hardly possible with a varied diet.

Vitamin B_7, also known as Vitamin H or Biotin, is an important cofactor in metabolism. A deficiency in Vitamin B_7 often results in certain forms of loss of appetite, so that feedback effects can occur and an already existing loss of appetite can be intensified. However, biotin can be produced by the body itself. The primary external sources are again yeast and liver. But bananas and walnuts are also rich in biotin, along with whole grain products.

Vitamin B_{11}, folic acid, also plays a role in a balanced diet. It has a certain preventive effect against arteriosclerosis (atherosclerosis), but above all, it seems to have a potentiating effect in combination with Vitamin B_{12}, the cobalamins. A deficiency in folic acid and Vitamin B_{12} can accelerate the course of dementia diseases (Alzheimer's). Folic acid is found in many leafy vegetables.

Vitamin B_{12}, Cobalamine, is a collective term for a whole series of similarly structured coenzymes that are of crucial importance in metabolism, particularly for cell division and blood formation. Due to its effect on the central nervous system, a deficiency in vitamin B_{12} should be avoided. Vitamin B_{12} is produced exclusively by microorganisms in the intestinal tract through complex mechanisms. Normally, the amount of cobalamins ingested through the diet is sufficient, provided that meat and offal are consumed. However, cobalamins are embedded in protein complexes [19], which must first be denatured—before entering the small intestine, so that protein-cleaving enzymes can digest the proteins and make the cobalamins available and biologically accessible. This process is a typical example of the relevance of biophysical-chemical processes that take place during the gastrointestinal passage. Vitamin B_{12} is only physiologically available if the protein complex that stores the vitamin is denatured beforehand (Fig. 1.12). However, this denaturation does not occur thermally, such as during cooking, but only through the action of acid. This loosens the bonds of the complex, releases the cobalamin, and passes it on to the intrinsic factor, which then allows binding to corresponding receptor proteins. Cobalamins can only be absorbed through the ileum (a part of the small intestine); otherwise, they are excreted. Normally, partial denaturation of

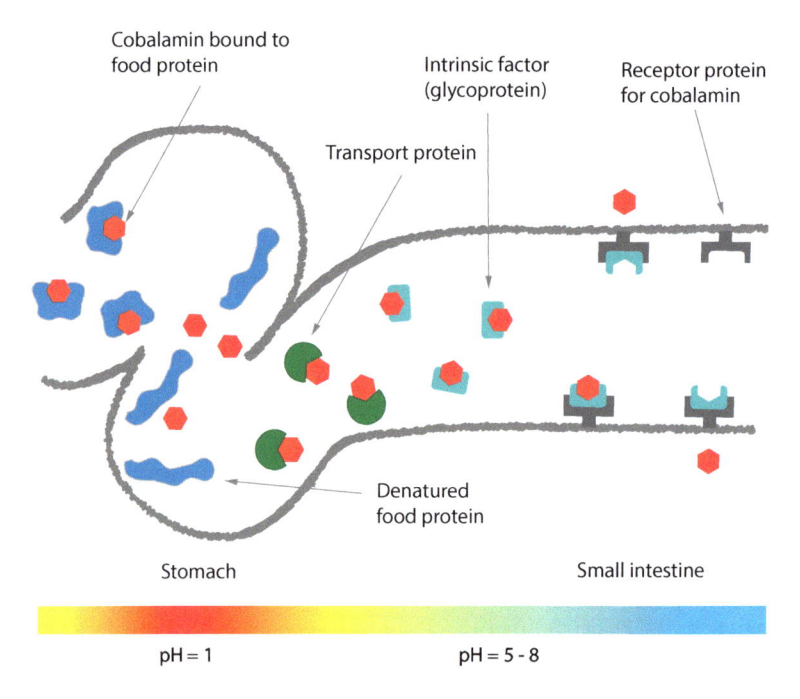

Fig. 1.12 A simple model for the uptake of vitamin B_{12}. It shows a unique interplay between protein physics and cobalamin. Initially (left), the cobalamin is bound to the food protein. In gastric acid, the bound cobalamin is released through a change in internal protein interactions (isoelectric point) and can then be bound by a transport protein in the less acidic environment. Only through further binding to the intrinsic factor (also a protein with suitable binding sites) can it be directed to the receptor centers

proteins occurs in the acidic stomach [20]. However, if acid blockers (proton pump inhibitors) are taken due to other diseases, insufficient denaturation can occur due to a lack of acid. The cobalamins are then not available and are excreted. The result is a vitamin B_{12} deficiency, and supplementation is necessary.

Vitamin C, Ascorbic acid, is the most well-known vitamin. It is easily absorbed through fruits and juices. A deficiency can be very easily compensated for. Sea buckthorn, sauerkraut, and fresh fruits and vegetables in all forms should be included in every person's diet.

The heat stability of vitamins during food preparation is often a particular focus. This point is addressed in Chap. 5 on the topic of "raw food".

1.6 Fat and Water—Solvents for Flavors and Taste

However, back to the immiscibility of fat and water. This is also important for the senses and thus for the perception of food. Fat and water separate taste and smell. All substances that we taste on the tongue—sweet, sour, salty, bitter, and umami— are water-soluble: sugar, acid (vinegar), salt, and phenols in bitter tea or coffee, and of course, glutamate and its natural form, glutamic acid, which is primarily responsible for the umami taste (a word creation from "umai" (うまい), Japanese for "delicious," and "mi" (味) for "taste"). The unfairly labeled as poison glutamate thus triggers the fifth basic taste.

Everything we smell—scents, aromas –, on the other hand, is largely fat-, oil-, and possibly ethanol-soluble [21]. Of course, also in organic solvents such as hexane, etc., but we limit ourselves here to food-grade solvents. What initially sounds like a mere physicochemical fact has consequences for our sensory perception when eating and drinking. This thermodynamic dividing line between taste and odor substances leads to us experiencing our enjoyment as we are used to. When chewing any food, we immediately perceive the taste on the tongue. Only the retronasal smelling of the released aroma substances from the food completes the taste. How sharp the separation between smell and taste is, is still best demonstrated by the kindergarten experiment. If vanilla sugar is tasted with the nose held closed, only the sweet taste of the sugar is perceived. Sugar dissolves in the saliva on the tongue and tastes sweet. However, the aroma substance vanillin hardly dissolves in saliva because it is not water-soluble; it evaporates quickly, and a few molecules of it buzz around in the nasal-pharyngeal area. The closed nose largely blocks access to the olfactory bulb, so the vanilla aroma is not perceived. After opening the nose and breathing, vanillin can reach the odor receptors. The vanilla aroma is smelled, and only then is the sugar perceived as vanilla sugar.

If you perform this experiment with cinnamon and, in parallel, with tonka beans instead of vanilla, attentive tasting with cinnamon reveals a slight temperature irritation away from the taste on the tongue, which increases the more cinnamon is taken. With the tonka bean, less so. These different stimuli of the trigeminal nerve system are added and further round off the sensory perception.

Vanillin is a pure odor substance; without any trigeminal stimuli, vanilla sugar cannot be recognized. Cinnamaldehyde, the main aroma substance of cinnamon, shows concentration-dependent, clear cold pain stimuli and can be recognized even without a sense of smell; the coumarin of the tonka bean, on the other hand, less or not at all. Sensory perception is thus very complex and goes far beyond the sense of taste, to "savor" food, to smell, taste, and consciously enjoy it.

Despite the complicated sensory perception, these simple considerations suggest that there are common preferences among all *Homo sapiens* across cultures: sweet, umami, and fat. The assumption of a universal, cross-cultural eating behavior is therefore obvious. This would be no wonder, as all humans have similar ancestors, the early hominids.

1.7 Taste as a Driving Force of Evolution

Looking at the history of humanity and focusing on the most important of all needs, food. Without food, survival is impossible. This was the driving force of existence. Food scarcity, far removed from today's concepts of hunger and diet, shaped early hominids more than anything else. Food had to be sought, and everything that was even remotely edible and served to satisfy hunger was consumed. The supermarket of early humans was their environment, their niche. Berries, fruit, herbs, roots, carrion, remains of torn animals, eggs, insects, small animals, accessible fish in rivers and lakes, by the sea. Everything was eaten raw, as fire was not yet controllable. The physiognomy of the hominids showed this: muscular jaws, long intestines with all kinds of enzymes and bacteria, so that the last macro- and micronutrient could be extracted from the meager meal and supplied to the physiology [22]. Actions were more instinctive, but still purposeful and forward-looking. The brains were smaller than today, but already large enough and developed enough to use the first tools, which were mostly used for food procurement and thus survival. Food was shared, passed on to offspring, as is still common among apes today. Early hominids had to be sure of the edibility of the food passed on, for example to the offspring, the children. Food control, a safety check, inevitably took place, but the only chemical testing laboratory of early humans was the mouth, nose, and eyes, in other words: the senses. How did it smell, how did it taste? Only with this could it be immediately decided whether the often unknown was suitable as food.

The function of the taste and smell senses becomes clear. These chemical senses served as food testing. Thus, important basic properties and relationships between taste and function of food quickly became apparent. Sweet-tasting, natural foods are generally not toxic, while bitter-tasting ones are more often, so caution is advised with bitter taste. Even slight acidity indicates that the food is rather safe and not toxic. At pH values below pH 5, the growth rate for pathogenic germs decreases rapidly. In addition, acidity causes saliva to flow, which is essential for swallowing.

But that cannot be all, because it was of little use to early humans if the stomach was full but the contents provided no nutrients. "Salty" also enhances saliva flows and where there is salt, there are usually other minerals as well. Minerals, i.e., ions, electrically charged atoms (or molecules) of different valence (charge strength), are, as already mentioned, the best biochemical and molecular biological switches. They are small, highly dynamic, and thus control many cellular processes. In addition, for example, calcium-phosphate complexes harden bones [23, 24], and what the hominids could least afford in their daily food procurement were muscular weaknesses and orthopedic problems, to put it casually.

How essential the taste test was can be seen from the fact that none of the raw foods were ever germ-free. Fire could not be used in these times. The hominids had to rely on what they tasted, smelled, and saw. Of course, this was associated with a high health risk, so the sense of taste had to be strongly developed in order to be able to eat all available foods. Evolution sometimes helped with this—and made slightly alcoholic substances, such as fermented fruits, edible for some species.

1.8 Alcohol Dehydrogenases—The Step Towards More Food in the Evolution of Early Humans

In fruits, berries, and other sugar-rich foods, a fermentation process (alcoholic fermentation) occurs as soon as yeasts can act. Wild yeasts are present everywhere in nature and are responsible for these fermentation processes as soon as the microorganisms infect the fruits. Orally ingested alcohol could not be broken down by the hominids. Paired with the typical fermentation taste and smell, it was repulsive to many species. Spontaneously fermented foods did not pass the taste and smell test. Alcohol dehydrogenases, so-called alcohol dehydrogenases, are responsible for the breakdown of alcohol. These are special enzymes that produce intermediate products from ethanol (which cause hangovers when abused), which can then be further processed physiologically. Today it is known that many monkeys and precursors of humans indeed had the ability to break down chemical alcohols, which are produced during digestive processes in the intestine, through so-called ADH4 variants, which are ineffective for orally ingested ethanol. Over time, certain chimpanzee species mutated and developed additional alcohol dehydrogenases, which allowed them to digest orally ingested alcohol, as shown in Fig. 1.13 [25, 26]. Fermented fruits and plants with their alcohol contents between two and four percent by volume could be used as food. This may sound irrelevant today, but it was crucial in the evolution for these chimpanzee species when considering the energy balance: carbohydrates (sugar) and protein 4 kcal/g, ethanol 7 kcal/g. The gain on the energy balance side is considerable. The slightly alcoholic fruits were thus a welcome addition to the energy balance. Of course, the amounts of alcohol in the fruits are much lower than in today's alcoholic beverages, but the advantage

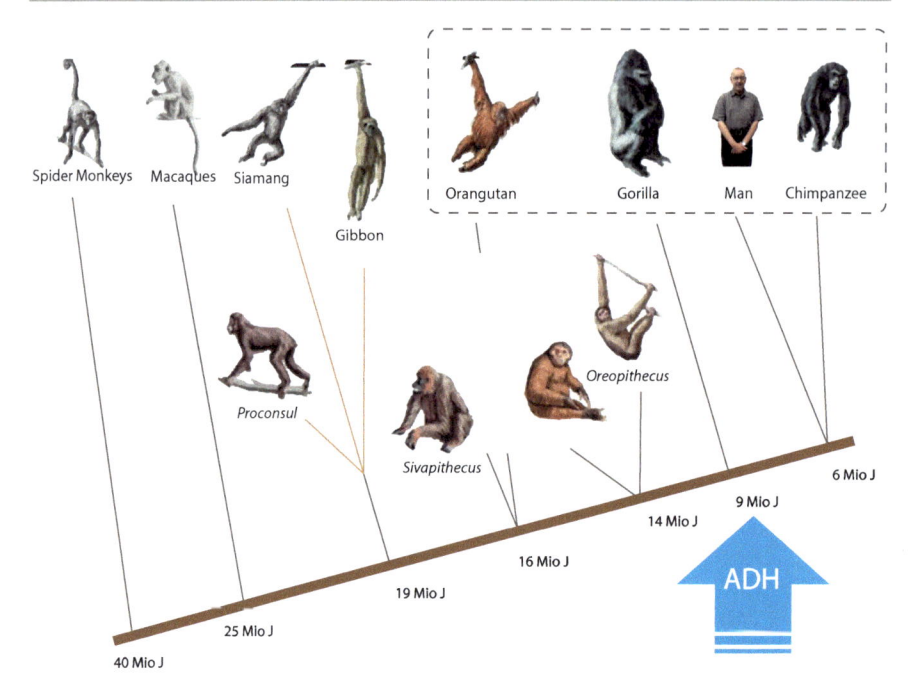

Fig. 1.13 Alcohol dehydrogenases were formed late in the course of evolution. They mainly affect chimpanzees and *Homo sapiens* 9 million years ago and thus before the use of fire. The framed species belong to the hominids

of being able to eat such fruits is considerable when considering the available food supply in the time before the control of fire. In addition, the fermentation process pre-digests the fruits and eliminates anti-nutritive components, such as fructose, oligosaccharides, and other indigestible components that cause irritation (see FODMAPs, Sect. 3.4, Fig. 3.33). Furthermore, the bioavailability of the food ingredients is increased.

What is less discussed in this context, however, is the detailed change in taste during this "wild" fermentation. Acid formed through lactic acid bacteria and the ubiquitous wild yeasts, causing the pH value to drop. In addition to receptors for sweet and bitter, acid receptors were already part of the basic human equipment for taste testing. But that is not all. The fermented fruits and other plants had to taste good in order to appreciate the advantage of the higher nutritional value. This is the first sign of the savory taste "umami", as umami is present in virtually every fermentation process of food, at least as an accompanying taste, since every food contains proteins, such as in the cell membranes of fruits. And this accompanying taste must not be perceived as disturbing, but had to withstand the taste test.

The ability to eat, utilize, and above all, enjoy mildly fermented foods can therefore be classified as the first revolution in the "becoming human" process.

1.9 The Control of Fire—The Beginning of Better Nutrition for Modern Humans

The second milestone in the nutrition of early humans was the control of fire and thus the beginning of food technology. Food was cooked, fried, and heated, which had several advantages for humans. The most important was the improvement of food safety. Food was pasteurized and sterilized during cooking, as the cooking temperatures were high enough. The second advantage was the denaturation of proteins in animal products and the softening of cell structures in hard vegetables or roots. In this way, the nutrients contained in them became much more biologically available. The supply of macro and micronutrients to humans improved significantly [27]. As simple as it sounds, this step of cooking and heating was crucial for human development. At the same time as the cultivation of cooking as a precursor to our modern cooking, the human body continued to evolve. Food no longer needed to be chewed as much, the jaw muscles and lower jaw receded and adapted to the food, as did the digestive tract. At the same time, brain volume increased, and fine motor skills improved. Tools could be better developed and constructed. *Homo habilis* and *Homo erectus* evolved steadily towards modern humans, to Neanderthals and *Homo sapiens*, as indicated in Fig. 1.14.

This argument is widely accepted and corresponds to the current prevailing opinion [28–30]. However, more can be inferred from this. Cooking and preparing animal foods with fire were so important to early humans that they took great risks for it. Large animals were hunted at the risk of life, and their meat and fat were recognized as valuable sources of energy and nutrition. However, there had to be high motivations for this, because to avoid the danger of becoming victims of animals like mammoths, it would have been enough to return to the original diet, eating and cooking roots, grasses, ancient vegetables, etc. Fire was usable. Apparently, a significant advantage was recognized in meat and animal protein, and measurable progress in human abilities due to the consumption of meat outweighed the risk. No wonder, as animal protein in meat is biologically rapidly available compared to the much less abundant proteins in plants and helped in muscle building, especially through the large supply of the amino acid leucine from animal foods. The greater supply of essential long-chain and non-plant fatty acids EPA and DHA contributed to the continuous development and performance improvement of the brain.

Moreover, there is clear evidence that the significant increase in brain volume is directly correlated with the improving diet and not with the social structures forming due to fire [31] and the intellectual learning pressure directly associated with these structures. Thus, the "Social Brain Hypothesis" [32] does not solely explain the development of the human brain. Reasons for the increase in brain volume are simply the provision of nutrients through the new cooking techniques at that time and the broadening of the food supply with protein-rich, animal nutrients. This is logical, as the brain mass must first be built from essential macronutrients, and

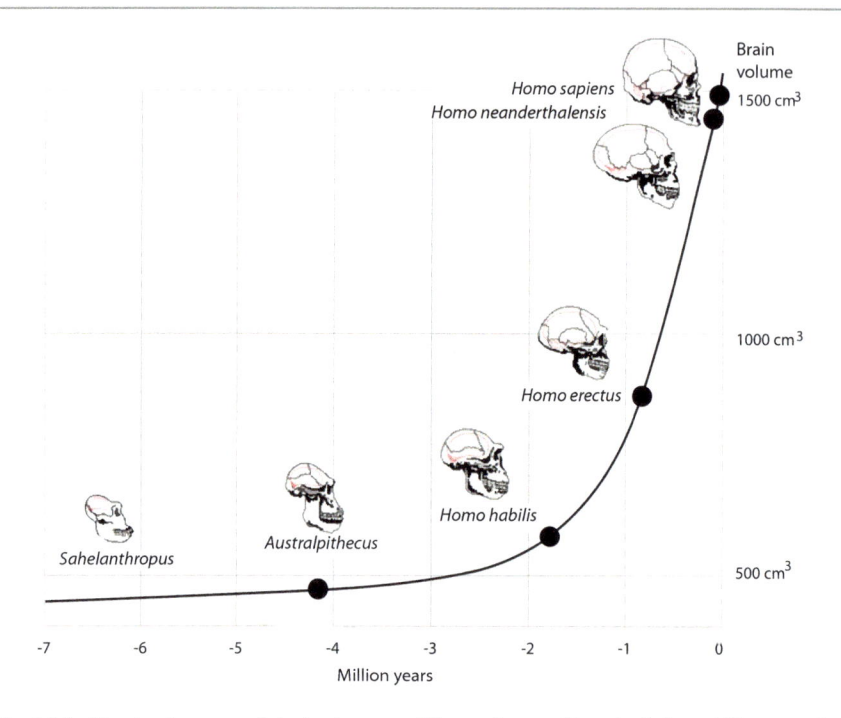

Fig. 1.14 The development of the brain over millions of years. Neanderthals and *Homo sapiens* interbred, and Neanderthal genes are still detectable in our genetic makeup today. The shape of the head and brain volume have changed little to this day. The developmental stages in the gray-shaded area are referred to as hominids, meaning precursors of today's humans

these must be supplied through an improved diet. The neurologically functioning brain is the organ with the highest energy consumption in the body and cannot be "physically-chemically-biologically-physiologically" nourished through social structures. For this, real, molecular matter in measurable quantities is necessary. Mental nourishment is a wonderful, but still philosophical image at the meta-level. To digest mental nourishment, gain insights, implement them, and systematically plan the next steps, the mass, capacity, and energy are needed first. Cognitive abilities can only grow on the molecular-biological basis created, the rapid and lasting networking of neurons. Put simply, this means: Only when neurons have networked in large numbers and form a learnable neural network can previously unknown connections be recognized, interpreted, and evaluated.

It is likely that cooking and preparing meat and animal products created a delicious taste that unconsciously outweighed all the risks, the hurdles of food preparation, the long cooking, the frying over fire with all the necessary tools. Apparently, the umami taste—the taste triggered by a glutamate-dominated amino acid mixture—was an important driving force for the motivation to hunt and cook.

1.9 The Control of Fire—The Beginning …

If it had not tasted good, this method of nutrition would not have been further developed, not even until today. The umami taste and thus the glutamate taste is a central point in the evolutionary history of humans and the long development of cooking.

Thus, the five basic taste qualities sweet, sour, salty, bitter, and umami (Table 1.2) and their function for human development is of great importance. The interaction of these five independent taste directions allowed our ancestors to assess and evaluate food—primarily for its edibility. However, the taste also had to be harmonious, because only when it was perceived as sufficiently good, the food was prepared a second, third time. If the experiences and taste remained good, perhaps even better than in the first eating and cooking attempts, the food and preparation method were included in the diet plan. A systematic Paleolithic cooking canon developed. In contrast, ideas and suggestions for a Paleo diet that are widespread in today's world [33] are not valid. The dietary forms are not objectively comparable. Popular recommendations that even go beyond this [34, 35] do not withstand rigorous scientific scrutiny.

The bitter taste reveals how close the "good" and the "bad" food can come to each other. Many bitter-tasting substances have beneficial effects in low doses, as shown in Chaps. 5 and 6. However, some bitter substances (such as strychnine) are highly toxic and cause severe damage to cells and often death. The systematic learning and interpretation of the bitter taste were therefore particularly important steps in evolution. To this day, this is evident, as bitterness is only tolerated and enjoyed by children at a later age.

Table 1.2 What primarily counts are the taste, the culinary and physiological purpose of the basic taste directions in evolution

Taste quality	Cell function	Evolutionary function	Trigger
Sweet	Glucose, energy	Nontoxic	Glucose, fructose, glycosides
Sour	pH value regulation, information transmission	Salivation, safe food ($3<pH<5$); warning of unripe fruits ($pH < 3$)	Protons
Salty	Switch function with multivalent ions	Physiological mineral balance	Sodium, calcium, magnesium
Bitter	Antioxidant effect in low doses; cell damage through antinutrients and toxic effects of bitter substances	Warning of poison	Phenols, polyphenols, bitter substances
Umami	Building blocks for muscle cells	Targeted protein intake	Glutamic acid, Aspartic acid, nucleotides

1.10 Glutamate Taste—Fire and Protein Hydrolysis

As has already become apparent at two points, it is time to devote more attention to the savory taste: Umami, the taste of glutamate or the taste that is mainly triggered by the glutamic acid, a proteinogenic acid, coupled with other molecular flavor enhancers. These terms often evoke skepticism, as skin itching, throat scratching, and other impairments are associated with them. Glutamate—the epitome of the China-Restaurant Syndrome [36], still stands for the evil of industrialized food production [37] and the harmfulness of eating itself [38]. However, none of this is scientifically tenable [39], the diverse biological function of glutamic acid in various animal and plant systems alone speaks against it. If glutamic acid actually triggered such pathological syndromes, it would most likely have been the end of the "cooking monkeys," at least the end of the role of cooking over fire in evolution. Entirely different dietary forms would have inevitably developed. Not only that, but without glutamic acid as a proteinogenic building block, our life, based on the respective genetics and the resulting proteins and their biological function, whether fauna or flora, could not have developed in the first place. In fact, brain function without the neurotransmitter glutamic acid is not possible. Therefore, the physiologically necessary concentration in the brain is orders of magnitude higher than it can ever be in food [40]. Even in plants, glutamic acid acts as a signal transmitter for communication and defense against predators when the leaves are injured [41]. It is much more likely that glutamic acid was the decisive driving force of evolution. It allowed the species that consumed wild fermented and cooked food to recognize the wondrous taste that only came about through these processes.

In doing so, cooking over an open fire, like grilling, was just one form of food preparation. But seemingly quite incidentally, cooking methods were developed that allowed for gentle and careful cooking, such as earth ovens, for example. Or cooking pits lined with leather from the skins of hunted animals and filled with water, which was heated with hot stones from the edge of the fire. The "cooking monkeys" [28] followed their intellect and taste, by giving in to the umami taste, which is most evident in fermented and heated foods. Both processes have a common root: the breakdown of proteins into fragments and their building blocks, the amino acids, as shown in Fig. 1.15. This process is called hydrolysis.

These taste-relevant hydrolyses, as shown in Fig. 1.15, only occur during processing, i.e., during cooking and fermenting. All amino acids have a taste—some taste bitter, some sweet, some slightly sour. For amino acids or short peptides to be detectable on the tongue at all, they must be free, i.e., first extracted from the long protein. Only then are they small enough to find a place on the taste receptors on the tongue. It is only the extensive hydrolysis of proteins that enables a comprehensive, deep taste experience as a reward for long cooking and fermenting (Fig. 1.16).

Apparently, several positive effects came together during cooking: The food became better, animal protein was physiologically much more available than plant protein, as can be seen in detail in Chap. 5, and the composition of animal protein, the mixture of essential amino acids, is closest to the needs of *Homo sapiens* and its

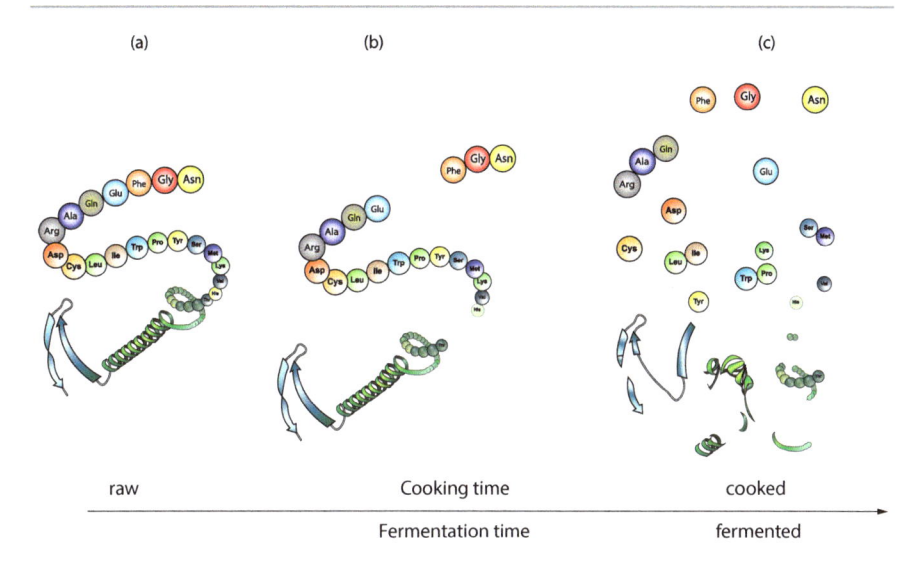

1.10 Glutamate Taste—Fire and Protein Hydrolysis

Fig. 1.15 Under heat and enzymatic catalysis, proteins (**a**) hydrolyze into fragments (peptides; **b**) and individual amino acids (**c**). Because only free amino acids and short peptides have a chemosensory (and bioactive) effect, taste is generated only through this process

Fig. 1.16 The cooking monkey (R. Wrangham) [26]. The driving force of the evolution of cooking and food preparation, with all the associated risks and challenges, was the umami taste, triggered by the molecule glutamic acid, which is hinted at in the chef's hat. *Homo sapiens* is not only "a knowing one" but also a "tasting one," as the root word *"sapio"* (Latin: I taste, I understand) correctly suggests

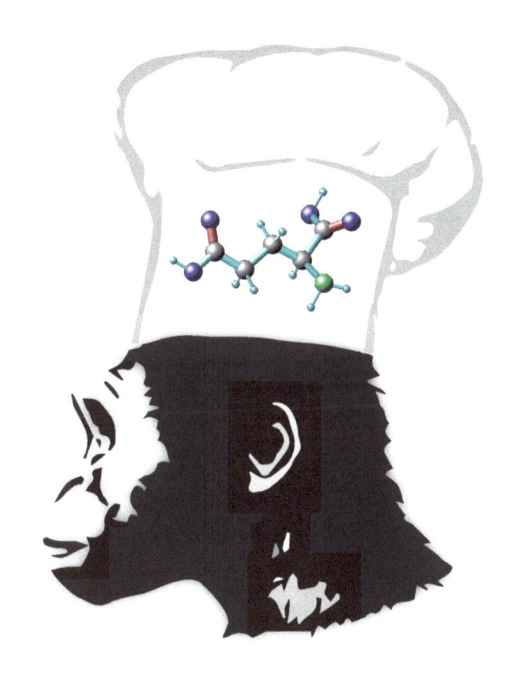

precursors. The volume of the brain increased, cognitive abilities improved, as well as the ability to grasp, analyze, and integrate complex relationships into living conditions. There are a whole range of good, physicochemical reasons why glutamic acid determines the umami taste. All taste triggers must be water-soluble because they must be dissolved directly by the saliva. Thus, all red-underlined fat-soluble amino acids in Fig. 1.19 are eliminated, as well as most essential amino acids.

The taste-triggering amino acids, therefore, cannot be essential. In fact, it would be fatal if only the hydrolysis products of such proteins that carry essential amino acids were tasted. A protein deficiency would be the result, as these are not equally present in all proteins and would thus be a poor indicator. Also, the hydrophobic and thus essential amino acids taste bitter. This taste must first be learned and accepted.

For biophysical reasons alone, the non-essential glutamic acid is present as an important building block in every protein. This is necessary for physical reasons alone, as glutamic acid is negatively charged and thus, along with aspartic acid, the only one that introduces a negative charge into proteins. It thus provides appropriate structural elements for protein folding and is largely responsible for their water solubility. Hypothetical proteins without the building block glutamic acid would have other, non-functional structures.

Glutamic acid can be physiologically (enzymatically) converted to another amino acid—glutamine. This polar amino acid is significantly less water-soluble and therefore performs other biological tasks. Through enzymes, glutamic acid and glutamine can be directly converted into each other on-site, i.e., glutamic acid becomes glutamine and vice versa, depending on biological requirements.

Glutamic acid does not react to toxic substances. It is almost heat-stable, and it does not form toxic Maillard products, such as the amino acid histidine, which reacts to histamine during fermentation or frying, which is toxic in high doses and can be responsible for a whole range of allergic reactions, including the symptoms of Chinese restaurant syndrome. This point is discussed separately in Chap. 4. However, it is also well known that acrylamide, which is formed in every Maillard reaction, is produced by thermal action from aspartic acid and glucose, as discussed in detail in Chap. 5.

There were thus very good factual reasons in evolution to choose glutamic acid as an indicator for protein taste. This alone shows that from today's perspective, natural and at the same time glutamate-free protein powders are not possible [42]. A common objection is that glutamate is artificial, while glutamic acid is natural. This is not true. Because the chemical structure of dissociated glutamate and glutamic acid is identical in form, symmetry, and shape, as shown in detail in Chap. 4 in Fig. 4.16 within the context of the chemistry of umami taste.

1.11 What Is Really in Plants—The Fundamental Problem of Raw Food

The already mentioned development of humans, the formation of brain mass and the associated increase in creativity, cognitive performance, and fine motor skills were clearly correlated with the improvement and change in diet. Cooking with fire was one of the main reasons for this. The plant-based diet and raw

1.11 What Is Really in Plants—The Fundamental ...

food provided the macronutrients sugar and carbohydrates as well as micronutrients such as minerals, trace elements, and vitamins as well as secondary plant substances such as phenols, etc. The drawback of many plants, however, is the proteins and their amino acid composition, which applies then as it does now. Although plants do contain proteins, they are present in significantly lower concentrations compared to animal products such as meat, milk, or eggs.

On the other hand, the mere presence of proteins in plants is only one aspect. Much more important was their accessibility to the early human organism. Extracting every macro- and micronutrient from plants during the gastrointestinal passage is difficult for modern humans. Our ancestors were therefore equipped with a completely different microbiome adapted to a purely plant-based diet in order to survive at all—both the jaw and the gastrointestinal length and structure were more geared towards securing nutrient supply through plant-based food in the past.

Understanding the background is provided again by looking at the structure of plants. Plants, like animals, consist of cells, which in plants are surrounded by hard cell walls. These protect the cell content as much as possible, meaning everything from the cell membrane, organelles, and nucleus to salts, secondary plant substances, and dissolved proteins (Fig. 1.17). The cell walls consist of cellulose, hemicellulose, pectin, and other non-starch carbohydrates, against which today's human enzymes can do nothing. Conversion into nutrients is not possible, so these components are considered dietary fiber.

Cell materials such as cellulose, hemicellulose, and pectin are not water-soluble and very resistant in their native associations, as they are present in raw form. Plants cannot escape from predators and are exposed to the full influence of the climate,

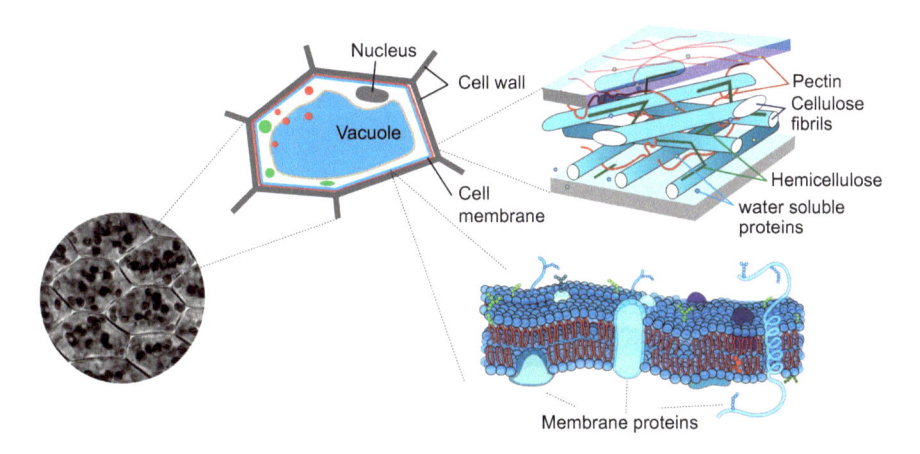

Fig. 1.17 The complex structure of plant-based food. On the left is a microscopic image of plant cells. The schematic representation of the cell shows the water-filled vacuole, the nucleus, and various functional components, such as droplets of essential oils and other substances. The hard cell wall made of various biopolymers provides the cell with stability and protects the membrane made of phospholipids, phytosterols (plant cholesterol), and membrane proteins located directly under the cell wall

strong mechanical stresses such as wind and hail, rain, e.g., a lot of water in a short time, large temperature differences between day and night, summer and winter, as well as high UV exposure from sunlight. The cell walls are therefore protected by biomaterials that can withstand extreme stresses. Cell walls are also insensitive to the common enzymes released by microorganisms, insects, and other predators.

Therefore, it was (and still is) important to prepare plant-based food in the mouth for digestion and the accompanying extraction of nutrients. This ensures that as many nutrients as possible, which are located behind the enzymatically difficult to dissolve cell walls, are biologically available. The old advice "well chewed is half digested" still applies today. In Fig. 1.18, it becomes clear what advantage the strongly developed jaws and powerful dentition of the plant-eating ancestors of humans had before the use of fire. The strong jaw muscles, a pronounced set of teeth, and long chewing allowed the exposure of many cells, whose content was already processed in the mouth and supplied to the digestive system with swallowing.

Then, that the chewing muscles and dentition were strongly developed in the early plant-eating hominids. This was and still is quite similar in herbivores, such as most monkeys. Consequently, after the use of fire, the chewing muscles and dentition changed. These pronounced jaw tools were no longer necessary. Cooked vegetables, roots, and meat have a soft texture, the cells are largely burst due to the effect of temperature, the hard cell material is "softened," and nutrients are

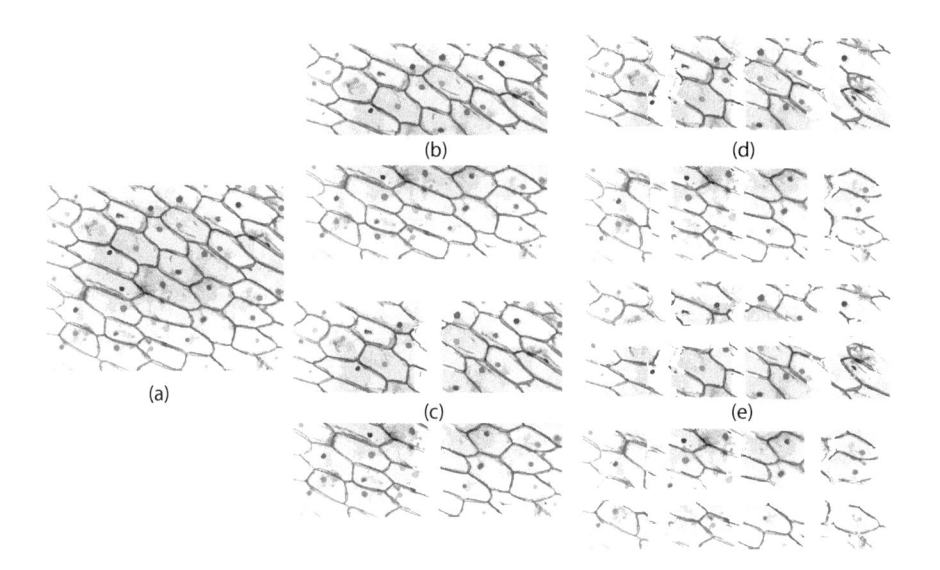

Fig. 1.18 During the chewing of raw plants, it is important to destroy as many cells as possible (**a–e**). Only in this way are the nutrients located within the cells biologically available. The more thoroughly chewed, the smaller the fragments (**d** and **e**), the more cells are destroyed, and the more nutrients are directly available. In the last image on the bottom right (**e**), all cells are opened, while on the far left (**a**), only the cells on the borders are opened

released much more easily. Pectins have partially lost their bonds, and the interactions of cellulose and hemicellulose are weakened. The required chewing work is significantly less compared to the raw state. The physiognomy of the "cooking monkeys" changed towards *Homo habilis* and *Homo erectus* [43]. These cell components no longer play a role in the energy balance of modern humans. In pure herbivores, however, these components must first be pre-fermented in special stomachs or intestinal sections. The appendix, cecum, still bears witness to this today. In *Homo sapiens*, the dietary fibers are fermented as much as possible in the large intestine. There, this process lowers the pH value and thus protects against pathogenic germs, and fermentation products serve the microbiome. This point will also be addressed several times in later chapters.

1.12 Decay, Fermentation, and Digestion: Intestines and Microbiome

Food components that cannot be enzymatically digested must be "fermented" in the intestine with the help of microorganisms such as bacteria, fungi, and viruses. This relationship is particularly evident in the changes in the microbiome during evolution [44]. Of course, food influences the microbiome, but this also happens in reverse: what humans can digest is eaten, what cannot be digested and fermented is not eaten or excreted.

Figure 1.19 shows the adaptation of the microbiome to the respective dietary specification. *Fibrobacter* are only present in pure herbivores to optimize nutrient yield. In *Homo sapiens*, this bacterial form is no longer necessary, as humans had already evolved into omnivores. To this day, the microbiome shows relationships that indicate a correlation between country- and ethnicity-specific diets [45]. However, not without a common denominator, because intestinal bacteria do similar things in the intestine as lactic acid bacteria, yeasts, and fungi do with food during conventional fermentation: they ferment much of what has not yet been broken down to produce intermediates that are important for metabolism. Gases and fermentation products are formed, which in turn serve as "food" for some microorganisms. In the work of Vital et al. [45], the function of a significant intermediate product, butyric acid (butyrate), is pointed out, which serves as a nutrient for intestinal cells and strongly influences the metabolic pathways and the immune system of our body.

In fact, the human microbiome is designed for diversity and can adapt to the available food within a very short time—an essential point, because *Homo sapiens* thus had the opportunity to survive despite times of scarcity, natural disasters, and other adversities. In this respect, the importance attributed to the microbiome in recent times is justified, even from this naive perspective. We can best maintain the diversity and function of the microorganisms in our microbiome through the food we consume, from raw, cooked, to fermented, from plant-based to animal-based, as *Homo sapiens* has been doing intuitively for millennia. The best functional food

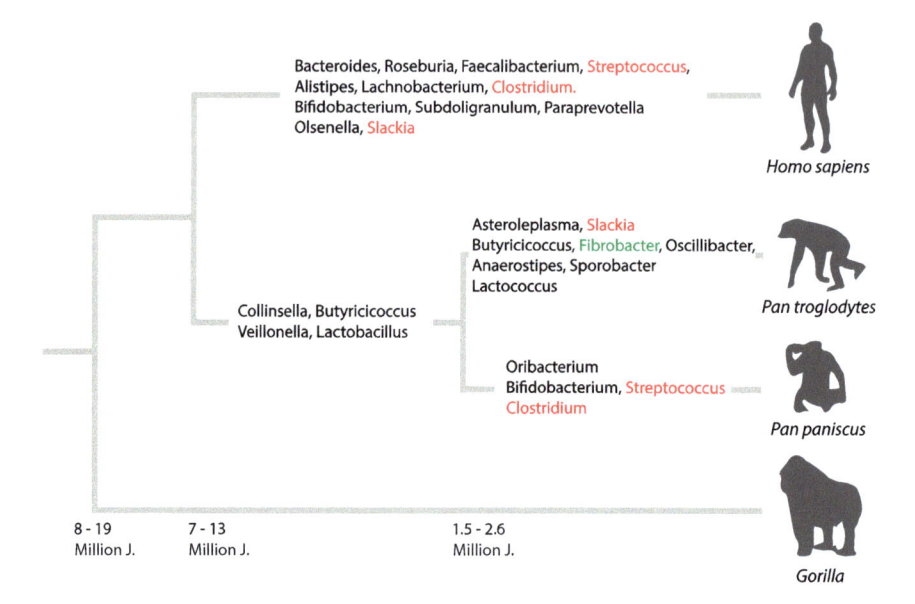

Fig. 1.19 The development of the microbiome based on an examination of the stool of certain relevant species. The red-colored bacteria are still present in *Homo sapiens* today, while the green-colored ones were classified as unnecessary during the course of evolution and the change from herbivore to omnivore and are no longer found in *Homo sapiens*. *Fibrobacter* are only found in pure herbivores and are no longer present in *Homo sapiens*

is still what fields, stables, forests, shrubs, trees, lakes, rivers, and seas naturally provide.

Apart from that, these considerations reveal an important fact: our food is ultimately a mushy, complex, structured molecular structure during intestinal passage. These molecules encounter enzymes and microorganisms in the tract from the oral cavity to the large intestine, which produce something from them, namely molecules that switch, trigger, stimulate, and thus keep our lives running. All living beings are therefore primarily "only" complex biomachines. The molecular diversity of food serves as fuel for us humans on a cellular length scale and below.

1.13 How Do We Digest Food? What Are Valuable Ingredients?

The fermentation of enzymatically indigestible food components in the large intestine using the microbiome is the last resort during the gastrointestinal passage to extract the last possible nutrients from previously indigestible substances.

Basically, these are secondary reactions. The macronutrients—fat, protein, and carbohydrates—must first be metabolized using much more effective and energetically favorable mechanisms that provide these valuable ingredients in a physiological form. These primary processes are primarily responsible for providing essential macronutrients and are therefore not carried out by bacteria, but by enzymes. The energy expenditure is therefore much lower. Enzymes—fat-splitting lipases, protein-splitting proteases, as well as amylases that split starch—are secreted by the pancreas, to name just the most important ones.

Enzymes are proteins that act as catalysts and have an active center on their surface. As soon as a suitable molecule approaches and binds, it is cleaved or otherwise modified. This means that each enzyme is responsible for exactly one group of molecules. For example, proteins cannot be cleaved by lipases. However, this also means that the molecules to be cleaved can dock directly onto the reactive center. Macro- and micronutrients that are still in intact (swallowed) plant cells with their cellulose wall are not accessible. Plant cell components cannot be enzymatically cleaved, nor are micronutrients, such as calcium or magnesium involved in the ionic pectin bonds, 100% biologically available in the small intestine (see Fig. 1.17 and 1.18).

1.14 The Evolutionary Advantage of Animal Foods

The fact that foods consist of a complex system of structured molecular groups such as proteins, fat, carbohydrates, etc., allows for a better understanding of their nutritional value in addition to their basic function in foods. This reveals new relationships that, in turn, allow for a deeper understanding of human nutrition.

The evolutionary advantage of hominids becoming omnivores and consuming animal foods when available is clearly evident. The protein density is extraordinarily high, the "exposed" proteins are easily available, in contrast to packaged plant proteins. In fact, the easy enzymatic accessibility of proteins for digestion plays a crucial role. Proteins consist of amino acids, which are chemically zwitterions and can therefore be most strongly influenced by the pH value. Amino acids change their electrical charge depending on the pH value. The main amino acids responsible for this are the basic amino acids arginine, histidine, and lysine, as well as the acidic amino acids aspartic acid and glutamic acid, as their protonation and deprotonation are most strongly influenced by the pH value. Proteins therefore have a so-called isoelectric point, a precisely defined pH value at which the total charge of the protein becomes neutral, which of course does not mean that each individual amino acid carries no charge. Only the sum of the charges of the amino acids present in the protein adds up to zero. In most physiological and food-typical proteins, this isoelectric point is around pH 5. If this is exceeded, the proteins become unstable, denature, and split at some amino acid bonds, also in the acidic environment of the stomach.

In the stomach, due to gastric acid and depending on stomach filling, pH values below 3 prevail. It is therefore quite acidic, and for most orally ingested proteins, the pH falls below the isoelectric point, causing the proteins to change their shape. This has significant consequences for protein utilization. In gastric juice, there are also acid-tolerant exception proteins that do not change up to a pH value of 1. This is pepsin, a mixture of similarly structured proteases that immediately exert their effect after the denaturation of food proteins and already break down proteins into peptides in the stomach. They show their highest activity at pH values between 1.5 and 3, precisely the pH range that occurs during and after eating when gastric acid (mainly hydrochloric acid, HCl) is buffered accordingly by food. Pepsins thus prepare long protein chains for further degradation by pancreatic proteases.

For this to happen, however, the corresponding peptide bonds must reach the active centers of the enzymes. For example, if whole muscle fibers are swallowed, the enzymes can only attack the surface proteins, and the yield is lower. Therefore, proteins are already pre-denatured by gastric acid so that pepsin (and also cathepsin present in the gastric mucosa) has a chance to bring its catalytic effect to bear.

1.15 The Advantage of Cooking for Nutrition and Bioavailability

From a molecular perspective, the advantage of cooking for human development is also revealed. Food is prepared for significantly better and more effective digestion through thermal processing. Nutrients, especially proteins, are prepared for direct access by enzymes. Plant cells become soft, the high temperature causes them to burst, plant proteins become much more available, and even antinutritive substances that serve the plant for defense are converted, in many cases to high-quality nutritive secondary substances. Membrane proteins, such as antinutritive lectins and other sometimes highly incompatible molecules, become valuable food because they lose their harmful effects through cooking. Not only through chewing, but also through cooking, the biological and physiological availability of nutrients is significantly enhanced. In addition, pasteurization takes place through heating, as already mentioned several times. The use of fire thus undoubtedly brought decisive advantages for the nutrition of early humans.

The question of when early humans began cooking and using fire is not definitively resolved. Evolutionary biologist Wrangham suspects that this happened 1.9 million years ago [27]. This is the time when the first fireplaces are dated, but it is unclear whether they were the result of bushfires, lightning strikes, or similar events. However, it is clear that the *Homo erectus* already used fire about 1–1.5 million years ago. This can be proven very clearly [46]. Further sites for fireplaces from about 1.5 million years ago can be found, for example, in the African region of East Turkana. An exact date cannot be determined, but this is not crucial. However, it is remarkable that the enlargement of the brain (see Fig. 1.14) occurred in parallel, which is supported by archaeological finds and modern analytics.

1.16 Digestion is Life, is Physical Chemistry

In the course of the previous discussion, it becomes increasingly clear what a central role molecular interactions, molecular structure, and the nature of macronutrients play. These are quantum physical, physicochemical interactions. Food is thus reduced to what it really is. Far removed from macro and meta levels, eating and digesting are nothing more than mundane physics, physical chemistry, biochemistry, or physiological chemistry.

In order for the brain to develop, it required an excellent combination of nutrients that could only be supplied through food during human evolution. On the one hand, the brain's energy demand is extremely high, so glucose, one of the main energy carriers for cell functions, must be supplied in sufficient concentration. This was aided not only by eating glucose-rich fruits and berries but especially by cooking starch-rich roots. Native starch is enzymatically indigestible. It is also called "resistant" starch and must be fermented in the large intestine or excreted partially undigested. The starch-splitting digestive enzymes (amylases) have no chance to interact with the densely packed, partially crystalline, highly branched polymers, as discussed in Chap. 3.

The digestion of cooked starch begins already in the mouth. In the saliva, there are salivary amylases, so-called α-amylases, which cleave starch molecules along their α-(1,4) glycosidic bond. They produce various fragments from starch, called oligosaccharides, with different chain lengths, referred to as maltodextrin by food technologists. Maltodextrin is formed after eating bread, potatoes, rice, and other starchy foods. Likewise, isomaltose—the less sweet sugar used by chefs and the confectionery industry, which causes less tooth decay [47]. Furthermore, the branching points of amylopectin, so-called limit dextrins, remain, which are in turn processed by other enzymes, just as maltoses are ultimately cleaved into glucoses by maltases (see Fig. 1.20).

An important aspect for carbohydrate as well as protein digestion is the effectiveness of the enzymes in the various pH ranges of the gastrointestinal tract (Fig. 1.21).

Fat digestion is also initiated via enzymes, lipases. The processes of fat digestion and fat absorption are somewhat more complicated, so we can only catch up on these discussions later due to the complex physical and physiological processes [48].

The individual steps of fat digestion are far more complicated than the elementary steps shown in Fig. 1.22. A whole series of physicochemical processes is necessary for this, which we can only understand after the following elementary considerations. In Chaps. 4 and 5, we will come back to this. First, we will stick to evolution and summarize the advantages that cooking brought to early humans.

1.16.1 The Advantages of Cooking are Measurable

The preparation of food using so-called cultural techniques (see Chap. 2) also plays a major role in the physiological processing and utilization of food. The

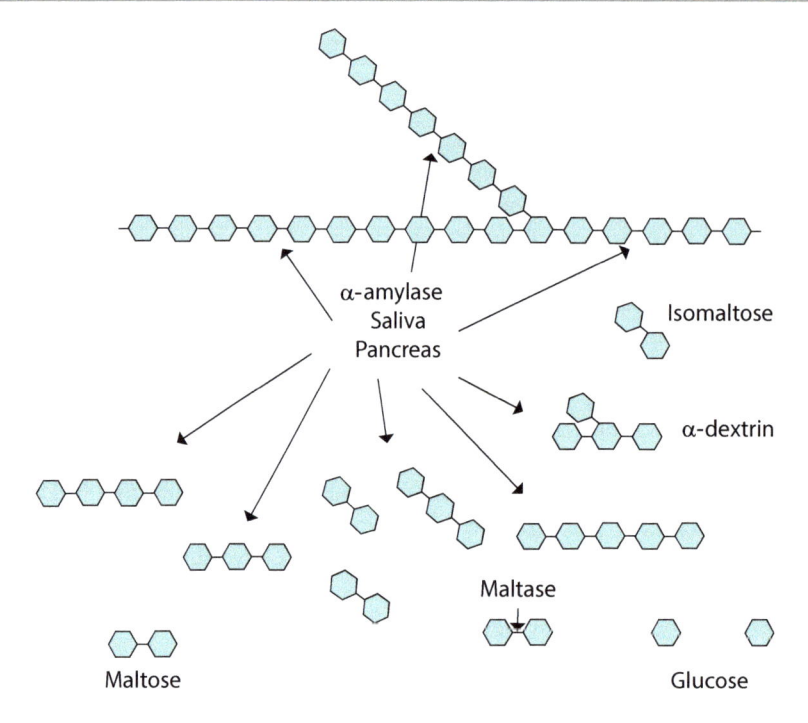

Fig. 1.20 The enzymatic digestion steps of cooked (swollen) starch in saliva and small intestine via amylases. Only after this can glucose be released through a whole cascade of reactions

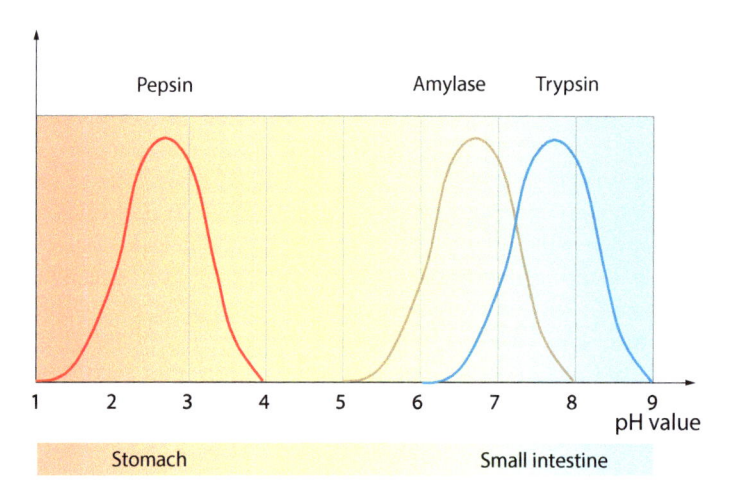

Fig. 1.21 The action of the three most important digestive enzymes in the various pH ranges during the gastrointestinal passage

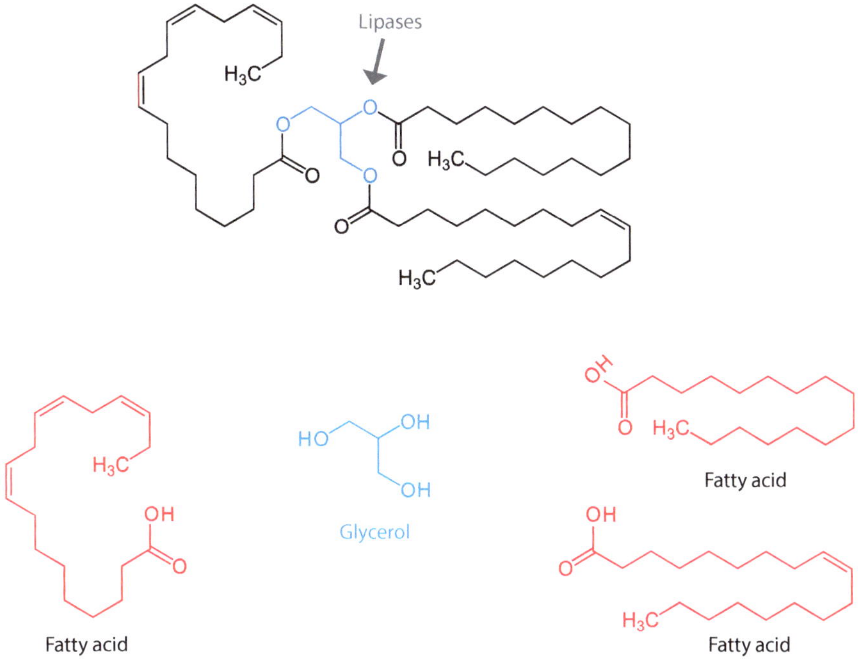

1.16 Digestion is Life, is Physical Chemistry

Fig. 1.22 Fats are cleaved from glycerol by lipases, resulting in free fatty acids and the trihydric alcohol glycerol (glycerin)

enzymatic cleavage and fermentation by the numerous microorganisms of the microbiome are not in vain, but cost energy, which is naturally provided by the ingested energy. How much energy must be provided by the body depends on the preparation of the food.

Intuitively, it can already be recognized that raw, unprepared, and unprocessed food requires the highest effort for the body, as can be seen in the strongly developed jaw muscles of early humans as well as gorillas. However, such statements do not need to be speculated upon, as they can be systematically verified through experiments [49]. Since reproducibility and verifiability are prerequisites for such experiments, the food used must provide reproducible results. Meat is suitable for this purpose, as the protein present in the muscles always has the same structure and is therefore easily accessible to protein-cleaving enzymes. Furthermore, for the experiment, living beings are needed whose digestive energy can be easily measured when they consume meat. The choice was therefore made for python snakes, which were fed whole meat, meat puree (pureed), cooked meat, and cooked meat puree with identical energy content and at a weight of 25% of their body weight each. The required digestive energy could then be deduced from the oxygen consumption in the following days.

The results are clear. Raw meat requires the highest energy (range I in Fig. 1.23). If the meat is pureed (II), the expended digestive energy already decreases by over 10%. Cooking whole meat is in a similar range (III), while cooking and pureeing result in a further significant reduction of the required energy to below 80%.

Although snakes are not comparable to humans, these results show an important aspect. Snakes do not chew their food but swallow it whole. Therefore, they are predestined for this experiment, as all the energy for the production of gastric acid and the provision of enzymes must be expended in the stomach and intestinal tract before the food is further processed by the enzymes in the small intestine. This energy is mainly provided by the molecule ATP, adenosine triphosphate, which is known from muscle work. It is a nucleotide with three phosphate residues that must be synthesized in the body cells in the mitochondria. In other animals, the energy expended for muscle work during chewing would significantly distort the result on oxygen consumption.

This makes it clear which molecular processes cooking takes away from digestion: thermal protein denaturation, collagen conversion to gelatin, and partial cleavage of proteins by heat. At the same time, the accessibility of proteins for enzymes is significantly increased.

1.16.2 Physicochemical *in-vitro* Intestinal Models

In humans, the digestive processes are more complex, so attempts are made to replicate digestion as accurately as possible using precise methods [50]. In fact, the gastrointestinal tract can be reconstructed [51]. The advantage of this *in-vitro*-method is that the effects of acidic gastric juices and pancreatic enzymes can be specifically observed. Food is introduced into the tube system that replicates the gastrointestinal tract. This then ends up in a glass flask that roughly corresponds to the stomach with appropriately acidified gastric juices (containing the enzyme

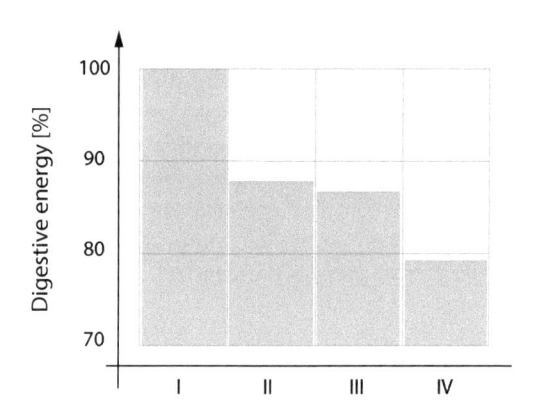

Fig. 1.23 The digestive energy expended by python snakes depends heavily on the form of meat food: I: raw and whole, II as raw puree, III cooked whole, IV cooked meat puree. The values are based on 100% for comparison

pepsin). Small samples can be taken and analyzed before the simulated stomach contents are transferred to the simulated small intestine. There, the pH value is adjusted to the higher values present in the natural small intestine, and the pancreas is added. This allows the changes in food molecules due to acid, enzymes, or bacteria to be tracked very precisely step by step.

With these intestinal models, it is also possible to observe very accurately the differences that occur during the digestion of cooked and raw meat [47]. The meat structure can be microscopically tracked after various stages of digestion, and the remaining fragments of different meat and muscle proteins can be precisely determined. Differences in the digestion time of differently cooked and prepared meat can also be detected. Furthermore, the different effects and contributions of the stomach and small intestine areas on protein decomposition can be examined more closely.

Even in processes involving pepsin in the stomach, conclusions can be drawn about digestibility and preparation, specifically the optimal cooking temperature, using these methods [52]. Raw meat shows only a few points of attack for the active center of pepsin to break down meat proteins. Pre-digestion in the stomach is not optimal, while an optimum is shown at cooking temperatures between 60 and 70°C. Of course, because the proteins have hardly lost any water, they are highly mobile and not yet completely clumped, the meat is tender [53] (see Fig. 1.24). This clumping only occurs at temperatures above 75°C [54]. At even higher temperatures, the proteins are strongly clumped and physically networked, the meat is tough and dry. Many peptide bonds can no longer be reached by the active centers of the enzymes. The meat becomes more difficult to digest again, as symbolically represented in Fig. 1.24.

This also corresponds to the early techniques used by prehistoric humans, who preferred to cook longer in earth ovens or pit cooking. The temperatures reached there are rather moderate at around 60°C, and the meat remains tender. Meanwhile, more umami taste is formed, while at the same time, the digestibility

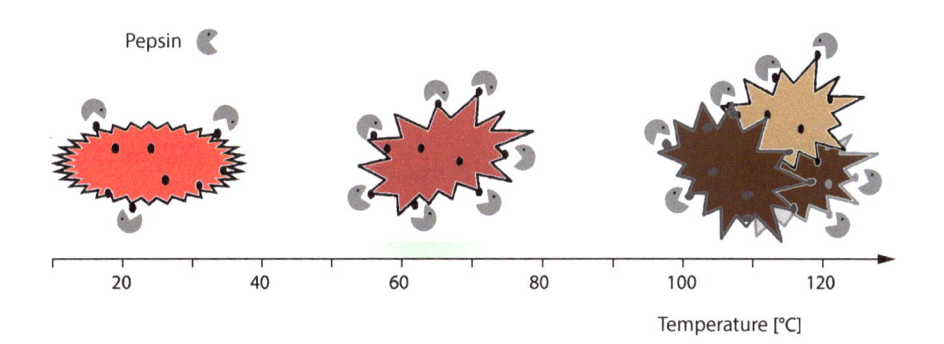

Fig. 1.24 Temperature and digestion in the stomach via pepsin. The active center of pepsin is represented as Pac-Man. The results of Bax et al. [48] clearly show that during the digestion of meat at temperatures below 70°C, the enzymes can perform their best work. The accessibility of the pepsin-sensitive peptide bonds is very high (green area)

and macronutrient yield are significantly better. *Homo sapiens* apparently sensed what was good for them and were rewarded for it.

The comparison of energy expenditure for raw, pureed, and cooked meat only provides information about the theoretical content of macronutrients. Since only these provide measurable energy, micronutrients do not, which is why comments on the destruction of vitamins or other micronutrients by grinding, pureeing, mechanical processing, or heating are misplaced here. Raw food advocates are therefore already at a disadvantage twice: they consume many nutrients but can only physiologically utilize a fraction. Furthermore, they have a much higher physiological energy expenditure in fermentation processes in the intestine to ultimately access fewer macronutrients. A good basis for a more comprehensive diet is therefore to eat both raw and cooked food.

1.17 Apes Would Choose Cooked Food

Great apes choose cooked food [55], even if they cannot prepare it themselves. The experiment with apes described below is based on the idea of using chimpanzees, bonobos, gorillas, and orangutans as model systems for Paleolithic hominids to determine their preferences. Raw and cooked apples, roots, and raw and cooked meat were chosen as food. It turned out that in many and especially statistically significant cases, cooked food, whether apples, potatoes, or meat, was preferred by many animals. Even chimpanzees, whose normal diet is predominantly meat-free, ate cooked meat with a significant preference. However, some animals and species preferred raw apples to cooked ones, as well as potatoes. Apparently, texture also plays a role. Thus, part of human evolution is reflected in this experiment.

The potential for food preferences has also been recently explored [56]. Chimpanzees cannot cook, but they were presented with cooked and raw food. It turned out: chimpanzees prefer cooked food. It seems that the chimpanzees realized the physiological transformation associated with cooking from the raw to the cooked state. Likewise, there was a strong willingness of the animals to wait for the cooked food, even though the raw form was already available. In addition, it became clear that the animals hoard raw food to "cook" it through ripening and spoilage at low temperatures. The purpose of cooked food was thus recognized by the animals.

Such research results occasionally give rise to the assumption that these ape species would "cook" or have the prerequisites for it. This is, of course, not correct, as there is no living being other than *Homo sapiens* that can control fire, cook something over or next to it, and then extinguish the fire again. This, quite apart from the many other achievements of humanity, distinguishes *Homo sapiens* from animals. Even if much of what humans do is not perfect, it is "counter-evolutionary" to anthropomorphize animals.

1.18 What the Body Wants—and How It Tells Us

With fire and the ability to digest alcohol, hominids were able to expand their food supply. The taste as a test and the body's sense of what was good for them were real "driving forces". Hominids became omnivores, eating everything. More meat, and thus more protein and fat, was on the menu, and the ability to eat fermented foods also helped. Only cooking made starch and complex sugars from roots and the precursors of legumes (pulses) physiologically available to some extent. The brain grew, and cognitive abilities increased. Food supply became predictable (Fig. 1.25).

Again and again, there are indications that other flavors besides umami played a role in evolution and ensured survival. For example, there are insects like fruit flies that taste "watery," which allowed them to find vital moisture [57]. There is evidence that rats and mice perceive a fat taste, as do some humans, when free fatty acids trigger receptors on the tongue. Starch taste is often mentioned when short-chain starches, the maltodextrins, trigger stimulus currents in receptors not responsible for the sweet taste. Even though these weak and unclear perceptions are not yet counted among the five basic tastes, they point to evolutionary conditions that ensured the survival of certain species. Mind you, we cannot live without water either. Apart from being essential for life, water is also the only solvent that dissolves all molecules responsible for the basic tastes and makes them tasteable at all. Fat is the macronutrient with the highest energy density—and solvent for

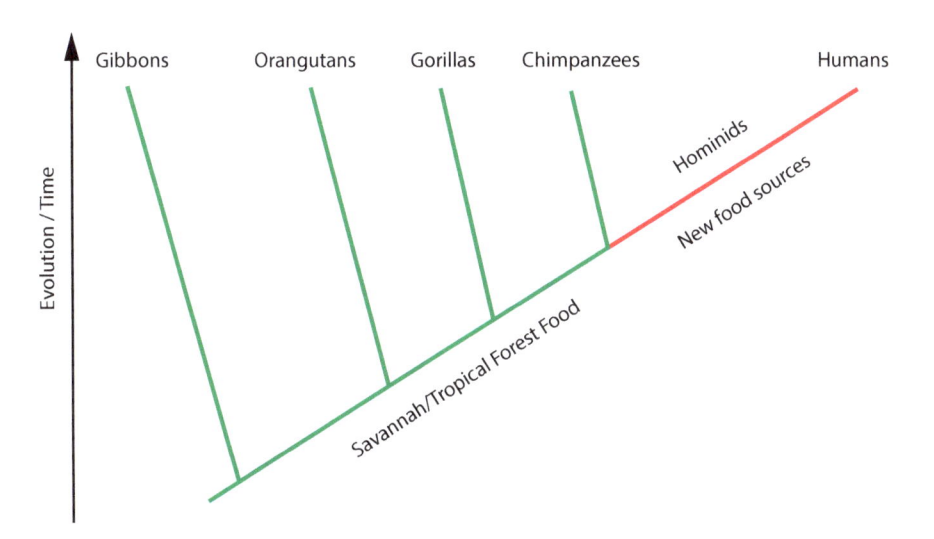

Fig. 1.25 Human evolution at a glance. The various ape species were bound to the food available in the tropical savanna, accessible in this niche. Only with the expansion of the food supply was "humanization" possible

aroma compounds. It is quite conceivable that there are lipases in the saliva on the tongue that already split fat in the mouth [58] and thus trigger weak positive sensations in some species [59]. The same is conceivable for a starch taste [60], which is not mediated via the conventional sweet receptors. Starch is already broken down into maltodextrin by amylases in saliva. This can signal further eating apart from sweet to access the glucose contained therein. In times of scarcity, these signals were vital for survival. In times of prosperity, saturation, and a certain excess, they are no longer important. All the more important is to listen to the body signals that express themselves, for example, through "appetite for..." or "desire for...". However, they can only be relearned with a pleasant feeling of hunger, and they should be listened to instead of consuming function-free products like power bars or detox drinks outside of meals. Even the supposedly healthy smoothie as a snack is calorie-rich and does not provide lasting satiety due to the strong shredding of the ingredients, but makes you a little thicker each time. Why this is the case is explained in detail at the cellular level in Chap. 6.

The hominids, the Neanderthals, and the first representatives of *Homo sapiens* certainly made no major mistakes in their evolutionary dietary choices, otherwise, we would not be sitting here in this world today. The greatest catastrophe in the development of humanity would have been if a few hominids sitting around the campfire had started to systematically refuse food as we do today; then things would have been bad for us humans. The red branch in Fig. 1.25 would not exist today.

References

1. http://www.faz.net/aktuell/politik/acrylamid-debatte-kuenast-kein-bratkartoffel-verbot-182976.html.
2. https://www.verbraucherzentrale.de/acrylamid; https://www.foodwatch.org/de/informieren/acrylamid/2-minuten-info/.
3. Madle, S., Broschinski, L., Mosbach-Schulz, O., Schöning, G., & Schulte, A. (2003). Zur aktuellen Risikobewertung von Acrylamid in Lebensmitteln. *Bundesgesundheitsblatt-Gesundheitsforschung-Gesundheitsschutz, 46*(5), 405–415.
4. Johnson, K. A., Gorzinski, S. J., Bodner, K. M., Campbell, R. A., Wolf, C. H., Friedman, M. A., & Mast, R. W. (1986). Chronic toxicity and oncogenicity study on acrylamide incorporated in the drinking water of Fischer 344 rats. *Toxicology and applied pharmacology, 85*(2), 168.
5. Jirgensons, B., & Straumanis, M. (2013). *Kurzes Lehrbuch der Kolloidchemie*. Springer.
6. https://www.sein.de/energetisiertes-wasser-universitaet-lueftet-geheimnisse-um-wasser/ .
7. Taubes, G. (2001). The soft science of dietary fat. *Science, 291*(5513), 2536–2545.
8. Vilgis, T. (2010). *Das Molekül-Menü*. Hirzel.
9. Gardner, H. W. (1989). Oxygen radical chemistry of polyunsaturated fatty acids. *Free Radical Biology and Medicine, 7*(1), 65–86.
10. Gonder, U. (2015). Kokosöl fürs Gehirn. *Zeitschrift für Orthomolekulare Medizin, 4*(4), 16–19.
11. Elmore, J. S., Cooper, S. L., Enser, M., Mottram, D. S., Sinclair, L. A., Wilkinson, R. G., & Wood, J. D. (2005). Dietary manipulation of fatty acid composition in lamb meat and its effect on the volatile aroma compounds of grilled lamb. *Meat science, 69*(2), 233–242.
12. Matthan, N. R., Ooi, E. M., Van Horn, L., Neuhouser, M. L., Woodman, R., & Lichtenstein, A. H. (2014). Plasma phospholipid fatty acid biomarkers of dietary fat quality and

References

endogenous metabolism predict coronary heart disease risk: A nested case-control study within the women's health initiative observational study. *Journal of the American Heart Association, 3*(4), e000764.

13. Zheng, H., Yde, C. C., Clausen, M. R., Kristensen, M., Lorenzen, J., Astrup, A., & Bertram, H. C. (2015). Metabolomics investigation to shed light on cheese as a possible piece in the French paradox puzzle. *Journal of agricultural and food chemistry, 63*(10), 2830–2839.
14. Kaleta, C., de Figueiredo, L. F., Werner, S., Guthke, R., Ristow, M., & Schuster, S. (2011). In silico evidence for gluconeogenesis from fatty acids in humans. *PLoS computational biology, 7*(7), e1002116.
15. Gosby, A. K., Conigrave, A. D., Raubenheimer, D., & Simpson, S. J. (2014). Protein leverage and energy intake. *Obesity reviews, 15*(3), 183–191.
16. Biesalski, H. K. (2015). *Mikronährstoffe als Motor der Evolution*. Springer Spektrum.
17. Bolland, M. J., Grey, A., Gamble, G. D., & Reid, I. R. (2014). The effect of vitamin D supplementation on skeletal, vascular, or cancer outcomes: A trial sequential meta-analysis. *The lancet Diabetes & endocrinology, 2*(4), 307–320.
18. Sattler, S. E., Cahoon, E. B., Coughlan, S. J., & DellaPenna, D. (2003). Characterization of tocopherol cyclases from higher plants and cyanobacteria. Evolutionary implications for tocopherol synthesis and function. *Plant physiology, 132*(4), 2184–2195.
19. Matthews, R. G., Kotmas, M., & Datta, S. (2008). Cobalamin-dependent and cobamide-dependent methyltransferases. *Current Opinion in Structural Biology, 18*, 658–666.
20. Carmel, R. (1997). Cobalamin, the stomach, and aging. *The American Journal of Clinical Nutrition, 66*, 750–759.
21. Vierich, T. A., & Vilgis, T. A. (2013). *Aroma, Die Kunst des Würzens*. Stiftung Warentest.
22. Cunningham, E. (2012). Are diets from paleolithic times relevant today? *Journal of the Academy of Nutrition and Dietetics, 112*(8), 1296.
23. Scherberich, J. E. (2008). Kalzium-Phosphat-und Knochenstoffwechsel. *Der Nephrologe, 3*(6), 507–517.
24. Pereira-da-Silva, L., Costa, A. B., Pereira, L., Filipe, A. F., Virella, D., Leal, E., & Serelha, M. (2011). Early high calcium and phosphorus intake by parenteral nutrition prevents short-term bone strength decline in preterm infants. *Journal of pediatric gastroenterology and nutrition, 52*(2), 203–209.
25. Milton, K. (2004). Ferment in the family tree: Does a frugivorous dietary heritage influence contemporary patterns of human ethanol use? 1. *Integrative and Comparative Biology, 44*(4), 304–314.
26. Dominy, N. J. (2015). Ferment in the family tree. *Proceedings of the National Academy of Sciences, 112*(2), 308–309.
27. Carmody, R. N., & Wrangham, R. W. (2009). The energetic significance of cooking. *Journal of Human Evolution, 57*(4), 379–391.
28. Wrangham, R. (2009). *Catching fire: How cooking made us human*. Basic Books.
29. Wuketits, F. M. (2011). *Wie der Mensch wurde, was er isst: Die Evolution menschlicher Ernährung*. Hirzel.
30. Wrangham, R., & Conklin-Brittain, N. (2003). Cooking as a biological trait. *Comparative Biochemistry and Physiology Part A: Molecular & Integrative Physiology, 136*(1), 35–46.
31. DeCasien, A. R., Williams, S. A., & Higham, J. P. (2017). Primate brain size is predicted by diet but not sociality. *Nature ecology & evolution, 1*(5), 0112.
32. Dunbar, R. I. (2009). The social brain hypothesis and its implications for social evolution. *Annals of human biology, 36*(5), 562–572.
33. Richter, N. (2014). *Paleo – Power for life*. Christian Verlag GmbH.
34. Hildmann, A. (2014). *Vegan for youth, Becker-Joest-Volk*. Hilden.
35. Greger, M., & Stone, G. (2016). *How not to die: Discover the foods scientifically proven to prevent and reverse disease*. Pan Macmillan.
36. Kwok, R. H. M. (1968). Chinese-restaurant syndrome. *New England Journal of Medicine, 278*, 796.

37. Grimm, H. U. (2014). *Die Suppe lügt: Die schöne neue Welt des Essens*. Droemer eBook.
38. Grimm, H. U., & Ubbenhorst, B. (2013). *Chemie im Essen: Lebensmittel-Zusatzstoffe. Wie sie wirken, warum sie schaden*. Knaur eBook.
39. Freeman, M. (2006). Reconsidering the effects of monosodium glutamate: A literature review. *Journal of the American Association of Nurse Practitioners, 18*(10), 482–486.
40. Farthing, C. A., Farthing, D. E., Gress, R. E., & Sweet, D. H. (2017). Determination of l-glutamic acid and γ–aminobutyric acid in mouse brain tissue utilizing GC–MS/MS. *Journal of Chromatography B, 1068*, 64–70.
41. Toyota, M., Spencer, D., Sawai-Toyota, S., Jiaqi, W., Zhang, T., Koo, A. J., Howe, G. A., & Gilroy, S. (2018). Glutamate triggers long-distance, calcium-based plant defense signaling. *Science, 361*(6407), 1112–1115.
42. https://supplementreviews.com/forum/index.php?topic=28361.0.
43. Walker, A. (1981). Diet and teeth: Dietary hypotheses and human evolution. *Philosophical Transactions of the Royal Society of London. B, Biological Sciences, 292*(1057), 57–64.
44. Moeller, A. H., Li, Y., Ngole, E. M., Ahuka-Mundeke, S., Lonsdorf, E. V., Pusey, A. E., & Ochman, H. (2014). Rapid changes in the gut microbiome during human evolution. *Proceedings of the National Academy of Sciences, 111*(46), 16431–16435.
45. Vital, M., Karch, A., & Pieper, D. H. (2017). Colonic butyrate-producing communities in humans: An overview using omics data. *MSystems, 2*(6), e00130-e00217.
46. Berna, F., Goldberg, P., Horwitz, L. K., Brink, J., Holt, S., Bamford, M., & Chazan, M. (2012). Microstratigraphic evidence of in situ fire in the Acheulean strata of Wonderwerk Cave, Northern Cape province, South Africa. *Proceedings of the National Academy of Sciences, 109*(20), E1215–E1220.
47. Palacios, C., Rivas-Tumanyan, S., Morou-Bermúdez, E., Colon, A. M., Torres, R. Y., & Elías-Boneta, A. R. (2016). Association between type, amount, and pattern of carbohydrate consumption with dental caries in 12-year-olds in Puerto Rico. *Caries Research, 50*(6), 560–570.
48. Ahrens, E. H., Jr., Insull, W., Jr., Blomstrand, R., Hirsch, J., Tsaltas, T. T., & Peterson, M. L. (1957). The influence of dietary fats on serum-lipid levels in man. *Lancet, 1*(943), 19–57.
49. Boback, S. M., Cox, C. L., Ott, B. D., Carmody, R., Wrangham, R. W., & Secor, S. M. (2007). Cooking and grinding reduces the cost of meat digestion. *Comparative Biochemistry and Physiology Part A: Molecular & Integrative Physiology, 148*(3), 651–656.
50. Guerra, A., Etienne-Mesmin, L., Livrelli, V., Denis, S., Blanquet-Diot, S., & Alric, M. (2012). Relevance and challenges in modeling human gastric and small intestinal digestion. *Trends in biotechnology, 30*(11), 591–600.
51. Kaur, L., Maudens, E., Haisman, D. R., Boland, M. J., & Singh, H. (2014). Microstructure and protein digestibility of beef: The effect of cooking conditions as used in stews and curries. *LWT-Food Science and Technology, 55*(2), 612–620.
52. Bax, M. L., Aubry, L., Ferreira, C., Daudin, J. D., Gatellier, P., Rémond, D., & Santé-Lhoutellier, V. (2012). Cooking temperature is a key determinant of in vitro meat protein digestion rate: Investigation of underlying mechanisms. *Journal of agricultural and food chemistry, 60*(10), 2569–2576.
53. Zielbauer, B. I., Franz, J., Viezens, B., & Vilgis, T. A. (2016). Physical aspects of meat cooking: Time dependent thermal protein denaturation and water loss. *Food biophysics, 11*(1), 34–42.
54. Siehe zum Beispiel Vilgis, T. (2010). *Das Molekül-Menü*. Hirzel.
55. Wobber, V., Hare, B., & Wrangham, R. (2008). Great apes prefer cooked food. *Journal of Human Evolution, 55*(2), 340–348.
56. Warneken, F., & Rosati, A. G. (2015). Cognitive capacities for cooking in chimpanzees. *Proc. R. Soc. B, 282*(1809), 20150229.
57. Cameron, P., Hiroi, M., Ngai, J., & Scott, K. (2010). The molecular basis for water taste in *Drosophila. Nature, 465*(7294), 91.

References

58. Voigt, N., Stein, J., Galindo, M. M., Dunkel, A., Raguse, J. D., Meyerhof, W., & Behrens, M. (2014). The role of lipolysis in human orosensory fat perception. *Journal of lipid research, 55*(5), 870–882.
59. Galindo, M. M., Voigt, N., Stein, J., van Lengerich, J., Raguse, J. D., Hofmann, T., & Behrens, M. (2011). G protein-coupled receptors in human fat taste perception. *Chemical senses, 37*(2), 123–139.
60. Lapis, T. J., Penner, M. H., & Lim, J. (2016). Humans can taste glucose oligomers independent of the hT1R2/hT1R3 sweet taste receptor. *Chemical senses, 41*(9), 755–762.

Recognizing Food, Learning to Eat: A Look into Evolution

Abstract

To this day, our diet is determined by fire. However, the settling down of humans after the Stone Age brought new foods into the diet, such as cereals, milk, and dairy products, as well as new techniques, like controlled fermentation. In order for milk to become digestible for humans, a point mutation in the DNA was necessary. Lactose tolerance was able to prevail in some niches.

2.1 Hunters, Gatherers, Energy Gainers

Our (food) culture is determined by fire to this day. Despite high-tech cooking systems, induction plates, microwaves, sous-vide cookers, programmable combi steamers, Ohmic heating, etc., not much will change, because all these methods only thermally alter the food, like fire. How exactly this happens is insignificant details that merely control the process from the outside but do not change anything essential at the molecular result. Although the heat transfer can be adjusted and precisely controlled in various ways with modern techniques, the cooking method is always the same: heat energy is transferred to food, and they react according to their composition to the associated temperature. Thermal cooking, regardless of the media—fire, radiation, or water baths—has only the melting of fats, the destabilization of cell membranes, the denaturation of proteins, and the softening of plant cell material as a result, depending on the temperature. Since the relevant cooking temperatures have not changed since then, it is no wonder that cooking techniques on a molecular scale hardly differ.

This outstanding cultural technique—cooking using the heat of fire—has proven itself to this day. No wonder, then, that today, despite high-tech, we still cook basically as the first attempts of hominids suggest. Cooking and food technology thus began with fire and the end of microbes—and therefore with food safety. The control of fire was tantamount to a "cultural revolution" in human history.

As can be seen using modern methods from archaeological finds [1], the dietary change from raw (and wildly fermented) to cooked and roasted was a crucial point for the further development of humans. As already noted in the previous Chap. 1, the proportion of animal food increased significantly with the use of fire, and thus also the proportion of essential macronutrient components, which until then played only a subordinate role.

Hunters and gatherers developed various strategies [2, 3] to ensure their survival. Challenges were climatic fluctuations, weather periods, or, depending on latitude, seasonally very different living conditions. Especially before settling down, securing food was correspondingly difficult.

2.1.1 Scavenging and Hunting—Nutrition before the Utilization of Fire

The consumption of carrion, insects, and other small animals, and eventually the targeted hunting of small animals by the first hominids were crucial steps in the systematic change of diet. Hominids increasingly differentiated themselves from chimpanzees and other non-hominid primates. The basic prerequisite for this was the ability to walk upright. For only in this way were the hands freed for new activities apart from locomotion. This was a decisive step for food procurement, as food could be transported in much larger quantities over longer distances and made accessible to members of growing social structures. Hunter-gatherers spread more and more and developed methods and systematic structures for food procurement that differed significantly from the food procurement abilities of chimpanzees, who hardly hunted animals, as meat is not primarily part of their dietary components. Meat-eating hominids and the first hunter-gatherers evolved about 2.5 billion years ago. However, hunting meant a great risk due to the lack of developed techniques at that time. The first protein and meat sources were therefore insects, carrion of killed animals, soft-bodied and small animals. This step can be compared to the development of alcohol dehydrogenase (see Sect. 1.8), as the meat not only provided the macronutrients proteins and fat, but also a whole range of vital micronutrients in physiologically highly available form, such as iron, zinc, and especially B-complex vitamins, which are hardly found in plant-based food. Even though raw meat is difficult to chew and thus requires energy itself, the advantage was evident.

2.1.2 Effort and Yield of Hunting and Gathering—The Energy Balance

As in Chap. 1 was already mentioned, the umami taste combined with the high biological availability acted as a reward and incentive for humans to take on the

2.1 Hunters, Gatherers, Energy Gainers

high risks of hunting larger animals. However, there is also an economic aspect, namely the energy balance for the hunters and gatherers. Hunting and gathering require energy, which in turn must be provided through food. And since the increase in human brain volume also required energy, there must have been a much greater energy intake in the form of macro- and micronutrients. Apparently, hunters and gatherers intuitively "maximized" their energy gain during food search [4]. This can be quantitatively recorded and more precisely defined within the framework of ethnological studies on indigenous peoples, such as the Aché in Paraguay or the San (!Kung) in Africa, historical data, and scientific analyses of archaeological bone finds. For this purpose, a recovery rate for energy during food search is defined in the research branch of ethnology [5]. For this, the energy, for example, the calorie amount of the whole food that was brought home, such as an entire pig or a deer, or the energy content of the amount of berries, fruits, roots, etc., is divided by the time required for their procurement. In doing so, not only the pure hunting times, digging times, and gathering times must be taken into account, but also the times for the distances that had to be covered for this must be included, because all this required energy, which the hunters and gatherers had to take in through food first. From this, an energy rate can be defined, which relates the energy expended in the search to the recovered energy through the corresponding food [6]. It clearly shows that the energy recovery rate is highest for animal food, as shown in Fig. 2.1. It is therefore not surprising that in the evolution of hominids animal foods rapidly gained in importance in order to ensure survival.

In this context, it is noteworthy that not only hunters and gatherers maximize energy recovery, but also animals, which especially allowed the omnivorous precursor of humans to survive and have an increasingly longer life due to the

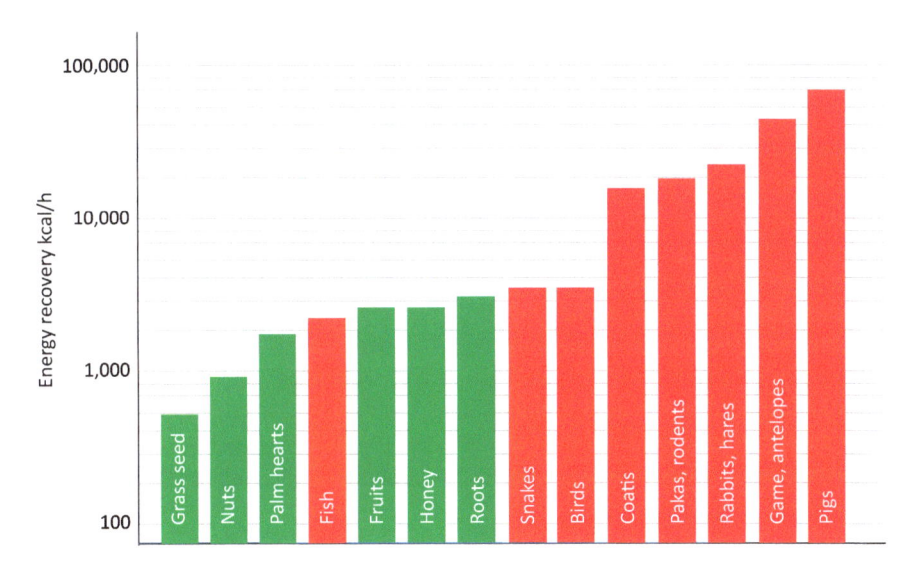

Fig. 2.1 The energy recovery in comparison. Red: animal foods, green: plant foods

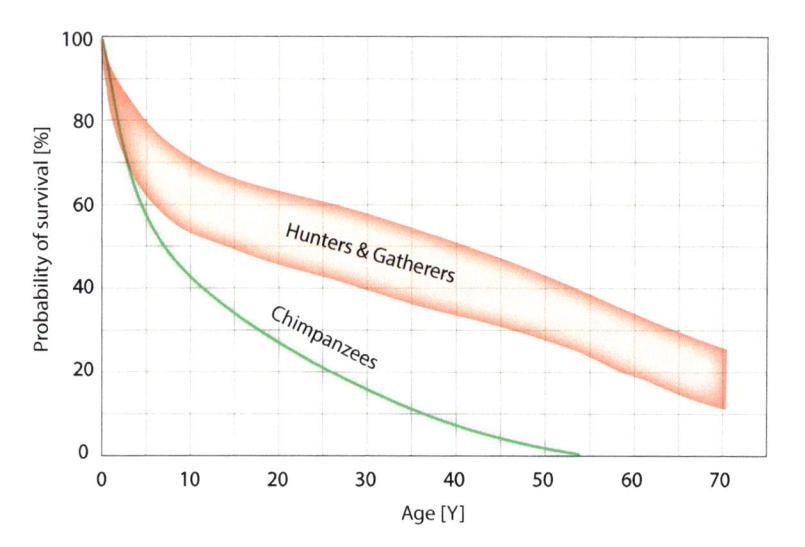

Fig. 2.2 Survival probability of chimpanzees (green line) compared to ethnic groups (red area) belonging to hunters and gatherers

constant optimized nutrition [7]. In Fig. 2.2, the advantage of hunters and gatherers becomes clear. The survival rate is significantly higher than that of chimpanzees, and the lifespan is also significantly extended. The ethnic groups used for analysis and considered as pure hunters and gatherers are Kiwi from New Zealand (upper limit) and Hadza from Tanzania (lower limit).

These high energy recovery rates from animal food can thus be considered as another driving force of evolution. The advantage of animal food for early humans during evolution is evident: it provided not only enough energy for the necessary activities for food procurement but also for the further development of humans.

With the higher nutrient density, especially of biologically available proteins and the associated essential amino acids and animal, often unsaturated fatty acids, along with the high energy recovery rates, the hominids took a first step towards the cultivation of food, which will be discussed in more detail later. This includes, above all, the targeted selection of food, which only became possible through increasing dexterity, refining fine motor skills, and growing cognitive abilities. In the early history of humans and when comparing chimpanzees and non-hominid primates with the first *Homo sapiens*, this becomes clear, as indicated in Fig. 2.3.

In order to obtain food such as larger game, appropriate tools and hunting systems had to be developed. Hunting and the conscious selection of food based on their energy production rates now allowed the hominids to develop early forms of an increasingly differentiating "food culture" compared to non-hominid primates.

How this development bore fruit for the hominids, the hunters and gatherers, is evident in a very early development of a culture that apparently defines itself fundamentally through food. Comparing the procurement of the daily energy

2.1 Hunters, Gatherers, Energy Gainers

Homo sapiens

- larger, hunted game
- small animals
- honey/extracts
- roots
- insects
- ripe fruits
- raw roots
- leaves

Chimpanzees / non-hominid primates

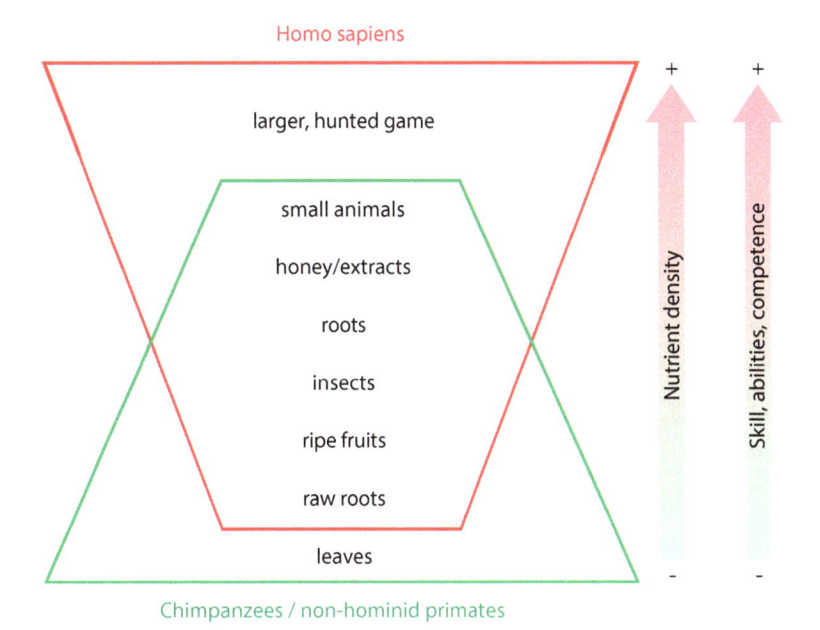

Nutrient density → +/−
Skill, abilities, competence → +/−

Fig. 2.3 The development of humans as a result of increased nutrient density and increasing motor skills is reflected in a systematic use of animal food, which was no longer attainable with simple means. The food of chimpanzees and early forms of humans overlapped significantly. The high proportion of animal food proved to be crucial for the better energy supply and development of *Homo sapiens* (see Fig. 1.25)

requirements of chimpanzees and non-hominid primates with that of hominids reveals astonishing results, especially when calculated over a lifetime, as shown in Fig. 2.4.

Fundamental differences become apparent in this comparison. Non-hominid primates and chimpanzees cover their daily needs without procuring more food than necessary. (Male) hunters and gatherers, on the other hand, procure significantly more food, measured by its energy content, than is necessary for their own needs from a certain age. This "overproduction" benefits the community to a great extent, apparently far beyond their own offspring. This reveals an early model of a generational contract, as hunters and gatherers must be provided for by the generations already in the overproduction phase during their young and later years of life. Social structures, therefore, inevitably formed relatively early in human history. Ultimately, these are the basic prerequisites for social structures and their targeted and controlled further development.

Non-hominid primates and chimpanzees, on the other hand, live a simplified life "from hand to mouth," as can be seen from the green curves in Fig. 2.4, while for *Homo sapiens*, food security through overproduction was in the foreground.

This is another indication of the early development towards a targeted "nutrition culture" based on food security (see Fig. 2.5). Only the procurement of

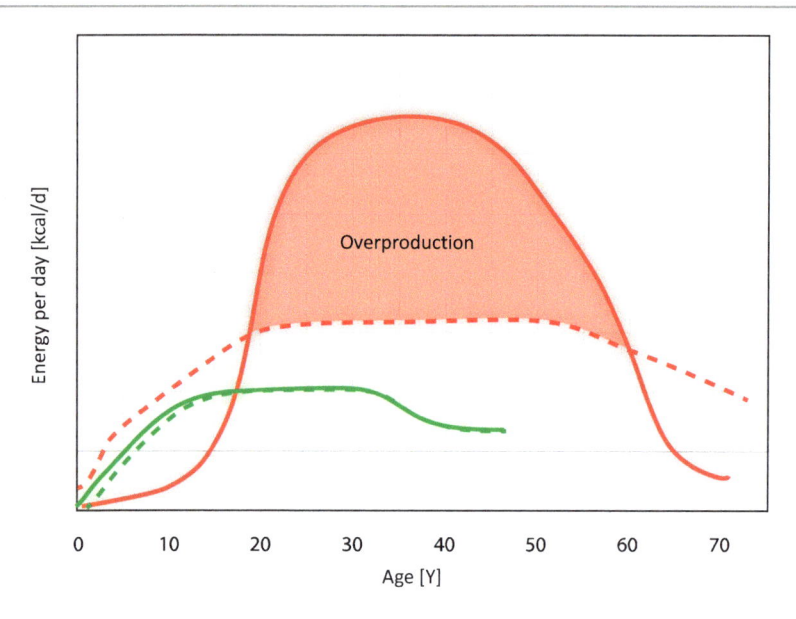

Fig. 2.4 The energy of food for the daily needs of chimpanzees (green) and hunters and gatherers (red) over a lifetime. The solid lines represent procurement, the dashed lines represent the estimation of self-consumption. Overproduction is only possible for hunters and gatherers—and therefore also for early hominids

high-calorie food in sufficient quantities enabled the development of *Homo sapiens* and later also the control over fire and targeted cooking, which, as already described in Chap. 1, significantly accelerated this path.

2.2 Animal Foods—High Energy Gain

As already noted, animal food played a major role in human development. The direct comparison shows this: While chimpanzees consume only 10-40 g of animal protein in the form of small animals per day, the amount for hominids/hunters and gatherers ranges from 270–1400 g per day. That complex learning processes are necessary for this more complex diet is known from simple experiments [8]. If animals are raised in captivity, they can be easily released into the wild if they are herbivores (plant-eaters). Their food search is simple enough to quickly learn the required knowledge in the wilderness. Carnivores, including meat-eating monkeys with complex food searches, cannot be easily released into the wild without risking failure in food procurement. This shows that a complex learning process within the species is necessary for learning appropriate hunting techniques [9]. In fact, this can still be observed in indigenous peoples today: A certain age and thus life experience is necessary for learning the techniques for higher productivity in food procurement, as can be seen in Fig. 2.4.

2.2 Animal Foods—High Energy Gain

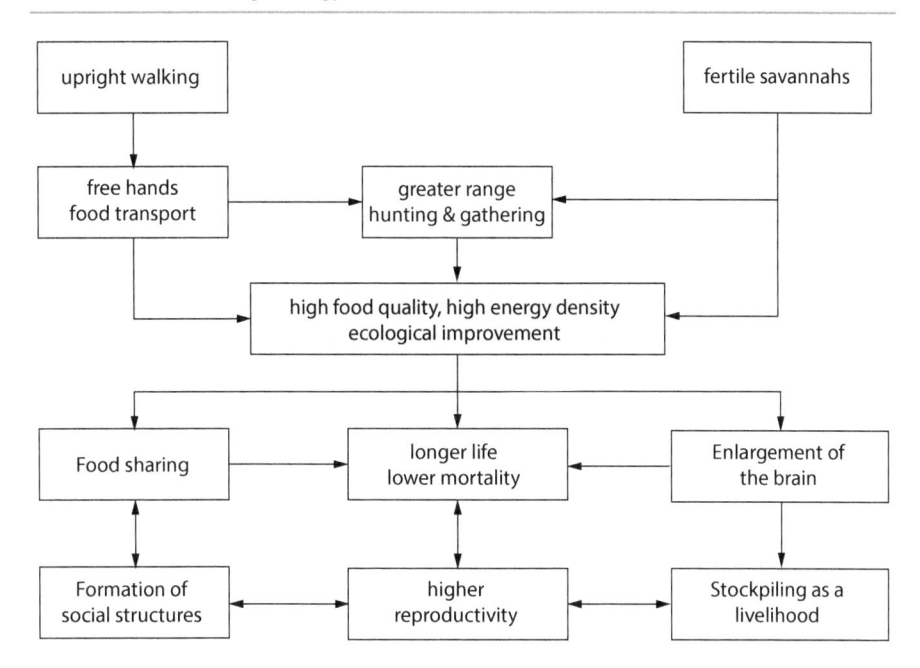

Fig. 2.5 The development of culture in human history is also determined by food. The food supply and the constantly improving handling of it are at the center and are responsible for the ecological and economic as well as the social structures

The hunting techniques of *Homo sapiens* already differed early on from those of other primates and animals. Hunting animals usually secretly ambush their prey or use simple pursuit techniques. Hominid hunters, as well as today's hunter-gatherers, take into account a wealth of information, making context-specific decisions both during the search and when encountering prey, and can quickly adapt to new situations and make new decisions. This has been deliberately referred to as the "beginning of science" [10]. The evolution to hominids and later to *Homo sapiens* was associated with great intellectual abilities and cognitive achievements [11]. Herbivores have not made this step to this day.

Hunted, killed, and fire-cooked wild animals were one of the best food sources available to hunter-gatherers. But not only their meat, but also fat and bone marrow were highly valued foods and had the highest priority in the diet of the Paleolithic [12]. Although hunting requires a high investment of time, high risk, and high skill, it is rewarded with a high energy return rate. Therefore, the development to hunter-gatherer in evolution can be considered one of the most significant steps that chimpanzees, who invest far less time and intellect in their food search, have not made. The resulting diet and the growing proportion of animal food are thus an essential point in human evolution. These anthropological approaches, in turn, suggest in line with previous findings that the self-imposed effort for food search and dietary changes were responsible for the increase in brain volume. Only then could social structures, responsibility, sharing of food, and the already mentioned generational contract develop. Only through food security could the reproduction rate increase and the offspring learn and educate themselves to later participate in higher energy returns.

So far, the most important facts on a macroscopic and "metascopic" level, which already show the development of hominids from an ethnological and anthropological perspective. However, the physiological advantage of tapping into new food sources only becomes apparent when considering the functions of the nutrients, which ultimately are the keys to understanding evolution, similar to how the umami taste and the associated glutamic acid were the driving force for the further development of cooking for hominids, Neanderthals, and *Homo sapiens*. Apparently, economic principles were systematically realized very early on.

2.3 Fat, Brain, Bone Marrow—Sources of Essential Omega-3 Fatty Acids

But it was not just the meat alone, as the hunted animal provided far more nutrients that played an essential role in development. The mentioned high energy extraction rates could only be achieved if the animal was eaten completely. Nothing could be wasted, every kilojoule counted. Thus, the tubular bones of the hunted animals were broken open and the bone marrow was eaten [13]. Bone marrow has a high fat content, the fatty acid spectrum is widely distributed and has a high proportion of previously inaccessible essential fatty acids, especially of the eicosapentaenoic acid found only in animals C 20:5 (n-3) and docosahexaenoic acid C 22:6 (n-3) (see Chap. 1). But also other polyunsaturated fatty acids and valuable animal *trans*-fatty acids (in ruminants) were consumed [14].

From Table 2.1, it can be seen how much fat is contained in the parts of wild animals. As mentioned, bone marrow represents a valuable source of long-chain essential fatty acids. Plant-based food does not contain these fatty acids, as shown in Table 2.2 for the late Paleolithic period.

The more frequent consumption of these animal fats in the Paleolithic period also enabled the development of hunter-gatherers into *Homo sapiens*. Table 2.2 also shows how advantageous a mixed diet of animal and plant food was and still is for the essential fatty acids for *Homo sapiens*.

Table 2.1 Averaged fat content of a wild animal. Measured by the essential fatty acids, the bone marrow is one of the most valuable parts of the animal

Tissue	Weight proportion (%)	Edible portion/ Animal (%)	Fat (%)
Muscle meat	45	90	2.3
Brain	0.5	1.0	9.0
Liver	1.9	3.8	3.3
Bone marrow	1.5	3.0	50.0
Fat, abdominal fat…	1.0	2.0	80.0

2.3 Fat, Brain, Bone Marrow—Sources of Essential ...

Table 2.2 Long-chain polyunsaturated fatty acids are found only in animal foods in adequate concentration. In particular, docosahexaenoic acid 22:6 (n-3), DHA, plays a central role in brain development in higher mammals

Fatty acid	From plant food (%)	From animal food (%)
Linoleic acid 18:2 ($\omega 6$)	4.28	4.56
α-Linolenic acid 18:3 (n-3)	11.4	1.21
Arachidonic acid 20:4 ($\omega 6$)	0.06	1.75
Eicosapentaenoic acid 20:5 (n-3)	0.14	0.25
Docosatetraenoic acid 22:4 ($\omega 6$)	0.00	0.12
Docosapentaenoic acid 22:5 (n-3)	0.00	0.42
Docosahexaenoic acid 22:6 (n-3)	0.00	0.27

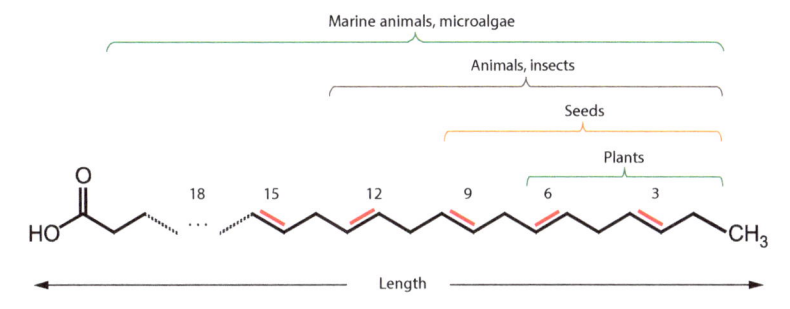

Fig. 2.6 The systematics of the possible occurrence of the positions of the double bonds (red) of Omega-3 fatty acids in mammals, insects, and plants. The chain length varies, with a maximum chain length of 24 carbon atoms C being relevant for human nutrition.

For the following explanations, the systematics of the fatty acids is of great importance, which results from the inherent difference between animal and plant foods, as schematically shown in Fig. 2.6.

It must be emphasized that, for physiological reasons, the triple unsaturated form (α-linolenic acid, ALA) can occur in land plants at most. All plant Omega-3 fatty acids are neither longer than 18 carbon atoms nor more unsaturated. These Omega-3 fatty acids are reserved for insects and animals. Higher chain lengths no longer occur in plants. The patterns and lengths of the Omega-3 fatty acids thus allow a universal and therefore generally valid distinction of the sources of these essential fatty acids. Plant and animal food can thus be fundamentally distinguished based on the fatty acids.

As already noted, animal Omega-3 fatty acids were (and are) indispensable for brain development. Above all, the long and more than triple unsaturated fatty acids have contributed significantly. Fatty acids from plants, which only have the triple unsaturated α-linolenic acid, are not sufficient for brain development alone. This is still the case today, and the significant effects of long-chain fatty acids on the brain growth of fetuses have long been known [15]. Nowadays, these fatty acids are usually attributed to fish, and their occurrence in bone marrow seems to be forgotten. The classic marrow dumpling soup, a fine bone marrow puree, or simply heated bone marrow on toast with salt are high-quality delicacies with lasting effects on our nutritional status. Moreover, Fig. 2.6 illustrates the advantages of insects as food. It is not only the high-quality protein that is currently coming into focus, but also the polyunsaturated fatty acids (which are also present in the cell membranes), which elevate insects to high-quality food.

In contrast to today's Western dietary habits, significantly more of these essential, long-chain animal fatty acids were consumed during the late Paleolithic period, which contributed considerably to the enlargement of the brain and the development of *Homo sapiens*, as clearly demonstrated by the analysis of bone findings. Residents near the coast in African habitats apparently had an even greater advantage. Long-chain essential fatty acids could be obtained particularly through fish. The development of hunter-gatherers accelerated [16].

2.4 Fatty Acids: Function, Structure, and Physical Properties

In this context, the question arises as to why specifically EPA- and DHA fatty acids were and still are so crucial for the evolution of humans, especially in the development of the brain. Why not more accessible fatty acids and why specifically long-chain and highly unsaturated animal fats? This apparently has molecular causes, including in the cell membranes, which are designed for a precisely defined composition of fatty acids depending on the function of the cells. In some cases, it is primarily physicochemical properties that can be regulated via fatty acids. The fatty acid composition in the neural cells of the brain is of particular importance in this regard.

Cell membranes, as we already learned in Chap. 1 for plant cells (Fig. 1.17), are so-called lipid bilayers made of phospholipids. These consist of a water-soluble head and two tails made of fatty acids. The molecule is thus "amphiphilic," meaning it loves both water and fat.

Aside from fat molecules, the triacylglycerols, which were already addressed in Chap. 1, fatty acids play a very significant role in these phospholipids. In contrast to fats, phospholipids carry only two fatty acids, as shown schematically in an example in Fig. 2.7.

Phospholipids can be constructed differently depending on the requirements of nature. The water-soluble head group can have different shapes, or the fatty acids can vary greatly. To understand basic questions, it is sufficient to initially consider simple models for phospholipids that only take into account the simultaneous fat and water solubility, as shown in Fig. 2.8. Nutritional scientific subtleties, which are determined, for example, by the exact shape and properties of the fatty acids, can be much better understood in context once the physicochemical basics have been worked out.

2.4 Fatty Acids: Function, Structure ...

Fig. 2.7 A phospholipid (**a**) consists of a water-soluble head (here a nitrogen-phosphorus group, a so-called choline group) and two fatty acids, which can be saturated (as on the right) or unsaturated (on the left). The space-filling model of the same phospholipid is shown in (**b**)

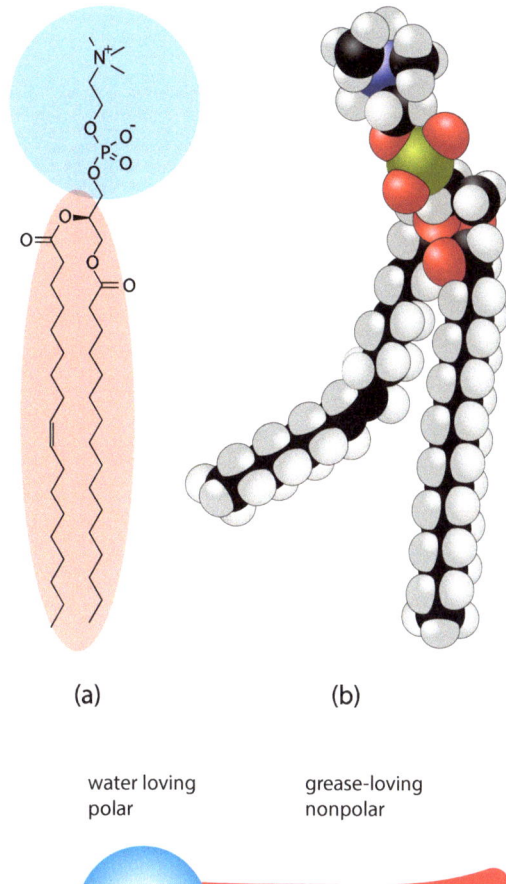

Fig. 2.8 A simple model of phospholipids: The hydrophilic head is shown in blue, the two fatty acids are red

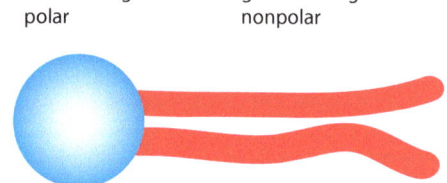

The purpose of this seemingly crude simplification is to demonstrate the "self-organization" of phospholipids. It is based on a simple thermodynamic principle: "Like attracts like," meaning, the fatty acids cluster together, as do the polar heads. The physical reason for this is interaction forces, in this case, dipolar interactions and van der Waals interactions via the fatty acids. In the cell membranes, the phospholipids organize themselves into double layers, as shown in Fig. 2.9.

Phospholipids are among the fundamental molecular classes that play central roles in all living systems, whether plant or animal. These lipid bilayers surround all natural cells that exist on this planet. And this is where the fatty acids come into play: They determine the mechanical properties of the membranes, their mobility and flexibility, or their temperature stability, as illustrated in Fig. 2.10.

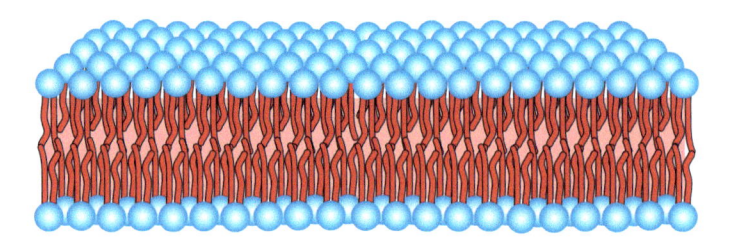

Fig. 2.9 The self-organization of surface-active molecules. The membrane consists of a double layer of phospholipids

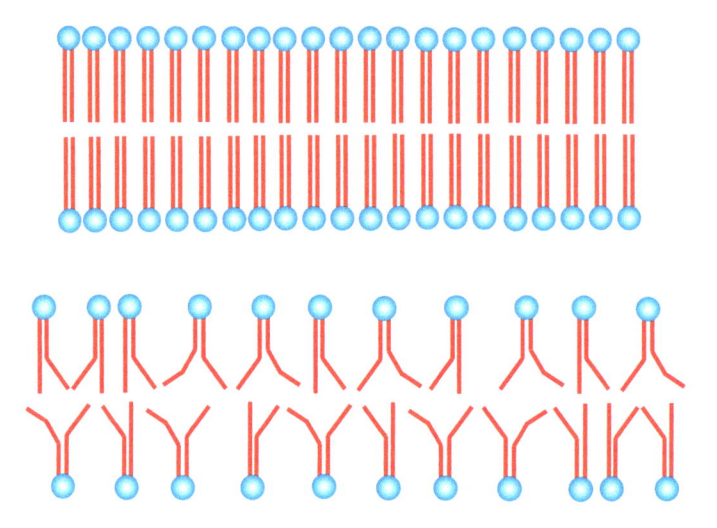

Fig. 2.10 Membranes made of different fatty acids. Top: A bilayer of phospholipids with saturated fatty acids, which can quickly arrange and crystallize. Bottom: A bilayer of unsaturated fatty acids, due to the kinks in the fatty acids, they arrange themselves only with difficulty, the membrane remains rather "liquid" (exaggerated representation)

These membranes must remain fluid and flexible in living, functioning cells, meaning that the individual phospholipid molecules must remain mobile within the membrane, as in a practically two-dimensional liquid. In the "solid", crystalline phase (Fig. 2.10top), this is not always the case, such as at low temperatures. Then the fatty acids will crystallize, the individual phospholipids are frozen in place, and the membrane breaks very easily. Cell death would be the consequence. The degree of saturation of the fatty acids is thus a biophysical method for precisely adjusting the physiologically required membrane fluidity and membrane flexibility [17]. But not only the degree of saturation, but also the exact location of the double bond [18] has an effect, especially in brain cells. This is precisely the crucial point in the function of omega-3 fatty acids, because their position of the double bond has a significant physical effect on local membrane fluidity, mainly because the double bond already begins at the third bond from the methyl end, i.e., at the fat-loving end (see Figs. 1.3 and 1.4). Thus, one of the basic tasks of unsaturated fatty acids in cell membranes is clear: They ensure fluidity and flexibility at very specific points, namely in the middle of the cell membrane, where the two layers meet at the fatty acids. They do this without changing the cohesion and mechanical flexibility—as well as the fluidity of the membranes—too much. Only omega-3 fatty acids can do this better than omega-6 fatty acids, whose first double bond is already closer to the hydrophilic choline heads.

This leads to a fundamental insight: Only a very balanced mixture of saturated and unsaturated fatty acids with variable saturation levels and variable chain lengths of the phospholipids creates the required flexibility, fluidity, and bending stiffness of the cell membranes. In addition, the human body temperature deviates only slightly from 37°C, so the mixing ratio of saturated and unsaturated fatty acids must not vary greatly for physical reasons.

However, membrane flexibility is not only generated by the different fatty acids, but also by cholesterol, as indicated in Fig. 2.11. Cholesterol is also located in the membrane and differs significantly in its chemical structure from the lipids in the bilayer. The OH group serves as a hydrophilic head, while the lipophilic fatty acids are significantly shorter. Cholesterol can thus be used to locally control flexibility. However, if essential unsaturated fatty acids are lacking, more cholesterol molecules must be incorporated into the membrane to maintain the physical properties. A membrane physics-adapted ratio of omega-6 and omega-3 fatty acids in the diet is therefore desirable.

2.5 The Omega-6/Omega-3 Ratio of Foods

The ratio of omega-3 and omega-6 fatty acids in cell membranes can be controlled to some extent through diet. A whole range of fatty acids in the phospholipids is not essential, and the body's own physiology produces them as needed. However, essential fatty acids must be obtained through food. This implies that it makes sense to select foods based on their omega-3/omega-6 ratio, in such a way that it is best adapted to the requirements of human physiology and the temperature of 37°C.

In contrast to today, hunters and gatherers consumed significantly more essential Omega-3 fatty acids. The ratio was usually close to n-6:n-3 = 1:1. In modern times, especially for people who consume industrially produced food, this ratio is more likely to be between 8:1 and 10:1. A balanced Omega-6/Omega-3 ratio is crucial for the development of the brain in newborns, for the same reason that it was relevant for brain growth during evolution: only a high level of long-chain, essential, and thus polyunsaturated animal fatty acids can supply and protect the brain to the required extent. This is where cell membranes come into play again. In contrast to other cell membranes in the body, the membranes of brain cells are equipped with a high proportion of docosahexaenoic acid (DHA) in the phospholipids, as schematically indicated in Fig. 2.11. They keep the membranes in an

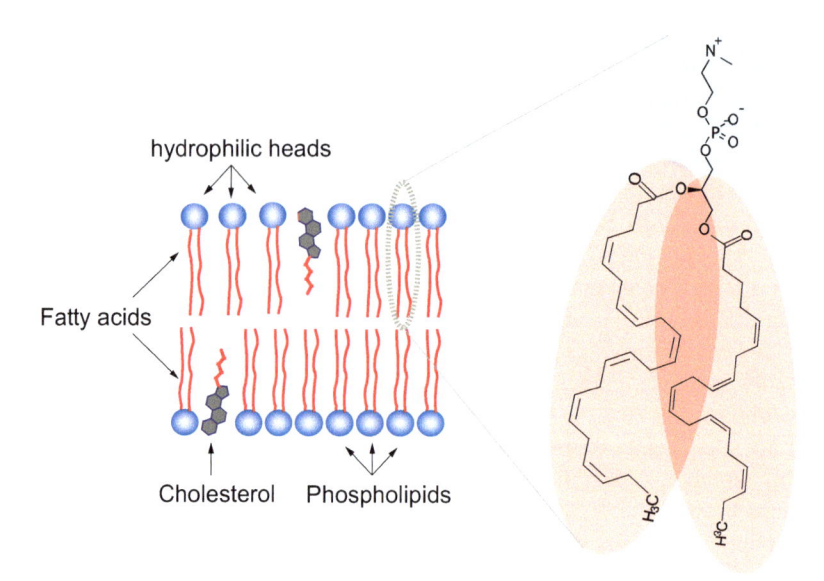

Fig. 2.11 Schematic representation of a small membrane piece, a double layer of phospholipids, whose two fatty acids stretch towards each other and whose hydrophilic heads form the outer and inner membrane of the cell. Some phospholipids in the brain (also known as the emulsifier "lecithin") carry a high number of omega-3 fatty acid docosahexaenoic acid (DHA) and/or eicosapentaenoic acid (EPA), as well as esterified (= bound to a fatty acid) cholesterol, which locally regulates the flexibility of the membrane

extremely flexible state, which is counterproductive for other cells—depending on their function.

Particularly in brain cells and naturally UV-rich light-exposed retinal cells in the eye, the molecular interplay of EPA and DHA shows another strength. Fatty acids protect cells from peroxidation where it is most important: in the brain and eyes through a special, but remarkable molecular process that can only be achieved through a balanced Omega-3/Omega-6 ratio [19]. In Chap. 1 (Fig. 1.5), the peroxidation of polyunsaturated fatty acids was already discussed. These oxidation processes and the associated release of highly reactive free radicals also naturally occur in membranes: the double bonds break, electrons, reactive fragments (free radicals), and highly reactive intermediates can then break further double bonds along the membrane, as indicated in Figs. 1.5 and 2.12. In this process, the cholesterol dissolved and distributed in the membrane can aggregate and form ordered crystalline regions. The membrane loses its biological function, becomes brittle, and permeable to unintended molecular components, damaging the cell content.

2.5 The Omega-6/Omega-3 Ratio of Foods

These facts can be directly demonstrated with elaborate methods of physics and physical chemistry using well-defined analytical methods. X-ray structure analysis and thermal methods can clearly identify the crystalline regions of cholesterol. They also manifest in a measurably reduced membrane thickness, as indicated in Fig. 2.12, as soon as the unsaturated fatty acids in the membrane are oxidized. The progress of the oxidation process in the membrane can also be detected using nuclear and electron resonance. Furthermore, the results show an important point that is often neglected in public ecotrophological and medical discussions: A efficient physiology is initially based on elementary physics and chemistry, which must follow the elementary interactions of self-organized "soft matter". If these physicochemical parameters are not correct, they must first be understood and repaired. Only then can symptoms be meaningfully treated.

The peroxidation in the cell membranes is such a textbook example that clearly shows us one of the functions and therefore the value of the animal long-chain, polyunsaturated fatty acids: They are necessary as effective antioxidants to maintain this process, which also occurs spontaneously. Good candidates for capturing free radicals are fat-soluble vitamins, as they can dissolve particularly well in the fatty acid-rich membrane layer, move freely within it, and unfold their effect. But also specifically shaped and polyunsaturated omega-3 fatty acids, such as eicosapentaenoic acid (EPA), are radical scavengers [20]. They have two advantages: On the one hand, their shape fits perfectly with the membrane structure, and on the other hand, they themselves are polyunsaturated and capture free radicals during decay. Due to their fivefold unsaturated nature, they do not form an excess, additional free radical even during complete decay. The fatty acid EPA is therefore the ideal blocker of radical formation in the membrane and thus a cell (membrane) protection, as shown in Fig. 2.13. This is particularly important in cells that are exposed to high stress (e.g., in the retina of the eye) or in brain cells that cannot be supplied with the usual repair mechanisms across the blood-brain barrier as with other organs of the body. Coupled with sufficient vitamin E (α-tocopherol), the progression of the oxidation process can be stopped.

But back to evolution: In fact, only a change in diet allowed these antioxidant processes to develop in this sophisticated form. For this, these long-chain, five- and sixfold unsaturated fatty acids were (and are) indispensable. Individual DHA

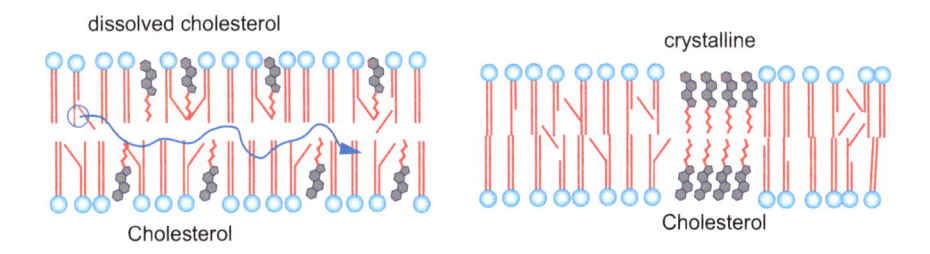

Fig. 2.12 Lipid peroxidation in membranes. A spontaneously forming free radical (circle) can migrate through the membrane (blue arrow on the left) and trigger further oxidation processes

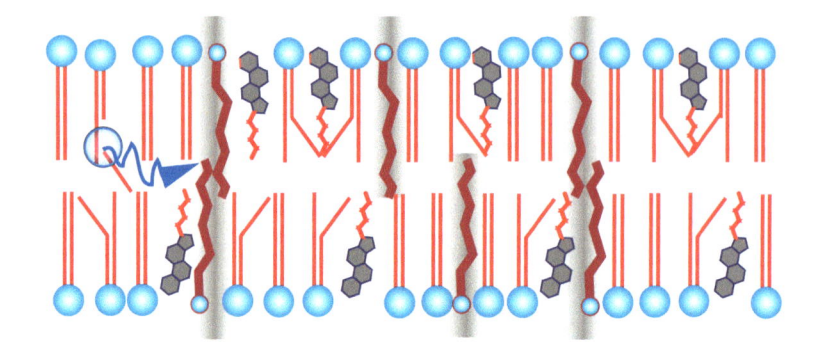

Fig. 2.13 Free fatty acids, e.g., EPA, represented by a thick dark red fatty acid, can be incorporated into the membrane at appropriate sites. There, they act as a blockade for spontaneously generated free radicals, stopping their cell-damaging diffusion. Reasons for this are their fivefold unsaturated form and their adaptive, precise length in different "conformations": They have four more carbon atoms compared to most membrane phospholipids (C 18-fatty acids)

and EPA can indeed be synthesized from α-linolenic acid by the body itself in complex, highly energy-consuming processes, but not to the extent required, as will be shown in more detail later [21].

In fact, DHA makes up about 40% of the fatty acids in the membrane phospholipids of the brain. Both eicosapentaenoic acid (EPA) and DHA have an effect on membrane receptor function, neurotransmitter formation, and metabolism. In particular, EPA also has an antioxidant effect. A ratio of about 1:1 between omega-6 and omega-3 fatty acids is optimal for cell metabolism, especially in the brain [22]. Again, this shows that the evolutionary shift to more animal foods in the development of *Homo sapiens* allowed the most important organs to be much better protected: the brain as a constantly evolving central computer and the vital eyesight. Everything else, such as the prevention of cardiovascular diseases, are desirable but essentially secondary side effects.

Even today, game is essentially an ideal food, also in terms of the balanced ratio of omega-3 and omega-6 fatty acids. Although the "lean" meat is always emphasized by nutrition science, this is only half of the truth. In fact, intramuscular fat is only present in small amounts, but the long-chain, polyunsaturated fatty acids are also found in the lipids of muscle membranes, which are inevitably consumed when eating meat [23]. Phospholipids can be easily utilized and digested by humans.

By the way, in many of these experiments with model membranes and laboratory-generated membranes of animal and human organisms, it is repeatedly shown that an excess of glucose accelerates lipid peroxidation [24]. Opposing claims [25] stand against these experimental results.

2.6 Offal—Forgotten and Valuable Foods

As already mentioned, when discussing meat consumption in evolution, one should not only focus on the sheer muscle meat. The scientifically correct terminology would be "eating animals," because the meat provides high-quality protein and thus a mix of essential and biologically easily available amino acids well-tuned to *Homo sapiens*. However, the animal provides much more in terms of high-quality "side programs," which have largely disappeared from the eating biography of humans in the 21st century due to increasing prosperity. Since the Paleolithic period, all offal was consumed. Especially the brains of hunted and killed animals, as they are packed with EPA and DHA. At the beginning of "humanization," it would have been a real sin not to eat the brain and to forego the polyunsaturated fatty acids. Brain was also part of the diet until the post-war period. This food, popular among our fathers and grandfathers, lost more and more importance with increasing prosperity and abundance. The bovine spongiform encephalopathy disease and the resulting BSE crisis finally put an end to the consumption of bovine brain. However, the brain, regardless of the animal, would still be a real superfood today and is superior to walnuts, chia seeds, and avocados solely due to the high concentration of essential fatty acids.

Occasionally, some people will still remember cod liver oil. This oil, derived from fish liver, prevented diseases such as rickets and had a whole range of other benefits. Of course, it is rich in vitamin E and, above all, relevant precursors to vitamin D_3. Cod liver oil is rich in the essential fatty acids DHA and EPA, but also valuable in terms of the micronutrients vitamin A, D, and E. Since, as already noted, wild animals incorporate the DHA/EPA protection into their cell membranes, it is no wonder what an outstanding function offal, especially the liver, had in nutrition during the evolution. This also enabled tribes whose niches had no access to coasts to ensure a basic supply of essential long-chain fatty acids. In view of today's perspective, it is downright negligent to banish liver and other offal from the diet. Although most livestock are now fattened with feed for high meat and milk production, liver is still an excellent source of vital micro- and macronutrients. Especially when the animal husbandry is appropriate, the animals are on pastures and kept as species-appropriate as possible.

Apparently, the meat, the unsaturated fatty acids, the favorable Omega-3/Omega-6 balance, the eating habits, and the exercise associated with hunting and gathering were responsible for the fact that no "civilization diseases" emerged, although civilization constantly progressed through food improvements and social structures. With one not insignificant difference: the boys and girls back then did not smoke or drink yet. The pollutants introduced into the body by this, nothing more than highly reactive molecules, provoke, among other things, the peroxidation in the cell membranes. A vegan dietary approach proves to be counter-evolutionary at this point. Quite apart from the fact that there would not have been the *Homo sapiens* in this form if it had remained strictly vegan [26]. Gradually, it becomes apparent which steps were first necessary until, from the various relatives of the early hominids, ultimately modern humans like Neanderthals or *Homo sapiens* could emerge, because in all stages of evolution, food always played a central role.

2.6.1 From Early Hominids to *Homo Sapiens*

The increased proportion of animal proteins and fats had another significant advantage that accelerated human evolution: women could wean their children much earlier than non-hominid primates [27]. While these primates nursed their offspring until the age of four or five years, *Homo sapiens* managed to cope with significantly shorter breastfeeding times. Children were weaned earlier, allowing women to become pregnant again more quickly. The advantage is evident in evolution: early humans multiplied rapidly, a high population ensured survival. In conjunction with the high energy return rate (see Figs. 2.1 and 2.4), this was of utmost importance for the growing groups of early hunters and gatherers. This shortening of breastfeeding time correlates highly with the enlargement of brain volume, the refined motor skills, and the strong increase in animal nutrition, especially considering the influences of polyunsaturated animal fatty acids, when observations of coastal indigenous peoples in South America are taken into account [28]. The advantage observed in this regard for carnivores can be very accurately predicted and understood through a universal model [27].

The change in the intestine due to the dietary change is also a clear indication. This is evident in the gradually increasing shortening of the intestine compared to raw food eaters and herbivores. On average, the intestine of plant-eating primates is significantly longer and can accommodate more volume, which is necessary to extract a maximum of macro- and micronutrients from raw food.

The changing diet in the Paleolithic during evolution consequently changed the hominids and only made the development of *Homo sapiens* possible. In this sense, "*Homo sapiens* is what he eats" and we are still what early humans ate. This long preparation enabled the next "revolution in evolution," the sedentism. The Neolithic Revolution began.

2.7 Sedentism—Fundamental Change in Food and Nutrition

Cooking not only made the food source of meat more accessible to the digestive system but also previously almost indigestible, starch-rich roots and tubers. It seems only consistent that such foods were cultivated, cultivated, and selectively bred, as well as animals like aurochs or precursors of pigs were cultivated, provided this was in line with the climate, soil fertility, and the possibilities of sufficient water supply for irrigating the cultivated areas.

This is to be seen as a fundamental change in culture: Agriculture replaced the culture of hunting and gathering partly between 20,000 and 10,000 years ago [29].

This, in turn, led to a change in diet: less meat was consumed, and more starchy roots, grass seeds, and "new breeds" were added to the menu. The systematic path to grain began its course.

2.7.1 Grains and Starch: Additional Food Sources

However, the sedentary lifestyle and the use of early grains and corn were accompanied by the first civilization diseases: As is known from bone and tooth findings, the first forms of caries diseases occurred. Using spectroscopic and nuclear-physical isotope methods, other problems can also be inferred in humans that evolution had not known until then. Thus, bone analyses indicate a deterioration of general health, with evidence of morbidity, poorer dental health, for example, a receding occlusion in the bite. Increased iron deficiency anemia, initial signs of bone loss, and a higher number of infections can now be detected with precise gene analyses and microbiological methods [12]. The food mix changed compared to the Paleolithic. Targeted livestock breeding began about 11,000 years ago after the last ice age when humans in the Neolithic began to domesticate wild sheep. With the domestication of animals and the cultivation of plants, the Neolithic Revolution began [30].

Although meat continued to be eaten, the meat and especially the fat of domesticated animals differed in composition compared to wild animals, as shown in Table 2.3 as an example.

Domesticated pasture-raised cattle still stood out with a favorable fat composition, even though they already performed worse compared to wild animals (Fig. 2.14). The n-6/n-3 ratio becomes significantly unbalanced, and the proportion of polyunsaturatedFatty acids are lower. Cattle that are fattened with grains and cereals have the least polyunsaturated fatty acids and a high n-6/n-3 ratio, meaning significantly more n-6 than n-3. The first domesticated animals were also much more "energy-rich" than wild animals. Domesticated animals no longer had to be hunted, reducing personal energy expenditure in this regard (even if it increased again with the cultivation of fields).

Table 2.3 The fundamental differences in the percentage of polyunsaturated fatty acids, the n-6/n-3 ratio, and the ratio of polyunsaturated to saturated fatty acids

Species	Polyunsaturated fatty acids (%)	n-6/n-3 ratio	Polyunsaturated/saturated fatty acids (%)
Giraffe	41	4	1.12
Antelope	32	2.5	0.67
African buffalo	32	3.4	0.83
White-tailed deer	16	2.9	0.33
Pasture-raised cattle	10	2.2	0.24
Feedlot cattle	**7**	**5.2**	**0.16**

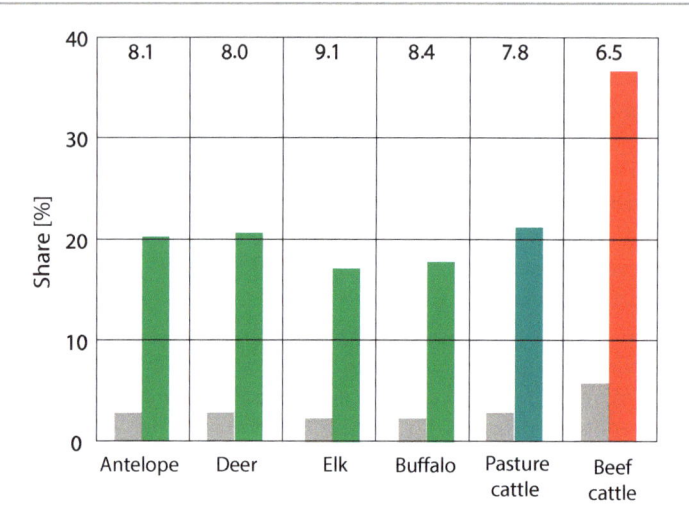

Fig. 2.14 Wild animals and the first domesticated animals already showed significant differences in terms of energy and fatty acid types. The upper row of numbers shows the ratio of weight to energy

Consequently, the "energy return rate" for protein and fat increased significantly compared to hunters and gatherers.

The change in energy density in pasture and fattened cattle when fed with protein- and starch-rich grains is striking. This trend can still be observed today. However, this also highlights an important point: It is not only the consumption of meat that makes up the diet, but also the type and quality of the meat. Meat from discounters or many butchers actually comes from animals fattened with (protein- and starch-rich) concentrated feed. Grazing animals are already more productive in terms of energy density, as well as in their fatty acid composition and the Omega-3/Omega-6 ratio.

2.7.2 The Fundamental Dietary Change

The transition from the culture of hunter-gatherers to farmers was more dramatic in evolution than it initially appears. Hunter-gatherers predominantly ate meat, even when enough plant-based alternatives were available [5], also enough of it to take in sufficient micronutrients. Dietary fiber from plant-based food proved useful in the shortening intestine over time to maintain bowel movement (peristalsis) and adapt to the conditions. At the same time, hunter-gatherers consumed so much unsaturated animal fat that the function of cell membranes could be optimized. Only with this diet could deficits be compensated, for example, the barely pronounced ability of the first hominids to synthesize sulfur-containing amino acid precursors from taurine [31].

Taurine, an aminosulfonic acid, is a metabolic product of the amino acid cysteine. For offspring in the womb, taurine is still essential today, as fetuses can only produce it after birth when the necessary enzyme is activated. In the womb, they depend on the supply from the mother. The mothers' diet was therefore of particular importance to ensure an adequate supply of taurine to the fetuses and thus a healthy development of the offspring. This directly links to brain size and the development of hominids to *Homo sapiens*, which accelerated significantly with the increased animal food of hunter-gatherers, as leading in taurine content are animal foods, especially liver, poultry, and fish such as mackerel and bream. Seafood and even cowmilk also stand out for their high taurine content [32]. Apart from the long-chain animal omega-3 fatty acids, these facts also speak for the great advantage of switching to primarily animal-based food during evolution.

Likewise, iron, the pigment heme, and porphyrin-iron-rich compounds in the appropriate (ionic) form could only be absorbed in physiologically sufficient amounts from the muscle dye myoglobin or the blood of animals [33]. Ethnologists and anthropologists assume that 56–65% of the food of hunters and gatherers was of animal origin to withstand the selection pressure. Basically, it seems grossly negligent to consider a vegan diet during pregnancy, as these fundamental conditions also apply to modern humans.

To this day, the iron present in the heme group is an excellent natural source for this micronutrient [34]. Unlike iron in plant-based food, heme iron is easily bioavailable, as will be discussed in more detail in Chap. 4.

2.7.3 Consequences of Sedentism—The Third Revolution in Evolution

The sedentary lifestyle led to a whole series of consequences in addition to the losses in population health, which only brought about a further development of food culture and above all new forms of storage. In particular, sedentary lifestyle allowed the development of new cultural techniques that could hardly exist among hunters and gatherers due to their constant search for food. Under the given circumstances, the population grew so rapidly that this can no longer be explained solely by the advantages of early weaning. Apparently, the establishment of social structures contributed to the population growth of tribes in the niches. People had to organize themselves according to abilities, new forms of civilization emerged, and agriculture was constantly improved. Crucial for the Evolution was the development of higher complexity lifestyles [35], as can be seen in the techniques and development in Fig. 2.15.

An important point is the energy return rate, which is very high among hunters and gatherers and is mainly due to the high proportion of fat and protein in animal food. It is initially unclear to what extent the sedentary lifestyle offers advantages for these macronutrients. The energy expenditure of tending, irrigating, and harvesting fields daily is not necessarily more energy-efficient than hunting and gathering. The caloric intake is also lower, as grains, insects, and vegetables have

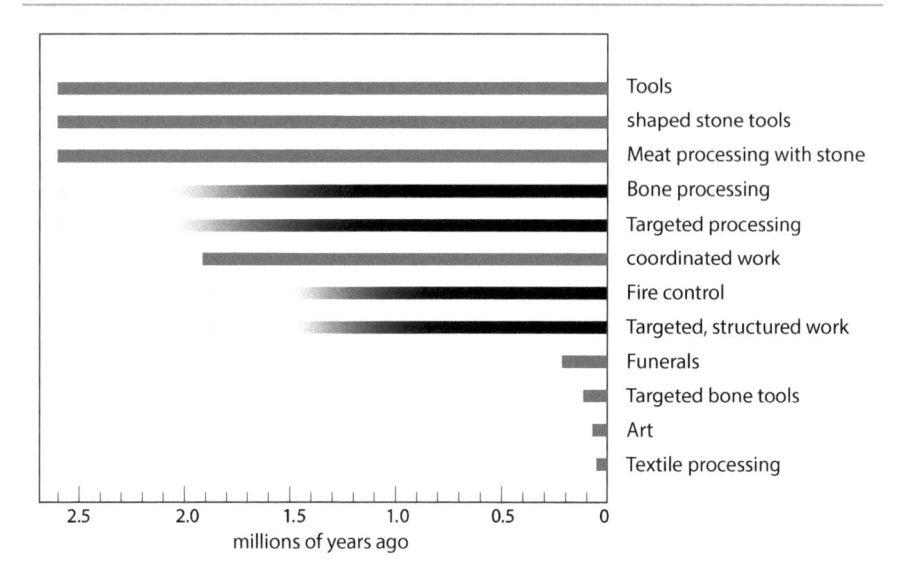

Fig. 2.15 Higher complexity in a sedentary ethnic group also determined the type of tools and the pre-planning of work. The leap is clearly visible in the very short bars in the times of the last 50,000 to 10,000 years

a significantly lower fat content than the meat of wild animals. This inevitably means that the missing energy had to be compensated for by carbohydrates and proteins. This already resulted in new forms of food economy in the Neolithic.

On the other hand, the sedentary lifestyle of a society also means more free energy resources, as there is no longer a need to roam around to find food, and people can thus devote themselves to the development of new techniques and agriculture. One advantage of becoming sedentary is evident: The age groups that did not contribute to high energy return rates and overproduction—the elderly, children, and adolescents (see Fig. 2.4)—could be provided for on-site and no longer had to be carried along with high energy expenditure, as was the case with hunters and gatherers when a change of niche was required. This results in a positive energy balance in favor of sedentary lifestyle, which at the same time enabled a more complex culture and the invention of new tools and techniques up to the development of storage facilities for food. Exchange transactions and early forms of trade in goods also contributed to better and more effective energy use. On the other hand, sedentary lifestyle also means higher conflict potential as well as risks of illness and infection, which become more likely in larger "living units".

Another important point is the choice of the area for settlement, which was of utmost importance for the sedentism. The environment had to be fertile, and so obviously so that people could settle there with good prospects [36]. People followed a "Garden of Eden principle" [37]. From this, archaeologists concluded the actual motivation for the beginning of sedentism. It was not the fundamental lack of food, but a comparatively abundant and predictable availability of food compared to the Paleolithic, supplemented by naturally given conditions, such as caves or protected rocks, for storing food.

Inevitably, technological progress, which had already begun with the use of fire long before, continued its course: Soon, with the ability to fire clay and earth, the use of cisterns and storage rooms began. In coastal areas as well as near rivers and lakes, food could be ensured all year round. Food for storage could be selected according to shelf life [38] and even meat and fish could be stored for up to a year in protected, dry places when dried and matured. The transition from the Paleolithic to the Neolithic is summarized again in a diagram in Fig. 2.16.

Also noteworthy is a fundamental difference between non-hominid primates and hominids: With the growth of the brain and the constantly increasing consciousness, more far-reaching religious inclinations of humans emerged with the increasing prosperity of the Neolithic. Hunter-gatherers develop spiritual rites, but no formalized religions. The handling of death, the awareness of death, of finite life, and the loss of fellow human beings, however, became much more pronounced in the Neolithic. Soon the first ritual burials took place, formal religions formed: Only the sedentary human created their gods.

The differences between the Paleolithic and the Neolithic are therefore fundamental and far-reaching. To this day, the differences can be seen in still existing groups of hunters and gatherers, such as the already mentioned !Kung. They built and still build only round, semi-open facilities to protect themselves from wind,

Tools:
flints, stones, bones,
spears

Tools:
Axes
spears

Tools:
stone blades, woodturning,
metal, hammers, chisels,
spears

Social Roles:
Men and women as
hunters, gatherers

Social roles:
Maternal role,
gathering, hunting

Social roles:
Collecting, farming both
Gender,
Spears

Food:
Meat, game, fish,
insects, berries, roots

Food:
Meat, insects, roots,
Grass seed

Food:
fruits, berries, breeding animals,
fish, Cereals, vegetables

Way of life:
Meat, game, fish,
insects, berries, roots

Way of life:
Meat, insects, roots,
grass seeds

Way of life:
Fruits, berries, breeding
animals, fish, cereals,
vegetables

Spiritual activities,
no religions

Religions,
Priest

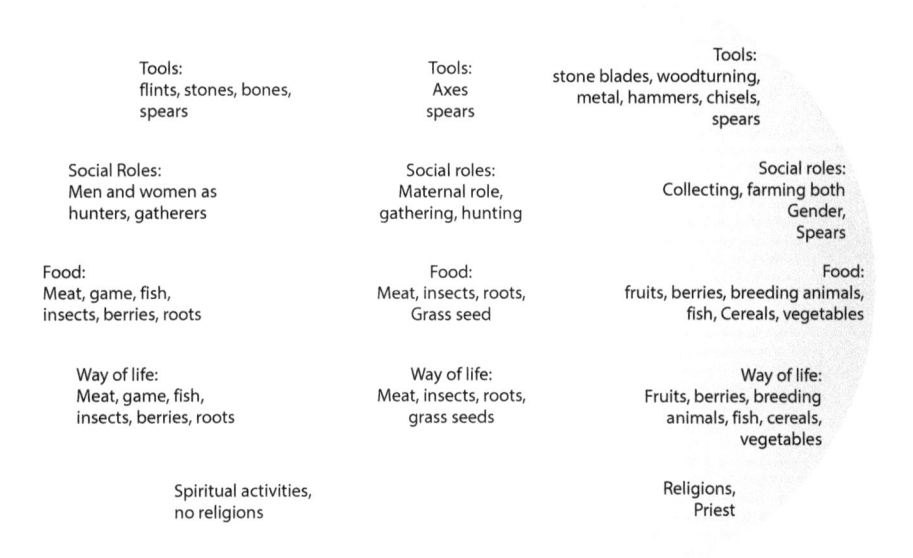

Fig. 2.16 The transition from the Paleolithic to the Neolithic is characterized by a fundamental change in all areas, not just in food

2.8 Lactose Tolerance—Point Mutation in DNA: Lactose Tolerance as a Result of Selection Pressure

rain, and cold, instead of houses, as has been customary among sedentary peoples for a long time. Like other hunting and gathering groups that still exist, they follow these basic principles to this day.

The sedentary lifestyle brought not only the targeted breeding of grains such as einkorn, spelt, and other early forms of wheat, as well as pseudo-cereals like millet, grasses, and vegetables, but also the domestication of animals. The idea of livestock, as they are called today, is much older than it is portrayed in modern times [39]. Livestock is the logical consequence of sedentary lifestyle, as the advantages of eating meat, cooking and frying with fire, and the taste drive umami persisted even in the transition from the Paleolithic to the Neolithic period. Wolves, wild sheep, and wild goats were domesticated, tamed, and became the first companions of humans. Domesticated wolves, or dogs, in particular, greatly changed life [40]. After taming, they were trained and helped the people of villages and communities as herding and shepherd dogs when sheep (Middle East), goats (Middle East), and cattle (Africa and Anatolia) were domesticated in various regions where they lived in their wild form. Animals and humans lived together for the first time in history. Soon, stables with urine drains were built, and animals were hand-fed by humans [41].

Sheep, goats, and cows became milk suppliers, a welcome but also problematic food, as the milk sugar could not be digested. Mammals, including humans, can generally only break down the disaccharide lactose during breastfeeding and split it into glucose and galactose (Fig. 2.17), which can then be supplied to human physiology.

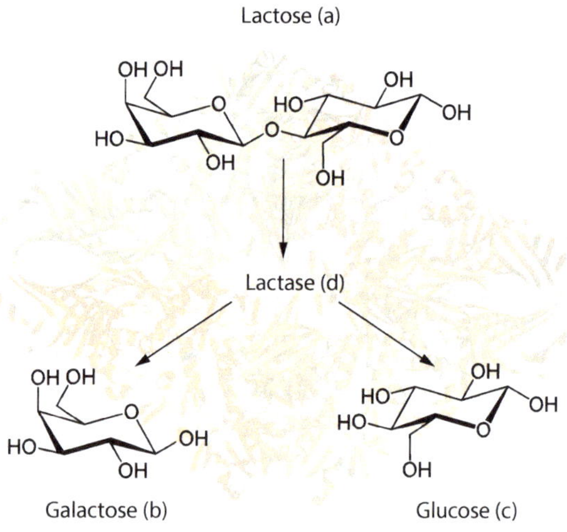

Fig. 2.17 From lactose (**a**) to galactose (**b**) and glucose (**c**)—in the presence of the enzyme lactase (**d**)

For this, the enzyme lactase (more precisely, lactase-phlorizin hydrolase) is necessary, which is normally no longer expressed in childhood. Lactase is produced in the small intestine via a specific gene in the offspring until weaning [42]. After that, the gene is switched off, and milk, as long as it contains lactose, is practically intolerable for almost all people. People all over the world are lactose intolerant, the splitting of milk sugar into glucose and galactose is not possible, instead, it is fermented by bacteria, which can lead to bloating, abdominal pressure, and diarrhea. It was different for some (Northern) European ethnic groups, specifically those who predominantly lived with dairy cattle. They developed lactase persistence after they gradually migrated to Europe with their cows and sheep and could consume milk as food. This genetic change took place within a few thousand years, a relatively short period of time for the timescales of evolution [43]. An adaptation effect is not possible on this short timescale; therefore, the evolutionary selection pressure must have been high, within which the gene responsible for lactase production mutated.

Since this example of gene modification is fundamental to understanding, it is worth taking a brief, highly simplified excursion into molecular biology at this point and focusing on basic relationships. The deoxyribonucleic acid (DNA) contains the genetic information, the genes. In their individual configuration for each person, it is also determined which proteins, which enzymes are formed by the cell. The DNA consists of a double helix; this is composed of nucleotides, which consist of phosphoric acid residues and sugar as well as four nitrogenous bases. The nucleotides are connected via the bases, as shown in Fig. 2.18.

The sequence of specific base pairs in Fig. 2.18 prevents the switching off of the lactase gene. Interestingly, exactly one base of a pair at a specific location in the lactase gene must be changed for a person to remain lactose-tolerant (Fig. 2.19). Considering that the genome of *Homo sapiens* contains 3.27 billion base pairs, a point mutation or a single nucleotide polymorphism at a specific location seems almost insignificant, but it also shows how essential the primary structure of DNA is. In the savannah in Africa, there were also such point mutations that allowed lactase persistence, but not to the same extent as in Europe.

In this context, the chemical mechanism of how milk sugar is split into glucose and galactose is also interesting. Enzymes are, as already explained, proteins specifically tailored for their catalytic tasks, so they consist of amino acids. For the separation of lactose, the negative charge of the glutamic acid is crucial. This, together with the environment of the active center of the lactase, is able to separate glucose from lactose in the first reaction step. For a short time, the remaining part of the milk sugar is bound to the glutamic acid. The lactose residue binds to a free glutamic acid site, while a polar water molecule is involved in hydrolyzing this intermediate product, and the lactose residue reacts to form galactose. These steps are indicated in Fig. 2.20. Glutamic acid thus plays an essential role in this purely biochemical process.

This fascinatingly demonstrates the role individual amino acids play in enzymes according to their physicochemical properties, in this case the negatively charged glutamic acid, which is also partly responsible for the umami taste.

Fig. 2.18 Basic structure of DNA: Two helices are connected via different nucleotides based on the bases cytosine (C), guanine (G), adenine (A), and thymine (T). The DNA is very long (approx. 2 m with over 3 billion base pairs per cell), so a huge number of different combinations are possible: The DNA of each person is thus individual

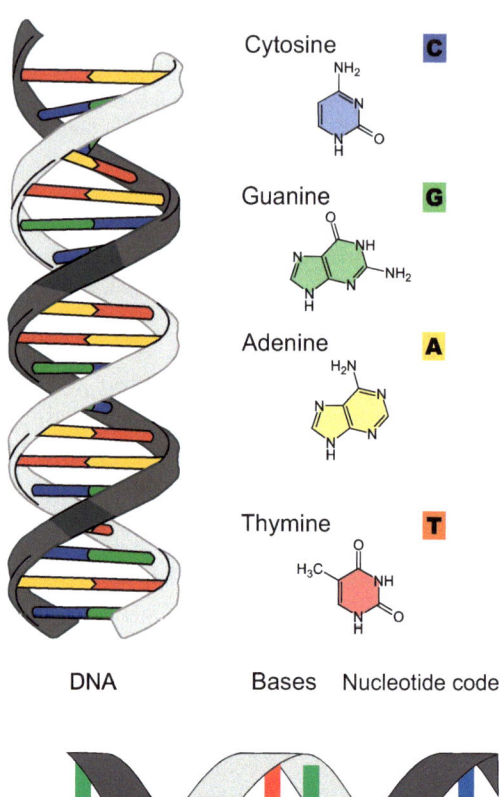

Fig. 2.19 The difference between lactose intolerance and lactose persistence can be precisely located at one base pair in the DNA. Such mutations are called single nucleotide polymorphisms (SNP). In the case of lactose persistence, this point mutation is a blessing

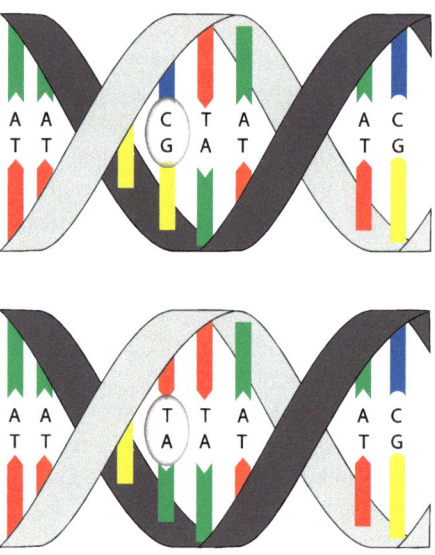

For the process shown in Fig. 2.20 to take place, the pH value must also be at the correct level during the gastrointestinal passage [44]. Lactase has its efficiency maximum at pH values around 6, which is why lactose cleavage occurs at corresponding locations in the small intestine. The physiological advantage of this

2.8 Lactose Tolerance—Point Mutation in Dna: Lactose ...

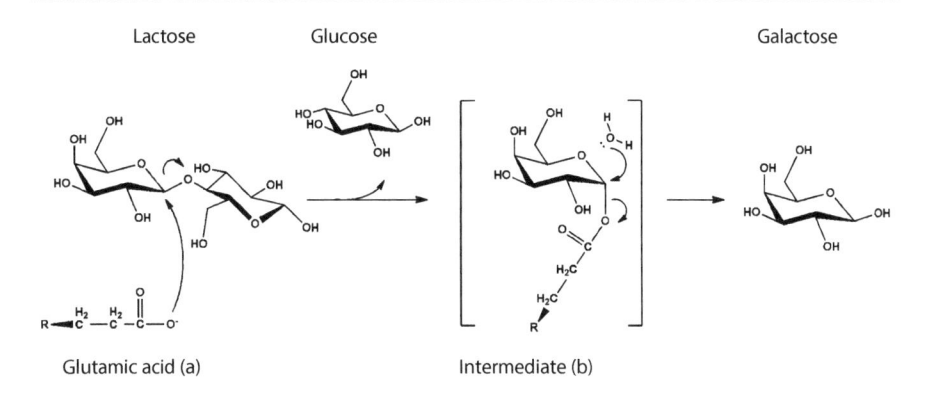

Fig. 2.20 The mechanism for splitting lactose into glucose and galactose: The negatively charged amino acid glutamic acid (**a**) is responsible for the cleavage of glucose, and in an intermediate step (**b**), the remaining residue is converted to galactose

separation is that glucose and galactose can be absorbed through channels by the intestinal mucosa, while lactose as such is not absorbed.

In the case of lactose intolerance, the milk sugar is transported further and reaches the large intestine, where it encounters the multitude of bacteria and microorganisms of the microbiome. There, milk sugar is fermented by specific (lacto-) bacteria. During this fermentation process, bacteria-specific fermentation products are formed. Small organic acids such as milk and acetic acid are formed, but they are not the problem, as the pH value in the large intestine is already lower and is thus buffered and temporarily increases intestinal movement. In any case, the lowering of the pH value due to the milk and acetic acid produced during fermentation would be a positive effect, as pathogenic germs would then have lower chances of survival. In principle, lactose even has a "prebiotic" effect. Furthermore, during fermentation, fatty acids are formed, but as by-products, fermentation gases such as methane, hydrogen, and the carbon dioxide typical of any fermentation are produced, which can cause cramp-like bloating and strong irritations when pressure increases. Like many polar molecules with sufficient OH groups, milk sugar also binds water and increases osmotic pressure. Water flows through channel proteins into the large intestine to reduce the high sugar concentration in the intestine and to provide pressure compensation. The volume of the large intestine content increases significantly, causing diarrhea, as schematically shown in Fig. 2.21. By the way, this simple principle also forms the basis for many electrolyte fluids used for bowel cleansing before surgery or colonoscopy.

This gene mutation during the Neolithic Revolution was thus a clear advantage. The food milk became a permanent part of the diet. Milk provides high-quality proteins, vitamins, and a healthy amount of fat, whose diverse fatty acid spectrum is unique. Many reports condemning milk as a food do not correspond to the facts. Therefore, milk and dairy products will continue to occupy us as valuable foods in the following chapters. We will examine the modern condemnations of (cow's) milk more closely in Chaps. 4 and 5.

78   2 Recognizing Food, Learning to Eat: A Look …

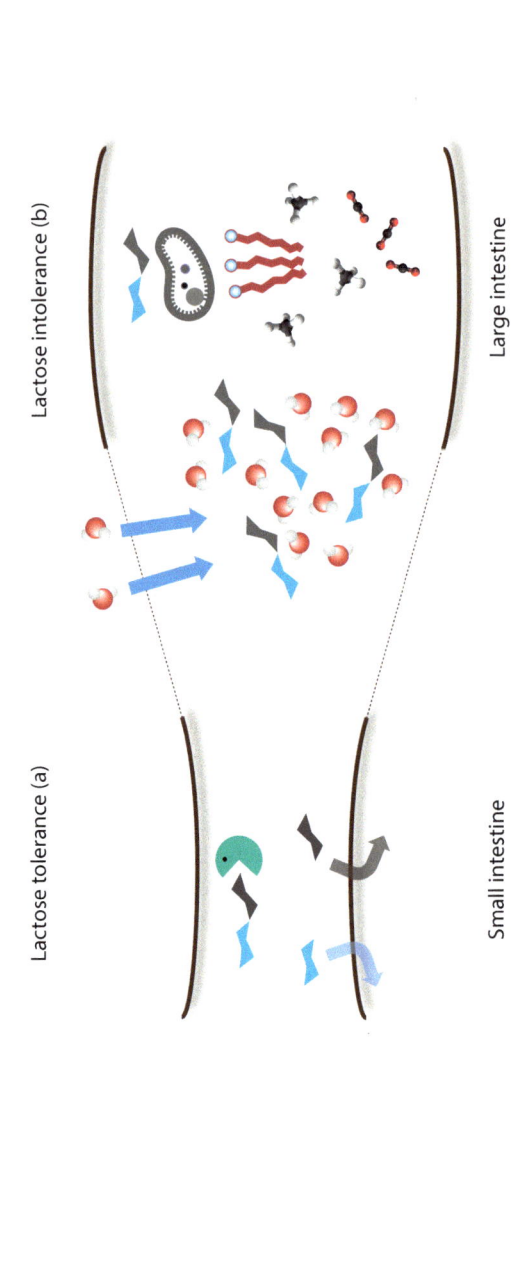

Fig. 2.21 In lactose tolerance, milk sugar is separated by enzymes, and the components glucose and galactose are absorbed by the intestinal mucosa (**a**). In the case of lactose intolerance, the milk sugar continues to move, with water (H_2O) flowing into the intestine for osmotic reasons (middle) and only in the large intestine (**b**) is the milk sugar fermented by the bacteria of the microbiome. This produces fatty acids, organic acids, and gases that cause bloating. The intestine expands, which manifests as pain (or cramp)

This point mutation is not trivial in the evolution and occurs widely in humans in Europe, but even there the distribution is not uniform. In southern Europe, lactase persistence is least pronounced, while in the north it is more pronounced. This gradient is also related to culinary preferences. In northern Europe, Finland, Sweden, Norway, England, northern France, and Poland, the dairy industry flourishes: butter, milk, and fresh cheese are among the most popular dairy products. Milk consumption in southern France, Italy, and Spain, on the other hand, is low. Cows are scarce there, also due to a lack of nutrient-rich meadows, and dairy products come from goat or sheep milk, as these animals can find food even in a sparsely vegetated landscape like the Garrique. Milk is mostly fermented into yogurt and cheese there—similar to about 10,000 to 20,000 years ago. This adaptation is highly selective and typical of a so-called niche culture [45], i.e., an active adaptation of the people living there to local conditions. This is also referred to as coevolution (gene-culture coevolution). Dairy cattle were bred, the animals were milked, the milk was drunk, as can be clearly demonstrated by archaeological finds together with gene-specific analyses. The consumption of milk increased the probability of survival and could secure the supply of macro- and micronutrients [46]. This is also a reason why indigenous peoples, such as the hunters and gatherers of the !Kung, still survive in their niche today. As long as food is secured there and the living culture can be preserved, there is no reason to leave the niche.

2.9 Fermenting and Fermentation—in the History of Food Culture

Lactose intolerance indicates that early dairy farmers milk fermented. Milk was soured or processed into yogurt- and cheese-like products [47]. Already around 10,000 years before our era, Indo-Europeans developed fermentation techniques to preserve food and protect it from spoilage [48]. The most durable form of milk is cheese with low water content, this food technology dates back to the 7th millennium BC. The targeted fermentation of fresh grape juice into durable wine also falls into this time period. Already around 5000 BC, the health benefits of fermented milk were documented in India, Mesopotamia, and Egypt. But already in the Neolithic period, sour products such as yogurt, kefir, curdled milk, etc. were made from fresh milk. These products offered the great advantage of microbiological food safety due to their low pH value and remained durable even without refrigeration. Controlled fermented foods thus ensure food supply to this day. Some essential developments, milestones, and types of fermented milk are summarized in Table 2.4.

There were thus two reasons for the souring of milk: on the one hand, of course, lactose intolerance, which is "as old as the Stone Age" [49], and on the other hand, shelf life. Fresh milk is only stable for a short time without adequate cooling. Soured milk, however, is. The souring of milk, which occurs spontaneously via lactic acid bacteria (lactobacilli), thus enabled its storage. Both requirements are strongly coupled on a molecular scale. The indigestible milk sugar is

Table 2.4 The history of fermented milk products

Product	Origin	Time Period	Remarks
Dahi	India	6000–4000 BC	Coagulated sour milk, also as a byproduct of butter production and ghee (clarified butter)
Chhash	India	6000–4000 BC	Buttermilk after butter production, drink
Khad	Egypt	5000–3000 BC	Sour milk spontaneously fermented in clay vessels
Leben	Iraq	3000 BC	Fermented milk with lactic acid bacteria, partially drained in cloths
Zabady	Egypt, Sudan	2000 BC	Yogurt, thicker consistency
Sour cream	Mesopotamia	1300 BC	Sour cream, spontaneous fermentation, acidification
Shrikhand	India	400 BC	Concentrated sour milk, spiced
Kumiss	Central Asia	400 BC	Mare's milk fermented with lactic acid and yeasts, contained CO_2, alcohol
Skyr	Iceland	870 AD	Sheep's yogurt with rennet

partially broken down by the lactic acid bacteria, and the fermented forms of milk, such as sour milk, curdled milk, or yogurt, contain significantly less lactose. On average, about 20% of the lactose is fermented by the lactic acid bacteria to lactic acid (and not, as often assumed, all sugars are fermented) [50]. The process is, as simple as it sounds, sufficiently complicated, as shown in the simplified diagram in Fig. 2.22. First, enzymes (β-galactosidase), represented by the scissors, cut the lactose molecule apart and release glucose and galactose. Only glucose can be converted by lactic acid bacteria to an intermediate product, pyruvic acid (pyruvate). Galactose is enzymatically converted to precursors of pyruvate in several intermediate steps, which in turn are converted to pyruvate and then to lactic acid in later steps. The enzymes required for this are also produced by the lactic acid bacteria, so this process is kept running. The resulting lactic acid lowers the pH value significantly: from the initial values of fresh milk, which are around pH = 6.7, to pH values below 4. Lactic acid bacteria can perform their work up to pH values just above 3, making them among the most acid-tolerant bacteria. The reason for the exceptionally long shelf life of fermented milk products is the low pH value of 3.5 to 3.2. Many germs responsible for the spoilage of fresh milk cannot multiply or survive at these acid levels. The functional proteins in the cell membrane of many bacteria, viruses, and fungal spores are already denatured at these low pH values. The germs die off, as also discussed in the corresponding fermentation processes in the intestine.

Both pyruvic acid and lactic acid are thus the fundamental factors for the food safety of "lactic acid-fermented" milk products, which humans have long used for their nutrition.

From this example, some conclusions can be drawn about our food culture: Lactic acid-fermented products, yogurts, and other sour milk products such as kefir,

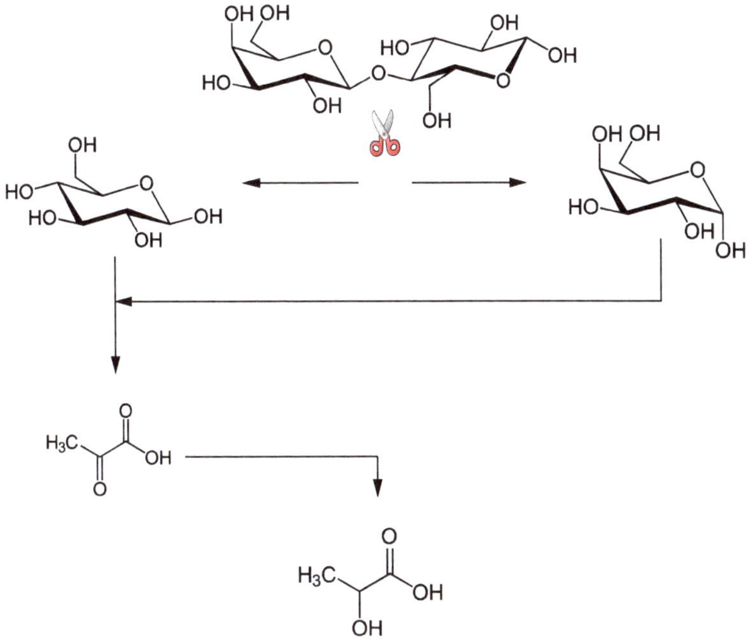

2.9 Fermenting and Fermentation—in the History ...

Fig. 2.22 From the disaccharide lactose (top) are derived glucose and galactose, from which pyruvic acid (bottom left) and finally lactic acid are formed (bottom center)

skyr, etc., are repeatedly attributed with "probiotic properties" because "good" lactic acid bacteria reach the large intestine and "support, expand, repair, and renovate" the "microbiome." While this is not fundamentally wrong, the various lactobacilli must first get there. The biggest barrier is the part of the gastrointestinal tract with the lowest pH value: the stomach. When fasting, the pH value is just above 1, which is too low even for acid-tolerant lactobacilli. Many die off. The drinking yogurt consumed on an empty stomach is therefore the least probiotically effective. Although the pH value in the stomach shifts to higher values depending on the amount consumed, it may still remain below 3. Only a few lactic acid bacteria make it to the duodenum. This is due to physical reasons, as the isoelectric point is exceeded for most membrane proteins, causing them to denature and lose their biological function. This means that cell function is no longer maintained. The yogurt for dessert is therefore, for physicochemical reasons, much more "probiotic."

The fermented milk also shows: The taste, acidity, and shelf life were decisive factors for the constant use of fermentation techniques. The higher digestibility of sour milk products and cheese as subsequent products also enabled the further development of fermentation techniques and provided Neolithic humans with useful tools for survival. Again, it was primarily the molecular changes, combined with taste, nutrition, and health benefits, that promoted these cultural techniques. The Paleolithic and Neolithic *Homo sapiens* was thus a being that unconsciously, through its taste and body sensation, developed precisely the right cultural techniques and occasionally passed on coevolutionary point mutations in certain niches under high selection pressure. This rich treasure left to us by our ancestors commands us to handle our simplest tools, our senses, carefully and to listen to them.

2.9.1 Fermentation of Vegetables

The sedentary lifestyle thus had far-reaching consequences for the ever-evolving food culture. Around 10,000 years BC, the people of the Neolithic period began their first pottery work. By using fire, not only could they cook and utilize food technology, but they could also burn, sinter, and process earth, clay, and sand into vessels, producing heat-resistant material that withstood high cooking temperatures [51]. But not only that, the containers could also be used for fermentation processes. There is chemical-biological evidence that alcoholic, beer- and wine-like beverages were produced using yeasts [52]. Even in the later "Ötzi," there are clear indications that fermentation with yeasts was already being deliberately applied during his lifetime. In many parts of the now-populated world, the idea of fermentation emerged with the systematic cultivation of cereals and the associated germination of seeds [53].

As mentioned several times before, the durability of fermented products is one of the strong arguments for this technique. Food could be stored, and the supply was ensured. It is therefore logical to also ferment conventional vegetables, roots, or even fruit. As already noted in Chap. 1, "decay" and the associated fermentation were a milestone in human history. Natural decay also points the way for fermentation. The crucial step was not to leave the processes involved to nature, but to control them so that the results would pass the taste and smell tests. Even during the sedentary lifestyle, these were the only control systems that humans had at their disposal. In fermentation, when germs and microorganisms are active, human senses are more challenged to reliably assess edibility.

The origin of fermentation actually lies in nature. Microorganisms such as lactic acid bacteria, yeasts, and fungal spores are found everywhere on plants and on human hands. As soon as the physicochemical conditions such as temperature or low oxygen content are right, spontaneous fermentation begins, as is still used today in Lambic beers [54] in Belgium or natural wine by brewers and winemakers [55]. Sourdoughs are also examples of spontaneous fermentation to this day: wild yeasts and Lactobacilli are applied with the hands or are found in bowls, flours, or are present in the room. The fermentation germs and their occurrence and frequency are therefore tied to location and time, as the season, temperature, and local parameters such as humidity determine the type and frequency of germs and thus the taste, aroma, and flavor (the sum of all impressions: taste, aroma, texture, visual and acoustic perception, e.g., the crunch of potato chips, the snap of fresh vegetables...) of the fermented preparations. Therefore, spontaneous, wild fermentation is currently gaining importance again, reflecting the signature of the chefs and the peculiarities of the region and the exact location.

Controlled fermentation starts precisely at this point. Because controlling the invisible microorganisms with the naked eye is much more difficult than controlling visible fire. The control of microorganisms begins with the careful handling of the food to be fermented, which must not have any decay spots and must be clean (washed). Otherwise, unwanted germs and their active enzymes would be introduced, leading to off-flavors and undesirable aftertastes.

In spontaneous fermentation, the food is selected, pre-sorted according to ripeness or initial decay, and washed, and then fermented under the prevailing conditions. Controlled fermentation with deliberately used starter cultures requires more technological knowledge and preparation. Although the first steps are identical, the fermentation process is strongly controlled. At the beginning of fermentation, the product to be fermented is inoculated with microorganisms. This targeted addition of microorganisms already determines a large part of the final taste and aromas. The inoculum can come from a preparation via spontaneous fermentation, as is usually the case at the beginning of a completely new process; alternatively, it can come from a previous fermentation. Both paths are shown in Fig. 2.23 via the arrows.

The natural and also spontaneous decay of food is shown for comparison on the far left in Fig. 2.23. Decay (a mixture of rot/anaerobic and decomposition/aerobic) occurs spontaneously in nature and, unlike fermentation, does not require cultural actions and interventions by hominids or *Homo sapiens*. Nevertheless, this process provides edible food in an intermediate stage, as already noted in Chap. 1. These fundamental differences will be important for later considerations, as they define, in addition to the handling of fire and cooking, a universal transition from nature to culture in human history.

The available oxygen content in the processes is of great importance for the effect and results of different microorganisms. Lactic acid fermentation and yeast fermentation are anaerobic, i.e., the bacteria and fungi work best under oxygen deficiency and lead to the desired results. In aerobic processes with cellular respiration, other processes take place and generally no preserving lactic acid is formed (Fig. 2.24).

Food-relevant controlled fermentation processes generally take place anaerobically, as the formation of lactic acid and alcohol serves preservation and food safety. However, the fermentation of alcohol to vinegar is an exception. Oxygen is required.

2.9.2 Asia—Advanced Culture of Fermentation

Above all, in the populated regions of Asia, targeted fermentation methods were already used very early in history [56]. The best-known example is Kimchi, a fermentation product based on (napa) cabbage with additional seasoning ingredients such as radishes, roots, etc., which can change depending on availability. How old Kimchi really is remains unclear to this day [57]. In India, too, vegetables have been fermented and fermented for a long time. Gundruk, salt-free fermented leafy

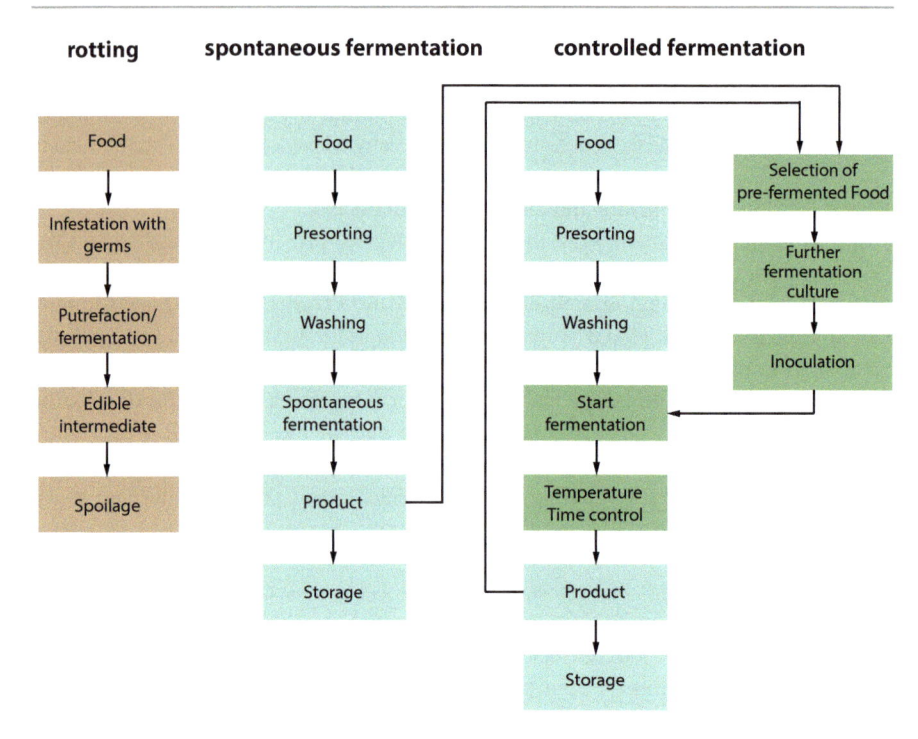

Fig. 2.23 Spontaneous and controlled fermentation. Control requires more cultural actions than spontaneous fermentation. There, the fermentation process is left to chance. The natural process of decay (rot) is shown for comparison. LM: food

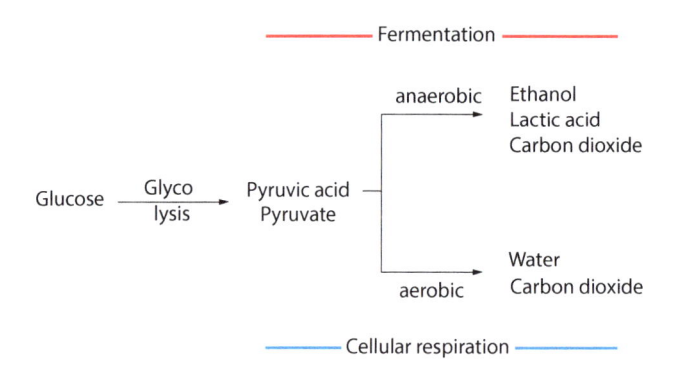

Fig. 2.24 Basic differences between aerobic and anaerobic microbiological processes

vegetables, is a traditional example from Nepal and the Himalayan region [58]. Salt-free fermentation is known in Japan as *sunki* [59]. But protein-rich foods such as fish have also been and continue to be fermented. The aim is always a longer shelf life. In many parts of Europe, Sauerkraut [60] is particularly well-known as a standard fermentation.

The procedure is very similar in all cultures. Vegetables are pre-cut, pressed in layers into fermentation containers, and mixed with about 2% of the weighing with salt. The fermentation container is sealed so that as little oxygen as possible can penetrate, but the carbon dioxide that forms can escape. Developing these techniques alone is a high intellectual achievement. It is known that this fermentation method has nutritive advantages in addition to shelf life and taste: Prebiotic and antinutritive substances of raw vegetables are enzymatically broken down during fermentation and partially converted into antioxidant substances. For example, the antinutritive, toxic in high amounts chlorogenic acid found in many raw vegetables is broken down into caffeic acid and quinic acid [61], which in turn can have antioxidant effects. So-called FODMAPs, this term stands for *fermentable oligo-, di- and monosaccharides and polyols* (fermentable oligo-, di, monosaccharides and polyols) are partially pre-digested during fermentation and partly converted into digestible sugars. Intolerances are less likely [62]. Lactose also falls under these FODMAPs. At the same time, nutrients, especially ionic micronutrients such as calcium and magnesium, are released and made physiologically available through fermentation [63]. The acidification also protects sensitive vitamins such as vitamin C, which has not already been pre-oxidized by the cutting process, from oxidation. To this day, sauerkraut, kimchi & Co are famous for their extremely high vitamin C content.

In this process, salt has several fundamental tasks. During fermentation, the pH value slowly drops to a value between 3.2 and 2.9. The critical phase is the initial period of fermentation when the pH value is still above 3.5 and harmful germs can still multiply strongly, as typical fermentation temperatures are between 25 and 35°C.

Until the pH value drops that far, no pathogenic germs should grow, as shown in Fig. 2.25. Therefore, during fermentation, the salt content must be sufficiently high to bind enough water in this critical phase and limit the growth of pathogenic germs. The osmotic effect of salt also helps to release bound sugar from the cells and cell proteins of the vegetables as fuel for fermentation. The bursting of still intact cells caused by this further destroys the cell structure. However, many sugars in vegetables are strongly bound in glycoproteins, saponins or glycosinolates (e.g., in the mainly used cabbage, radish, and onion plants for fermentation), which must first be enzymatically released to be available to the lactobacilli during fermentation.

At this point, the fundamental difference between salt-free fermentation, as in Gundruk or the Japanese *sunki*, becomes clear. Without salt, it is very likely that additional germs, such as those transferred from hands or the human body to the vegetables, will grow at the beginning of fermentation. These individual notes can be further enhanced by rubbing the vegetable leaves over the arms and face beforehand. What appears unhygienic is still hygienic. In the end, the low pH value prevails, inactivating pathological and harmful germs or keeping them within tolerable limits. However, there are strong odor differences in salt-free fermentation. The pungent smell of Gundruk and *sunki* is unpleasant to many people. In contrast to Kimchi, Sauerkraut & Co, salt-free fermentation also produces

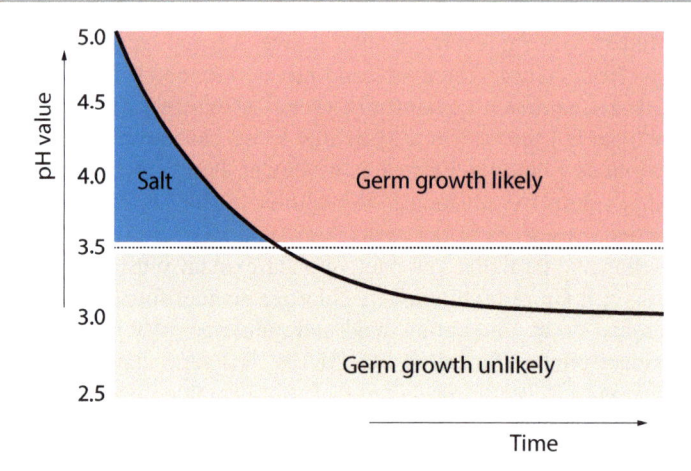

Fig. 2.25 The salt added during fermentation prevents the growth of unwanted, harmful germs during the critical phase, as long as the pH value is not sufficiently reduced

typical decomposition odors reminiscent of decay or animal carcasses. The biogenic amines, 1,5-pentanediamine (cadaverine) and butane-1,4-diamine (putrescine) are formed during fermentation from amino acids derived from plant cell proteins. They are also known as corpse poisons, although the name hardly reflects the chemical and biological properties of the two odorants.

Cadaverine is formed from the essential amino acid lysine, while putrescine is formed from arginine via the intermediate product ornithine. In very high doses, cadaverine and putrescine, like most biogenic amines, are toxic (see, for example, histamine), but they are also an essential part of our physiology. Both substances help significantly with digestion in the intestine and are also present in most human secretions, where they are responsible for the biologically stabilizing effect, as is exploited in cellular "anti-aging," as shown in detail in Chap. 6.

2.9.3 Fermented Beverages, Hot Drinks

In beer (grain) and wine (grapes), yeasts ferment sugar [64], producing alcohol, which in early times was at least digestible for some people. Wild yeasts fermented germinated or cooked grains. Yeasts cause the fermentation of sugar from the starch of grains to produce alcohol. In the case of wine, free sugars from grapes and fruits are metabolized by yeasts. These beverages were highly welcome, as the low alcohol content was well tolerated, the drink tasted good, was also reasonable durable, and was not infested by germs. The beverage culture thus began already in the Neolithic.

Ethnic groups that could not digest alcohol due to the absence of alcohol dehydrogenase developed a tea culture. Tea itself is again fermented but had to be brewed. These drinks are also safe: pouring hot water over them ensures germ-free conditions and—far more importantly—the steeping extracts a whole range of phenols and polyphenols (tannins) that have strong antioxidant and antibacterial effects. At the same time, tea, like wine, has an astringent trigeminal effect. Tannic acids in teas and wine cause this dry feeling in the mouth, which is also recognized as positive in practically all regions of the world: tea culture in Asia, beer and wine in Eurasia and Europe, astringent cocoa drinks in South America. Apparently, two fundamental aspects play a role in the development of all cooking techniques: safe food and appealing sensory properties. In addition to taste and smell, the trigeminal nerve also plays an obvious role in the development of cooking techniques.

2.9.4 Miso, Fish Sauce, Soy Sauce & Co—Brilliant Examples of (Complete) Fermentation and Flavor

In today's Europe, targeted fermentation using Lactobacilli was practiced during the Bronze Age. The best-known historical example is the fish sauce Garum [65], which was used by Greeks and Romans as a universal seasoning sauce. Fish were salted along with their innards—and thus also the enzyme-active pancreas as well as the liver as a fatty acid supplier—tightly packed in barrels and fermented for a certain period of time. The process teaches a variety of typical basic principles of fermentation. We have already encountered salt as an osmotic component. The retention of the innards has two reasons: fish liver has an exceptionally high density of nutrients. The pancreas, on the other hand, provides all enzymes relevant for degradation: lipases, proteases and glycosidases, which are capable of cleaving and thus digesting fats, proteins, and carbohydrates. During the fermentation of the fish, several processes take place in parallel. On the one hand, the pancreatic enzymes cleave fish proteins, fats, and the carbohydrate components of the glycoproteins into peptides, amino acids, and sugars. On the other hand, lactic acid bacteria ferment sugars to lactic acid in the oxygen-poor areas in the center of the barrels, which in turn lowers the pH value. The large molecules are thus broken down into smaller ones, the consistency changes, and the fish turn into a spicy, salty-sour sauce that contains only indigestible solid residues for the enzymes. This seasoning sauce has a long shelf life due to its acid content and high salt concentration.

In this context, the course of the pH value in the Garum sauce [66] is interesting, which is roughly depicted in Fig. 2.26.

The course of the pH value shows the processes in the emerging fish sauce. In the first ten to 20 days, the pH value drops steadily due to the progressive lactic acid fermentation, while in the remaining time the acid increases only slightly. During this time, however, a more or less complete enzymatic conversion of macronutrients, color pigments, and connective tissue takes place. Proteins are cleaved, fats are converted. New taste substances and aromas are formed. Even if the fermentation temperatures are between 15 and 30°C, enzymatic and oxidative browning occurs: The sauce turns brown and resembles, in color and taste, a

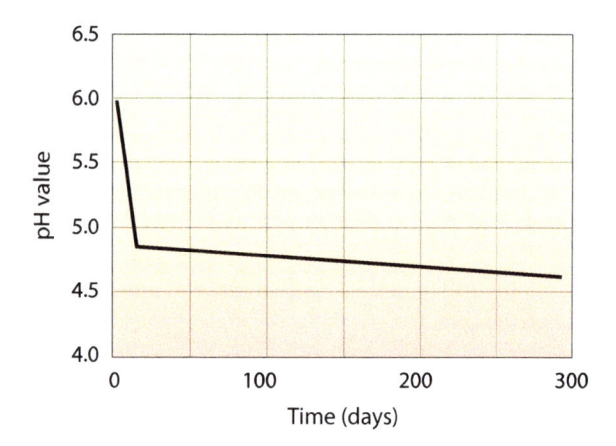

Fig. 2.26 The pH value during the fermentation of a fish sauce (type Garum)

gravy. This is precisely the central point: umami taste is created and an incredible mouthfilling sensation, mouthfulness mouthfeel (kokumi) develops. The fermentation of protein-rich foods can thus be represented in a similar scheme as shown in Fig. 1.15 already for cooking. With one difference: The proteins are not broken down by the long temperature exposure, but enzymatically. Fermentation is thus a "molecular cooking at low temperatures" [67].

A similar system can also be seen in soy sauces and miso pastes. There, the koji fungus (*Aspergillus oryzae*), the is an edible mold fungus, which grows, for example, on the substrate rice. This fungus performs a similar function to the pancreas of fish in the production of fish sauce: It releases proteases, lipases, and amylases as well as enzymes that cleave proteins, fats, and starch, provided they are cooked or steamed so that the active centers of the enzymes have access [68]. Mainly because of the starch-splitting amylases, this fungus is suitable for the hydrolysis of protein- and starch-rich seeds such as soybeans, rice, and cereals. But the koji fungus is also suitable for the production of mixed forms, such as miso preparations, which consist of fish and soybeans [69] and which, due to the animal amino acid spectrum, have a significantly expanded aroma formation and, in addition to salty and bitter, an intense umami taste [70]. Miso describes the product of a mixed fermentation of lactic acid fermentation and targeted enzymes via the malted koji rice, rice inoculated with the noble fungus *Aspergillus oryzae*. For this purpose, rice of a certain variety is soaked in fully desalinated water at room temperature for twelve hours and then steamed for one hour at 90 °C.

Steaming at 90°C is essential: On the one hand, the polished rice grains cook evenly, and no hard layer forms due to cooking in calcareous water. Furthermore, the crystalline starch melts and can absorb water. Starch, amylose and amylopectin are present in hard, twisted (amylose), and partially crystalline forms (highly branched amylopectin) in the uncooked state. These molecules are densely packed into hard starch granules. The enzymes have no access to these hard structures. Only steaming softens the grains, allowing the molecules to bind water and attach enzymes. Only then can enzymes such as amylases break down the starch into sugars.

2.9 Fermenting and Fermentation—in the History ...

Subsequently, the rice is cooled to 35°C and inoculated with Koji spores and fermented at 35°C for 48 hours. The fungus initially colonizes the surface of the rice grains but gradually penetrates them. The cell tissue becomes brittle, and the enzymes are activated. The resulting product, Koji rice, is therefore very enzyme-rich in application and is used, for example, in sake brewing and occasionally in baking. For the fish miso, fish meat is minced, vacuum-packed, and steamed at 90°C for one hour. It is then pressed to a water content of about 55% and mixed with Koji and salt (mixing ratio 5:5:1) and ground. The mixture is then allowed to ferment for one year at temperatures between 25 and 30°C. The procedure for soy miso is similar: The soybeans are soaked for twelve hours, then steamed at 90°C for one hour, and subsequently ground in the same mixing ratio with Koji and salt.

The addition of the enzyme-rich Koji fungus is also crucial for the development of taste and aromas. The highly active enzymes, especially proteases and amylases, ensure a high proportion of soluble sugars, peptides, and amino acids in the first 14 days of fermentation, with aroma formation starting rapidly. Comparative experiments of fish and soybean fermentation show a significant increase in soluble sugars and peptides during this phase. Without the addition of Koji fungus, no significant increase in enzyme activity is recorded in the same period. So definitely something that can also be prepared at home, provided you have the patience. In Fig. 2.27, a very simplified scheme for the enzymatic processes is shown. It becomes apparent how practically all components of the starting preparation accessible to the enzymes are converted into aromas.

During fermentation, the formation of a well-balanced aroma spectrum can be observed, as shown in Fig. 2.27. The brown coloration of the miso also occurs through enzymatic browning. This process initially causes the progressive browning by converting phenolic components such as ferulic acid, polyphenols, etc., but during the course of fermentation, it also produces roast-like or even bread crust-like smells when sugars and amino acids react with each other to form deliciously fragrant "Maillard products" slowly during the long fermentation times. Free sugars provide caramel aromas, but fruity and earthy aromas are also formed in abundance. Aromas and smells that have been shaping humans since their interaction with fire. Highly fermented sauces and pastes thus fit into the familiar smell pattern. It is therefore no wonder that fermentation found its place in the diet of *Homo sapiens* beyond food safety.

The explanation is relatively simple: During roasting, amino acids and sugars are thermally stimulated to react with each other. As shown in Fig. 2.28, sugars and amino acids are energetically raised during roasting by temperature (thermal energy) like in an elevator and can then easily overcome the energy barrier (left in Fig. 2.28). Enzymes, on the other hand, significantly lower the high potential barrier due to their catalytic effect, which can then be overcome even at low temperatures (right in Fig. 2.28). Amino acids and sugars react to form odorants that are similar to those produced during roasting. Soy sauces, Miso pastes and gravy sauces therefore not only look similarly dark brown, but also form similar aroma compounds in many cases: Both smell very similar. The classic enzymatic browning, as is known for many polyphenol-rich, raw fruits and vegetables, plays

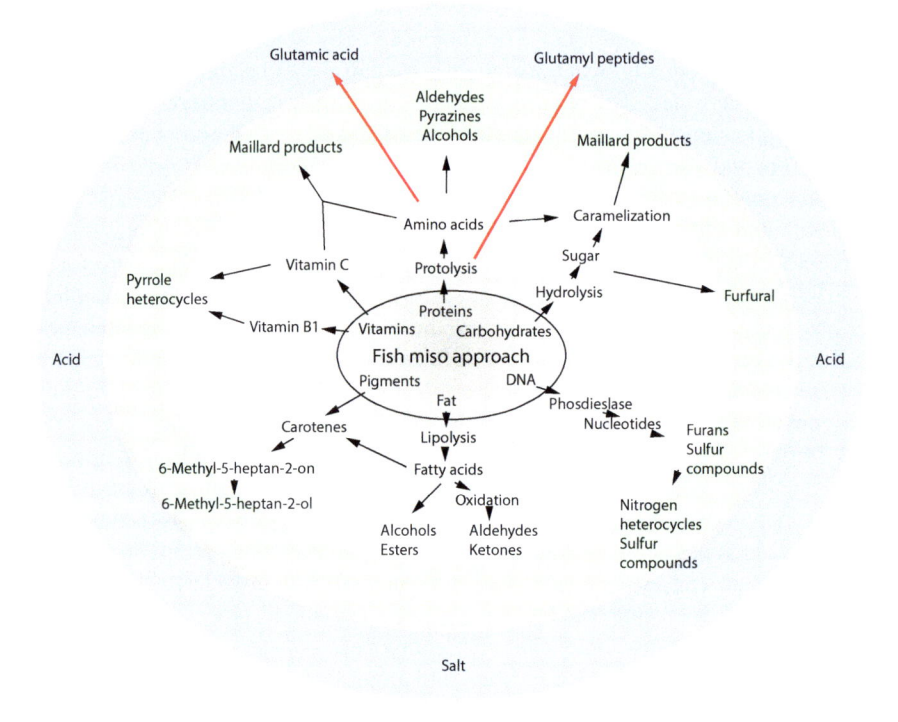

Fig. 2.27 The conversion of proteins, fats, pigments, vitamins, etc., during the miso fermentation of a fish miso approach [69]. The "aroma formation ring" fo the miso is shaded green, blue designs the "taste ring." The high proportion of free glutamic acid defines the typical umami taste of long-fermented foods

a subordinate role in these cases. Here, existing diphenols are oxidized to quinones via polyphenol oxidases, which produce a brownish coloration. Polyphenol oxidases are completely deactivated when steaming legumes.

An important aspect is, as always, the taste. The added salt does not change during fermentation, so it remains. Miso therefore fundamentally tastes salty. Lactic acid bacteria make the miso paste sour at the beginning of fermentation. However, a significant amount of free glutamic acid also forms, which is considered the main trigger of umami taste. Soy sauces, fish sauces, and miso pastes therefore represent a typical savory umami taste. In fact, this is associated with the reaction inertia of glutamic acid in conventional food preparations such as roasting, braising, baking, and fermenting. Glutamic acid does not form odor-active Maillard products during fermentation, nor browning substances or roasted flavors during roasting. Thus, it withstands common kitchen techniques and unfolds its full umami taste potential on the tongue. As already elaborated in detail in Chap. 1, the umami taste was the driving force of the Paleolithic era. It is therefore not surprising that with

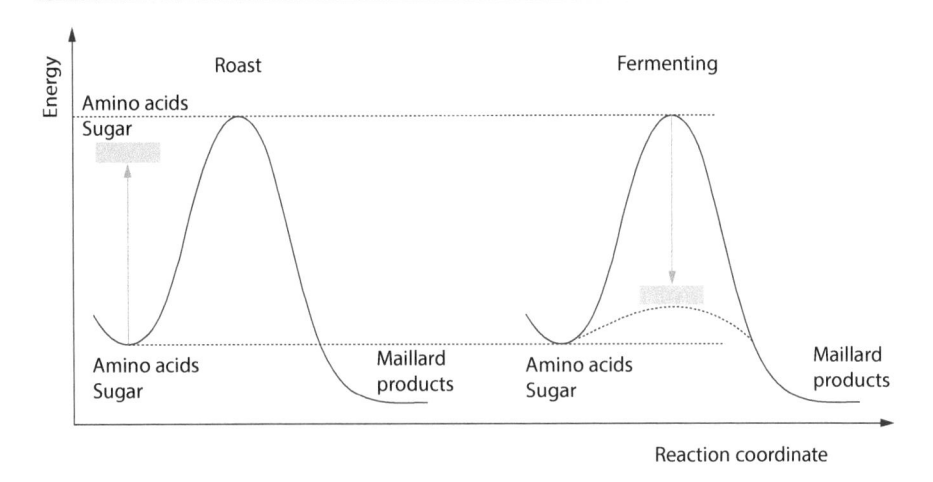

Fig. 2.28 Sugars and amino acids react to form Maillard products. At the molecular level, the difference between roasting/grilling/braising and fermenting is smaller than one might think

the control of microorganisms in the Neolithic and in modern times, fermentation as a fundamental process technology never lost its importance and is still cultivated with high-tech methods in all forms for the sake of delicious taste.

2.9.5 Safety and Advantage of Fermented Products

In the final analysis, this also reveals the great importance of the basic taste "sour". Mild acids with pH values between 4.5 and 3.5 stand for safer foods. Fermented vegetables thus met the main criteria: good taste and shelf life, as well as, depending on the fermentation duration and type, a more or less strong development of the umami taste. The cultural technique of fermentation thus fulfilled the most important prerequisites for its further development in cooking and food culture: Similar to cooking on fire, fermentation contributed to the preservation of life and further development. Incidentally, fermented products had a whole range of advantages for the nutrition of people in all epochs. Many nutrients become easily available through the multitude of chemical and biochemical reactions. In addition, as already hinted at in various places, a whole range of highly antioxidative substances with anti-inflammatory properties are formed, mainly from phenolic and glycosidic substances that are antinutritive in their raw state.

These include, for example, Nattō (see Chap. 5), the specially fermented soybeans, which have been part of the food culture in Asian and especially Japanese cuisine since 200 BC. To do this, the soybeans were cooked, then slightly dried again and fermented on rice straw. As a result, the cooked soybeans were inoculated with a Bacillus, which is now called *Bacillus subtilis natto*. The mere idea of cooking the beans and drying them again shows the cultural achievement behind this product and the research achievements and process optimizations that people were already capable of in the Neolithic era.

The fermentation takes about one day at room temperature. On the surface of the beans, a slimy-looking coating forms, which is perceived as unusual for European tongues. During this process, proteins are also broken down by the enzymes of the Bacillus, and even polyglutamic acid is synthesized, a polymer whose basic units consist only of glutamic acid. Since this polymer is electrically negatively charged at (almost) every basic unit, the water-soluble macromolecule binds many water molecules around itself and is thus responsible for this highly viscous, thread-forming slime on the surface of the fermented bean. In addition, a whole range of micronutrients are formed during the fermentation process, which in turn are available to human physiology, especially vitamin K_2, but also vitamins of the B-complex, water-soluble enzymes such as nattokinase, and longer-chain fructose-based dietary fibers. The aroma formation is also typical: in Nattō, heterocyclic pyrazines also dominate, whose odor attributes are described as nutty, roasted, earthy, burnt. Here, too, there is an aromatic proximity to typical temperature-induced roasting aromas, even if their concentration in Nattō is significantly lower.

Thus, as with cooking, many aspects come together during fermentation: smells good, tastes good, tastes like umami, and benefits humans through a variety of essential micronutrients. Only fermentation potentiates the concentration of relevant antioxidant properties, so it is no wonder that miso pastes, soy sauces, sauerkraut, kimchi & Co were developed and cultivated in all cultures. Consequently, these basic techniques remained integral parts of human nutrition over the millennia.

2.9.6 Kokumi— Perpetual Umami Companion for Millennia

But the fermentation—and also the braising—is responsible for another secret that we taste on our tongues. This is also a reason why these methods have been carried from the Paleolithic and the Neolithic to the present day. In Fig. 2.27, so-called glutamic acid and γ-glutamyl peptides appear (which are shown on the taste circle for simplicity, although this is not strictly correct). Both during long cooking and during fermentation, small fragments form from the proteins, which are responsible for the mouth fulness mouthfeel, "kokumi". They always consist of a glutamic acid and one or two other, hydrophobic amino acids. In soy, fish, and also in miso pastes, it is predominantly peptide pieces such as glutamyl-valyl-glycine (γ-Glu-Val-Gly), i.e., fragments consisting of a glutamic acid (Glu), a valine (Val), and a glycine (Gly) [71]. In miso pastes, for example, γ-glutamyl-cysteinyl-glycine (γ-Gl-Cys-Gly) and the dipeptide γ-Glu-Val are also added [72]. These protein fragments provide an extraordinarily high mouthfeel, apart from the five basic tastes, by stimulating calcium-sensitive sensors on the tongue that have long been overlooked [73].

This is another important point, because both cooking and fermentation show: Umami, unlike all other tastes, apparently never comes alone. Although glutamic acid dominates the results of these two cultural techniques, the evolutionary taste is always accompanied by glutamyl peptides, which are held responsible for the moutfulness mouthfeel. Umami and kokumi are thus inseparable partners (as long as one does not resort to conscious tasting with glutamate) and accompanied the cooking hominids up to the *Homo sapiens* for over a million years through the evolution. Humans are also driven by chemosensory means.

2.10 Germination as a Universal Cultural Technique for Food Enrichment

In addition to fermentation, the use of germinated seeds from cereals, pseudocereals, and legumes also became established and was used by most ethnic groups of the Neolithic period [74]. This is evident in noodles based on germinated and fermented millet in the late Neolithic, both in Africa [75] and Asia [76]. Germination not only increases the nutrient content of millet (as well as other cereals and pseudocereals), but also breaks down non-nutritive substances that serve plant defense, and above all, generates physical properties that allow the starch to stick and gelatinize. This significantly improves the processability of the doughs.

Germination thus served as a value-adding refinement of seeds, which are not edible raw due to their high lectin content, especially in legumes. From a molecular perspective, germination is a significant change in molecular structure, but at the same time, it produces a variety of new nutrients. During the germination process, antinutritive lectins, which can cause blood cells to aggregate at high concentrations, decrease dramatically. Germinated beans and their sprouts thus become edible without heating. Hard plant substances in the seeds, such as phytates and pectins, are partially enzymatically converted by enzymes released from membranes during germination and synthesized into nutrients (primarily for the germinating plant). Starch is also converted to sugars by amylases. Taste and aroma change significantly. During germination, the vitamin content increases, and protein-degrading enzymes (proteases) are released, but also enzymes that convert amino acids. As a result, the proportion of essential amino acids cysteine, tyrosine, and lysine increases significantly. Previously non-existent proteins and enzymes are formed, and the water-binding capacity increases. At the same time, the new proteins exhibit strong surface-active properties: The emulsifying ability and the capacity for foam formation increase significantly, as the proteins become partially water-soluble.

In contrast to fermentation, however, germination usually does not involve fermenting bacteria. Acidification, a strong umami taste formation, or a dramatic aroma conversion towards fire flavors does not occur. A significant advantage of germinated seeds is their use as natural biomachines in the production of bread, beer, and products that, like noodles, have a higher processing degree.

It is fascinating to recognize how advanced early food technology was, and to this day, the method and the resulting foods are the basis of enjoyment and daily cuisine [77]. But we must not forget that the foundations for this were laid millions of years ago. Early hominids and the early forms of modern humans were already sensory-driven beings.

2.11 The Origin of Human Chemical Sensing

The essential prerequisites for this can be found in the physiology of humans and their taste perception [45]. Above all, the ability to taste umami laid the foundation for the evolution of humans. Even in the phase before the utilization of fire, hominids consumed carrion. However, carrion is "ripened meat" and thus should be understood as a form of decay and therefore fermentation. Free glutamic acid and Maillard precursor products formed, the carrion tasted umami, and consuming it was recognized as an advantage. Also, because it could be chewed and digested with less energy expenditure than fresh meat due to the beginning of decay. Later at the fire, the umami taste was preserved. Roasting and cooking produced free glutamic acid, the umami taste became more intense, delicious flavors were added, which far surpassed the taste of "ripe carrion." The use of fire was obvious anyway: after bush and grass fires in the savannah, early humans found already burned, "grilled" meat and consumed it. Thus, animals that perished in the fire were probably the first grilled delicacies of early humans. It was precisely the sensory perception of the intense umami taste combined with the roasted flavors that led to a significant imprint [78] and control of food search. Meat-eating hominids thus had an expanded food supply and a significant advantage over plant-eating apes such as chimpanzees and other primates that do not respond to free glutamic acid with taste stimuli [79]. Biesalski [46] also points out another advantage: the expansion of taste preferences and the associated expansion of the food spectrum made it easy for the emerging human to access all essential micronutrients, "especially the critical micronutrients iron, zinc, folic acid, vitamin A, D, and B_{12}."

2.12 The Cultural Imprint of Universal Molecule Classes—Beloved Flavors

All lactic acid fermented vegetables have a very similar taste and a likewise typical smell. The few examples mentioned already show that fermented and fermented foods were developed in all cultures and ethnic groups as early as the Neolithic: yogurt, curd, kefir, cheese, beer, wine, vinegar, kimchi, soy sauce, early forms of miso, to name just a few.

This basic taste accepted in every culture, as well as the similarities in the aroma profile, suggest a deep imprint on humans that has occurred through a few universal processes: cooking and fermenting. With garum, miso pastes, fish sauces, soy sauce, nattō, etc., smells and aromas dominate that are very reminiscent of roasted and cooked, as has already been described in several places in this chapter. Humans apparently experienced an imprint through odorous substances, which, together with the typical umami taste and "sour" for "safe," determined the cooking culture. This imprint continues to this day.

Only in a very modern era were fermented products banned from the canon: in classical French cuisine and in the kitchen style defined as Nouvelle Cuisine, it was considered necessary to avoid "the fermented" on plates. This is incomprehensible in the light of the evolution of cooking humans. However, for some time now, this technique has been experiencing outstanding importance in our time of avant-garde kitchens. The new regional kitchen styles brought a significant boost to fermentation, as the sometimes highly restrictive product selection suggests recognizing fermentation not only as an opportunity for new aroma and taste variability but also, as in the times of refrigerators, for preserving food. In the best sense, this is not only *"back to the roots"*, but also a reflection on the elementary cultural techniques that are still the basis of all nutrition today—since the first hominids to *Homo sapiens.*

It is no coincidence that people still like braised, roasted, grilled, and fermented dishes today. Despite all fears and warnings about grilled or burnt food. Not infrequently, the almost black grilled sausages, the dark baked bread, and the almost black grilled peppers are eaten as antipasti, despite urgent warnings about acrylamide [80] and warnings from health experts [81]. This illustrates how strong the sensory imprint is on the food culture and the developed cooking techniques. Experience also shows that *Homo sapiens* continued to evolve as long as he trusted his taste—despite the health risks he faced when eating from today's perspective.

2.13 Culinary Triangle and Structuralism: The Universal Basis of Human Nutrition

At this point, a foray into cultural studies is appropriate, more specifically into ethnology and Structuralism. From a completely different perspective, similar conclusions are drawn. There is nothing else but raw, cooked, and fermented. All foods that we consume to this day are defined based on this foundation. This was demonstrated by cultural scientist and ethnologist Claude Lévi-Strauss [82], when he introduced his ideas of structuralism into culinary arts. He translated the universality of food states—raw, cooked (or done), and rotten/fermented—into a triangle with these corner points (Fig. 2.29). At the same time, he defined the transitions between nature and culture on one axis and from unchanged to changed on the other axis. Cooking, as a process led by humans, is thus recognized as a cultural achievement.

The term "raw" refers to the initial structure of each food in its original, biologically natural state, be it the crisp apple freshly picked from the tree, fresh carrots with soil, or the living animal, even René Redzepi's still living seafood offered for consumption at the "Noma" restaurant [83]. Any further process, even washing

Fig. 2.29 The culinary triangle according to Lévi-Strauss (inside), expanded by the symbolism of molecular processes (outside). The transition from nature to culture was only made possible through human control of temperature and microorganisms

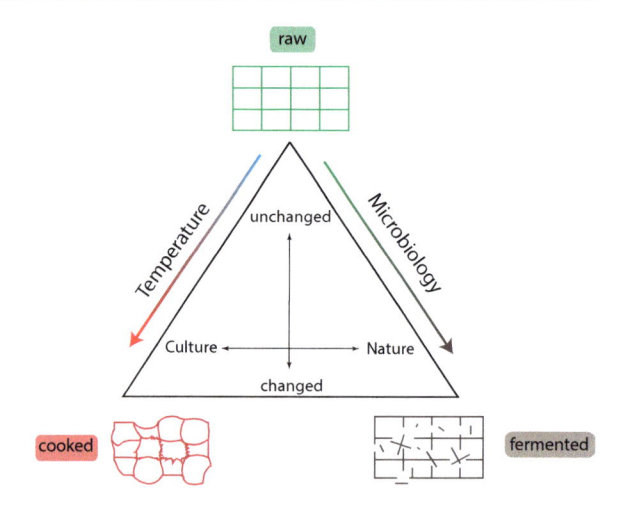

the carrots, the lettuce, or wiping the apples, is already considered a cultural act in cultural studies, leading slightly away from the raw state. If the raw is left to nature without cultural intervention, it will rot. However, cultural measures by chefs allow the rotting process to be controlled—then it is called fermentation. Cooking and fermenting are therefore to be appreciated as fundamental cultural actions dedicated to the continuation of life.

These abstract considerations of the culinary triangle can also be scientifically justified when considering the changes in molecular structures caused by specific process techniques [84]. Fermenting and cooking are different techniques in terms of process. Cooking is usually achieved with a change in typical thermodynamic parameters such as temperature, pressure, or volume. At the fire, stove, or grill, this is always the temperature. In doing so, the protein structures change first, and the texture of the food changes. Initially, physical changes in the food are in the foreground. Fermentation always involves microorganisms, such as lactic acid bacteria, yeasts, etc. This allows for enzymatic, chemical reactions to occur. The changes at the molecular level are of a different nature.

2.13.1 The Molecular-Evolutionary Variant of the Culinary Triangle

Apart from cultural-scientific discussions, this triangle takes on a much more concrete meaning when interpreted under the molecular aspects of the previous explanations on cooking and fermenting, as shown in Fig. 2.30. For simplicity's sake, we will limit ourselves to proteins, as their components, especially glutamic acid, are responsible for the umami taste.

In the immediately raw state, the molecular components of food are in their native form. Proteins, chain molecules made up of individual amino acids

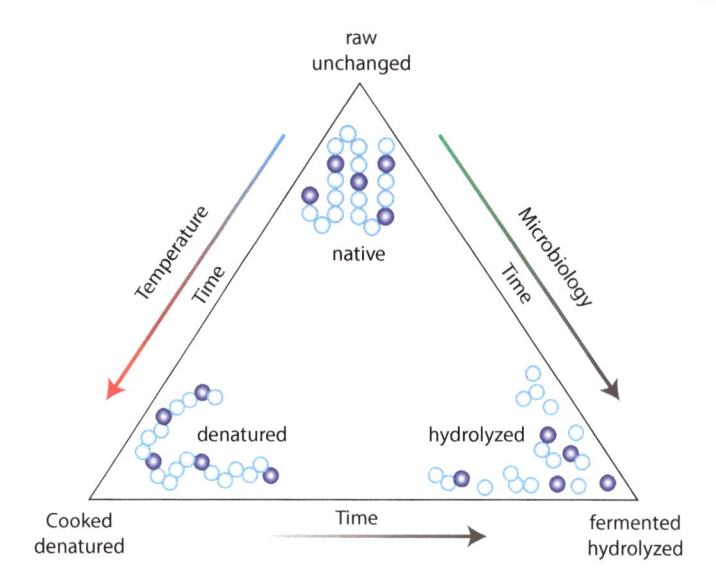

2.13 Culinary Triangle and Structuralism: The Universall …

Fig. 2.30 The culinary triangle in the focus of taste formation. Both long cooking and long fermentation break down (hydrolyze) proteins. In the lower right part of the triangle, the umami taste is anchored

(represented as spheres), are folded, retaining their (in most cases) original structure. When the "raw" is transformed into the "cooked," this occurs with an increase in temperature. The thermal energy unfolds the proteins (and other structural polymers), causing them to lose their native shape, and the texture of the food changes. When food is fermented, i.e., altered by enzyme-releasing microorganisms, proteins are gradually broken down into fragments until individual amino acids are present. This includes glutamic acid (colored dark blue), which is responsible for "umami," while some small fragments of two or three amino acids are responsible for "kokumi" (mouthfeel), provided they still carry a glutamic acid. Fermented sauces, fish sauces, soy sauces or miso pastes are therefore always savory and mouth-filling. Long cooking, as in stock or sauce bases, also leads to protein fragments and free glutamic acid, resulting in umami and kokumi. Different paths, one goal: the intense taste of well-being. Since time immemorial, all fundamental cooking processes have aimed for the umami taste. At this point, cultural sciences and natural sciences can be brought together, despite different approaches, formulations, and scientific methods.

2.13.2 Cooking Cultures in the Light of the Culinary Triangle

In fact, there is no traditional and artisanal cooking method that cannot be anchored in the culinary triangle. In cultural studies, the necessary culinary actions are primarily used as a criterion. For example, the contrast between raw and cooked is often divided into the weaker contrast of raw to fried and the stronger raw to boiled. In the case of boiled dishes, a vessel and water are added as mediators in the cooking process, which is why boiled dishes are culturally higher than fried ones. Smoke and air, or steam, are also used to describe cooking techniques with the triangle. Washing vegetables and fruits as well as selection (e.g., by ripeness) are comparatively simple cultural actions. However, this leads to contradictions and many discussions in cultural studies. These can be easily resolved if the molecular changes associated with the respective cooking technique are scientifically justified [85], as has already been done in Fig. 2.30.

The consistent expansion of the triangle of "cultural actions" and the connection with the changes triggered by them, such as gentle temperature increases during steaming or low-temperature cooking, marinating, pickling, and fermenting with knowledge of the molecular changes, lead to a new cooking landscape and many intermediate stations on the way from raw to cooked and raw to fermented, as shown in Fig. 2.31.

Therefore, all modern cooking techniques can be located in the culinary triangle. Low-temperature cooking using sous-vide technique [86] leads to cooking states between raw and cooked, as some proteins remain in their native raw state at the usual temperatures between 45 and 60°C [87]. The culinary triangle also retains its validity in connection with biotechnological developments in the sense of sustainability when novel proteins, such as Quorn, are produced from fungal spores, bacteria, or other bioreactors [88]. These novel foods can be located, similar to aged meat, fresh between raw and fermented. Even extracted proteins

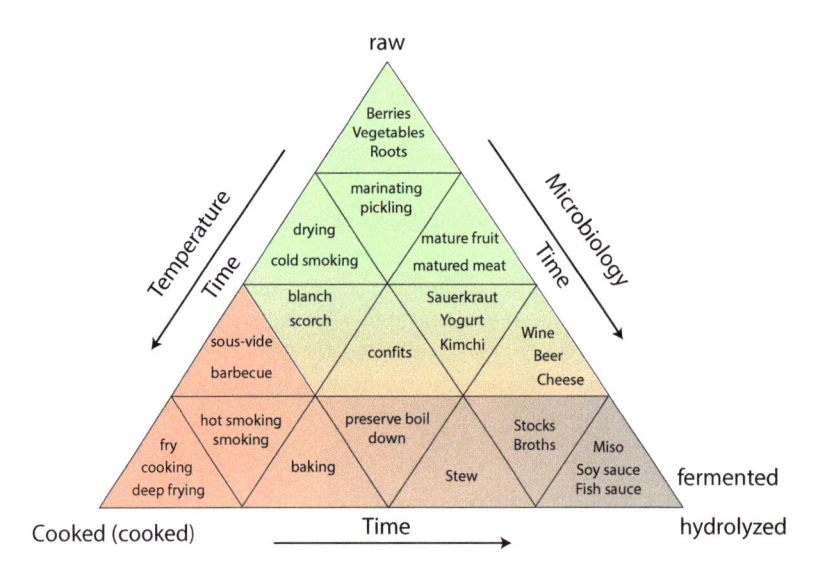

Fig. 2.31 All current cooking techniques can be located in the culinary triangle when the molecular changes and states are taken into account. The upper, still greenish areas can be defined as "pseudo-raw" to initiate a better definition of the term "raw food" (see Sect. 5.8)

from algae, soybeans, or other foods are still within the triangle. In the end, only the state and taste of the food count, which are exclusively defined by molecular parameters. Novel industrial process methods, on the other hand, find little space in the culinary triangle, as detailed in Chap. 5.

In cultural studies, the term "raw" is defined somewhat differently. There, even cleaning the carrot or washing the apple is considered a cultural act, and the washed food is, strictly speaking, no longer raw. Therefore, we will expand the term "raw" to accommodate the molecular aspects that are the only measurable and objective criteria for us. To this end, we introduce the term "pseudo-raw," which allows for cultural actions but is still minimally changed at the molecular level, as shown in the greenish area in Fig. 2.31. The expanded "raw" thus becomes the connecting part of unheated but fermented foods, such as yogurt, sauerkraut, or kimchi. Above all, however, "raw" and "pseudo-raw" offer a multitude of new possibilities for taste and aroma variety, new textures with captivating mouthfeel, and last but not least, new approaches to plate design, which we will revisit later in the book (Chap. 6).

The culinary triangle also illustrates a comprehensive roadmap of the various processes of kitchen craftsmanship. It also shows how systematically certain paths were chosen through cultural techniques to achieve umami taste. This applies to all cultures of the world; the desire for umami taste is universal.

To illustrate this, we look at the triangle in the simplified form from Fig. 2.30. It becomes clear how far-reaching the universal idea of the culinary triangle is. Braising and slow cooking first lead to the denaturation of proteins, then cooking or braising for a long time, and through the hydrolysis of proteins, the perceptions of umami and kokumi emerge, as represented by the light brown arrows. In long-fermented products such as miso, soy and fish sauces, the ingredients are first denatured by cooking or steaming, then the temperature is maintained—an initial hydrolysis sets in. Subsequently, the preparation is cooled to the fermentation start temperature, then inoculated with koji cultures and fermented for a long time, as shown by the dark brown arrows. In the case of kimchi or sauerkraut fermentation, the ingredients are washed and cut. This has only a minimal effect on the molecular structure. Then the lactic acid bacteria intervene and initiate a weak hydrolysis under acidification. The umami taste is less pronounced. Roasting is interesting. First, proteins are denatured until they hydrolyze on the roast surface or when grilling with the highest heat contact in the crust and, with further temperature increase, point towards umami. In this way, practically all preparation techniques can be represented as process paths in the culinary triangle.

It is important in this context to consider the direction of the arrows. They are not reversible. Once denatured, proteins do not fold back. Also, the amino acids and peptides of hydrolyzed proteins do not find their way back to a complete protein. The processes are irreversible. The representations of the results in Figs. 2.31 and 2.32 impressively show how cultural studies, ethnology, and anthropology meet with the basic natural sciences of physics, biology, and chemistry. This is only offered in this pronounced form by culinary studies.

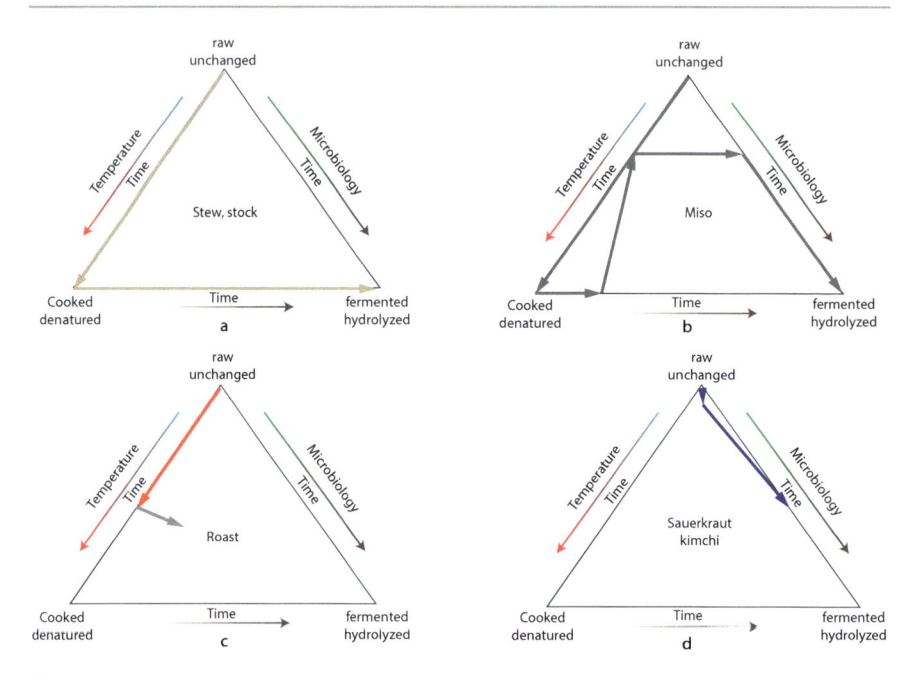

Fig. 2.32 The classical pathways of evolution towards umami in the light of the culinary triangle. **a**) braising/braising fond, **b**) miso, soy and fish sauce, **c**) roasting(grilling), **d**) sauerkraut and kimchi

2.14 Conclusion

The universality of the culinary triangle, the pursuit of umami, can thus be reconciled not only chemically, sensorially, and food technologically but is also linked to the sensory perception of the human species: *Homo sapiens* is apparently a chemically-sensorially driven being. This statement is increasingly confirmed by the development of food culture. Indeed, it is fascinating how strongly the "chemical senses" influenced survival, culinary skills, and cooking techniques. This is also one of the decisive characteristics of *Homo sapiens* and its developed culture: No animal consciously commits these cultural acts, controls fire, or even microorganisms (Fig. 2.33). These are the fundamental characteristics of humanity since the harnessing of fire. And ultimately, this is the clear dividing line between humans and animals, despite the high degree of overlap in the genetic structure of humans and other higher mammals.

Fig. 2.33 Fire, agriculture, and animal husbandry, as well as fermentation, proved to be milestones and fundamental cultural techniques of human evolution with significant impacts on nutrition. Whether this will also apply to today's technical revolutions remains to be seen

References

1. Jones, K. T., & Metcalfe, D. (1988). Bare bones archaeology: Bone marrow indices and efficiency. *Journal of Archaeological Science, 15*(4), 415–423.
2. Speth, J. D., & Spielmann, K. A. (1983). Energy source, protein metabolism, and hunter-gatherer subsistence strategies. *Journal of Anthropological Archaeology, 2*(1), 1–31.
3. Milton, K. (2000). Hunter-gatherer diets – A different perspective. *Amercian Journal of Clinical Nutrition, 71,* 665–667.
4. Winterhalder, B. (1981). *Optimal foraging strategies and hunter-gatherer research in anthropology: Theory and models* (pp. 13–35). Hunter-gatherer foraging strategies: Ethnographic and archaeological analyses.
5. Hawkes, K., Hill, K., & O'Connell, J. F. (1982). Why hunters gather: Optimal foraging and the ache of Eastern Paraguay. *American Ethnologist, 9*(2), 379–398.
6. Cordain, L., Eaton, S. B., Miller, J. B., Mann, N., & Hill, K. (2002). The paradoxical nature of hunter-gatherer diets: Meat-based, yet non-atherogenic. *European journal of clinical nutrition, 56*(S1), 42.
7. Kaplan, H., Hill, K., Lancaster, J., & Hurtado, A. M. (2000). A theory of human life history evolution: Diet, intelligence, and longevity. *Evolutionary Anthropology: Issues, News, and Reviews, 9*(4), 156–185.
8. Stanford, C. B. (1999). *The hunting apes: Meat eating and the origins of human behavior.* Princeton University Press.
9. Pruetz, J. D., & Bertolani, P. (2007). Savanna chimpanzees, *Pan troglodytes verus*, hunt with tools. *Current Biology, 17*(5), 412–417.
10. Liebenberg, L. (1990). The art of Tracking – The origin of science. David Philip Pub.
11. Gibson, K. R. (1986). Cognition, brain size and the extraction of embedded food resources. *Primate ontogeny, cognition and social behaviour, 93*–103.
12. Cordain, L., Watkins, B. A., Florant, G. L., Kelher, M., Rogers, L., & Li, Y. (2002). Fatty acid analysis of wild ruminant tissues: Evolutionary implications for reducing diet-related chronic disease. *European Journal of Clinical Nutrition, 56*(3), 181.
13. Frassetto, L. A., Schloetter, M., Mietus-Synder, M., Morris, R. C., Jr., & Sebastian, A. (2009). Metabolic and physiologic improvements from consuming a paleolithic, hunter-gatherer type diet. *European journal of clinical nutrition, 63*(8), 947.
14. Eaton, S. B., Eaton III, S. B., Sinclair, A. J., Cordain, L., & Mann, N. J. (1998). Dietary intake of long-chain polyunsaturated fatty acids during the paleolithic. *The Return of w3 Fatty Acids into the Food Supply, 83*, 12–23. Karger Publishers.
15. Crawford, M. A., Williams, G., Hassam, A. G., & Whitehouse, W. L. (1976). Essential. *The Lancet, 307*(7957), 452–453.

16. Crawford, M. A., Bloom, M., Broadhurst, C. L., Schmidt, W. F., Cunnane, S. C., Galli, C., & Parkington, J. (1999). Evidence for the unique function of docosahexaenoic acid during the evolution of the modern hominid brain. *Lipids, 34*(1), S39–S47.
17. Barton, P. G., & Gunstone, F. D. (1975). Hydrocarbon chain packing and molecular motion in phospholipid bilayers formed from unsaturated lecithins. Synthesis and properties of sixteen positional isomers of 1, 2-dioctadecenoyl-sn-glycero-3-phosphorylcholine. *Journal of Biological Chemistry, 250*(12), 4470–4476.
18. Stubbs, C. D., & Smith, A. D. (1984). The modification of mammalian membrane polyunsaturated fatty acid composition in relation to membrane fluidity and function. *Biochimica et Biophysica Acta (BBA)-Reviews on Biomembranes, 779*(1), 89–137.
19. Borow, K. M., Nelson, J. R., & Mason, R. P. (2015). Biologic plausibility, cellular effects, and molecular mechanisms of eicosapentaenoic acid (EPA) in atherosclerosis. *Atherosclerosis, 242*(1), 357–366.
20. Mason, R. P., & Jacob, R. F. (2015). Eicosapentaenoic acid inhibits glucose-induced membrane cholesterol crystalline domain formation through a potent antioxidant mechanism. *Biochimica et Biophysica Acta (BBA)-Biomembranes, 1848*(2), 502–509.
21. Domenichiello, A. F., Kitson, A. P., & Bazinet, R. P. (2015). Is docosahexaenoic acid synthesis from α-linolenic acid sufficient to supply the adult brain? *Progress in lipid research, 59*, 54–66.
22. Simopoulos, A. P. (2011). Evolutionary aspects of diet: The omega-6/omega-3 ratio and the brain. *Molecular neurobiology, 44*(2), 203–215.
23. Cordain, et al. (2002). Fatty acid analysis of wild ruminant tissues: Evolutionary implications for reducing diet-related chronic disease. *European Journal of Clinical Nutrition, 56*(3), 181.
24. Self-Medlin, Y., Byun, J., Jacob, R. F., Mizuno, Y., & Mason, R. P. (2009). Glucose promotes membrane cholesterol crystalline domain formation by lipid peroxidation. *Biochimica et Biophysica Acta (BBA)-Biomembranes, 1788*(6), 1398–1403.
25. https://www.welt.de/vermischtes/article155694524/Voelliger-Bloedsinn-dass-Zuckerkonsum-Diabetes-ausloest.html.
26. Mann, N. (2007). Meat in the human diet: An anthropological perspective. *Nutrition & Dietetics, 64*(s4).
27. Psouni, E., Janke, A., & Garwicz, M. (2012). Impact of carnivory on human development and evolution revealed by a new unifying model of weaning in mammals. *PLoS One, 7*(4), e32452.
28. Smith, E. K., Pestle, W. J., Clarot, A., & Gallardo, F. (2017). Modeling breastfeeding and weaning practices (BWP) on the coast of Northern Chile's Atacama desert during the formative period. *The Journal of Island and Coastal Archaeology, 12*(4), 558–571.
29. Larsen, C. S. (2003). Animal source foods and human health during evolution. *The Journal of nutrition, 133*(11), 3893S-3897S.
30. Reed, C. A. (1984). The beginnings of animal domestication. In Mason, I. L., & Mason, I. L. (Eds.), *Evolution of domesticated animals*. Longman.
31. Chesney, R. W., Helms, R. A., Christensen, M., Budreau, A. M., Han, X., & Sturman, J. A. (1998). The role of taurine in infant nutrition. *Taurine, 3*, 463–476. Springer.
32. Yamori, Y., Taguchi, T., Hamada, A., Kunimasa, K., Mori, H., & Mori, M. (2010). Taurine in health and diseases: Consistent evidence from experimental and epidemiological studies. *Journal of biomedical science, 17*(1), S6.
33. Henneberg, M., Sarafis, V., & Mathers, K. (1998). Human adaptations to meat eating. *Human Evolution, 13*(3–4), 229–234.
34. Battaglia Richi, E., Baumer, B., Conrad, B., Darioli, R., Schmid, A., & Keller, U. (2015, June). Gesundheitliche Aspekte des Fleischkonsums. *Swiss Medical Forum, 15*(24), 566–572. EMH Media.
35. Rafferty, J. E. (1985). The archaeological record on sedentariness: Recognition, development, and implications. *Advances in Archaeological Method and Theory, 8*, 113–156.

References

36. Fitzhugh, B., & Habu, J. (2002). *Beyond foraging and collecting. Evolutionary change in hunter-gatherer settlement systems.* Springer.
37. Binford, L. R. (1980). Willow smoke and dogs' tails: Hunter-gatherer settlement systems and archaeological site formation. *American antiquity, 45*(1), 4–20.
38. Nickel, R. K. (1975). Paleoethnobotany of the Koster site. *The Archaic Horizons, 20*(67), 75–77.
39. Nieradzik, L., & Schmidt-Lauber, B. (2016). *Tiere nutzen. Ökonomien tierischer Produktion in der Moderne.* Studien Verlag.
40. Turnbull, P. F., & Reed, C. A. (1974). The fauna from the terminal Pleistocene of Palegawra Cave, a Zarzian occupation site in northeastern Iraq. *Fieldiana Anthropology, 63*(3), 81–146.
41. Benecke, N. (1994). *Der Mensch und seine Haustiere. Die Geschichte einer jahrtausendalten Beziehung.* Theiss.
42. Burger, J. (2007). Die Milch macht's! Die ersten Bauern Europas und ihre Rinder. *Journal Culinaire, 4*, 32–35.
43. Curry, A. (2013). The milk revolution. *Nature, 500*(7460), 20.
44. Fallingborg, J. (1999). Intraluminal pH of the human gastrointestinal tract. *Danish medical bulletin, 46*(3), 183–196.
45. Gerbault, P., Liebert, A., Itan, Y., Powell, A., Currat, M., Burger, J., Swallow, D. M., & Thomas, M. G. (2011). Evolution of lactase persistence: An example of human niche construction. *Philosophical Transactions of the Royal Society B: Biological Sciences, 366*(1566), 863–877.
46. Biesalski, H. K. (2010). Ernährung und Evolution. *Biesalski, H. K., Bischoff, S. C., Puchstein, C.: Ernährungsmedizin. Nach dem neuen Curriculum Ernährungsmedizin der Bundesärztekammer, 4*, 4–19.
47. Burger, J., & Thomas, M. G. (2011). The palaeopopulationgenetics of humans, cattle and dairying in Neolithic Europe. *Human bioarchaeology of the transition to agriculture*, 369–384. Wiley.
48. Farnworth, E. R. T. (Ed.). (2008). *Handbook of fermented functional foods.* CRC Press.
49. Tong, P. (2013). Culturally speaking: Lactose intolerance: A condition as old as the Stone Age. *Dairy Foods Magazine, 26*. https://works.bepress.com/phillip_tong/46/.
50. Hertzler, S. R., Huynh, B. C. L., & Savaiano, D. A. (1996). How much lactose is low lactose? *Journal of the Academy of Nutrition and Dietetics, 96*(3), 243–246.
51. Caviezel, R., & Vilgis, T. A. (2017). *Koch- und Gartentechniken. Wissenschaftliche Texte und Erläuterungen.* Matthaes.
52. Cavalieri, D., McGovern, P. E., Hartl, D. L., Mortimer, R., & Polsinelli, M. (2003). Evidence for *S. cerevisiae* fermentation in ancient wine. *Journal of molecular evolution, 57*(1), S226–S232.
53. Dineley, M., Dineley, G., & Fairbairn, A. S. (2000). *Plants in Neolithic Britain and Beyond.* Oxbow Books.
54. Van Oevelen, D., Spaepen, M., Timmermans, P., & Verachtert, H. (1977). Microbiological aspects of spontaneous wort fermentation in the production of lambic and gueuze. *Journal of the Institute of Brewing, 83*(6), 356–360.
55. Howard, C. (2017). Skin contact whites: Perhaps amber is the new'orange'. *Wine & Viticulture Journal, 32*(3), 21.
56. Tamang, J. P. (Ed.). (2016). *Ethnic fermented foods and alcoholic beverages of Asia.* Springer.
57. Jang, D. J., Chung, K. R., Yang, H. J., Kim, K. S., & Kwon, D. Y. (2015). Discussion on the origin of kimchi, representative of Korean unique fermented vegetables. *Journal of Ethnic Foods, 2*(3), 126–136.
58. Swain, M. R., Anandharaj, M., Ray, R. C., & Parveen Rani, R. (2014). Fermented fruits and vegetables of Asia: A potential source of probiotics. *Biotechnology research international, 2014*. https://doi.org/10.1155/2014/250424

59. Tomita, S., Nakamura, T., & Okada, S. (2018). NMR-and GC/MS-based metabolomic characterization of *sunki*, an unsalted fermented pickle of turnip leaves. *Food Chemistry, 258*, 25–34.
60. Pedebson, C. S. (1961). Sauerkraut. *Advances in food research, 10*, 233–291. Academic Press.
61. Couteau, D., McCartney, A. L., Gibson, G. R., Williamson, G., & Faulds, C. B. (2001). Isolation and characterization of human colonic bacteria able to hydrolyse chlorogenic acid. *Journal of applied microbiology, 90*(6), 873–881.
62. Shepherd, S. J., Halmos, E., & Glance, S. (2014). The role of FODMAPs in irritable bowel syndrome. *Current Opinion in Clinical Nutrition & Metabolic Care, 17*(6), 605–609.
63. Hotz, C., & Gibson, R. S. (2007). Traditional food-processing and preparation practices to enhance the bioavailability of micronutrients in plant-based diets. *The Journal of nutrition, 137*(4), 1097–1100.
64. Haaland, R. (2007). Porridge and pot, bread and oven: Food ways and symbolism in Africa and the near East from the neolithic to the present. *Cambridge Archaeological Journal, 17*(2), 165–182.
65. Landi, M., Araújo, A. F. F., Bernardes, J. P., Morais, R., Froufe, H., Egas, C., Oliveira, C., & Lobo, J. (2015). Ancient DNA in archaeological garum remains from the south of Portugal. In Oliveira, C., Morais, E., & Cerdán, A. M., *Chromatography and DNA analysis in archeology*. Município de Esposende.
66. Vieira, M. M. C. (2008). Garum: Recovering of the production process of an ancestral condiment. *Congresso do Atum, Vila Real de St. António*.
67. Vilgis, T. (2013). Fermentation – molekulares Niedrigtemperaturgaren. *Journal Culinaire, 17*, 38–53.
68. Chancharoonpong, C., Hsieh, P. C., & Sheu, S. C. (2012). Enzyme production and growth of Aspergillus oryzae S. on soybean koji fermentation. *APCBEE Procedia, 2*, 57–61.
69. Giri, A., Osako, K., Okamoto, A., & Ohshima, T. (2010). Olfactometric characterization of aroma active compounds in fermented fish paste in comparison with fish sauce, fermented soy paste and sauce products. *Food Research International, 43*(4), 1027–1040.
70. Giri, A., & Ohshima, T. (2012). Dynamics of aroma-active volatiles in miso prepared from lizardfish meat and soy during fermentation: A comparative analysis. *International Journal of Nutrition and Food Sciences, 1*(1), 1–12.
71. Kuroda, M., Kato, Y., Yamazaki, J., Kai, Y., Mizukoshi, T., Miyano, H., & Eto, Y. (2012). Determination and quantification of γ-glutamyl-valyl-glycine in commercial fish sauces. *Journal of agricultural and food chemistry, 60*(29), 7291–7296.
72. Van Ho, T., & Suzuki, H. (2013). Increase of "Umami" and "Kokumi" compounds in miso, fermented soybeans, by the addition of bacterial γ-glutamyltranspeptidase. *International Journal of Food Studies, 2*(1).
73. San Gabriel, A., Uneyama, H., Maekawa, T., & Torii, K. (2009). The calcium-sensing receptor in taste tissue. *Biochemical and biophysical research communications, 378*(3), 414–418.
74. Mäkinen, O. E., & Arendt, E. K. (2015). Nonbrewing applications of malted cereals, pseudocereals, and legumes: A review. *Journal of the American Society of Brewing Chemists, 73*, 223–227.
75. Adebiyi, J. A., Obadina, A. O., Adebo, O. A., & Kayitesi, E. (2018). Fermented and malted millet products in Africa: Expedition from traditional/ethnic foods to industrial value-added products. *Critical reviews in food science and nutrition, 58*(3), 463–474.
76. Lu, H., Yang, X., Ye, M., Liu, K. B., Xia, Z., Ren, X., Cai, L., Wu, N., & Liu, T. S. (2005). Culinary archaeology: Millet noodles in late Neolithic China. *Nature, 437*(7061), 967.
77. Kwon, D. Y., Jang, D. J., Yang, H. J., & Chung, K. R. (2014). History of Korean gochu, gochujang, and kimchi. *Journal of Ethnic Foods, 1*(1), 3–7.
78. Rolls, E. T. (2000). The representation of umami taste in the taste cortex. *The Journal of nutrition, 130*(4), 960S-965S.

References

79. Hellekant, G., & Ninomiya, Y. (1991). On the taste of umami in chimpanzee. *Physiology & behavior, 49*(5), 927–934.
80. https://www.tagesspiegel.de/weltspiegel/gesundheit/vergolden-nicht-verkohlen/370008.html.
81. http://www.taz.de/!5163400/.
82. Lévi-Strauss, C. (1972). *Le cru et cuit*. Plon.
83. http://www.deutschlandfunkkultur.de/bestes-restaurant-der-welt-essen-oder-nicht.954.de.html?dram:Article_id=284017.
84. Vilgis, T. (2013). Komplexität auf dem Teller – ein naturwissenschaftlicher Blick auf das kulinarischen Dreieck von Lévi-Strauss. *Journal Culinaire, 16,* 109–122.
85. Vilgis, T. (2017) Evolution—culinary culture—cooking technology. In van der Meulen, N. Wiesel, J., & Reinmann, R. (Eds.), *Culinary Turn, Aesthetic Practice of Cookery.* Bielefeld: Transcript.
86. Tzschirner, H., & Vilgis, T. (2014). *Sous-vide: Der Einstieg in die sanfte Gartechnik.* Köln: Fackelträger.
87. Zielbauer, B. I., Franz, J., Viezens, B., & Vilgis, T. A. (2016). Physical aspects of meat cooking: Time dependent thermal protein denaturation and water loss. *Food biophysics, 11*(1), 34–42.
88. Wiebe, M. G. (2004). Quorn™ Myco-protein – Overview of a successful fungal product. *Mycologist, 18*(1), 17–20.

Consequences of Early Industrialization on the Molecular Composition of Food

Abstract

The term industrialization today strongly suggests altered food. The industrialization of food and agriculture inevitably began with sedentism, and the resulting production and trade forms had a major impact on human nutrition. The effects of industrialization on the molecular composition of food are of crucial importance. Using simple examples—animal fats, milk, and bread—these are introduced and the causes explained.

3.1 Neolithic—The Modernization of Food

The origins of the industrialization of food and agriculture can be traced back to the Neolithic period. The sedentary lifestyle and the resulting forms of production and trade had a significant influence on human nutrition. Although this has little to do with the modern concept of industrialization, food has been systematically manipulated and genetically modified through breeding since early times. Natural products and food were modernized and adapted to humans and the environment, as already described in the previous chapter.

Until the industrialization, as known today, the path was still long, but at this point it is useful to look at the change in nutrients on a rough scale, because with the sedentary lifestyle and the industrialization that took place thousands of years later, human nutrition fundamentally changed [1]. This shows a fundamental change (Fig. 3.1) exemplified by the intake of the two vitamins C and E (curves A and C in Fig. 3.1) from the available food. While in the epochs of the hunter-gatherers and in the Neolithic the content of these vitamins in the diet was constantly high, and even increased slightly with agriculture, it seems to decrease with the beginning of industrialization, and most strongly in recent history.

This development is even more evident in the case of dietary fat. The total amount of fat consumption (Curve B) increased only slightly until 1800 years ago, which is due to the increasing prosperity. In the course of the industrialization of food, the availability of animal products also increased. The fat content in the energy input of human nutrition thus steadily increased, and in parallel, the input of saturated fats (Curve D), which is not surprising, as stearic acid (C 18:0) is the most common fatty acid in many foods, whether animal or plant-based. If the intake of total fat increases, the proportion of saturated fats increases accordingly. Fig. 3.1 also shows two essential aspects: since the industrialization of food in the 19th century, the ratio of Omega-3- to Omega-6-fatty acids has changed compared to the Paleolithic period. While the ratio was about 1:1 for a long time in human history, the proportion of Omega-6 fatty acids (Curve E) exceeded that of Omega-3 fatty acids (Curve F) towards the end of the 19th century. As already noted in Chap. 2, the proportion of hunted animals, which have a significantly higher Omega-3 fatty acid content, decreased in animal foods. At the same time, the proportion of Omega-3 fatty acids in farmed animals decreased due to fattening. Feeding with cereal residues and other cultivated feedstuffs has long changed the fat spectrum of farmed animals—and is not fundamentally a problem of today [2, 3].

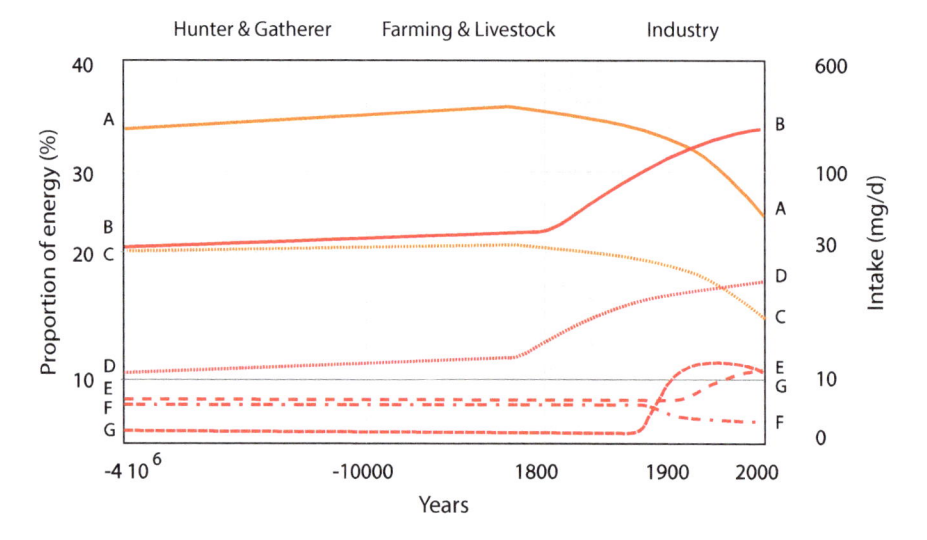

Fig. 3.1 The change in food and the intake of micronutrients Vitamin C (Curve A) and E (Curve C) as well as fats—B: Total fat, D: saturated fats, E: Omega-6 fats, F: Omega-3 fats, G: *trans* fats—over the course of the most important epochs of humanity. The light vertical shading defines the beginning of the Neolithic period (10,000 to 9,000 years ago) and the beginning of the industrial age (around 1900). Since the vertical scales are not linear, this representation is only schematic

3.1.1 Natural Fats, Industrial Fats, *trans* Fats

What is striking in Fig. 3.1 is also the increasingly growing proportion of *trans*-fatty acids over time. Even in the diet of hunters and gatherers or early farmers, *trans*-fatty acids were inevitably present. All ruminants produce them in their metabolism, so they are found in their meat, but especially in the milk and dairy products. They are also present in fish. However, they can also be formed from unsaturated fatty acids by heating. This chemical process is hardly avoidable at certain temperatures. The intake of certain *trans*-fatty acids has reached alarming levels only due to the fat-hardening processes developed some time ago. Originally liquid oils were hardened, making them solid—like margarine—and only then industrially processable in plants. These processes were developed in 1901 and revolutionized industrial fat processing in a certain way just a few years later. The reason for this is purely physical: In *trans*-fatty acids, the double bond is not in the bent *cis* form, but in a *trans* form. The fats therefore fit more smoothly into a crystal form, and the melting point becomes higher (see, for example, [4]). The melting points of the hardened fats *("shortenings")* can thus be precisely regulated, resulting in a whole range of process technological advantages in the production of cookies *(shortcakes)* or other convenience food.

Much is reported in the nutrition and medical literature about the negative effects of *trans*-fatty acids, but the reasons and explanations are less often found. Especially since no nutrition-related negative influences of *trans*-fatty acids from milk and dairy products of ruminants such as cows, sheep, and goats on physiology are known. However, using fat physics and fat biology, an insight into the world of differences between *trans*-fats from fat hardening and the *trans*-fats of ruminants can be recognized. As in most cases, small details in the molecular structure are responsible for a whole range of physiological phenomena and thus make them more easily understandable. For these considerations, the most important representatives of fatty acids that show fundamental differences are shown in Fig. 3.2.

Fatty acids with 18 carbon atoms are among the most common representatives for cell biological reasons and are found in (almost) all foods in larger quantities, even in coconut oil they are represented with over 6% on average as oleic acid. Therefore, for the following, fatty acids C 18 will be considered as an example to better understand the basic physical properties. The unsaturated version C 18:0, in Fig. 3.2 at the top, is the basic version. The monounsaturated oleic acid (C 18:1, 9) is, for example, dominant in olive oil, with the double bond occurring at the 9th position, symmetrically, regardless of whether it is counted from the methyl end or from the ester group. The corresponding *trans*-fatty acid, elaidic acid (C 18:1, 9t) shows the fundamental difference in shape: The *trans*-double bond is at the 9th position, exactly in the middle of the carbon chain. In the case of vaccenic acid (C 18:1, 11t), the *trans*-double bond is at the 11th position (counted from the ester group), closer to the methyl end of the fatty acid. This point proves to be a small but essential difference.

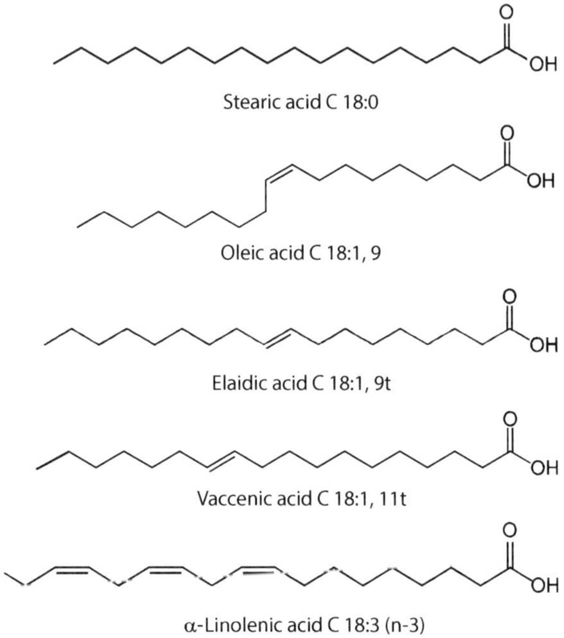

Fig. 3.2 The most important versions of the C18 fatty acid. *trans*-fatty acids are the elaidic acid and the vaccenic acid. The associated omega-3 fatty acid is the α-linolenic acid (ALA)

Stearic acid C 18:0

Oleic acid C 18:1, 9

Elaidic acid C 18:1, 9t

Vaccenic acid C 18:1, 11t

α-Linolenic acid C 18:3 (n-3)

All carbon-carbon double bonds, in contrast to simple carbon-carbon bonds, are not freely rotatable. This is a fundamental difference, because single bonds, like two spheres connected by a simple rod, can be easily rotated around their own axis. When they are chemically doubly connected, the second rod acts as a stiffening element and prevents rotation. In the *cis* form, kinks are therefore built into the chain, so that these fatty acids cannot be arranged as closely as the saturated fatty acids C 18:0 and are thus forced into different crystal forms (polymorphs). Fats that consist mainly of more unsaturated fatty acids therefore remain more liquid even at slightly lower temperatures.

Since industrial catalytic fat hardening for fat production uses vegetable oils, elaidic acid is one of the most common *trans*-fatty acids and is classified as harmful. Why, then, is vaccenic acid not problematic, but even positively rated [5, 6]? From a physiological point of view, vaccenic acid also serves as a natural precursor for conjugated linoleic acid. But this alone does not yet explain the big difference to elaidic acid. Here, too, the molecular structure plays the main role. The structural difference between the *trans*-vaccenic acid from the milk fat of ruminants and the elaidic acid from fat hardening is considerable, even if it does not appear so at first glance. Whether the double bond is at the 11th position or at the 9th position in the chain makes a big difference in terms of the mobility of the molecule. If the non-rotatable double bond is in the middle, the molecule becomes very stiff. At the 11th position, closer to the free end, the molecule remains more mobile in the direction of the ester group. This makes vaccenic acid structurally closer to omega-3 fatty acids. In these, the double bond is at the 15th position (the 3rd position in the omega counting system). The *trans*-vaccenic acid has a similarly high melting point (44°C) as elaidic acid, but the behavior between the phospholipids in the cell membrane and fat particles is more decisive for physiology. This will become clearer when considering the incorporation of *trans*-fatty acids into cell membranes, when their fluidity and flexibility are defined by the type of fatty acids in the phospholipids. Similar considerations have already been discussed in Chap. 2 in Figs. 2.9 and 2.10 for saturated fatty acids.

When phospholipids with *trans* fatty acids instead of *cis* fatty acids are incorporated into the membrane, the stiffness of the membrane increases, and it loses flexibility and pliability. As the number of *trans* fatty acids increases, the flexibility and thus the function of the cell membrane are severely restricted, which can lead to health problems in the long term. This is discussed in more detail in Sect. 3.1.4.1 (Fig. 3.12). Before that, a more detailed excursion into the physics of fluid membranes is necessary, as already mentioned in Chap. 2. Membranes surround all animal and plant cells and ensure their biological function. For this to work smoothly, cell membranes must comply with certain biophysical laws. In addition to the properties of fatty acids, cholesterol also plays a decisive role.

3.1.2 Cholesterol and Cell Membranes

Most people will have heard of the "good" and "bad" cholesterol (or cholesterin), the good HDL (High Density Lipoprotein) and bad LDL (Low Density Lipoprotein). But what exactly is behind this? The basic understanding of many properties of HDL and LDL begins with the contrast between water and fat. Fats are not soluble in water, but our body fluids are more or less colored water or physiological saline solution: blood, lymph fluid, "stomach and intestinal juices". Minerals, salts, and many proteins can be dissolved and transported in them. However, water does not dissolve triacylglycerols and only very limited free fatty acids, not in the physiologically required amounts. So there is only one possibility for effective fat transport: fats and fatty acids—and also cholesterol—must first be packaged in water-soluble shells so that they can travel through the veins with the blood and be unpacked again at the appropriate place, in order to deliver them to their intended destination. For these transport containers, all animals and humans have developed two different nanoparticles, LDL and HDL, which differ fundamentally in size and density and thus their physical properties.

So far, in the designations LDL and HDL there is no mention of cholesterol, but only of fat or lipids (lipo-) and protein. However, the cholesterol molecule fulfills a fundamental function in the cell membranes (Chap. 2), because it has a weakly water-soluble head and a short fat-soluble chain, as shown in Fig. 3.3. The chemical structure of the triterpene cholesterol shows (Fig. 3.3) a large part of the physicochemical properties: cholesterol is surface-active due to the weakly polar OH group and the water-insoluble residuesurface-active. The steroid structure, chemically composed of connected ring structures, is relatively rigid, while the very short fatty acid is more flexible. In addition, the molecule is significantly

Fig. 3.3 The cholesterol molecule has a polar, water-soluble hydroxyl (OH-) group, a very rigid, water-insoluble steroid residue, and a water-insoluble aliphatic (fatty) chain

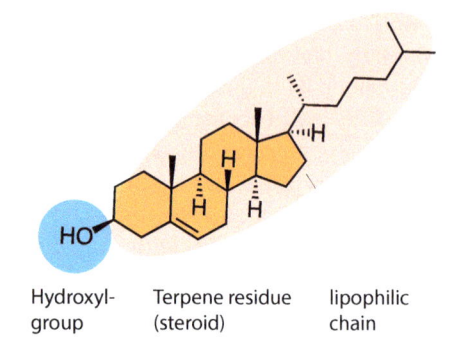

Hydroxyl-group   Terpene residue (steroid)   lipophilic chain

shorter overall than the phospholipids in the membrane. It is therefore a very special "space filler" with a stiffening function, but only for one "half" of the lipid bilayer. Cholesterol stiffens locally, i.e., only where a cholesterol molecule is currently located—without negatively affecting the global bending ability of the membrane. At other cholesterol-free sites in the membrane, the bending ability of the bilayer does not change. This suggests that cholesterol acts as a local switching element for the flexibility and fluidity of the membrane, whenever it is physiologically necessary, for example in the immediate vicinity of membrane proteins or other functional units anchored in the membrane. Cholesterol is thus to a considerable extent necessary for controlling biological and biophysical functions of the membranes.

Due to its surface-active properties, the cholesterol molecule is easily incorporated into the cell membrane. Apart from the structure of cholesterol, the molecule size is also of crucial importance. The length of the hydrophobic groups is such that the cholesterol molecule can fit into the gaps created by the higher space requirement of unsaturated fatty acids, as schematically shown in Fig. 3.4 [7, 8].

The incorporation of cholesterol results in a whole range of positive effects for membrane stability [9]. On the one hand, breaking points are repaired due to the dipolar interaction of the head groups, and on the other hand, the fluidity of the membrane is locally changed. Despite still high fluidity due to numerous unsaturated fatty acids, the cohesion of the membrane is maintained. Furthermore, cholesterol prevents phase separation of different phospholipids from fatty acids of different degrees of saturation. It is easy to imagine that phospholipids, which consist mainly of short-chain, saturated fatty acids, would prefer to cluster together and separate from phospholipids made of unsaturated, long-chain fatty acids, according to the principle "like attracts like." This would create two phases in the membrane layers, which would have different bending properties and fluidities. Such heterogeneous structures would not be particularly conducive to the function of the membranes. Cholesterol thus acts as a phase mediator between the differently structured phospholipids in these membrane bilayers. Even more: cholesterol molecules group phospholipids around themselves, forming clusters and thus new

3.1 Neolithic—The Modernization of Food

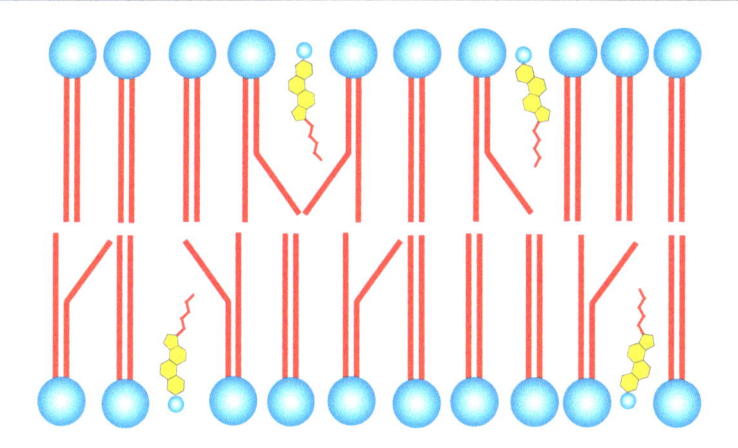

Fig. 3.4 Cholesterol molecules insert themselves into the membrane. They fill gaps that are present due to the structure of unsaturated fatty acids, thereby increasing membrane stability. The stiffening and sealing of the membrane are only achieved on the hydrophilic sides of the membrane, while the center of the membrane remains mobile

phases, which are of great importance for membrane function, especially when proteins are incorporated into specific areas of the membrane, where designated properties corresponding to the functional protein are required [10].

In fact, the influences of cholesterol on the fluidity of the membrane are temperature-dependent. The temperature dependence allows cholesterol to multitask in the cell membrane. At higher temperatures, cholesterol can restrict the free mobility of the fatty acid chains of the phospholipids and thus selectively stiffen the bilayer. Temperature fluctuations can thus be better compensated. At the same time, the cell membrane permeability (permeability) for small molecules, such as water, also changes. At low temperatures, on the other hand, cholesterol has a plasticizing effect in the membrane: through interaction with the fatty acid chains of the phospholipids, cholesterol can prevent membranes from crystallizing and maintain their mobility and flexibility. Cholesterol is therefore by no means harmful per se, but a molecule with a designated biophysical significance—which is why the body's own biochemistry laboratory produces in in high concentrations.

For cell function, cholesterol in its molecular form is biophysically indispensable. It should not be forgotten: The physiological cholesterol requirement of a 70 kg person per day is approximately 0.5–1.5 g. Only about one-third of this comes from food, the rest is provided by biochemical syntheses in the liver. There are indeed cells in humans that synthesize cholesterol on site, e.g., in the intestinal mucosa, in the steroid hormone-forming glands in the skin, to use it for vitamin D synthesis (see Chap. 5). However, the major part is synthesized in the liver and must be transported from there to many cells of the body. As we will see in later chapters, cholesterol is even a kind of universal molecule for various physiological processes (see

Chap. 5). For example, it serves as a precursor for the synthesis of steroid hormones and bile acids, without which no fat digestion would be possible, as shown in more detail in Chap. 4, and it is also essential for the formation of vitamin D.

3.1.3 Phytosterols

These basic physical functions are not specific to animals and humans, but must also be maintained in plant membranes. This is achieved through chemical similar sterols, which differ only marginally in the aliphatic chain. The rest is identical. So what are these phytosterols, which are occasionally attributed positive and health-promoting effects? For this, there is first a search image in Fig. 3.5: Which is the evil animal cholesterol—A, B, C, or D? And which is the good plant phytosterol? The answer is molecule B.

Upon closer inspection, there are fundamental differences in the various sterols, but these can only be found in the very short carbon chain. In cholesterol (B in Fig. 3.5), this consists of five carbon atoms, all of which are freely rotatable. This is similar in molecule A, which is structurally closest to animal cholesterol. Molecules C and D, however, have more complicated tails. In C, for example, the rotatability is restricted by a double bond, and in D there are additional steric

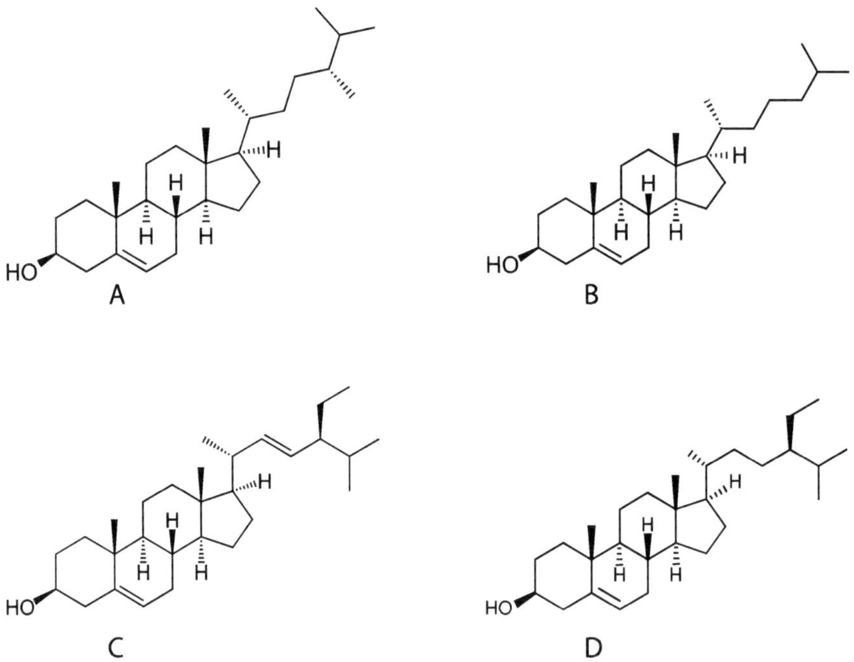

Fig. 3.5 Some plant sterols and one animal sterol. The animal cholesterol is molecule B. A: Campesterol, C: Stigmasterol, D: β-Sitosterol

3.1 Neolithic—The Modernization of Food

hindrances due to branching. Why is this the case in phytosterols but not in cholesterol? Quite apart from the different function of plant cells and animal cells, the temperature in animals is largely constant. Plants, on the other hand, are exposed to strong temperature fluctuations, so several differently structured sterols are needed, which, due to the structure of the hydrophobic tail, can influence membrane flexibility to different degrees in order to maintain its function. Plant sterols, therefore, have no immediate function in animal or human membranes due to these structural elements. Therefore, it is questionable from a biophysical point of view to assume that the regular consumption of phytosterols as a dietary supplement would have many positive effects.

For the specific cell function, it is important that the total cholesterol concentration in the membranes remains constant. Each cell requires a precisely adapted composition of phospholipids and cholesterol for its function. On the other hand, the cell membrane is constantly renewed, and thus the cholesterol is also exchanged. Excess and therefore currently unneeded cholesterol is neutralized—or more precisely, apolarized—so that it no longer has a place in the membrane. To do this, it is esterified and a fatty acid is anchored to the OH group. The esterified cholesterol is then neither polar nor surface-active, but highly fat-soluble, as can be seen in Fig. 3.6. Although the molecule is significantly larger after the esterification, it remains completely fat-soluble. Therefore, it can be transported very easily with the fatty acid thermodynamically together with the other fats via the larger LDL particles through the blood. For this purpose, physiology uses exclusively the supposedly suitable LDL for purely physical reasons.

3.1.4 Cholesterol, LDL, and HDL—What Is Good, What Is Bad?

The nanoparticles HDL and LDL are initially nothing more than very specific packaging materials adapted to their respective requirements, in order to effectively transport fats, cholesterol and phospholipids through the blood. To understand this more precisely, the fat must be followed on its way from the liver into

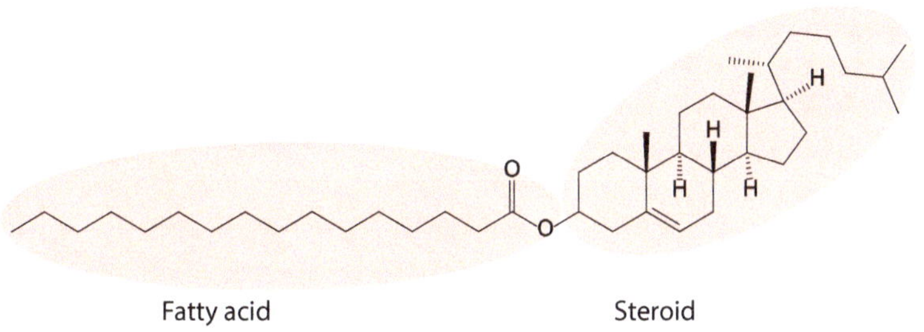

Fig. 3.6 Esterified cholesterol is no longer surface-active, but 100% fat-soluble. The fatty acid does not have to be saturated, as exemplified in this illustration; in general, the composition of the cell membranes reflects the entire fatty acid spectrum in the esterified cholesterol as well

the blood, to the cells and back again. The function of lipoproteins reveals further properties that can only be solved by various proteins, such as size and the necessary packaging and opening mechanisms, which must be naturally different for LDL and HDL. But what exactly do these fat transport droplets look like? To understand the mechanism, one must not forget that fat does not simply dissolve in water (blood) and certainly cannot be transported. Consequently, the fat released from the liver must first be emulsified so that it can be safely packaged in particles at all. All surface-active molecule types that are physiologically available are used as emulsifiers: phospholipids and cholesterol. This is practical because these two molecule groups must be transported to the peripheral cells anyway. Triacylglycerols and esterified cholesterol can be packaged in them. The physically compelling basic structure of the particles is shown in Fig. 3.7.

From the liver, initially up to 1000 μm large, unstable fat droplets are secreted, in which the esterified cholesterol is dissolved. These VLDL particles (Very Low Density Lipoproteins) are initially much too large and too unstable to be transported without disturbance. In these VLDL particles, the components to be packaged are highly unstructured, so they require a lot of space on a large volume, thus the density is very low. Fats, as well as free and esterified cholesterol, are therefore constantly repackaged, sorted, and organized until the size of the packaging particles reaches its absolute minimum and the highest possible density under the given physical conditions. The phospholipids, which must be transported to the cell membranes anyway, serve as packaging material. In addition, non-esterified, free cholesterol is incorporated into the surfaces of the droplets, analogous to the cell membrane. Phospholipids and non-esterified cholesterol form a "monolayer" (Monolayer; a "half double layer", i.e., a single-layer membrane, compared to the membrane) around the fat droplets, as shown schematically in Fig. 3.7.

Fig. 3.7 The basic structure of all animal fat particles without proteins. Neutral fats (triglycerides or triacylglycerols) and esterified cholesterol are located in the fat core. Free cholesterol and phospholipids are located as emulsifiers on the surface of the droplets

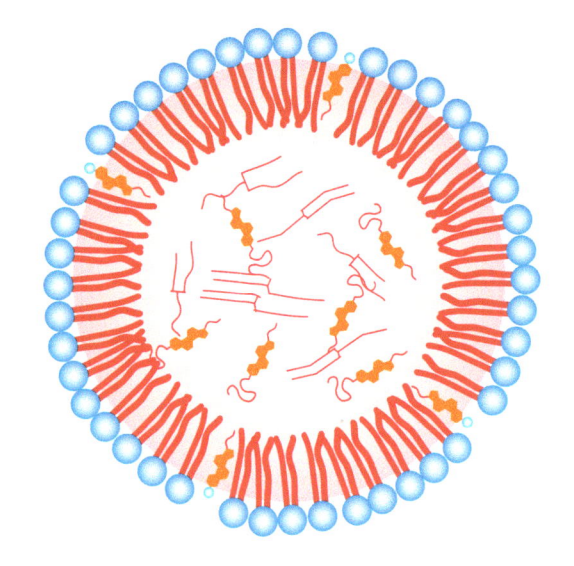

3.1 Neolithic—The Modernization of Food

At the core of the droplet are all neutral fat molecules, namely triacylglycerols and esterified cholesterols, while phospholipids and free cholesterol form the shell. The polar heads of the phospholipids and free cholesterol thus ensure the high water solubility of the droplets. Fat and cholesterol are therefore "emulsified" in the blood. However, it is known that such emulsions stabilized only by phospholipids (and cholesterol) are not particularly stable. The phospholipids are much too weak as emulsifiers. Furthermore, it is known from physics that larger particles are far more unstable than smaller particles due to the volume-surface ratio. Thus, the VLDL particles must be reduced in size and additionally stabilized. To do this, proteins are used in addition, which not only have biological tasks but also contribute to the stability of the fat particles. If the shape and size of the proteins match the dimensions of the particles, they are literally wrapped by the corresponding proteins, which significantly increases stability. To do this, it is useful to visualize the size ratios of the lipoproteins. In Fig. 3.8, these are illustrated.

The size and density ratios initially reflect the physics of these lipoproteins. The VLDL particles have a specific weight of 0.8 g/ml due to the loose packing, which is slightly below the specific weight of fat of 0.9 g/ml. The fat core is thus very large and there is hardly any protein present, whose density is much higher than that of fat. VLDL particles are therefore stabilized only by phospholipids and free cholesterol . HDL particles, on the other hand, have a specific weight of 1.063–1.21 g/ml, which is above that of water (1.0 g/ml). Consequently, the protein

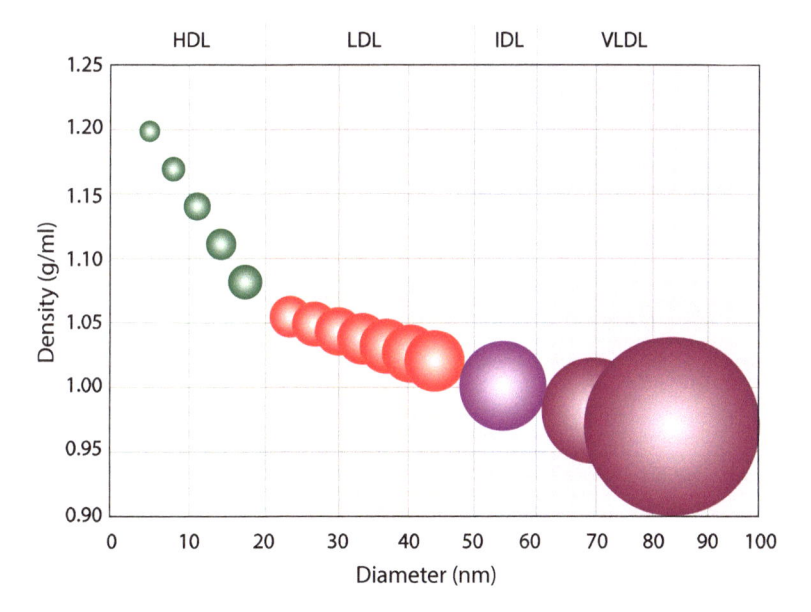

Fig. 3.8 Size ratios of lipoproteins in the body. The smallest particles are HDL (5–15 nm), they have the highest density. Next are the LDL particles (20–50 nm) with significantly lower specific weights. VLDL have the lowest density and the largest diameters. IDL stands for intermediate lipoproteins, which are formed during the first steps of the packaging process

content is significantly higher compared to VLDL. This is also illustrated by a more detailed analysis of the substances involved in the particles: it shows how strongly the compositions of the different particles differ. The various particles HDL and LDL are therefore not *per se* "good" or "bad". Their size corresponds exactly to the purpose of their tasks, and that is initially not pathological or physiological, but biophysically determined. This is precisely what can be seen in the composition of the particles, which is shown in Table 3.1.

Upon closer inspection, the interaction of the various components is already apparent. The large, VLDL released from the liver particles carry most of the fat and cholesterol esters. As the particle size decreases, the triacylglycerol content decreases, while the protein content increases. In fact, the VLDL particles are much too large to be stably transported through the bloodstream. For physical reasons, the stability of the particles decreases with increasing size. Smaller particles are more stable, similar to very large soap bubbles, whose shape fluctuates greatly, while small bubbles remain stable in the spherical shape. Therefore, these large VLDL particles are repackaged into more stable systems, according to Fig. 3.8 via intermediate stages such as IDL (intermediate lipoproteins).

It is striking that cholesterol esters are most common in LDL particles. However, this corresponds exactly to the packaging rules of fats. On the one hand, both free and esterified cholesterol molecules must be supplied to cell physiology from the liver. Some of the free cholesterol molecules find sufficient space in the phospholipid layer that surrounds all particles (Fig. 3.7). The large cholesterol esters, however, must be transported to the cells in the fat core, where the ester is cleaved if necessary. The LDL particle size is thus a physically optimized compromise between capacity, stability, surface, surface tension, and transport properties.

In order for the transfer and release of the components packaged in the LDL particles, such as phospholipids, free and esterified cholesterol to work, they must dock at the designated receptor proteins on the cells, essentially "anchoring" themselves so that the contents of the LDL particles can be transferred to the cells. At

Table 3.1 The average composition of the various cholesterol particles defines their size and function. For the sake of completeness, chylomicrons (CM) are listed. The particles with specific weights around 0.8 g/ml transport the dietary fats absorbed in the intestine, bypassing the liver via the lymphatic system into the bloodstream

Particle	CM	VLDL	LDL	HDL
Size (nm)	200–600	100–60	25	7–15
Total fat content (%)	99	91	80	44
Triacylglycerols (%)	85	55	**10**	6
Cholesteryl esters (%)	3	18	**35**	15
Cholesterol (%)	2	7	**11**	7
Phospholipids (%)	8	16	**23**	30
Protein (%)	2	4	**21**	50

3.1 Neolithic—The Modernization of Food

the same time, the particles must be equipped with very specific opening mechanisms so that their contents become accessible to the cells. This task is most conveniently accomplished with proteins that are predestined for this purpose by their surface amino acid pattern (of course, local biophysical processes also play a role, but we will leave these out for the sake of simplicity). Therefore, the anchor proteins (receptor proteins) must be of a very specific structure and shape so that they can recognize the packaging proteins around the LDL particles. Only then are the anchoring sites released for docking, thus maintaining the basic biochemistry of cholesterol metabolism. This class of proteins is known as apolipoprotein B, one of the most important being apolipoprotein B100, whose structure and properties have also been well studied with models and simulations [11].

Even after this consideration, LDL is still not "evil," but simply essential for life. Size and density are determined by physics and adapted to function and physiology [12]. The insight here is that individual physiology cannot easily outsmart the physicochemical laws that are determined by structure and interaction. Individual differences in cholesterol metabolism or even diseases must therefore be sought at a completely different level.

The reverse transport of non-esterified cholesterol from the cells is carried out via HDL-particles (the "good" cholesterol) with a significantly smaller diameter. This is also physically sensible, as the non-esterified cholesterols are smaller due to the missing fatty acid, require less space, and can be packed more densely. It is no wonder, then, that the least cholesterol esters can be found there, but more phospholipids from the cells and far less free cholesterol, which easily finds space on the surfaces. However, these return flows must be characterized by very high stability, which is why they are significantly smaller, but still large enough that they cannot pass through any unforeseen locations. This also requires special stability enhancers. Other apolipoproteins are responsible for this, mainly those of type AI, which differs significantly in its properties from type B100. Also, the HDL particles no longer need to release their contents at functional locations but are processed in the liver in a different way.

Therefore, the apolipoproteins of type AI are designed to provide maximum stability and do not carry docking and release sites, like the apolipoproteins of LDL-particles. Apolipoproteins AI encircle the HDL-particles like precisely fitting belts, as shown in Fig. 3.9, which only works up to a certain size of lipoproteins [13], and are designed with their amino acid sequences on the surface to interact with as few biological surfaces as possible. The reverse transport should proceed largely undisturbed. Deposition on the surfaces of the veins (epithelium) is therefore unlikely for physical reasons alone. The returning cholesterol is not stored in the liver but is recycled into bile acids. These play a very special role in emulsifying and digesting fat that we ingest through food. More on this later, but it already shows very clearly how universal and effective biochemistry, or physiological chemistry, works: with similar molecular structures that are suitable for many applications.

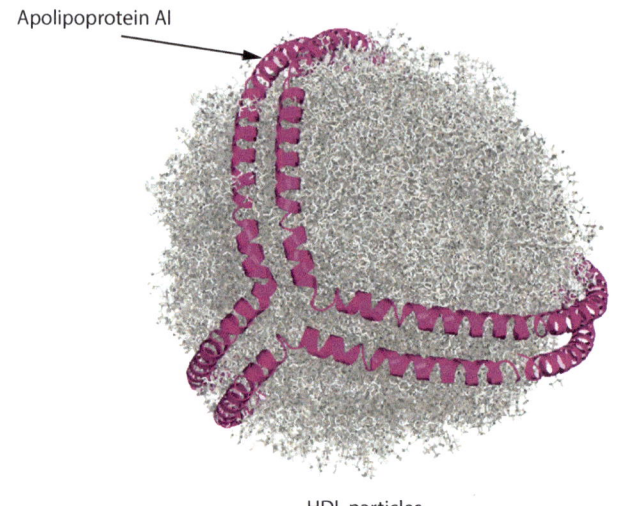

Fig. 3.9 Basic structure of an HDL-particle. The apolipoproteins encircle the phospholipid- and cholesterol-stabilized particle like a tight belt. Together with the phospholipids, high stability is guaranteed

These simple considerations show much that is hardly found in medical guides: Above all, physical and chemical properties determine far more than physiology the basic rules for behavior, size, and structure of LDL- and HDL-particles.

3.1.4.1 When Does LDL Become "Bad"?

Despite the good news from physics and chemistry so far, it is of course not to be denied that LDL particles can indeed have negative consequences and can cause inflammation and deposits—plaques—in the veins. However, a whole series of further processes is necessary for this, which primarily concern the apolipoproteins of type B100 and not the cholesterol molecule itself [14, 15], as roughly summarized in Fig. 3.9. The "evil-becoming" of LDLs takes place in several steps. For this to happen, oxidations, lipid losses, or sialic acid transfers must first occur in native LDL particles. In the process, the LDL particles become slightly more densely packed and shrink a little. Consequently, the apolipoproteins surrounding the particles must follow suit in their folding, initially changing their shape slightly. The LDL particles become slightly negatively charged, with LDL turning into LDL(−), where the bracketed (−) represents the negative charge. With further oxidation, the apolipoproteins fold over, directing the electrically negative charge inward and turning hydrophobic β-sheets outward. β-sheets generally tend to aggregate and are thus perfect sites for the aggregation of such LDL(−) particles. Larger and larger aggregates are then formed, as shown in Fig. 3.10.

The large aggregates can form on mucous membranes and at the boundary layer in the veins due to the electronegative interactions, as mucous membranes

3.1 Neolithic—The Modernization of Food

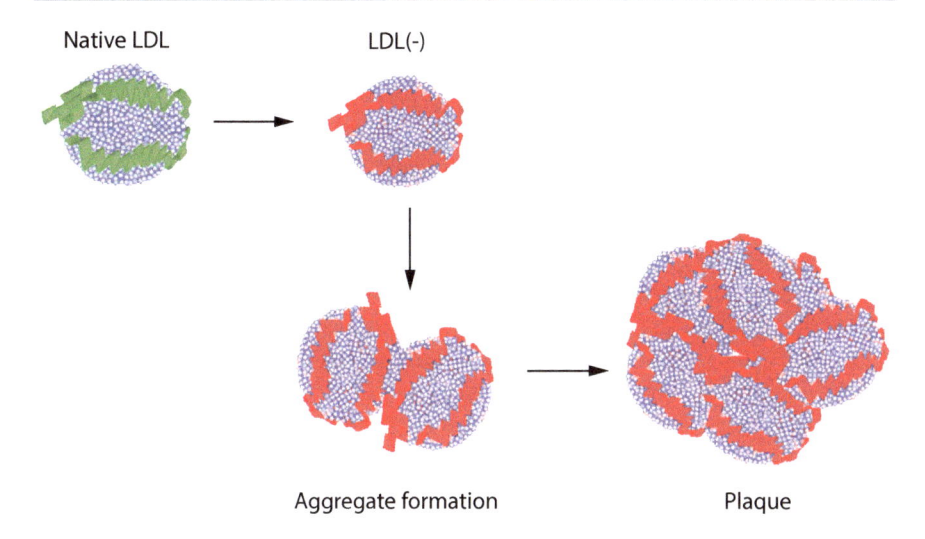

Fig. 3.10 The LDL particles become "evil". First, negatively charged LDL(−) particles are formed, which are slightly smaller and whose apolipoprotein B100 has been refolded to a higher proportion of β-sheets. This is followed by aggregate formation, which can lead to inflammatory plaques

have a high proportion of glycoproteins, to which the aggregates can attach. On the one hand, this is because the lipoproteins on the surface are structured in such a way that they can strongly interact with the surface glycoproteins, and on the other hand, because they have a higher inertia due to their mass and can only diffuse slowly. Due to their larger diameter, they also have more contact possibilities with surfaces, making adsorption easier. With the simultaneous formation of antibody complexes, an inflammation-prone plaque formation is more likely under these conditions.

The details of aggregate formation have been thoroughly investigated [16]. In this context, the mechanism of aggregation can be described precisely. The altered, partially denatured apolipoproteinsB100 of the different LDL particles join together via their β-sheets and form sufficiently strong connections that resemble those of amyloid formation, as observed in Creutzfeld-Jakob disease, suspected as its cause, and discussed in more detail in this chapter. This is not surprising, as many physical processes at the molecular level are universal and usually follow similar physical principles. Biophysical processes are also universal, as will be shown in more detail in Sect. 3.3.4.

At this point, it is worth realizing how high the purely physical hurdles of pathological plaque formation really are. Therefore, not every measurement of a cholesterol level above 200 (an arbitrarily set limit) must be pathological, but can be within the range of genetic variation. In light of these considerations, it becomes clear how ineffective cholesterol-lowering drugs are if they merely reduce the formation of the cholesterol molecule in the liver, rather than influencing the plaque-forming processes dominated by physical-chemical pathways. Moreover, these processes also show how unrealistic it is to control cholesterol levels through dietary restrictions. The influence of food on these processes seems to be marginal indeed. Not from the macroscopic perspective of laboratory measurements, but for molecular-physical reasons, especially considering that HDL and LDL particles, for example from eggs, are completely broken down, digested, and metabolized during the gastrointestinal passage and do not, as often assumed, pass directly from the stomach into the blood and be diagnosed there as "good or bad cholesterol." The proposed effect of so-called statins, which primarily inhibit the formation of the free cholesterol molecule in the liver, may serve as a mere symptomatic treatment, but contributes little to the actual cause-fighting. Packaging mechanisms—the sum of all makes the difference.

In this context, the packaging mechanisms of lipoproteins HDL and LDL are worth a closer look, providing insight into saturated, unsaturated, and *trans* fatty acids. As already noted, specific molecules, such as fats, esterified cholesterol, and phospholipids, must form compact particles that should meet as many physiological requirements as possible. For this purpose, the already mentioned apolipoproteins are used. However, there are more than just the two previously mentioned, stability and functionally important apolipoproteins ApoAI and ApoB100. With apolipoprotein E, additional packaging helpers are available [17]. These proteins are capable of packaging lipoproteins and supporting the transition from VLDL to LDL and HDL, as schematically shown in Fig. 3.11.

The various apolipoproteins help to sort and organize the diverse mixture of molecules according to physical interactions and to shape and size the droplets so

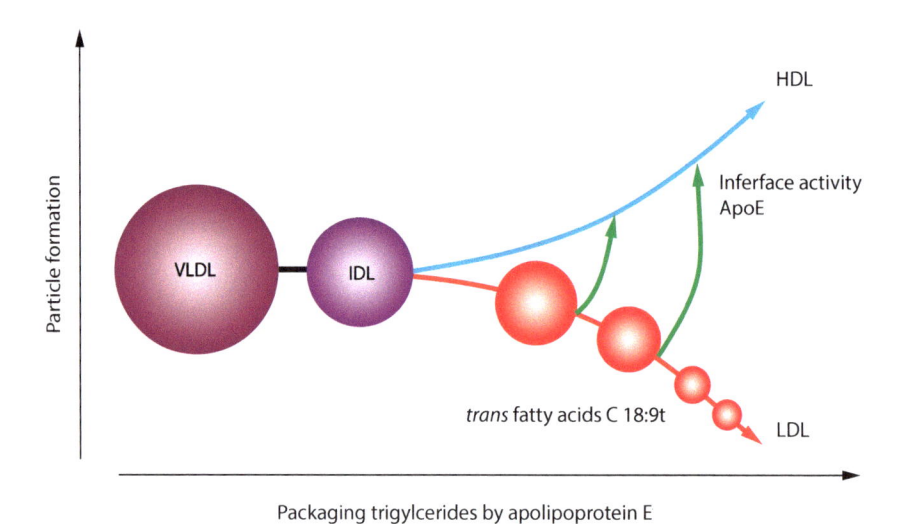

Fig. 3.11 Which fats are packaged in what way and in what sizes is regulated by apolipoproteins (ApoE). In this case, the *trans* fatty acids, if present, also play a significant role

that the LDL particles can be functionally enclosed by apolipoproteins B100 and the HDL-particles by apolipoproteins AI. In particular, the class of apolipoproteins E is significantly involved in fulfilling this required minimal biophysical prerequisite. The molecular supply of fats, fatty acids, cholesterol esters, etc., to some extent determines the ratio of LDL and HDL.

Over these molecular concepts, it also becomes apparent what influence, for example, *trans* fatty acids can have on the size of lipoproteins. This is shown in an exaggerated (and therefore unrealistic) model in Fig. 3.12. It is assumed that the phospholipids consist exclusively of fatty acids of the *trans* fatty acid C 18:9t. The hypothetical lipoproteins made of *trans* fatty acid phospholipids could not be reduced to correspondingly small radii, as required for HDL particles. An excess of *trans* fatty acids therefore disproportionately promotes LDL particles. This has been observed in various studies. *Trans* fatty acids tend to induce LDL particles with a larger radius. Now, however, it is clear why. Due to the strong rigidity in the middle of the fatty acids, they can be packaged into smaller droplets for purely geometric reasons. The causes for this are once again purely physical-chemical in nature.

Phospholipids containing *trans* vaccenic acids show a reduced influence on the curvature radius due to the overall higher flexibility of the free rotation of the carbon-carbon bonds of C 18:11t near the polar head group. Moreover, it is known that the *trans*-vaccenic acid is also converted in human physiology by enzymes to conjugated linoleic acids (CLA), which in turn have positive properties [5]. In fact, the *trans* vaccenic acid is the only known precursor of conjugated linoleic acid. Anxiety about this *trans* fatty acid in milk, dairy products, and butter is therefore not warranted.

3.1.5 Margarine, Fat Mixtures & Co

What can be said about margarine with this knowledge? Basically, these artificial products praised as "healthy fats" are rather superfluous from the perspective of human nutrition. Analog butter is nothing more than hardened vegetable

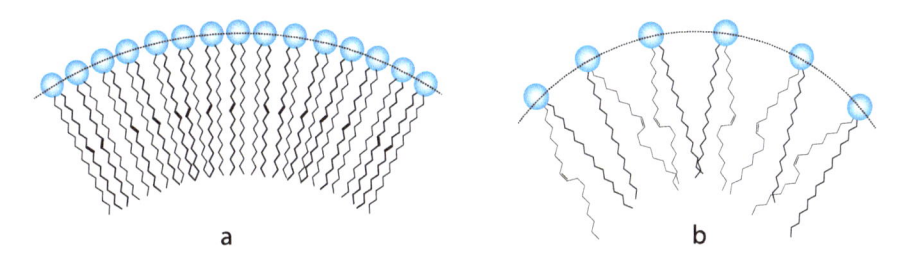

Fig. 3.12 If phospholipids consisted exclusively of the harmful C 18:9t fatty acids (**a**), only lipoproteins with large radii could be produced. Phospholipids containing unsaturated fatty acids inevitably lead to droplets with smaller radii (**b**)

fat. Unsaturated fatty acids become saturated. The manufacturing process is now designed in such a way that complete hardening hardly produces any or no *trans* fatty acids. Nevertheless, "valuable" vegetable margarine remains an illusion and self-deception: A saturated C 18:0 fatty acid is a saturated fatty acid and is therefore to be evaluated in the same way, regardless of whether it comes from a pig, cow, or after hardening from sunflower or rapeseed oil. Even if it was previously in an olive or another "healthy" oilseed, it is irrelevant for physiology where it comes from. Only the molecular property counts. This is described with C 18:0 and is completely independent of origin. The origin is therefore irrelevant as soon as the triacylglycerols are separated from the pancreatic enzymes into the fatty acids.

Our human metabolism is, after all, 100% animal, which should never be forgotten. Of course, the physically-chemically altered fat of plant margarine is "cholesterol-free" (but we now know more precisely what that means). However, this is only an apparent advantage, as the discussion about dietary cholesterol has long been off the table for the general population. The industrial plant fat, often enriched with vitamins, phenols, and other ingredients from dietary supplements, can be saved, including seemingly health-promoting claims, provided a varied diet according to the culinary triangle is followed.

At this point, it is again useful to look at evolution. *Homo sapiens* did not harden fats but used them in the form in which they occur in animals and plants. Depending on the type of fat, he used them for appropriate purposes. Butter can be spread at not too high temperatures, as can animal fats. Vegetable oils are liquid and were also used.

3.2 The Early Economic Models Using the Example of Agriculture

At the beginning of the Neolithic period, the ability of humans to drink was not particularly pronounced, as discussed in detail in Sect. 2.8. Animals were rather kept for milk processing and meat supply. Only after the development of lactose tolerance was milk discovered as a food. Quite apart from the expanded nutrient supply, it was obvious to adapt the keeping of animals to the new requirements. The early dairy industry was born [18]. Archaeological finds indicate the beginning of the systematic use of milk as far back as the 8th millennium BC, as evidenced by typical fatty acid residues from milk fat on clay vessels [19]. Of course, there were regional differences due to the respective climatic conditions. However, it is certain that between the 8th and 5th millennia BC, the use of milk began. The milk of sheep, goats, and cows was divided between humans and the offspring of the animals, which inevitably affected animal husbandry. Animal management changed, and economic questions far removed from the energy input of hunters and gatherers (Chap. 2) were asked for the first time. Inevitably, it had to be weighed whether to keep animals for food or as milk suppliers to best secure the life of the tribe [20]. An optimized mixed economy was thus necessary, as milk

3.2 The Early Economic Models Using the Example of Agriculture

and meat were directly linked. On the other hand, milk proved to be about four times more efficient in terms of energy input and energy consumption when converted and compared to the energy input in hunting, which probably also contributed to the significant increase in life expectancy in the Neolithic period [5]. In the course of optimization, targeted herd management began [21] and sophisticated slaughter rhythms were developed depending on milk optimization, age of the mother stock, and meat production.

It was recognized early on which patterns of animal husbandry were suitable for the respective living conditions. This is most evident in the case of sheep (or goats) [22]. During the Neolithic period, the highest commandment was to produce sustainably. The stock of the herd and its use were a large part of securing the livelihood as well as the capital base of a tribe and thus an economic value. This is schematically shown in Fig. 3.13.

There is a model of animal husbandry depicted that focuses not on overproduction, but on sustainability. The upper bar shows the relevant season for sheep in months. Sheep are born in February. A small portion of the lambs are separated from the mother sheep earlier to use a portion of the milk, represented by the thin black bar, for humans, while the majority of the lambs grow up with the mother sheep until weaning (represented by the blue bar) before being driven to the summer pastures. There, newborn lambs are separated from the mother after a certain time, the sheep are milked, as shown in the black bars. A certain measure of mostly male lambs, depending on the size of the herd, is used for meat production. The slaughtering is shown in the lower part of the diagram (red bars). The slaughtered milk lambs are clearly under one year old. Mother sheep are usually slaughtered between the ages of three and six years. Their number and age are also determined by the size of the herd. This model was (and is) applied by Kurdish mountain farmers in present-day Iran [23].

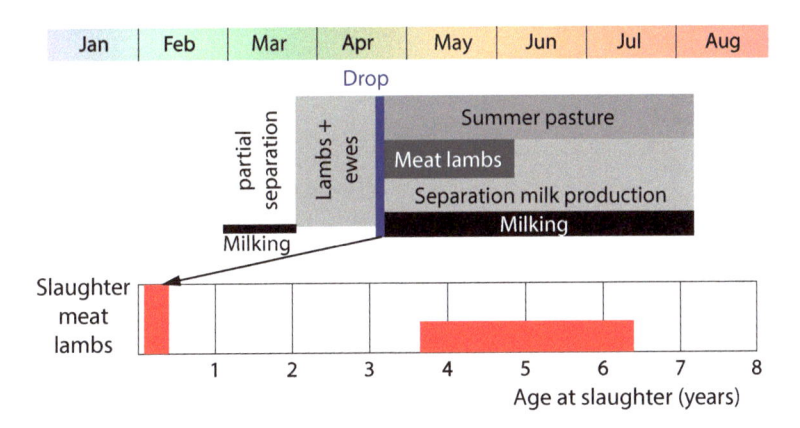

Fig. 3.13 The mixed economy model is a typical animal husbandry pattern at the beginning of the Neolithic period. Male young animals also served as meat lambs when too many rams were born

An alternative management model that focuses less on sustainability and more on stockpiling, which quickly leads to overproduction, is shown in Fig. 3.14. Although this model respects a similar seasonal cycle, it allows for an early separation of a large part of the lambs from the mother sheep. In this way, milk can be produced over a longer period. At the same time, in this model, male animals are slaughtered before weaning and used for meat supply. In both models, the slaughtering age of the animals determines the livestock population.

The animals in Neolithic times were what are now called dual breeds: neither pure meat nor pure dairy animals. In this case, the first model (Fig. 3.13) proved useful. For larger herds and specialization in pure dairy farming, the modern model proves to be practical.

These models are still used today. Small farmers or organic farmers follow the first model, regardless of whether it is cows, sheep, goats, or even chickens for egg production (instead of milk). The stock is maintained, the young animals are only separated from the mother animals after weaning (if at all). Young animals are slaughtered depending on their health condition or economic situation and demand. Male animals are cared for in the production of goat's milk, for example, and slaughtered and eaten or directly marketed when they reach a sufficient weight. The mixed farming model is much closer to the needs of the people to be supplied. Strong overproduction beyond manageable stockpiling is not planned and would not be possible for many farmers. The living animals are pure operating capital due to the milk production. Only what can be processed, eaten, and traded is killed. Therefore, even today, extraordinary pieces of goat bucks or rams can be found in agricultural areas, producer markets, or organic markets in cities. Depending on the region and eating culture, unusual parts such as tripe, testicles, udders, and the more familiar offal such as liver, kidney, and heart are available. Unlike today, sustainable food production used to be the prerequisite for economic viability and the basis of nutrition.

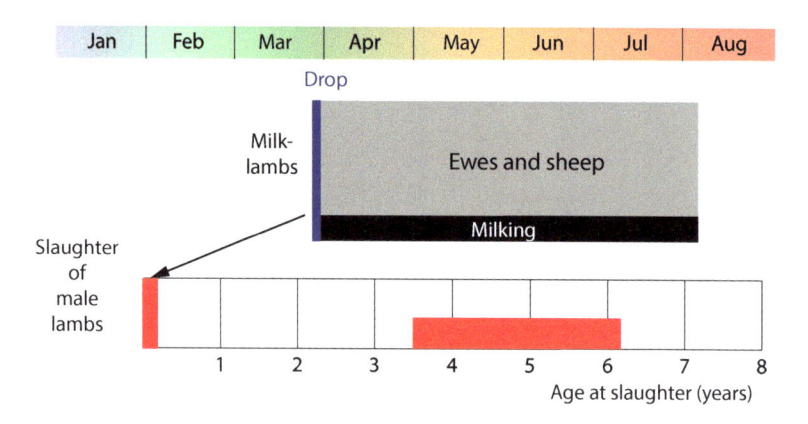

Fig. 3.14 A modern model allows for high overproduction from a critical number of animals

3.3 Industrial Animal Production and BSE (Bovine spongiform encephalopathy)

3.3.1 Contemporary Industrialized Agriculture

The goods in supermarkets must be inexpensive (which, in fact, offal would be), suitable for mass consumption, and conform to the price structure dictated by corporations. The second model is therefore the simplest version of complete industrialization of animal production, as expressed, for example, in the shredding of chicks. In factory farming, male animals are killed immediately after birth or hatching if it is not economically worthwhile to raise them for consumption, for example, because it costs too much feed or the highly bred animals do not produce marketable meat. These two highly simplified agricultural models immediately show why this is possible under silence or even acceptance in large parts of the population: The supply of mass production is not based on the actual needs of people, but unconditionally submits to the estimates of trade analyses. No wonder, then, that as a result of the supposedly abundant supply, too many goods are not eaten and have to be thrown away.

Overproduction is ultimately also responsible for the fact that not all products and all parts of the animals have to be eaten. Thus, today's society can afford not to eat or use offal or supposedly inferior pieces. There is enough of the so-called prime cuts that are easy and quick to prepare. The argument of the industrial meat industry that these pieces are eaten in China or other countries is questionable. The fact that it is no longer necessary to eat all parts of an animal increases the demand for fillet, fat-free or supposedly healthy parts, and factory farming, thus leading to an exaggeration of the modern model. Current attempts by chefs or Slow Food and other movements to use the "Second Cuts" are commendable, but do not reach the masses to curb industrial livestock farming and meat production. Phenomena of this magnitude of abundance have never existed in any epoch of human history and remain limited to industrialized nations. From the perspective of nutrition, the value and nutritional value of food, and their molecular composition, the current eating behavior appears incomprehensible and many "civilization diseases" avoidable.

3.3.2 The Early Beginnings of Industrialization

The Neolithic model of a sustainable mixed and meat economy was developed to preserve the stock, and it proved to be useful when adapted to the respective regional conditions and seasonal conditions. Today, people in the affluent society have a choice: everybody can buy the products of the modern model, including overproduction, but no one is forced to do so. There are many opportunities to buy our food from direct marketers and producers. Unlike in the supermarket, one is much less inclined to buy pointless products that supposedly represent an attractive offer. These economically cheap foods are only possible with the modern economic model and unnaturally bred animals. They do not reflect their value creation, let alone the work on the farms. It is also our Western eating habits that summon factory farming. This can be conclusively traced using the example of industrial chicken farming. Mass poultry farming did not arise solely from greed for profit. The egg was no longer recommended as a food in the 1950s and 1960s due to cholesterol, which curbed private consumption. Shortly thereafter, the first suspicions spread that red meat made people sick, whereupon the wealthy world population turned to white meat, i.e., chicken and turkey. Demand increased and the vicious circle began: poultry was produced en masse to satisfy demand. The poultry was bred with unnatural, oversized breasts, and fattening with high-calorie feed led to rapid growth of the meat in a short time. The short life of the poultry until slaughter became torture. The barns became more crowded, a sick chicken could infect countless other chickens in the confined space, and medication and antibiotics had to be added to the feed prophylactically and for economic reasons. The consequences are clear: economic models like these cause great harm to animals and humans.

3.3.3 Animal Food for Ruminants?

The occurrence of mad cow disease BSE (Bovine spongiform encephalopathy) [24] is an example, as it shows the impact of misguided animal husbandry forms in all consequences and at all levels. All basic principles of species-appropriate animal husbandry were turned upside down. Cattle were fed the bone meal of their own species as an additional protein source for fattening. The pure herbivores were given protein-rich animal food as fattening feed. The idea seems obvious at first glance, because it should not matter where the amino acids ultimately come from. If it were not for the prion problem and the fed animal feed were not contaminated. This neurological disease, very similar to Creutzfeldt-Jakob disease (CJD), can actually be understood on a physical basis [25]. The triggers are not viruses or bacteria, but spontaneous refolding of proteins, i.e., purely physical processes that are very similar to those at the edge of the partial unfolding of apolipoproteins B100 in LDL particles, which are also observed and ultimately belong to the main triggers of plaques. In the case of neurological diseases, proteins in the brain are refolded.

3.3.4 Prion Hypothesis—Physical Infections

Proteins are specifically folded chain molecules made up of the 20 proteinogenic amino acids, as illustrated in Chap. 1. How they are folded is mainly determined by the sequence of amino acids, some of which are hydrophilic (water-loving) and some hydrophobic (water-repellent), but lipophilic (fat-loving). This competition of interactions forces the protein chains to fold into a very specific shape, which is determined solely by physics. Some of the amino acids fit well together,

3.3 Industrial Animal Production and BSE

others do not, and still others repel each other strongly if, for example, they have the same electrical charges. However, the repulsion is spatially limited because the equally charged amino acids are fixed in the protein chain at very specific positions. This results in many competing folds. However, evolution ensures that completely unfavorable folds, which would have no biological functions, no longer occur. Such biologically nonsensical amino acid sequences were discarded and banished as non-functional during evolution. Most proteins, therefore, always find their way into the most favorable fold predetermined by nature, even if there may be occasional traps along the way and the proteins get stuck in intermediate folds. However, they find their way out of these through molecular fluctuations, as the only conclusive criterion for this selection is provided by physics: The state, i.e., the fold, of the lowest energy is sought. This is a basic principle of all physical systems, including proteins. The state of the lowest energy is always sought. Only this energetically lowest ground state guarantees the accurate function of the protein.

Physically, one can imagine this as if someone were standing on a mountain and rolling a ball down. As long as there are no obstacles such as very deep or very wide holes into which the ball could fall and get stuck, it will quickly roll into the valley, to the lowest point. This is then the absolute minimum of the mountain of potential energy. Something similar happens with proteins as well. On the way to the perfect fold, various intermediate states, or "relative minima" as side valleys in the mountain model, are reached, which do not yet represent the absolute minimum but are already very close to the perfect fold. Essential elements are pre-folded, which is precisely why proteins linger in these valleys for a certain time, for example, to form biologically relevant elements such as α-helices or β-sheets. The rest is then pulled along, and the path to the energy minimum, a protein with its assigned function, is thereby facilitated, as schematically shown in Fig. 3.15.

These processes are regulated through the physical examination of the many interactions between individual amino acids: through the contrast of water and fat solubility of amino acids. Some of them like each other more, some less, some even hate each other, while others repel each other when they are oppositely charged. The optimal folding must be found between all these opposing and competing conditions. In this process, water mediates hydrogen bond interactions between some unfavorable amino acid interactions—until the most optimal and energetically lowest folding is achieved. In theoretical physics, this is often referred to as frustration-free foldings. However, helices and beta sheets are equally necessary for the function of the protein. Due to the necessary fine-tuning, the energetic distance of some possible foldings to the perfect one is not that large. It is conceivable that under very specific circumstances, refoldings are possible for some proteins: helices randomly become beta sheets, as shown very schematically in Fig. 3.16.

This can lead to health problems, as already described for evil becoming LDL particles, because the extended β-sheets have a strong tendency to form aggregates. However, several such refolded proteins must be present for this

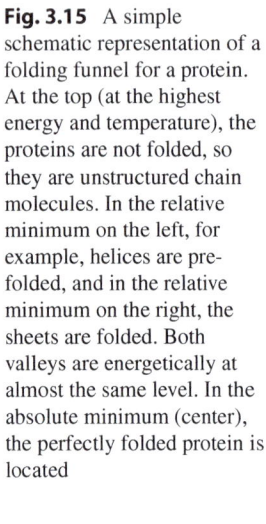

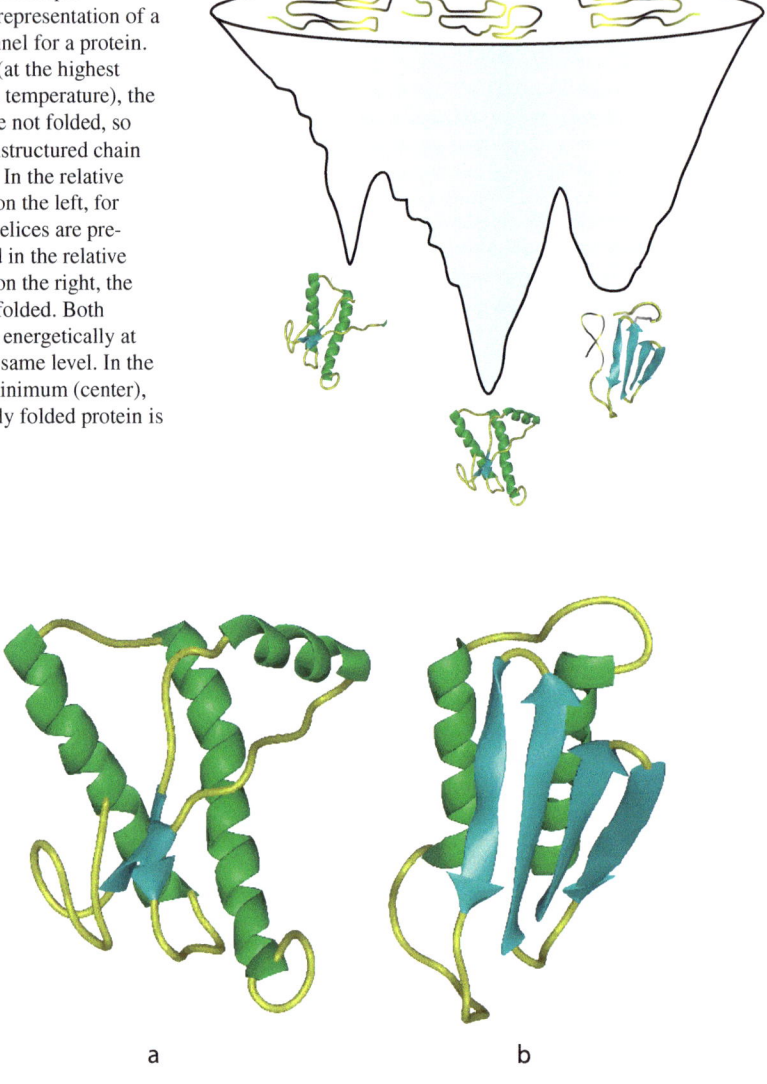

Fig. 3.15 A simple schematic representation of a folding funnel for a protein. At the top (at the highest energy and temperature), the proteins are not folded, so they are unstructured chain molecules. In the relative minimum on the left, for example, helices are pre-folded, and in the relative minimum on the right, the sheets are folded. Both valleys are energetically at almost the same level. In the absolute minimum (center), the perfectly folded protein is located

Fig. 3.16 Example of a refolding of proteins. From parts of the light green α-helices **a**, blue-green β-sheets **b** are formed. The structure (b) is the prion.

to happen. But it is enough if an already refolded protein comes close to a still original protein, because then the native proteins experience different electrostatic forces. However, since the two minima are energetically very close to each other, these small changes in interactions are sufficient to refold the still original protein. This is part of the prion hypothesis, when refolded proteins, called prions, can trigger diseases. Infection is often mentioned, but this contagion is virus- and bacteria-free. It only happens through a physical process and can be observed very precisely in laboratory experiments [26]. The refolding is induced by changed physical forces. Very similar to what is known from elementary physics: electric fields can generate dipoles or charge shifts and thus initiate new processes on molecular scales.

Up to this point, the unfolding processes would not be particularly bad, as a few unfolded prions do not yet cause a disease on their own. The decisive effect is of a collective nature. Due to their predominantly hydrophobic properties, beta sheets have a high potential for aggregation. They therefore form fibrillar structures and long aggregates, which are called amyloids in medicine. As a result, fewer and fewer beta sheets are forced to surround themselves with free water (in the physiological environment as saline solution). And this corresponds to a further significant minimization of energy. Aggregating into a long amyloid is energetically much more favorable than the sum of the energy of non-aggregated prions. This is shown schematically in Fig. 3.17.

In doing so, the physical nature of prion formation becomes clear. Two different areas are important for prion diseases. The blue-colored funnel leads each individual protein through the two intermediate valleys into the deepest, native state. Therefore, only the interactions of the individual proteins are relevant there. However, proteins from the native state can spontaneously refold back into the

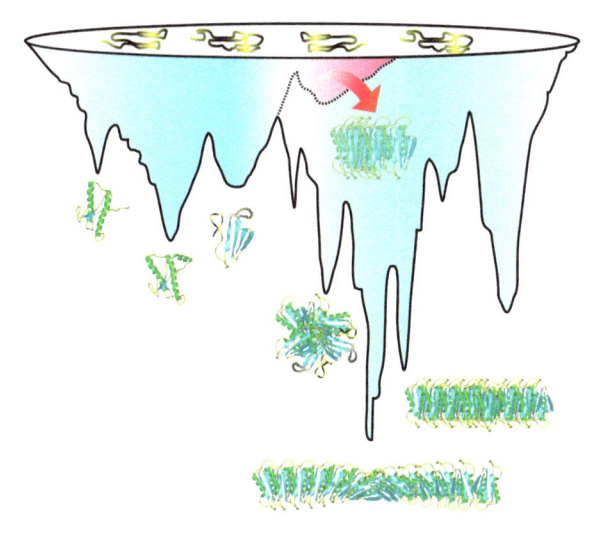

Fig. 3.17 The physical infection of proteins is a "many-particle problem". On the left (colored blue) is the folding funnel from Fig. 3.15. When proteins fold spontaneously, they increasingly occur in beta sheet-dominated structures and induce further unfolding (red arrow). At higher concentrations (blue-green colored area), these can aggregate into small aggregates (below red arrow). If these nuclei are large enough, large aggregates with different structures can grow from them. Long amyloids can be found in the deepest minimum. The amyloid minimum is much deeper than the minimum of the perfectly folded protein. If enough proteins are unfolded, they will aggregate into amyloids.

β-sheet-dominant intermediate structures. To do this, they jump from the minimum of the blue-colored area into the higher valley to the right of it (Fig. 3.17). If other original proteins are nearby, they will fold with a certain probability. However, this happens in the blue-green colored area, where interactions between different proteins must be taken into account. Thus, for an unfolded protein to access others, as already described, they must come so close that they are within the range of electrostatic and hydrophobic forces and new hydrogen bond connections can be established. If this happens frequently, they can nucleate into an aggregate. Therefore, the barrier between the blue and blue-green areas is lowered. When the nucleated aggregates reach a critical size, they can grow continuously. Fibrils and large aggregates form. Each additional unfolded protein can attach itself to these fibrils. The fibrils become longer and longer, causing damage to the normal tissue. These processes are particularly relevant for proteins in the nervous system. Therefore, the spinal cord, nerve pathways, and especially the brain are severely affected. Both in Creutzfeldt-Jakob disease and in mad cow disease, these amyloid structures are responsible for the onset of dementia and other neurological malfunctions. Unfortunately, no way has been found to date to stop this physical infection in related diseases in time.

For all these processes, neither chemical nor viral or bacterial processes are responsible. All this happens only because of the potential given by the protein and the physical conditions. Therapies, therefore, also look for ways to inhibit or prevent this fibril growth. The cause of the occurrence of mad cow disease is still unclear. The assumption that it is due to prions introduced through animal feed has not been clearly proven to this day. BSE as well as CJD also occur spontaneously [27], as the energy minima are close together and are thus not separated by very high barriers. Molecular fluctuations may already be sufficient to trigger such cascades. This by no means implies that cattle should be fattened with animal feed. This has nothing to do with species-appropriate animal husbandry. Just as little as feeding carnivorous pets with vegan food.

3.3.5 The Well-Intentioned Attempt to Impose Omega-3 Fat on Sheep

Far less risky was the attempt to improve the fat of sheep as already briefly mentioned in Chap. 1. Based on the view that animals contain too many of the saturated and thus unhealthy fatty acids, agricultural scientists agreed to enrich the feed of sheep with linseed oil, which contains the essential α-linolenic acid (ALA), and also with fish oil, which contains the essential fatty acids EPA and DHA [28]. Furthermore, combinations of linseed and fish oils and algae were also added to the feed. The idea was to modify the natural fat of the animals through the feed towards healthy fat. Meat consumption was to become healthier in this way.

The fatty acid spectrum of the intra- and extramuscular fat of the animals actually changed due to the oversupply of unsaturated fats compared to grass: More unsaturated fatty acids were incorporated into the fat of the animals, as was shown in the analysis of the fatty acid spectrum after slaughter. Depending on the feed additive—linseed, fish oils, algae or combinations of fish oils and microalgae—even increased EPA and DHA can be enriched in the fat of the lambs. The most effective were feed additives from fish oils and algae, in this case, the concentration of essential fatty acids EPA and DHA increased considerably. The goal was achieved. But the meat no longer tasted good, and thus the price for the enrichment is much too high. Simply put: The lamb meat became rancid quickly after slaughter. The reason for this is already known from Chap. 1: the oxidation of unsaturated fatty acids. In the process, too many fragments of the many unstable double bonds are formed, and thus precursor compounds to these off-flavors.

This is not surprising, because the species-appropriate food for sheep consists of grass, herbs, and plants, whether the animals graze on dikes, in fields and meadows, or in the barren mountainous regions of the Mediterranean. An excess of highly unsaturated fatty acids has no biological or physiological place in the plant-eating ruminants at their constant body temperature, as this experiment by agricultural engineers shows: The imposed fat does not correspond to the natural and physiologically required compositions of land animals. The meat becomes aromatically inedible. It is unhealthy for the animals in any case, because an excess of polyunsaturated fatty acids means a multitude of free radicals for the metabolism of land animals. And if they were not slaughtered so early, they would soon succumb to the resulting oxidative stress (see Fig. 1.5) and the resulting cell damage. At the same time, the proportion of *trans* fatty acids of various isomers of C 18:1 increased noticeably when feeding with fish oils and algae, as is usual in ruminants. The best-known of these is the already mentioned vaccenic acid C 18:1, 11t.

Since the Neolithic period, grazing animals have been fed exclusively grass, hay in winter, or later, when technology allowed, silage (fermented grass/hay). Thus, these principles were violated in two ways in the attempt to make animal fat "healthier." Although flaxseeds are plant-based, they are not the first choice of food for ruminants. It is even more critical with fish oils, which definitely do not belong to the diet of sheep. As we already know, the fatty acid pattern of animals is determined by genetics, the evolutionary biology of sheep, and thus their living conditions. The function of animal membranes and body and subcutaneous fats corresponds exactly to the natural requirements, such as body temperature and climate. Different sheep breeds, therefore, have a composition of fatty acids adapted to this, which is mainly determined by the food the animals find in the pastures. Grasses, meadow herbs, and flowers they eat have exactly the composition of fats adapted to the soil and climate. Particularly species-appropriate animal husbandry of land animal breeds takes place on native soil. Organic farmers make extensive use of this idea, and young breeders are happy to return to old breeds in their operations—and not without reason.

This example demonstrates very impressively what herbivores can do and what we humans cannot. The very limited ability of humans to produce the longer-chain, more unsaturated fatty acids like EPA and DHA from polyunsaturated plant fats such as ALA is indeed physiologically predetermined. Herbivores—especially ruminants—cannot take up DHA or EPA from their natural plant-based diet, unlike omnivores. However, grasses can only synthesize α-linolenic acid (ALA) in their metabolism at most. Consequently, herbivores must produce the polyunsaturated fatty acids, which are also necessary for them, themselves. This occurs in several processes in the various stomachs and digestive systems that are essential for ruminants. In the rumen, therefore, the green fodder is pre-fermented to break down the hard cell material such as cellulose, hemicellulose, and other polysaccharides, making the cell contents and the fatty acids of the phospholipids in the cell membranes available. The fats of the feed are converted for the physiological requirements of the ruminants through this pre-fermentation. A whole series of enzymes is necessary for these changes, such as desaturases—they can incorporate double bonds into saturated fatty acids—or elongases—these can extend shorter fatty acids. Just as much as is needed. Thus, DHA is produced in the mammary glands for the milk [29], to pass on these fatty acids to the offspring, lambs, dairy sheep, and calves.

If an excess supply of DHA and EPA is forced through unnatural food, these polyunsaturated fats are packed into the fat deposits and muscles. There, the fatty acids can also oxidize more intensively after slaughter and cause off-flavors. By the way, these cuts of meat would not be harmful for consumption. However, the usual taste and aroma expectations are not met, and the enjoyment falls by the wayside.

3.3.6 Intensive Fattening, US-Beef—Other *trans* Fats

The influence of feed on the taste of meat is therefore obvious. The fatty acid spectrum of the fats in the meat changes significantly. This is also evident in modern (industrial) fattening, especially when comparing US beef under intensive fattening and beef cattle that were in the pasture and exclusively ate green fodder. This shows the direct effect of the feed on the *trans* fatty acids and the placement of the double bond. To do this, one must first take a closer look at the essential processes for the two most important fatty acids from fattening and green fodder in the rumen, which are shown in a highly simplified form in Fig. 3.18. Lipases cleave the fatty acids from the feed until they are present individually, separated from the glycerol. The triple unsaturated ALA of the green fodder is converted by isomerases to shift the double bond. The *cis* double bond at the 9th position remains, while the *cis* double bond at the 12th position becomes a *trans* double bond at the 11th position. There are usually two cases, as the isomerases can shift the *cis* double bond of the 15th carbon atom either as a *cis*- or *trans* double bond to carbon atom No. 13. This results in so-called conjugated linoleic acids. Most of these are again removed by reductases, taking away one double bond and thus feeding it to the pathway on the right.

In Fig. 3.18 on the right, it is shown how, starting from linoleic acid (a doubly unsaturated omega-6 fatty acid), as found in nut oil seeds, especially in the concentrate corn and in sunflower seeds, so-called conjugated linoleic acids(CLA)

3.3 Industrial Animal Production and BSE

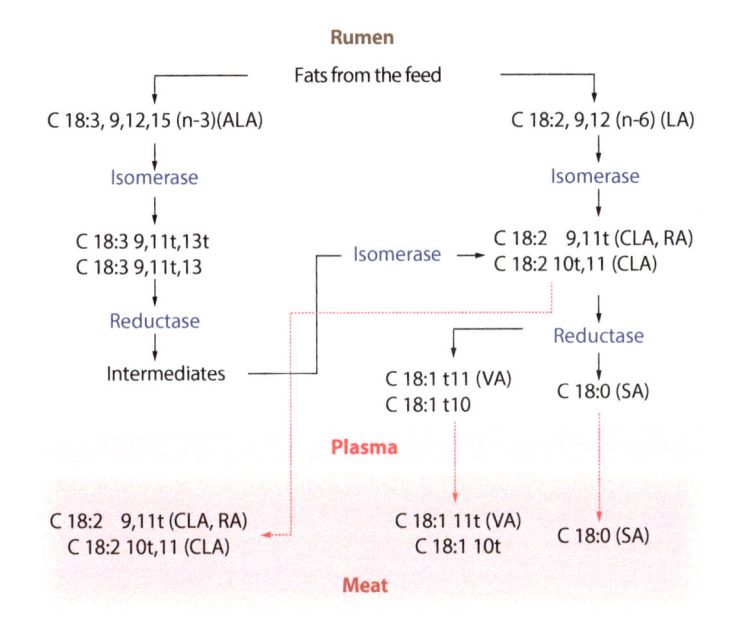

Fig. 3.18 Simplified representation of the natural hydrogenation of unsaturated fatty acids and the synthesis of conjugated linoleic acids (CLA) in the rumen. Involved are isomerases, which rearrange the double bonds, and reductases, which can remove double bonds

are formed via isomerases. Likewise, when α-linolenic acid (shown on the left in Fig. 3.18) is altered, which is synthesized into CLA in muscle tissue through a variety of enzymatic steps.

These conjugated linoleic acids can already enter the meat (as subcutaneous fat or intramuscular fat) via the blood plasma and accumulate there. One of them, referred to as C 18:2, 9,11t, is also called rumenic acid (RA), because rumen is the technical term for the first stomach chamber of ruminants. From some of these conjugated linoleic acids, the saturated stearic acid C 18:0 (SA), the *trans* fatty acid C 18:1, 11t, known as vaccenic acid (VA), but also a *trans* fatty acid with its *trans* double bond at the 11th carbon atom is formed. This one is formed—in addition to the elaidic acid (C 18:1, 9t) known to us from Fig. 3.2—increasingly also during industrial fat hardening. Similar to elaidic acid, the *trans* fatty acid C 18:1, 9t also has its *trans* double bond very close to the center of the fatty acids. Like elaidic acid, this *trans* fatty acid has a strong stiffening effect and therefore exhibits "unhealthy" properties that have already been described (Fig. 3.2 and 3.12), while vaccenic acid no longer shows these properties. However, vaccenic acid is not formed during catalytic fat hardening (hydrogenation); it is largely reserved for ruminants through their biohydrogenation in the rumen.

This universally valid scheme, however, does not yet provide information about the frequency of fatty acids in intramuscular fat after fattening. It is likely that the frequency of different fatty acids in animal fat depends on fattening and thus on feeding. Do the animals receive more green fodder, meaning more ALA, or more concentrated feed, which, in addition to the intended protein content for muscle building, also has a different fatty acid pattern and emphasizes the linoleic acid of seeds and grains more? In fact, feeding experiments show a clear difference in the fatty acid pattern between grazing animals and fattened animals [30]. Detailed studies [31] therefore show clear results, for example, the ratio of Omega-3- to Omega-6 fatty acids is significantly more favorable with soilage and grazing. In addition, cattle from intensive fattening have a much higher proportion of the *trans* fatty acid C 18:1, 10t, as can be clearly seen from Fig. 3.19.

The vaccenic acid serves as a precursor to conjugated fatty acids, while other *trans* fatty acids do not. Consequently, *trans* fatty acids whose double bond does not occur at the 11th carbon atom cannot be metabolized into CLAs. They must be transported via the lipoproteins LDL and HDL. Therefore, the earlier argument from Fig. 3.11 applies here: The further the double bond is from the center of the fatty acid, i.e., well below position 9 or further above, the less influence it has on the curvature of phospholipid monolayers and thus on the radius of the LDL particles (see Fig. 3.12). *trans*-fatty acids with 12t or 13t as well as 6t or 7t (the gray boxes in Fig. 3.19) can still be packaged into HDL particles, while *trans* fatty

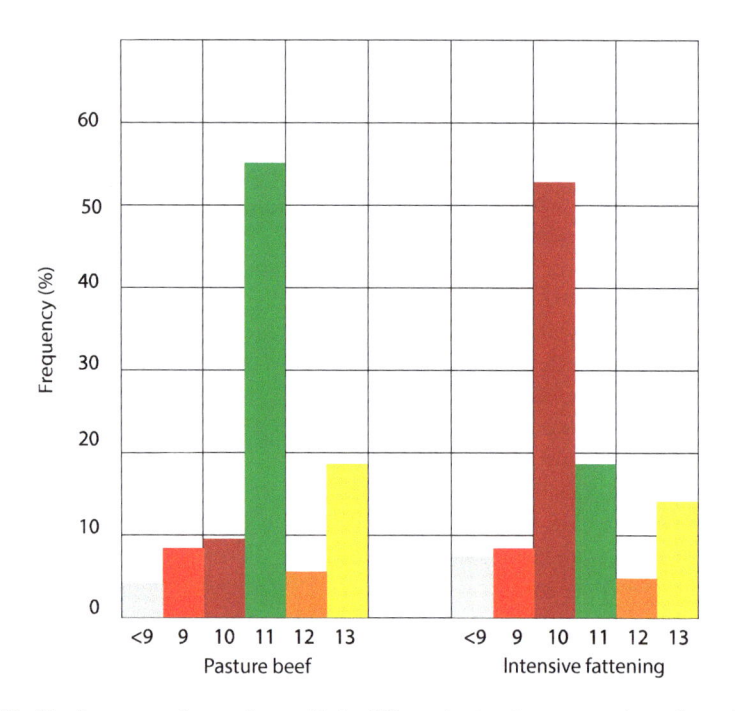

Fig. 3.19 The frequency of trans fatty acids in different husbandry systems depends on the husbandry and feeding [31]. The number on the horizontal axis describes the position of the *trans* double bond, 11, thus vaccenic acid is present. Intensive fattening with feed high in linoleic acid apparently favors the formation of non-metabolizable *trans* fatty acids

acids of 9t or 10t are significantly worse. The curvature of the HDL lipoproteins would be disturbed, therefore LDL particles are favored by these fatty acids, as already indicated in Fig. 3.12.

3.4 Staple Food Bread: Physics, Chemistry, Nutrition

3.4.1 Grains

The question: "Which came first—beer or bread?", remains largely unanswered. Much speaks for beer, as during storage, grain often became damp and began to germinate. Wild yeasts are everywhere, and the vessels needed for storage provided an ideal fermentation space [32, 33]. However, there is recent evidence that bread is much older than previously thought. As archaeological excavations have shown, a transitional culture, the Natufian culture in Jordan, already baked sieved mixtures of ground flour from grass seeds and water over fire to make unleavened flatbreads [34]. This transitional culture is still counted among the hunters and gatherers but was already sedentary. Analytical studies of charred and preserved bread remains show that for this "Paleo bread," precursors of grains such as barley, einkorn, oats, and the seeds of the club-rush *Bolboschoenus glaucus* were used. The grains were ground, sieved, and kneaded into a dough, which was then baked. Sedentism and the beginning of agricultural culture with the ancient grains and early grass breeding allowed for the production of bread from ground grains. People quickly learned to systematically use grain. These possibilities quickly arose from the ability of germination: The grass seeds produced grasses with usable seeds. From these observations, early breeding research emerged. Soon, the people of the early Neolithic learned to develop appropriate methods to cultivate grasses. The path of cereals and pseudocereals took its course and was reflected in the human diet [35]. According to current knowledge, the origin of agricultural practices lay in a broad belt of Southeast Asia, including the areas of present-day southern Turkey, Palestine, Lebanon, and northern Iraq. There, a great variety of wild grains still exists today, and the climatic conditions for growth were excellent. Thus, wheat (*Triticum*) was created, and the first forms of barley (wild barley, *Hordeum spontaneum*) were frequently collected by local residents.

In Southeast Asia, there is already evidence around 7800 BC for a gradual transition of hunters and gatherers to grain cultivation, which gradually became a significant part of their energy intake and brought forth new cultural techniques. Soils were worked with simple tools, grain was ground to obtain precursors of flour. One to two thousand years later, around 5000 BC, hunted animals made up only about 5% of daily food, while grain and livestock became a considerable part of daily nutrition. In present-day Egypt, an agricultural-based civilization developed in the 5th millennium BC: People became specialists in the cultivation of wheat, flax, or barley, which was also used for beer production. Systematic agriculture was born.

The early wild cereal species of the genera *Triticum* (wheat) and *Hordeum* (barley) were genetically diploid, possessing male and female genetic traits and bearing only a few seeds. Abundant harvests were not to be expected at the beginning. However, the diploidy revealed a remarkable variation in the protein and starch content of the cereal grains, which could be specifically used for breeding. The beginning of early agriculture and the use of irrigation systems allowed the survival and spread of polyploid grains, reducing genetic variations but making cultivation more predictable and thus plannable. This targeted breeding testifies to the great achievement of early farmers in recognizing and utilizing this.

The first stable breeding of polyploid grains dates back to around 6000 years BC. Genetic variability and selective choice were necessary to adapt the grains to different environmental conditions [36]. Today, archaeological finds and genetic analyses show that the wheat variety *Triticum turgidum* var. *dicoccoides* was crossed with *Triticum tauschii* to create *Triticum aestivum*, the precursor of our modern wheat. *Triticum aestivum* is a hexaploid wheat with 42 chromosomes compared to the 14 chromosomes of einkorn (small spelt), *Triticum monococcum*. It is no miracle that such a genetically powerful cereal replaced all existing wheat varieties: Currently, there are 20,000 varieties of *Triticum aestivum* for professional wheat cultivation. The same applies to emmer (two-grain, *Triticum dicoccum*) [37], when attempts are made today to adapt cereals to climatic conditions, such as the high drought of some regions. It is also not surprising that modern methods such as Next Generation Sequencing, CRISPR-Cas9 and genetic engineering are used today [38] to make breeding processes more effective.

Neolithic breeding research was primarily focused on growth and prosperity, so that the homogeneity and productivity of cereals improved, but it quickly became apparent how the processability of the varieties varied. Increasingly, the properties of the flours and doughs produced from them came to the fore. After harvest, the cereals had to be easily ground, and good adhesive properties were necessary for the doughs so that good, durable breads could be baked from them.

Glue is the appropriate keyword. Most dough properties, kneading, and baking are defined by the "glue protein." The grain needs good glue so that bread can be baked at all. Other grains and many pseudo-cereals do not store this protein in the grain or only to a very limited extent. In addition to selection for growth and climatic conditions, another criterion was added: processability and thus the proportion of the glue protein gluten. Selecting for high and also less weather-dependent gluten content is still practiced today. Thus, in the last 200 years, active genetic selection and genetic manipulation have increased the gluten content of the original *Triticum*. These findings are rightly considered a great cultural achievement and show how persistently and systematically cereal breeding and research were carried out empirically without laboratories and precise analysis methods.

Higher gluten yields, however, had another aspect in the early years of agriculture. It was also the significantly higher protein content that made wheat more valuable compared to other grasses. Starch provides a lot of energy in the form of glucose, but wheat protein offered those essential amino acids that secured high nutritional standards for *Homo sapiens*. An aspect that was only recognized late and that seems to no longer appear in today's discussion about gluten-free. However, the gluten as a protein-rich nutrient did not play an important role in all regions of the world, as only a few climatically favored cultivation areas could breed highly developed wheat forms. In large parts of Asia, rice was cultivated, while on the American continent corn grew, and in Africa, millet and its relatives such as sorghum. Rice, corn, and millet do not contain gluten as storage protein. Therefore, no leavened bread was prepared with them, as the fermentation did not work particularly well due to the lack of cohesion, and the doughs could not withstand the dough rise. The majority of the Neolithic world population did not live on wheat bread. The high spread of wheat only occurred much later.

From these facts alone, a whole series of today's relevant relationships becomes clear. The adhesive properties of proteins determine processing, taste, and texture. It also becomes clear that gluten-free or low-gluten grains and pseudo-cereals are suitable for the production of other foods than grain varieties with a high gluten content. The molecular properties are crucial. Nature primarily defines these. People had recognized this and drawn the appropriate conclusions. Nowadays, many thousands of years later, the molecular backgrounds are largely known, are disseminated through scientific publications, the internet, and popular media, but apparently, the very modern human is unfortunately only rarely able to draw logical conclusions from them. Reason enough to return to the molecular and nutritional physiological questions about gluten-free baking apart from necessities in Chap. 4 in detail.

3.4.2 What can be learned from the germination of grain

What exactly lies behind the millennia-old empirically researched dough and baking technology? The most instructive approach is to take a closer look at the germination, as the first breeders did in the Neolithic period. The seeds germinate underground with sufficient moisture and appropriate temperatures. The grain swells, absorbs water, and sprouts form. This germination and sprouting process requires energy, which, as long as the germinating seed is underground, cannot come from sunlight through photosynthesis. The grasses therefore store energy in the seeds in the form of storage proteins, gluten (amino acids), densely packed starch (glucose), and more or less fat. These are the very macronutrients that all organisms need: During germination, new proteins, including enzymes, must be formed, and plant cells for the cotyledons must be created. A complete biochemical machine is set in motion, and glucose is the main fuel that the germinating grain takes from the starch. Special enzymes, amylases, are needed to break down the starch, lipases for fat degradation, and proteases to obtain the individual amino acids from the storage proteins, which are then assembled into new proteins and enzymes. To make this germination process as efficient as possible, the biochemical equipment of the seeds was gradually adapted over time, depending on the origin and climatic conditions. It is no wonder that the enzymes balance each other out to incorporate more or less phospholipases in order to also obtain free fatty acids—and thus convertible germination energy—from the lecithin. Only these biochemical helpers make it possible for a plant to thrive from the seed. In this sense, the plant already largely determines what is technologically possible in terms of food, as the dough and baking processes require precisely these facilities provided by the grain. Since the Neolithic period, humans have developed breeds over thousands of years that are well known today. All this information is still stored in the DNA of modern wheat. It should therefore come as no surprise that the wheat genome contains far more genes than that of *Homo sapiens* [39].

3.4.3 Wheat Bread and Aids—Industrial and Natural Methods?

Bread has been one of the staple foods of humans since the cultivation of grasses. However, bread became much more than a food; it became a cultural and religious symbol. In Christianity, it even becomes the body of Christ, the epitome of purity. In times of industrialization, bread seems to be moving further and further away from this reference to purity. While bread used to consist only of cereal flour, water, salt, and possibly (wild) yeasts, nowadays, complicated-sounding ingredients give rise to discussions. So-called flour treatment agents are common in industrial bakery products, including ascorbic acid (E 300), ascorbates such as sodium-l-ascorbate (E 301) and calcium-l-ascorbate (E 302), lecithin (E 322), guar gum (E 412), mono- and diglycerides of fatty acids (E 471), lactic acid esters of mono- and diglycerides of fatty acids (E 472b), wine and acetic acid esters of mono- and diglycerides of fatty acids (E 472f), sucrose esters of fatty acids (E 473), sugar esters (E 474), polyglycerin esters or polyglycerol esters of fatty acids (E 475), sodium stearoyl-2-lactylate (E 481), calcium stearoyl-2-lactylate (E 482), and l-cysteine (E 920) [40]. This list of stabilizers, release agents, emulsifiers, and even the amino acid cysteine is indeed frightening for many—firstly, because they are not expected in bread, and secondly, because many of the chemicals are also labeled with an E-number.

At the latest since the industrialization of bread, we have been confronted with the question of the effects of additives in a daily consumed food. Additives are generally identified as causes of intolerances [41], without, however, developing an understanding of why industrial baking processes use them in the first place or where the additives come from [42]. In most cases, the biochemistry of the grain is the starting point for the use of such aids, as they are present in the grain or are formed during germination. What is urgently needed, however, is the presentation of the relationships, so that it becomes clear what the additives really do, how food technologists come up with the idea of using them, in what quantities they occur, and what they can and cannot do during digestion.

The basis for all breads is flour made from ground, sifted, and largely husk-free grain, along with water, an appropriate amount of salt, and a leavening agent, depending on the choice and type of bread, yeast or sourdough. From a scientific perspective, this corresponds to a relatively simple model system, but it is also

3.4 Staple Food Bread: Physics, Chemistry, Nutrition

already the basis for a deeper understanding of the noted list of additives. These additives are not always used, but only under certain conditions, as artisan bakers can understand from their own experience. To do this, the basic ingredients of flour, water, salt, and leavening agent must be examined more closely. For a basic understanding, however, it is not enough to look at the components only at the macroscopically visible and tangible level, but the focus must be directed at the invisible to the naked eye, microscopic dimensions. Only at the level of molecular dimensions do the processes and mechanisms of flour, water, salt, and leavening agent fully unravel [43]. Even the observations made above about germination suggest that even the most natural and original organic bread contains more than just the macroscopically visible ingredients mentioned.

Before baking, the dough must first be prepared. To do this, flour, a certain amount of water, and some salt are mixed [44]. From the powdery, almost dry flour, an elastic, highly stretchable material is produced, reminiscent of rubber or chewing gum. Without this elasticity, it is hardly possible to shape and bake a fluffy bread with a pleasant crumb. This well-known process reveals a deep insight into the molecular world of flour. Flour is made up of a multitude of components that the grain supplies. A small selection of these is shown in Fig. 3.20. In addition to the energy stores for germination—protein and starch—flour also contains a whole range of other additives that primarily serve the grain but also prove useful

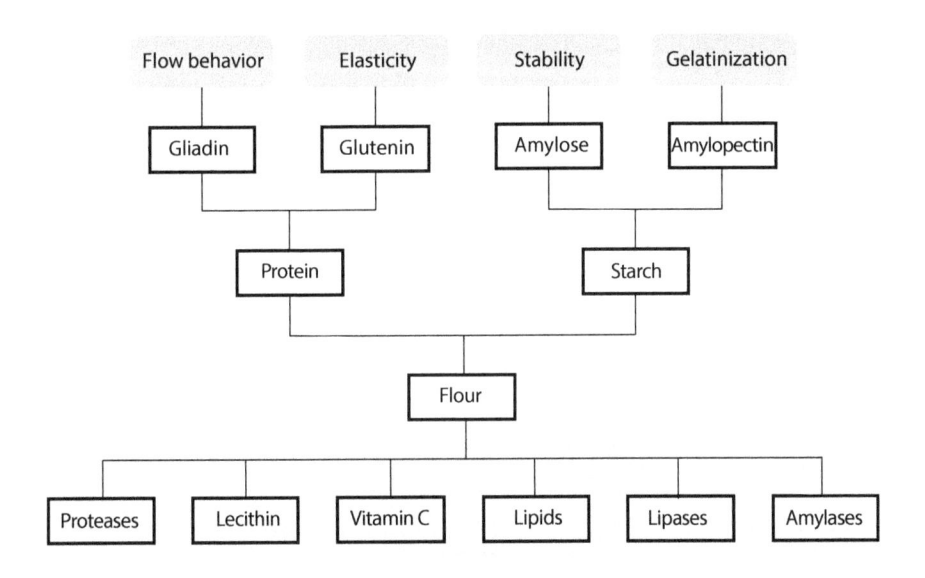

Fig. 3.20 The most important components of (natural) flour include a whole range of functional substances. The macronutrients protein and starch, which also perform important baking technological tasks, are colored green. A few additives contained in the flour are colored red at the bottom, including enzymes that break down proteins, fats, and starch (i.e., proteases, lipases, and amylases), as well as emulsifiers (lecithin and other lipids) and ascorbic acid (vitamin C)

in the dough and during the baking of bread. These include enzymes, as well as emulsifiers or lipids, such as the phospholipids found in plant cells, as well as other free fatty acids and the fat derivatives mono- and diglycerides. The latter are therefore found on many lists of additives. All foods contain such molecular compounds for biophysical and biochemical reasons.

In all wheat-like flours, even in the ancient wheat varieties emmer and einkorn or spelt, gluten forms a highly elastic dough when kneaded with the addition of water. The processability of flour into dough depends, as already indicated, on the quantity and quality of the protein. Gluten plays an outstanding role in this and shows the unique potential of this protein and how difficult it is to replace it.

3.4.4 Dough Properties, Dough Processing, and Gluten

In fact, the adhesive protein of wheat with its extraordinary properties plays a crucial role in dough processing. To understand the reasons, it is sufficient to consider adhesive proteins as more or less long pearl chain molecules, some of whose amino acids are water-repellent and others are water-loving, as schematically shown in Fig. 3.21.

Behind the collective name Gluten, there are different proteins, of which only two types are really important for processing. On the one hand, the relatively short Gliadins, and on the other hand, the significantly longer, low- and high-molecular Glutenins [45]. The precise form a protein takes during kneading and processing—whether elongated, tightly coiled, or folded into regular structures such as helices—depends on the type and sequence of amino acids it is made up of. But also on the environment in which the protein is located. In the water-rich dough, it is thermodynamically unfavorable for a large number of adjacent hydrophobic amino acids to stretch out, as they do not like water. They therefore clump together as closely as possible according to the principle "like attracts like" to protect

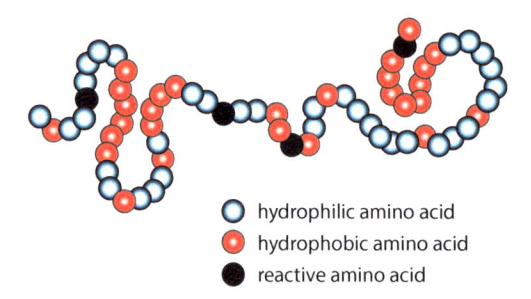

Fig. 3.21 Schematic model of an unstructured protein chain. The pearls represent the amino acids, depending on their solubility in water in different colors. Also, the amino acid Cysteine, which can chemically react with each other to form permanent cross-links (black circles), is incorporated into wheat proteins in the right places. Cysteine and the partial water solubility of certain amino acids are functionally important for dough and bread

3.4 Staple Food Bread: Physics, Chemistry, Nutrition

themselves from the unwanted water. The hydrophilic regions of the protein, on the other hand, like to surround themselves with water and stretch out more in it. During kneading, the protein chains must arrange themselves in such a way that as few thermodynamically unfavorable interactions as possible are forced to occur (molecular frustration) [46]. The physics of protein folding and protein structure is therefore characterized by a high number of compromises [47]. Furthermore, some amino acids (Arginine, Lysine, and Histidine) are electrically positively charged, while others (Aspartic acid and Glutamic acid) are electrically negatively charged. Since different charges attract and like charges repel, but both groups are highly water-soluble, there are further molecular competitive situations in the dough that must be resolved gradually according to physical laws during kneading. Therefore, a highly stretchable, elastic dough is formed gradually, as indicated in Fig. 3.22.

Only weakly interacting parts, such as the polar blocks of the protein chains, will knot and entangle [48], as is the case in many polymer solutions and melts, which largely explains the tough, viscous, and elastic behavior of polymers. Different electrical charges come together due to strong attraction, while the hydrophobic sections in the protein, for example, form aggregated β-sheets [49]. These very simple model ideas already show the high stretchability of the dough: If we pull very slowly, the hydrophobic clumps dissolve first, as they have the least cohesion. Then the electrostatic complexes dissolve, for which higher forces are already necessary, and with even higher deformations, the loops and knots gradually dissolve.

It is precisely these properties that matter in doughs that undergo fermentation, specifically sourdough fermentation involving wild yeasts, lactic acid, and acetic acid bacteria during fermentation. The microorganisms involved enzymatically

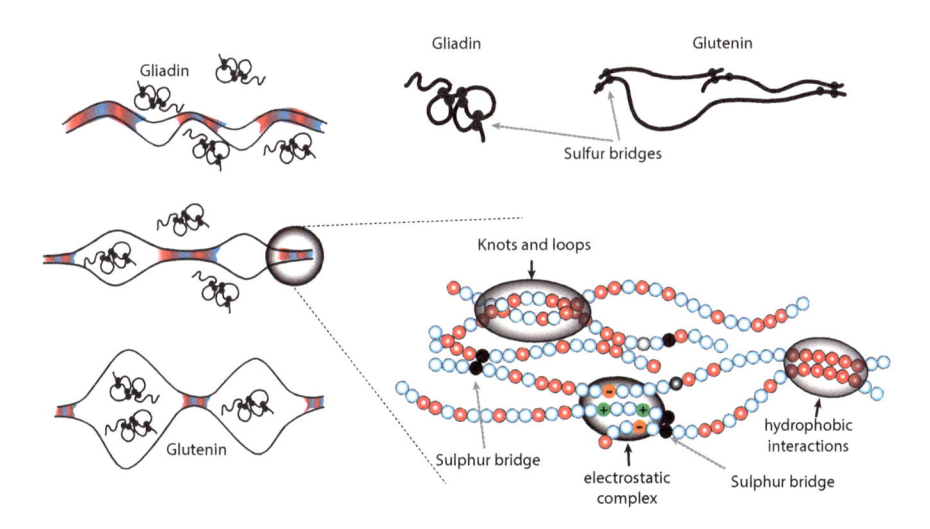

Fig. 3.22 Various structural elements in Gluten that form after the addition of water and kneading: Entanglements and knots, electrostatic complexes, and hydrophobic associations

digest a portion of the available sugars and produce carbon dioxide, which provides dough loosening. For the adhesive properties to be sufficient and for these dough cohesion properties to occur, a number of molecular properties must come together.

In particular, the acidification of the dough offers another advantage: The charges of the amino acids change, and the overall charge of the glutenins weakens. Consequently, electrostatic complexes (see Fig. 3.22) play a lesser role. The glutenin chains can intertwine better. This allows for further dough improvement, making breads lighter and airier.

3.4.5 Yeast and Sourdough—Dough Leavening, Fermentation

After kneading, the resting phase follows. Yeast and added basic sourdoughs now need nourishment. Yeasts and Lactobacilli metabolize glucose (grape sugar) and produce carbon dioxide (CO_2) in the dough, which inflates the dough and thus provides loosening. But where does the sugar come from? Flour brings glucose, but most of it is present in starch, as starch molecules are huge associations of glucose molecules. The Amylose is a linear chain molecule made up of linked glucose units. Amylopectin is also a chain molecule, but these molecules are highly branched and very tightly packed. The leavening agent is thus firmly bound and not available to the yeast cells.

The solution to the dilemma are enzymes Enzymes, amylases, which each wheat grain contains for good biological reasons. After all, the wheat grain should germinate and then grow. For this, it needs energy in the form of glucose during the germination phase, which it takes from the starch. During dough production, various amylases break down the starch and make it available to the Yeast in the form of glucose. If the flour still has enough amylases, the nutrient supply for the yeast is ensured. Even during kneading with water, some amylases are activated and produce glucose, mainly from the few amylose molecules that have been released from the starch grain during kneading. These are easily accessible to the enzymes. During fermentation, this process continues, and amylases continuously produce glucose, so the yeast cells are able to produce a lot of gas and thus guarantee the desired dough loosening. Once the yeast gets going, it also releases amylases (and proteases) that promote the starch degradation process.

The enzymes act as molecular scissors, which, depending on the type of amylase (there are several of them), can cut starch chains apart or separate the branching points of amylopectin to expose maltose, which can be cleaved into glucose by maltases or α-glucosidases (Fig. 3.23). However, this can only work if the flour provides enough amylases. However, this is not always guaranteed, because cereals are a natural product and are therefore subject to the usual fluctuations, depending on rainfall, sunshine, temperature progression during growth, etc.

The strong production of fermentation gases by the microorganisms ensures the desired dough loosening. The dough significantly increases its volume. The glutenin network must withstand these high deformations. Good adhesive properties

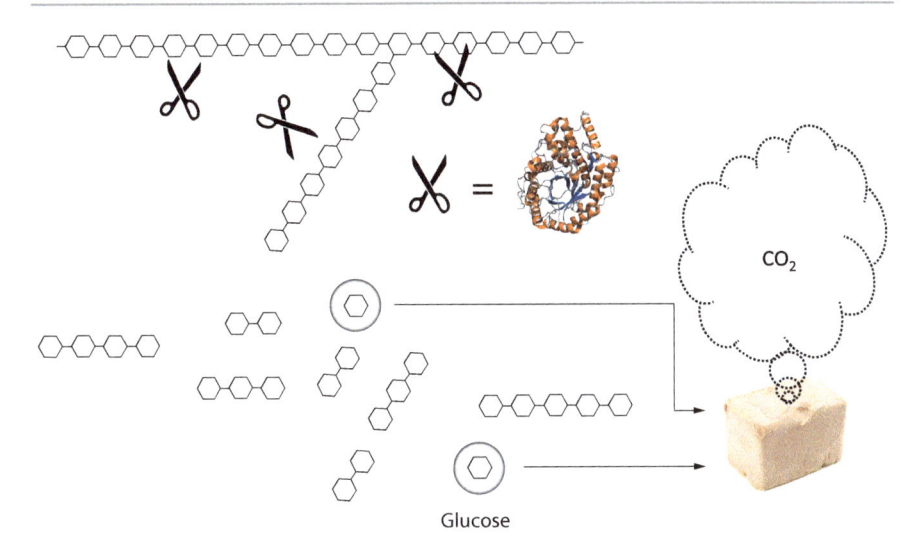

Fig. 3.23 Enzymes, here amylases, are proteins with cutting function, represented here as scissors. Different amylases can cut starch, i.e., molecular chains of glucose, at specific points until individual glucose molecules are present (shown circled). Only these serve as food for the yeast, which can then produce carbon dioxide (CO_2) for dough leavening

are expressed on a molecular scale through wide-meshed glutenin networks. These imply high elasticity, high tear resistance, and high extensibility on the macroscopic dimensions. The molecular chains must not tear during the leavening process, as shown in Fig. 3.23.

In tthe case of poor adhesives, for example when only a small amount of high-molecular-weight glutenin is formed, fewer entanglements can form, and the networks tear apart more easily. The dough can no longer withstand the high elongation. The network hangs "by the molecular thread" (Fig. 3.24, right), while the parts above and below the bubbles are hardly stretched. The dough then does not rise far enough. A strong glutenin network is also important during baking, as the gas bubbles expand more under temperature increase than the dough mass, putting even more strain on the network. Even then, the network must not tear, as the structure of the bread would be lost. This also applies to the pan breads made from wheat flour that have developed in many cultures, such as some flatbreads—Pita (Israel), Lavash (Iran), Yufka (Greece), Chapati, Naan, or Roti (India)—as exemplified in Fig. 3.25.

In these unleavened breads, only (highly refined) wheat flour and water are kneaded (in some regional areas, oil is also added), and the dough is then baked in thin flatbreads in very hot pans. The loosening of the dough is caused by the rapidly evaporating water and the associated strong gas expansion. Even with the quickly occurring deformations, the gluten must not tear before the baking process is completed. This is because the structure is solidified under temperature via disulfide bridges, as indicated in Fig. 3.26.

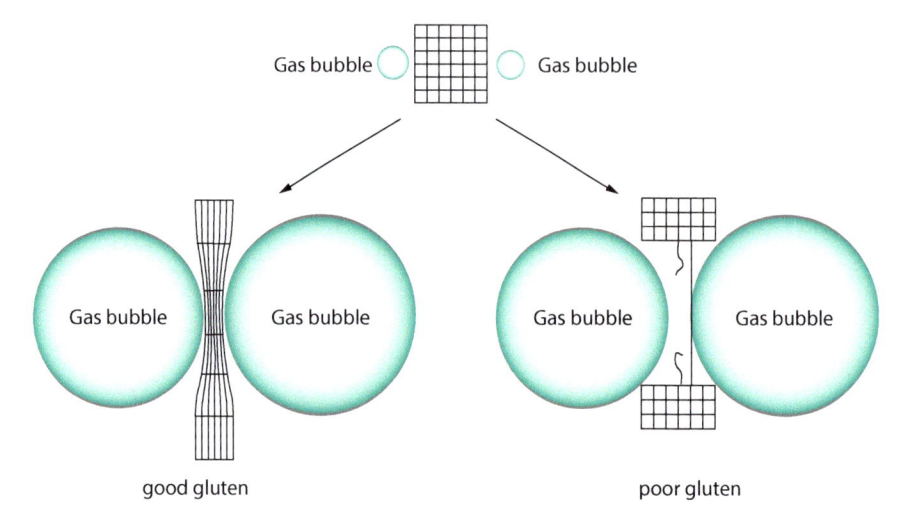

Fig. 3.24 The difference between good (left) and poor (right) adhesive properties of a bread dough is also evident during proofing when the glutenin network is heavily stretched

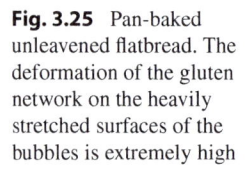

Fig. 3.25 Pan-baked unleavened flatbread. The deformation of the gluten network on the heavily stretched surfaces of the bubbles is extremely high

3.4.6 Heating of Gluten: Vulcanization of the Network

When heating the gluten, the gluten structure created during kneading is fixed. The reactive sulfur-containing amino acid cysteine comes into play, as it can chemically react (in Fig. 3.21 it is represented as a black circle). Cysteine is characterized by a side group of sulfur and hydrogen that can oxidize. Then the hydrogen is released and sulfur becomes reactive. When two cysteines meet, they can react at temperatures from about 65-70°C to form a sulfur bridge, a so-called disulfide bridge, and thus form a permanent chemical bond that is very stable. Cysteine can therefore cross-link proteins with itself or with other protein chains, as schematically shown in Fig. 3.26.

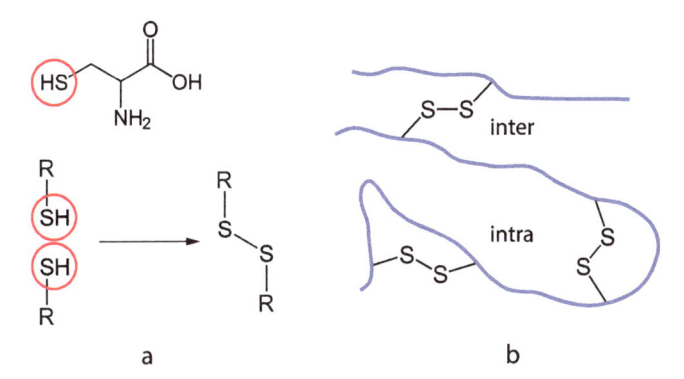

3.4 Staple Food Bread: Physics, Chemistry, Nutrition

Fig. 3.26 The formation of disulfide bridges: The hydrogen sulfide group -(S–H)- reacts to form sulfur bridges -(S–S)- (**a**) and can thus cysteine within a protein (intra), but also connect two adjacent proteins (inter) with each other (**b**)

The disulfide bridges are formed mainly during heating, i.e., baking, of the bread. At temperatures between 65°C and 75°C and an exposure time of about five minutes, almost all disulfide bridges are formed [50]. The difference between the various wheat proteins comes into play here. In the short gliadins, consisting of only about 300 amino acids, the reactive cysteine is predominantly in the middle of the protein chains and is already cross-linked. For cross-linking, the sulfur group in cysteine must be free and must not have been cross-linked before, in the native state, as in gliadin (Fig. 3.22). In gliadin, all cysteines are already cross-linked to disulfide bridges in the native state. This gives the protein a special stability. Therefore, gliadin is not directly involved in network formation in breads. Gliadin is thus only squeezed as a self-crosslinked globule between the large network meshes of glutenin.

Quite different with the longer, up to over 800 amino acids long glutenins. In these, the reactive groups are located at the ends and are therefore predestined for network formation in the dough. The ratio of gliadins and glutenins (Fig. 3.27) in the flour is crucial for the mechanical properties and thus for the crumb responsible. While the glutenins with their cross-links provide elasticity, the small, self-crosslinked gliadins act more like a soft lubricant and thus lead to high extensibility. The softness, elasticity, and extensibility of the baked bread crumb can be attributed to the interplay of gliadin and glutenin [51, 52].

Fig. 3.27 already illustrates how the ratio of gliadin to glutenin affects the network properties and thus the dough properties: If a lot of gliadin is present in a flour variety, it can even prevent the chemical reaction of two cysteines and thus the formation of disulfide bridges. The gliadin/glutenin ratio thus also defines the dough and baking properties.

These detailed considerations of the properties of gluten were necessary for several reasons. On the one hand, they show the fundamental influence on the

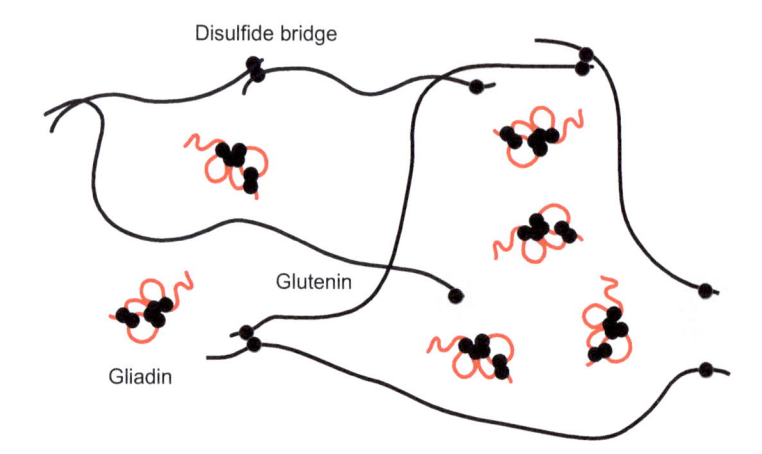

Fig. 3.27 Gliadin is already self-crosslinked and is squeezed between the meshes of the crosslinking glutenin. The still free cysteines are indicated with gray shading. For the reaction and the formation of disulfide bridges, they must come close enough

properties of dough processing and baking results. In addition, it becomes clear why leavened bread works well only in elastic doughs based on gluten, as hardly any other gluten-free cereals and pseudocereals are suitable with their proteins to withstand strong dough leavening. The long, high-molecular glutenins brought forth these cultivation techniques. Paired with the low pH value of sourdoughs, these inventions were a blessing. Sourdoughs were durable, cultivable, and the breads were also characterized by long shelf life. This was in demand in the Neolithic: processability with simple means, good acidification properties, and reasonable shelf life of the breads. These facts taken together show the worthwhile advantages of a lengthy breeding of *Triticum* towards increased gluten content.

3.4.7 Starch and Water Management in Baked Bread

Once the leavened dough has risen properly after proofing and final proofing, the breads are placed in the oven at high temperatures. Various processes take place here: On the one hand, physical changes in the proteins and starch, and on the other hand, chemical reactions set in at high heat, including the Maillard reaction, which flavors the bread.

Under heat, the protein network changes, it tightens even more, as significantly more cross-links via disulfide bridges can form when water partially evaporates or is also redistributed. Because when the temperature rises to 65–70 °C, the highly branched crystalline starch, amylopectin, melts. It is ready for water absorption, water can penetrate, and the starch becomes pasty. The water molecules are driven directly into the widely protruding starch branches after the amylopectin has melted and are bound there, which later, after baking and cooling, manifests itself as residual moisture in the bread and through the softness of the crumb of the fresh bread. The linear amylose can indeed form knots, entanglements and loops à la glutenin, but the polymer is not suitable for immediate and effective cross-linking. The highly branched amylopectin molecules cannot even form loops. They face each other like gears in a gearbox, but due to the high degree of branching, they cannot come too close or even intertwine and knot into a network, as shown in Fig. 3.28. Nevertheless, they fill the space available to them as best as possible, bind the water, and form a starch paste, they pasty or, anglicized, gelatinize.

Starch therefore forms a tough, pasty paste when heated and not a highly elastic gel. However, it binds a lot of water in hydration shells, as indicated by the blue shading in Fig. 3.28 [53]. The starch binds the water, the highly branched amylopectin molecules provide a moist paste for a loose cohesion and for the high water binding and thus the moisture retention of the baked bread. The elasticity of the baked crumb is mainly generated by the cross-linked gluten. No wonder that gluten-containing wheat- or wheat mixed breads, with whatever variations, are still a successful model for human nutrition today due to this sophisticated interplay of the macronutrients protein and starch.

3.4.8 The Taste of Bread

There is another reason for the millennia-long success of bread, namely its taste. This may sound trivial, but it is not, because the taste of baked bread, apart from the later added salt, has a considerable proportion of umami and sweet. In Chap. 1 the umami taste as a decisive driving force of cooking and roasting in the Paleolithic was already discussed in detail. Wheat breads seamlessly join as a consistent development in the Neolithic Revolution. Breads are baked at high

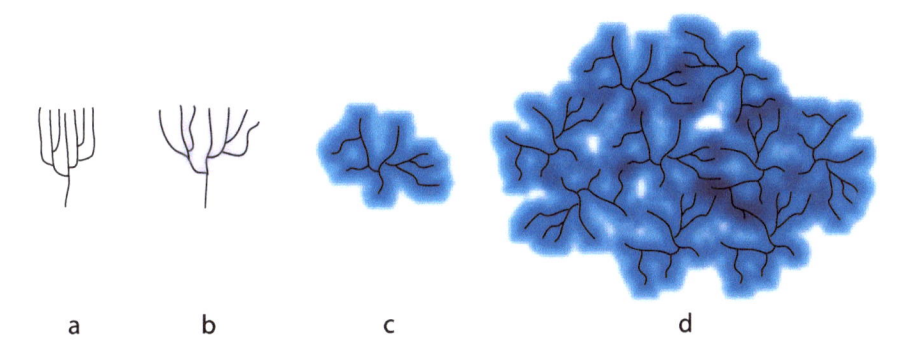

Fig. 3.28 The pasting of amylopectin. Before the crystals melt (**a**), no water can be absorbed. When melting (**b**), water absorption begins. The completely melted amylopectin binds a large amount of water (**c**). Swollen amylopectin cannot come too close due to the high degree of branching. Several molecules keep themselves at a distance because of their long branches, forming a starch paste (**d**)

temperatures in fire, in ovens of all kinds. However, these high temperatures only affect the crust, which quickly browns and emits characteristic, beguiling roasted aromas. Inside the baking bread, there is a high water content, which gradually evaporates a little. As long as water evaporates, the temperature cannot rise above 100 °C. However, it takes a certain baking time for the temperature to reach this level in the center of the bread. During this time and at temperatures up to 80 °C, the amylases are active and gradually release glucose from the starch, the more the starch melts. In the temperature range between 65 °C and 80 °C, amylases continue to be active and produce the preferred taste "sweet" in the form of glucose, which is immediately noticeable on the tongue and not only after the action of salivary amylases in the mouth.

The umami taste of bread, especially in the crust, is a direct result of gluten. Gluten protein contains a high proportion of glutamic acid, the amino acid that is largely responsible for the umami taste. During fermentation, a small part of the gluten is already cleaved in the dough by the enzymes of the lactobacilli. The peptides, fragments carrying a glutamic acid, begin to contribute more and more to the taste and mouthfeel with each step of further shortening.

At the bread surface, the cleavage process of proteins is greatly accelerated at high temperatures, as evidenced by the rapid browning. Browning substances and roasted aromas are formed from free, cleaved amino acids and free glucose molecules. Only free glutamic acid hardly participates in this Maillard reaction, but remains unchanged in the crust. Since it is ionic, i.e., negatively charged, it can combine with salts, which are present in every food, to form glutamate, which in turn dissolves in saliva and triggers a distinct umami sensation. No wonder, because the storage protein of wheat, gluten, contains up to 30% glutamic acid [54], which, as already noted, is also of great importance for the elastic properties and the partial swelling capacity in water. The umami taste of bread is also anchored in the crust. Paired with the roasted aromas and the sweetness, bread seamlessly fits into the taste patterns that *Homo sapiens* has preferred since the emergence of humans by the fire. Meat was given an adequate partner in the diet of early farmers with bread. In wheat-free regions and countries of this world, other cultural techniques developed to get the most out of the available grain.

3.4.9 The Early Regional Bread Culture

The previous explanations also make it clear why it is not possible to produce leavened breads with gluten-free grains such as rice, corn or pseudo-cereals such as millet: They cannot withstand the drive and the associated deformation. Yeasts and other leavening agents have enough sugar available, but the gas bubbles escape as they grow. There is no long and sufficiently elastic protein present that encloses the growing bubbles and prevents their growth or escape from the dough. Therefore, a completely different grain and bread culture developed in the regions where no wheat grew. The gelatinization of starch is sufficient for the cohesion of simple, less airy flatbreads made from gluten-free pseudo-cereals and water. A large part of the protein storage in corn is so-called zein proteins (zein and glutelin-2 (zeanin)) [55], which exist in various variations. In contrast to the largely unstructured proteins in gluten, the proteins in corn are present in extremely stable helix-dominated structures, as shown in Fig. 3.29 for example.

This compact structure and the strong hydrophobicity, which is given by the frequent occurrence of the amino acids alanine, proline, and valine, illustrate the main problem with corn proteins. It hardly unfolds, and the majority of corn storage proteins are not water-soluble [56]. Simple processing steps, as in the production of wheat dough, are not possible. They also cannot be easily denatured by heating. Another problem is the bioavailability of the amino acids, especially that of the essential amino acid valine. The protein, when added to the diet in its native form, is neither water-soluble nor denatured in the acid of the stomach. Proteases cannot easily reach the corresponding interfaces. The protein remains enzymatically indigestible, and the amino acids are not biologically available. In this form, these storage proteins contribute little to human nutrition.

Due to their strong hydrophobicity, zein proteins are only soluble in ethanol, ether, or glycol [57]. However, these solvents are of little use for the processing of corn, and they were even less useful for early humans in South America. Nevertheless, zein shows a significant increase in water solubility and thus the bioavailability of amino acids at alkaline pH values around 11.5. At alkaline pH values, zein denatures, dissolves much better in water, can even be processed, but tastes strongly alkaline and therefore soapy. However, the early farmers of the Aztecs already used these properties 1500 years BC in South America when the cultivation technique of nixtamalization was developed.

3.4.10 Nixtamalization—Ancient Grain Technology of Gluten-Free Cereals

The protein spectrum in corn is less diverse than in wheat. In addition to α-, β-, and γ-zein, very short-chain glutelins are crucial [58], which have nothing in common with glutenin and its structure-forming properties. Glutelins are already

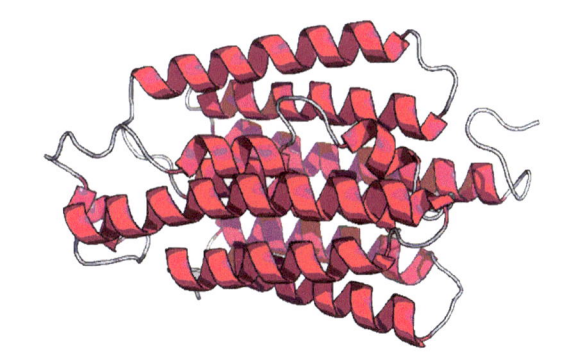

Fig. 3.29 Typical storage proteins in corn (zein) are highly structured proteins. Their secondary structure is predominantly dominated by helical helices

strongly linked to themselves in their native state in the corn kernel via sulfur bridges. Above all, the zein proteins of gluten-free corn require special treatment in order to be able to bake it at all, so that preparations such as tortillas (i.e., flatbreads) can be successful.

In this process, fresh corn kernels are boiled in lime water (made with ash) and soaked for several hours. (Slaked) lime raises the pH value of the soaking water to alkaline, the corn swells in lye, and the cohesion of the corn proteins is weakened in the pH range between 9 and 11. The zein proteins can partially unfold with the glutelins, dissolving the intra-disulfide bridges at pH values above 9, and later form a weak network after grinding and kneading. The treated corn kernels are then washed. They can then be ground wet and shaped into flatbread dough pieces. During baking, these proteins form a weak network, with the glutelins, which are also soluble in the alkaline range, supporting network formation. The corn starch, which gelatinizes simultaneously during baking, is then sufficient for the cohesion of the tortillas. However, leavened breads with leavening could not be baked with the nixtamalization process, as the proteins are not long enough.

The effects can be demonstrated strictly scientifically and systematically when zein and corn starch are cooked with lime (in this case, with pure calcium oxide) in different ways [59]. These experiments clearly show how the thermal denaturation of the zein proteins is adjusted when cooked with lime. While no thermal denaturation can be observed in native zein proteins, it is sufficient to hold the zein proteins at a high pH value at 61 °C for 20 minutes to denature them. The resulting pastes and doughs then also show significantly better mechanical properties. The doughs are more elastic and bind water together with corn starch very well.

Another advantage of nixtamalization is the breakdown of antinutritive substances that serve to ward off predators, such as chlorogenic acid. The main problem with corn is the high proportion of phytic acid complexes, which bind minerals and also vitamin B_3 (niacin) via their hydroxyl groups and no longer release them.

This shows a more fascinating connection: Many of the long-developed early cultural techniques related to food and food preparation are the results of very specific physical-chemical properties of natural products. What an enormous and noteworthy research achievement was made by cooking and eating people over 3000 years ago! The corn with its poor protein properties required a special chemical treatment through a systematic increase in pH value. This allowed for better processability on the one hand, and better availability of macronutrients and some micronutrients on the other, in the case of corn, niacin. Only these food technological methods ensured survival based on corn. In regions of the world where corn was not a staple food, nixtamalization was not developed. In wheat regions, these techniques are unnecessary, as the gluten develops during kneading simply by contact of the flour with water, and its broad amino acid spectrum is biologically immediately available.

In this context, reference should be made to deficiency symptoms such as pellagra, which can be found in regions that predominantly rely on corn as a staple food. Pellagra is a typical niacin deficiency disease, with symptoms including pigmented, burning or itching skin areas, as well as diarrhea, vomiting, and neurological failure symptoms. However, conventional pellagra only occurs in combination with a tryptophan deficiency, and the exclusive consumption of corn not treated by nixtamalization, due to the already mentioned amino acid pattern, promotes the development of pellagra.

The corn proteins, zein and glutelins, are significantly less valuable than the wheat protein mixture gluten. In all early and late Neolithic wheat-growing areas, such deficiency symptoms are not known from archaeological finds and analyses. No wonder, as gluten is one of those plant proteins that can contain all essential amino acids. A blanket demonization of gluten [60] and wheat [61, 62] is therefore, from the perspective of molecular nutritional sciences, a rather questionable trend. This will be discussed in more detail in Chap. 5.

3.4.11 A Look Into the Modern World of Additives Using Bread as an Example

3.4.11.1 Additives in Industrially Produced Breads

Wheat and its ingredients are subject to more or less large fluctuations like all natural products [63]. As with all products growing in fields and gardens, there are good and bad years. Every year after the harvest, the question arises how much gluten the wheat has, what the enzyme status is, and what its baking properties will be. These were essential questions for early farmers. Today, in the age of technical-scientific possibilities, help can be provided. For example, weaknesses can be compensated for by simply mixing wheat from different regions or vintages, by adding amylases, or by adding gluten.

Wheat contains its own enzymes, vitamins, emulsifiers, etc., but in fluctuating amounts. If the flour lacks sufficient amounts of these, additives can be used to compensate. Without these aids, a standardized and largely automated baking process would not be possible.

First, it is important to distinguish between the added substances. Furthermore, a distinction must be made between artisanal and industrially produced bread. A baker who kneads his dough daily in the bakery has a sense of how much water he needs, whether he should mix in another flour or change mixtures for mixed breads. Industrial machines (so far) are not that intelligent; they always require a very clearly defined flour-water ratio. But that's not all, because the adhesive properties, the starch content, the proportion of lipids and lecithin, and the enzyme status are crucial for the results to be reproducible after mixing and fermentation. Customers of mass-produced baked goods ultimately want an identical bread. Fluctuations are tolerated just as little as higher prices. If the flour does not have good or sufficient gluten, gluten is added until the flow properties of the dough, the elasticity, and the crumb stability are correct. The enzyme status can be determined in the same way. If it is too low, amylase can be added, which is not harmful to humans. In artisanal bakeries, amylase is used less often; instead, enzyme-active baking malts are added, which similarly enrich the flours with amylases. The germinated and dried grain is an amylase supplier and chemical factory *per se*; every beer brewing process depends on it. Germinated grains release significant amounts of amylases during germination, which then support the wheat's own amylases. Amylases thus help on two levels: They release maltose for the yeast, increasing dough leavening. Later during baking, di- and trisaccharides react with amino acids to form Maillard products, which create the characteristic bread aroma.

The function and properties of these additives are well known. If ascorbic acid and ascorbates is added to industrially produced bread, which the wheat already contains in varying amounts, is neither a problem for health nor for ethics. It is not much anyway, as any overdosing far beyond the natural amounts would seriously disturb the proofing process.

3.4.11.2 The Effect of Emulsifiers

Emulsifiers are usually used in emulsions such as creams and mayonnaise for stabilizing the oil droplets in the aqueous environment. In this process, the emulsifiers attach themselves to the many interfaces between oil and water. Therefore, it seems counterintuitive at first glance to add them to bread doughs, in which fats and oils do not play a significant role. However, bread like any food is strongly dominated by interfaces. The fluffy bead crumb, for example, consists mainly of interfaces that form between baked dough and the many gas bubbles. The bubbles take on the role of oil, as air is, like oil, hydrophobic. Different flours contain varying proportions of the emulsifier lecithin. This is normal, as lecithin is one of the most important components of all cell membranes, whether animal or plant. A multitude of phospholipids (lecithin) self-organize, forming dense and self-contained cell membranes and separating extra- and intracellular spaces. This self-organization occurs for purely physicochemical reasons, as lecithin, a mixture of phospholipids, has a water-soluble head group and two fatty acid tails that are exclusively fat-soluble. Fat and water are thus united in one molecule. Just right for the foam structure of bread: The hydrophilic part is located in the water-rich dough, while the fatty tails of the lecithin protrude into the gas bubbles. Other natural lipids such as mono- and diglycerides, which, contrary to many opinions [64], occur in every animal and plant fat metabolism, also help with stabilization. Again, it is obvious to supplement emulsifier-poor flours from critical years with lecithin to ensure a controlled baking process. Such ways of thinking are not widespread among traditional bakers, as their individual clientele accepts fluctuations in qualities more readily than customers who buy baked goods of industrial origin in supermarkets or baking factories at very low prices.

3.4.11.3 L-Cysteine—Molecular Manipulations in the Gluten Network

The sulfur-containing amino acid L-cysteine is essential for the structure and stability of the adhesive network, as shown in Fig. 3.25 and 3.26. The high-molecular glutenin, in which the cysteine is located more at the ends of the long protein chains, defines the mesh size of the network during cross-linking and thus the elasticity of the crumb and consequently the mouthfeel when eating the bread. Poor gluten is characterized, for example, by a high gliadin content or low concentrations of high-molecular, but higher concentrations of low-molecular glutenin. These are not good prerequisites for the stretchability and elasticity of the doughs. However, there is a physically obvious trick to increase the mesh size of the network and thus the stretch and load capacity of the adhesive network: The addition of very low concentrations of the amino acid L-cysteine during dough preparation has many effects. It reacts during kneading and heating with some cysteine-occupied ends of the glutenin molecules, so that these cannot react directly with neighboring other proteins. The consequence is clear (Fig. 3.30): The protein network becomes more open-meshed, stretchable, and can better withstand the drive during fermentation despite less high-molecular glutenin.

Free cysteine amino acids, as they occur in small amounts in every flour, combine with those in the chains and prevent further closing with other chains, making

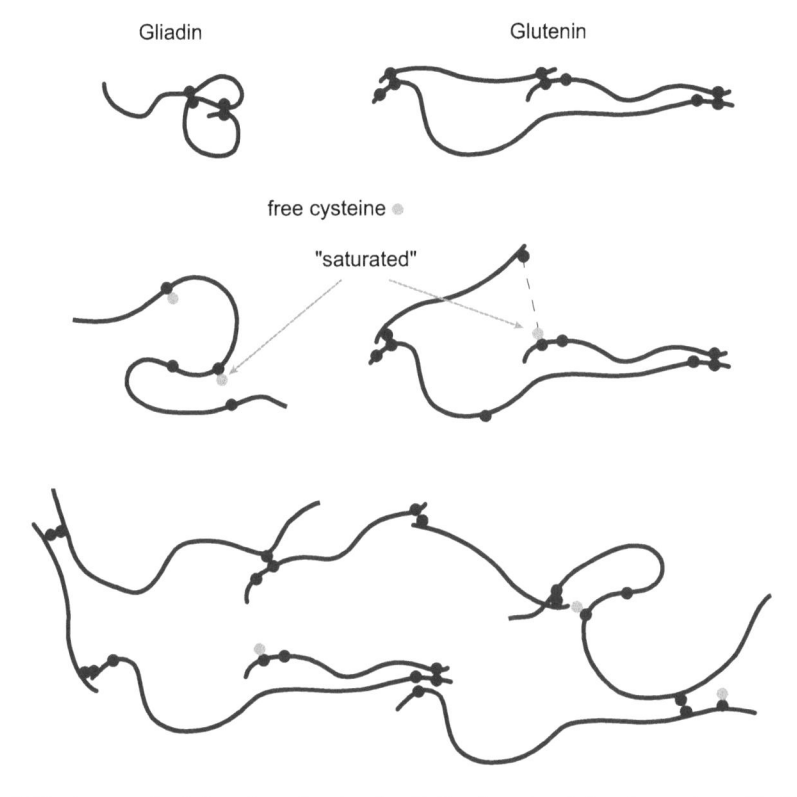

Fig. 3.30 A network of glutenin molecules. In gliadin, the network-forming amino acids (black circles) are more internal. They network themselves, in glutenin the cysteines are at the ends, forming disulfide bridges (sulfur bridges) that hold the network together. If free cysteine is added (gray circles), they saturate bound cysteine, the protein chains can no longer network there, the network remains more open-meshed. Even gliadin can then contribute to the elasticity

the gluten softer and more deformable. It is already clear: Apparently, the number and effect of permanent connections between the chains and thus the adhesive properties can be controlled by vanishingly small additions of cysteine (in ppm—*parts per million*).

The amino acid is obtained from cysteine-rich proteins, which are otherwise rarely used. This includes all structural proteins that get their shape through sulfur cross-linking, such as proteins in skin, hair, including animal hair and pig bristles. Headlines like "Hog's bristles and human hair in bread" [65], are of course nonsense. Cysteine from human hair must not be used in food. And the claim that pig bristles are in bread is, of course, also a completely unqualified assertion. For the function and properties, it does not matter at all where the amino acid cysteine comes from. The utilization of pig bristles and poultry feathers would even be entirely in line with the ancient, but currently practiced again, nose-to-tail movement, so even a very good idea. Even the cysteine from human hair would theoretically be no problem. With the tons of hair that accumulate in hairdressing businesses, a natural raw material source practically ends up in the trash. The same applies here: Only the structure determines the function.

3.4.11.4 Non-natural and Wheat-Alien Additives

The previously discussed additives used are in harmony with the diversity of molecules that wheat already contains and are therefore harmless to health. However, the industrialization goes many steps further, as a whole range of flour treatment agents have nothing to do with the ingredients of the cereals, which cannot all be discussed in detail here.

First and foremost are non-natural, i.e., synthetic, emulsifiers such as tartrate esters, sugar esters, lactate esters, etc., which are not provided by nature but are repeatedly used in industrial baked goods. However, the basis for these non-natural emulsifiers is natural fats (edible fats) and likewise natural products such as tartaric acid, acetic acid, lactic acid, or various sugars and sugar alcohols, which are esterified with mono- and diglycerides. Depending on this, one speaks of tartrate esters (**DiA**cetyl **T**artaric **E**ster of **M**ono- and Diglycerides, DATEM), acetate esters (ACETEM), or sugar esters. The principle of the structure is shown in Fig. 3.31.

The emulsifiers on the left side in Fig. 3.31 are natural; they occur in nature and are used by plants and animals through metabolism. On the right are examples of artificial (better synthetic) emulsifiers. They do not occur in nature.

The question is, of course, justified why the "*Homo industrialis*" comes up with the idea of using natural or synthetic emulsifiers in a staple food like bread. As already mentioned, wheat or other cereals and pseudo-cereals bring a certain amount of natural emulsifiers with them. The two natural emulsifiers represent, from a physicochemical point of view, two extremes. Mono- and diglycerides are only limitedly water-soluble due to their fatty acid dominance and the very weakly polar glycerol residue group. They have only one or two free OH groups at the positions where the fatty acids are missing. Thus, these emulsifiers are highly fat-soluble. In contrast to lecithin. Its head group of choline and a phosphoric acid is

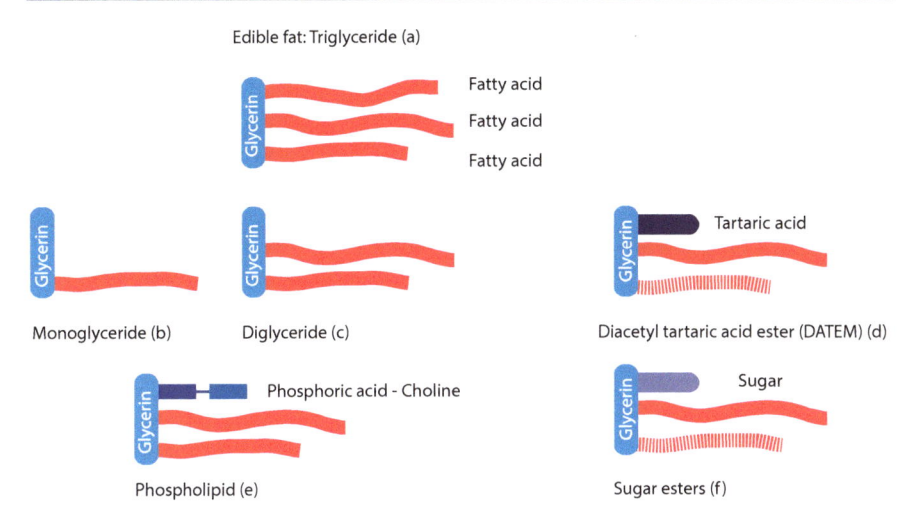

3.4 Staple Food Bread: Physics, Chemistry, Nutrition

Fig. 3.31 The structure of various emulsifiers. Above is a triglyceride (triacylglycerol; **a**), a fat with the glycerol or glycerin backbone and three fatty acids. When fatty acids are enzymatically separated (by lipases), mono- and diglycerides are formed (**b, c**), lecithin replaces a fatty acid with a phosphoric acid and a choline group (phospholipids, **d**). Synthetic emulsifiers replace a fatty acid with a water-soluble molecule, for example, tartaric acid (**e**) or various sugars (**f**) or sugar alcohols.

highly polar and therefore highly water-soluble. Thus, the hydrophilic head group itself pulls fatty acids with 18 or 20 carbon atoms into the water, while the fatty acids of the mono- and diglycerides pull the weakly polar group of glycerol into the fat. The properties and modes of action of the two emulsifiers are therefore completely different.

For the baking industry (and other applications), it is therefore desirable to have emulsifiers with properties between these two extremes, i.e., head groups with different water solubilities, in order to meet the multitude of interactions in bread dough at the physicochemical level. Then, the interfaces between gluten–air or the differently hydrophilic-hydrophobic amino acids, the different properties of the surfaces of the starch granules can be mediated more finely. This is precisely achieved by synthesizing fats with head groups whose water solubility is different. In this way, emulsifiers with a very targeted **H**ydrophilic-**L**ipophilic-**B**alance (expressed by the so-called HLB value) can be created [66].

Different emulsifiers are useful from a physical point of view already during kneading, as the bread dough is always a multiphase system, the interactions of the individual components of which are in competition with each other. The opposing interactions between hydrophilic and hydrophobic have an effect on the formation of the crumb structure during baking. The bubble structure of the crumb is solidified by the baking and the associated processes such as water withdrawal, starch gelatinization, protein cross-linking, etc. In the case of dough without additives, at home or at the artisanal (organic) bakery, these tasks are taken by are the lecithins present in the flour, as well as mono- and diglycerides. Depending on the flour properties, additional emulsifiers can be added to industrial flours.

The interactions of the components are also strongly temperature-dependent and change during baking. Each increase in temperature causes an increase in thermal energy and thus a weakening of non-covalent bonds: hydrophobic and hydrophilic bonds become weaker. On the other hand, additional permanent disulfide bonds between neighboring gluten molecules are formed. The strength is increased, and the bubble structure hardens. At the same time, the small phospholipids remain mobile and maintain their emulsifying effect. They also positively influence the crumb structure at higher baking temperatures. Due to the hydrophilic areas of the emulsifiers, water droplets are also better retained in the dough. Water in the dough means higher moisture during baking. This also benefits a uniform crumb structure and aroma formation. Especially in wheat with poorly developed gluten properties, emulsifiers can help. They compensate for deficits that the wheat brings with it in poor harvest years. The emulsifiers ensure good dough volume and good baking properties.

The dough yield is also larger with the help of emulsifiers. This baking technology term describes the measure of the amount of dough that is produced when 100% cereal products are mixed with a certain amount of water or other bulk liquids. As always, industrial bread can be eaten despite the additives, but it doesn't have to be, depending on personal preferences. When deciding whether adding emulsifiers is good or bad, facts help more than belief. Even artificial substances like DATEM are harmless in the amounts used.

I can't tolerate that—about "Slow Baking", bread machines, and industrial bread.
Recently, there have been increasing reports of intolerances to cereals and bread [67]. This refers to irritations such as mild intestinal rumbling or flatulence, which are associated with the consumption of bread [68]. Not whether wheat makes you dumb, as some popular book titles suggest, but whether wheat actually promotes disease, is the subject of various scientific disciplines [69]. Whether modern gluten "conglutinates" the intestine can be examined in detail using serious scientific methods [70]. The latter is, of course, not universally valid, as the high-quality proteins of gluten are broken down into peptides in the stomach by the enzyme pepsin and then cleaved into peptides and amino acids by the pancreatic enzymes trypsin, chymotrypsin, and peptidases, which are then supplied to the physiology. However, this process depends on the individual enzyme equipment of each individual. It is written that in the past, and even earlier, in the Neolithic, everything was better and modern wheat is highly overbred [71], so that we can no longer digest it [72]. However, this does not correspond to the scientific data situation [73]. Humans have been breeding wheat since time immemorial, wisely and purposefully, so that they can bake and consume bread.

Many irritations have less to do with the storage proteins, some with the baking process and the wholemeal flour, and with completely different classes of molecules [74], which are in focus in detail in Chap. 5 when it comes to the question of gluten-free products. While people have long been careful to separate the chaff from the wheat and to take as exclusively as possible the macronutrients starch and protein from the endosperm, the doctrine of whole food nutrition held the view that postmodern humans consume too few dietary fibers and should therefore eat the whole grain [75]. This is positive for the microbiome in many respects. However, whole grain bakers buy, in addition to soluble and insoluble dietary fibers such as the macromolecules cellulose, pectins, hemicellulose and cellular mucilage (hydrocolloids) and minerals, a whole range of new ingredients that can cause irritations, gas, and rumbling during the gastrointestinal passage in people with a predisposition for such issues. The enzymatically indigestible components are fermented in the large intestine, depending on the state of the microbiome, to a greater or lesser extent. This is basically a desired process beneficial to the microbiome, but it also leads to some irritations.

Grain does not want to be eaten, neither by humans nor by the predators in the field. The grain protects itself from this, as well as from drought stress and intense sunlight. When it gets too hot and dry, plants have to make do with little water, and they must not be smashed by hail or storm. The grain provides for this, and it grants special protection to its offspring, its seeds, which we consume as grain. This happens in several stages. Each grain is therefore packed in several layers of the aforementioned macromolecules. That is why grains also contain many divalent ions, known as minerals such as calcium and magnesium, which strengthen the cohesion, for example in pectin, through ionic bonds and later negatively affect the baking properties, as all whole grain bakers experience with their own dough when it tears during baking or does not form a crumb with good consistency.

Following this are the seed coat, seed shell (testa), hyaline membrane and the aleurone layer with proteins and lipids. At the top sits the germ with all the structures for the first growth processes, i.e., proteins, enzymes, which are specially protected. Only then does one penetrate to the actual substance: the flour body with the storage proteins gluten and the starch granules.

For the protection mechanisms and defense against predators, lectins are incorporated (which we will discuss in more detail in a completely different context), for effective frost protection by means of freezing point depression and as "water retention molecules", small and medium-sized sugars (oligosaccharides and -fructoses) are available. Depending on the plant, these include fructoses, lactoses, fructans, galactans, and sugar alcohols (polyols). With their help, the plant can manage its water balance even in stressful situations. These small molecules, which are hidden in significant amounts in and around the cells, are summarized under the term FODMAPs (**F**ermentable **O**ligo-, **D**i-, **M**onosaccharides **A**nd **P**olyols) summarized.

The common point of these various substances with different chemical structures is their resistance to enzymes from the pancreas during the gastrointestinal passage. They can therefore, unlike starch, not be broken down into their individual components and must first be degraded and fermented by enzymes released by certain bacteria, fungi, or viruses of the gut flora of the microbiome. Fermentation is indeed the correct term, because just as yeast ferments sugar into alcohol and carbon dioxide (CO_2) or lactobacilli ferment lactose into lactic acid and CO_2, this happens in the intestine. The resulting gas cannot escape through the intestinal wall and accumulates in the intestine, taking up a large volume, which can lead to bloating. With the multitude of intestinal bacteria, not only the odorless CO_2 is produced, but also methane (CH_4) and sulfur gases, which smell like rotten eggs and are perceived as fecal odors. As the pressure increases, they find their way out through flatulence. This is not dangerous. Sugar and sugar alcohols (sorbitol, mannitol or xylitol) have an additional effect: they have different osmotic effects, i.e., as the sugar alcohol concentration in the intestine increases, water is bound. At the same time, water flows into the intestine through osmosis. The stool volume increases, as does the intestinal movement (peristalsis). Such polyols can therefore have a stool-promoting effect in sensitive people. This is not dangerous either, but very unpleasant.

Some bread consumers react to such FODMAPs and complain about bloating and flatulence. And this has been increasing in recent years. Of course, this is usually attributed to the "evil" industry and the baking mixes with their many additives. Even the German weekly newspaper *Die Zeit* suggests in its magazine that there are no good bakers anymore. In a bread today, so many additives are baked in that even the ducks in the city park won't eat it anymore. As a remedy, simple home baking with bread machines was suggested [76]. This, too, is a misconception.

One reason for the poorer tolerance of bread compared to ancient times is indeed the FODMAPs in whole grain flours. These molecules are not more prevalent in plants and seeds than they used to be, but they remain—unlike in white flour—in whole grain bread. This is because many industrially baked breads are no longer fermented long enough. The fermentation times are often much too short. Yeast breads "rise" only for a short time, and there is no time for the classic three-stage sourdough in fast-bake shops. However, this would be important, as during this fermentation, FODMAPs are fermented by the microorganisms of lactobacilli into aromas or flavor precursors. They therefore do not need to be fermented in the intestine and do not cause any irritation. The acidification has another advantage: The low pH value stimulates the production of enzymes in certain sourdough microorganisms that break down antinutritive substances such as phytin and phytic acid.

Bread is usually well-tolerated even in modern times, provided that the germs of sourdough and yeasts are given enough time to break down these antinutritive substances [77]. In Fig. 3.32, the measured degradation of FODMAPs for a classic whole wheat yeast bread is shown. It can be seen that the concentration of all three different types of FODMAPs—Raffinose (black) as an indigestible trisaccharide, Fructose (red) and Fructans (green)—have decreased to an irrelevant level after 4.5 hours of proofing. It is also striking that the increase in fructans and fructose occurs during a one-hour proofing. Usual fermentation times for yeast doughs of about one hour even increase the proportion of some FODMAPs. This is easily understandable, as many fructoses are initially glycosidically bound to other molecules, which can be detached in the initial phase of fermentation. Thus, fructose

3.4 Staple Food Bread: Physics, Chemistry, Nutrition

Fig. 3.32 The decrease of some wheat-relevant FODMAPs with the proofing time in hours. The black curve describes Raffinose, the red curve Fructose and the green curve Fructans [77]

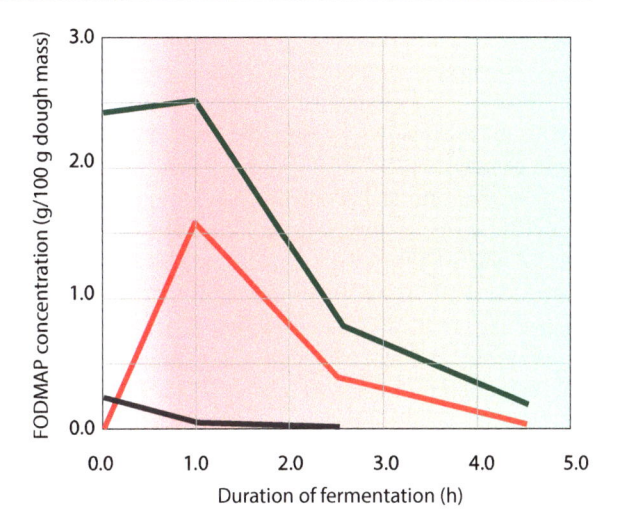

is only released during short fermentation times and is only broken down by the enzymes of microorganisms after long periods.

In this context, it is interesting to look at the molecular structure once again to better understand these things. The FODMAPs are very systematically structured, as shown in Fig. 3.33. They consist of the sugars Glucose (Glc), Fructose (Fru) and Galactose (Gal), which is already known from Lactose. Raffinose, Stachyose, and Verbascose, all also known from various vegetables and legumes, including soybeans, are merely extensions of sucrose by one galactose unit each. The chemical bond between the galactoses as well as between galactose and glucose is different from that between glucose and fructose. Many people do not have enzymes

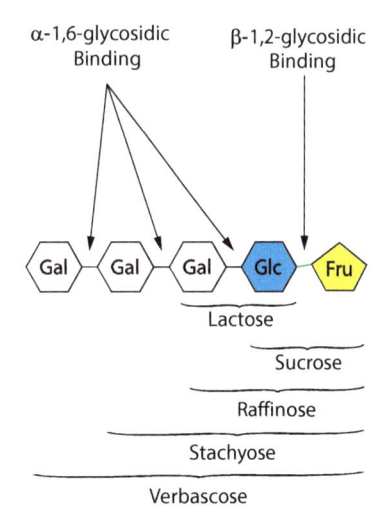

Fig. 3.33 The chemical systematics of some indigestible and bloating sugars

for their cleavage (Chap. 2). Consequently, these sugars must be fermented, resulting in bloating due to the fermentation gases produced.

However, microorganisms from yeasts and sourdough produce enzymes to cleave these units. As a result, they no longer end up in the intestines of bread and grain eaters.

Therefore, rapid baking machines, with which one can bake "healthy" whole grain bread, help only a little. What these machines cannot do is to sufficiently reduce the FODMAPs with too rapid fermentation. Furthermore, it should be noted: In white flour, there are hardly any FODMAPs present. Therefore, this is always, as far as intestinal rumbling is concerned, more tolerable for some people.

However, FODMAPs and their effects are not particularly bad or even life-threatening. It should also not be forgotten that many representatives of FODMAPs were considered prebiotic just a few decades ago (which they still are today). In fact, FODMAPs train the microbiome and expand it through their presence and their need to be fermented, by adding more bacteria and fungi. This is basically positive, as the systematic increase of acid-tolerant lactobacilli contributes to a lowering of the pH value in the colon and protects against a strong colonization of pathological germs that can cause inflammation. Not every flatulence that occasionally bloats the intestine is an irritable bowel sydrome or even an allergy, but rather a positive and natural sign that something good is happening in the microbiome. These facts will be addressed again in other contexts later on, as FODMAPs have a prebiotic effect and contribute in a very special way to the preservation of the microbiome.

3.5 Food Spoilage and Preservation Methods

3.5.1 The Desire for Preservation

Early preservation methods such as drying and fermentation ultimately accelerated industrial processes. This approach reflected people's desire to have food safely available beyond the season. Preservation methods also offered strategic advantages. Rulers provided security for the people and ensured armies their supply. Political and military power were also driving forces of an early industrialization of food. Drying methods were early preservation methods and allowed Huns and other steppe peoples extensive rides and conquests [78]. Thus, it is no wonder that Napoleon Bonaparte, as early as 1795, when he was appointed General of the Army, offered a prize for the development of preservation measures [79]. The supply of the army was ensured after the Parisian chef and confectioner Nicola Appert succeeded in achieving a longer shelf life by heating and bottling in sealed bottles [80]. The prize was his, and noteworthy in the laudation was not only the verifiable shelf life, but also a satisfactory preservation of taste and seasoning over several years. These works were forgotten until they were picked up by Louis Pasteur, further developed scientifically, and published in 1866 [81]. The path to the canned food was thus not far, as advances in metal processing allowed hot and thus pasteurized food to be quickly sealed airtight in metal cans. The biochemical durability, the preservation of taste for a long time, and safe transport were thus ensured [82].

3.5.2 Drying, Canning, and Sterilizing

The early developed methods such as drying, Pasteurization, optional acidification, and subsequent banishment of oxygen are still important foundations of all preservation techniques, be it on a small, private scale or on an industrial scale. For a long time, preservatives and methods have been used: For millennia, the smoke of fire was used to preserve, fermentation was also one of the first achievements in preservation, and new methods were constantly developed. Already 1200 years ago, fish were dried, preserved, and eaten at the appropriate time by the Vikings in Haithabu [83], fish were and are heavily salted, as is common today with stockfish [84, 85]. These methods are also applied to meat. With the handling of acetic acid bacteria, weakly alcoholic beverages could be fermented into vinegars, which in turn still serve as preservatives today [86]. Another long-practiced form of preservation is cooking in fat followed by marinating in the cooking fat; to this day, gourmets enjoy confit ducks, fish, or vegetables.

With the availability of sugar, new preservation methods came into fashion. Already 2000 years ago, archaeological traces were found that indicate the preparation of jam-like preparations. The first written mentions of canned, heavily sugared fruit preparations date back to the Middle Ages.

With the Industrialization, the post-war period, and advanced food technology, these ancient cultural techniques receded into the background. The refrigerator allows for strict temperature and humidity control to this day, even sensitive vegetables and meat have a longer shelf life. Furthermore, modern shock freezing methods [87] enable almost non-destructive freezing of water-rich foods for several months. Vacuum sealing, which involves a significant removal of atmospheric oxygen, also gained increasing importance. Or even freeze-drying, in which (shock-) frozen foods are deprived of water through sublimation processes under vacuum.

The development of various methods in the respective cultures is, of course, related to the local living conditions, the food supply, and the respective preferences. A second argument should not be forgotten: the close coupling of the developed methods to the prevailing type of bacteria, which naturally also depends on the respective climate zones. This is also evident in the different temperatures at which bacteria are killed or inactivated. Various types of bacteria can grow in the entire temperature range between 0°C and 100°C. For food, only the range between 0°C and 75°C is relevant. Below 0°C, food-relevant bacteria cease to multiply, and above 75°C, virtually all disease-causing germs are harmless and destroyed. An exception is the special bacterium *Clostridium botulinum* from the family of Clostridia, which (usually in canned food) produces neurotoxic proteins and whose spores only become harmless at temperatures above 120°C (Fig. 3.34).

Since the temperature increase affects the disintegration of cell membranes and the denaturation of proteins and enzymes in the cell membrane, the duration of temperature exposure is also important. Time and temperature play a crucial role in killing pathogens. Germ reduction does not occur immediately but only after a certain time. It follows an exponential law. Therefore, heat must always be applied for a certain time until the number of germs is reduced to a certain level, and the remaining germ count can no longer cause any damage [88]. This can be quantified using the decimal reduction values, the D_T values. The decimal reduction value, $D_{T=t\,min}$, describes how many minutes (t) are necessary at a temperature (T) to reduce the number of germs to one-tenth of the initial value.

Another difficulty is that for many germs, not only time and temperature are crucial, but also the environment in which they grow. Thus, the time required for the decimation of staphylococci at 60°C varies depending on water activity and thus the moisture of the food, ranging from one to 20 minutes. While the decimation of listeria in skim milk at 62°C takes only two minutes ($D_{62=2\,min}$), it takes up to an hour at the same temperature in meat ($D_{62=60\,min}$). It is indeed not easy to provide accurate values for killing in every food. Simple rules of thumb are listed in Fig. 3.2 (Table 3.2).

The example of meat and milk shows how several hurdles must be overcome to kill germs. Time, temperature, and water activity are just three of several such

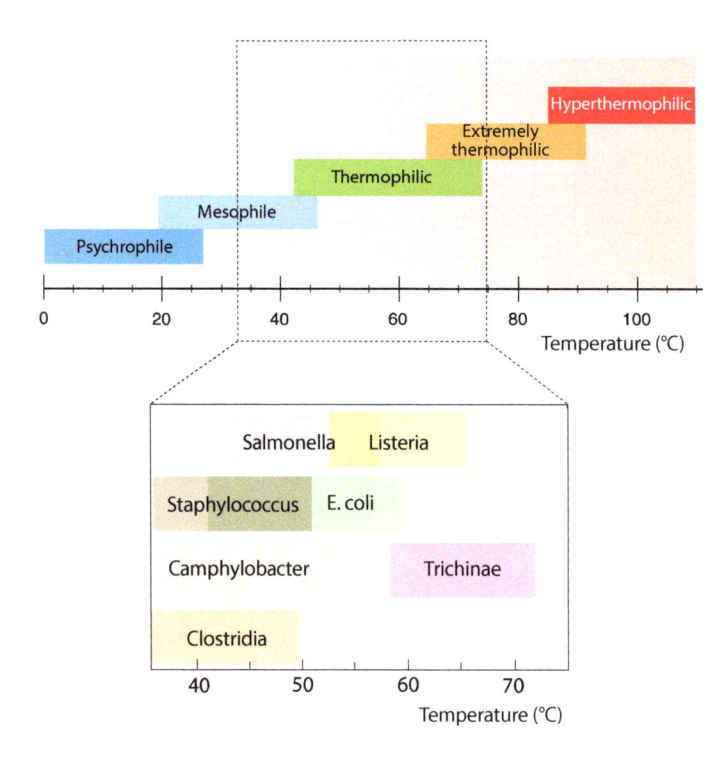

Fig. 3.34 The most important food-relevant germs that are problematic in food. Above 75°C, the germs mentioned here are harmless. *Clostridium botulinum* is not explicitly considered in this illustration

3.5 Food Spoilage and Preservation Methods 165

Table 3.2 Sterilization of germs depending on time and temperature

Germs	Temperature in °C	Time in minutes
Listeria Staphylococci Salmonella	62	30
Viruses Molds Yeasts	80	30
Bacteria *Clostridium botulinum*	120	15
Fungal spores	100–120	30

hurdles. Humanity has quickly learned from serious mistakes that led to poisoning, illness, and death how to make food safe.

3.5.3 Hurdle Concept and Barrier Theory

All these seemingly completely different cultural techniques are based on a strong principle: germs must overcome at least one, if not several hurdles in preserved foods in order to multiply at all [89]. The possible hurdles (Fig. 3.35) are: heating to kill germs, cooling and reduced oxygen supply to prevent germ growth, a low pH value to kill germs, a reduction of water activity to deprive germs of their basis for life, and, if all that does not help or is not possible due to certain food requirements, the addition of preservatives.

Food safety has become a duty to guarantee food safety. However, the discussions about raw milk and raw milk cheese also show societal discourses [90]. It is therefore no wonder that with increasing industrialization and the ever more advanced knowledge of chemistry, preservatives were developed [91], some of which are antibacterial but are also not particularly healthy when ingested in higher doses over a longer period of time through preserved food.

Due to the progress of basic research and food technology, the shelf life of food has been greatly extended [92]. Be it through preservatives such as ascorbic acid

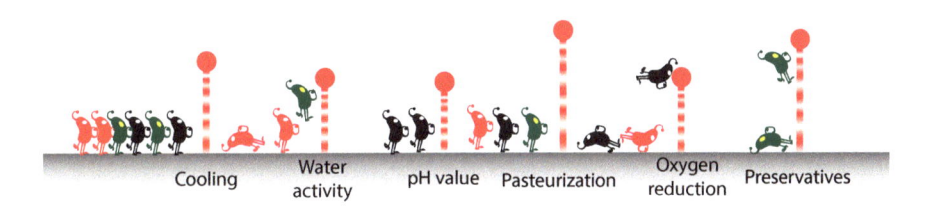

Fig. 3.35 The hurdle theory in food preservation knows six common hurdles to make as many different bacteria as possible fail

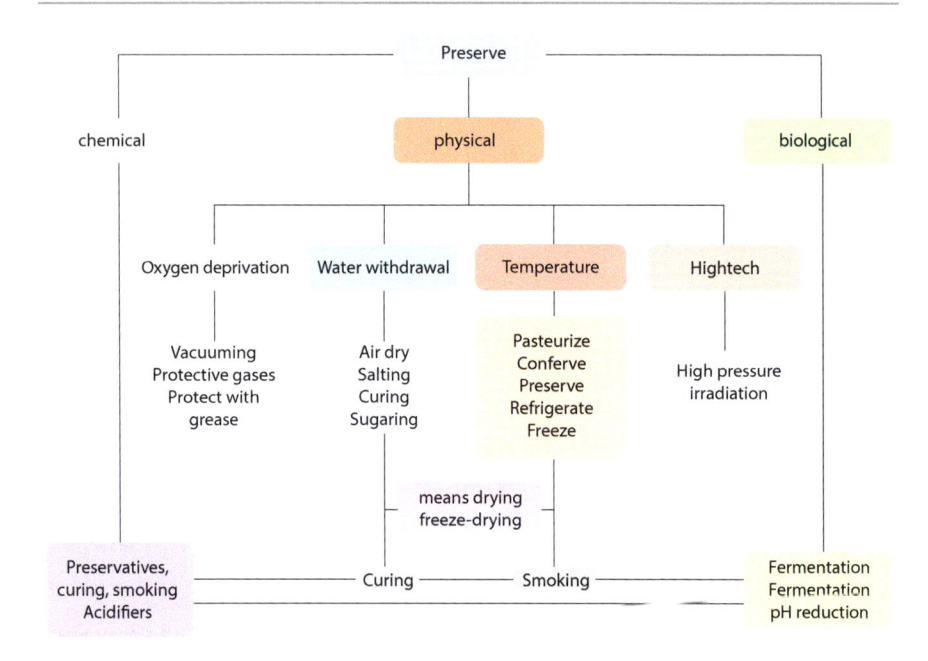

Fig. 3.36 Various preservation possibilities include a wide range of physicochemical methods

(vitamin C), through acids such as malic acid, acetic acid, lactic acid or citric acid, or through phosphates, rosemary extracts (carnosic acid and carnosol), natamycin (an antibiotic), benzoic acid or even butylhydroxytoluene. These are substances that are repeatedly criticized and whose health problems are discussed. The last three mentioned preservatives play hardly any role in industrial food production due to lack of acceptance. Therefore, purely physical methods such as high-pressure treatment (pascalization) or irradiation with UV, beta, or gamma rays are increasingly coming to the fore [93]. The variety of different methods for increasing food safety is roughly summarized in Fig. 3.36.

It is not always easy to clearly separate the individual processes. This becomes apparent in the case of curing. Despite the fundamentally chemical nature of curing, physical phenomena related to the ionic effects of salts are at the forefront. Smoking also, depending on the smoking temperature, not only causes chemical changes but also brings about a series of physical and physicochemical effects.

3.6 Misunderstood Preservation Methods

Rarely are the "how and why" of preservation methods questioned in popular publications [94]. Therefore, the effects and methods of the following three examples will be examined more closely.

3.6 Misunderstood Preservation Methods

3.6.1 Example of Misunderstood Industrialization—Liquid Smoke

Since the harnessing of fire, smoke has been used as a chemical preservative mixture to preserve food. However, smoke has strong health effects in addition to its preservative and aromatic properties. Real smoke consists not only of a multitude of molecules but also of solid particles, partially burned fine dust particles, soot particles, heavy oils from polycyclic aromatic heterocycles (PAH), micro- and nanodroplets (see e.g. [95]) to be removed.

Today, modern techniques make it possible to filter smoke and process it in such a way that only the ingredients that are actually necessary remain. These would include, for example, formaldehyde for preservation, carboxylic acids with their antibacterial properties, and typical odorants from smoke for aroma. For this, no solid particles, no carcinogenic polycyclic hydrocarbons, or even fine dust particles are needed. To achieve this, the smoke must be passed through water or oil and run through various filters, centrifuges, and other chromatographic separation processes. The purified residual smoke is then liquid—which leads to gross misunderstandings. It is referred to as "artificial" or "fake" or even more harmful than "real" smoke [96, 97], as well as by some organic associations [98] and enjoyment organizations [99]. As a result, liquid smoke is banished as industrial smoke, and old traditions are adhered to instead.

3.6.2 Example of a Misunderstood Water Binding with Polyols—Sorbitol

Water-binding auxiliaries, which are used to reduce water activity, are repeatedly criticized [100]. These include sugar alcohols, including those polyols that we have already encountered as FODMAPs.

The idea of water binding via such sugar substitutes is inspired by nature, as in plants, for example, sorbitol is incorporated. Of course, nature takes into account two physical conditions: there must not be too much glucose, fructose, or sucrose. Due to osmosis, cells would burst even at low concentrations, and the plant would suffer damage even without frost. At the same time, water should still be strongly bound. Therefore, molecular structures deviating from conventional sugars must be incorporated, such as sorbitol, which binds significantly more water, as symbolically represented in Fig. 3.37.

It turns out that trehalose, a disaccharide from fungi, and sorbitol have a significantly better water-binding capacity compared to sucrose [101, 102]. The decisive factor is not only the number of OH groups but also their accessibility. The examples of sucrose and trehalose show the different orientation of the OH groups concerning the sugar axes. Sucrose has axially (a) and equatorially (e) oriented OH groups, while trehalose has only equatorially oriented OH groups, which are more easily accessible and can more easily gather water molecules around them.

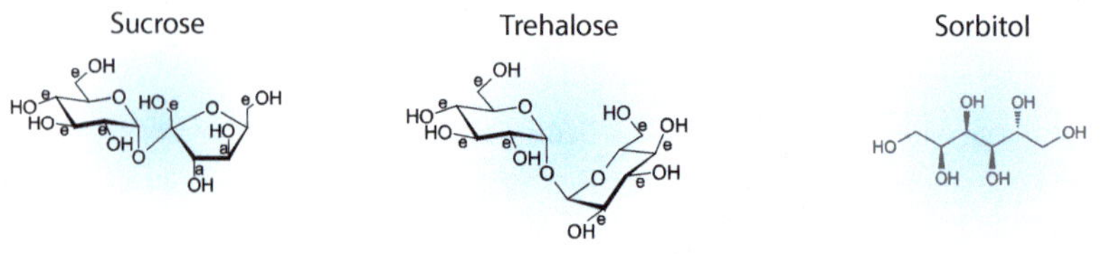

Fig. 3.37 Hydration shell and water-binding capacity of Sucrose, Trehalose, and Sorbitol

The structure of sorbitol, a sugar alcohol, is very simple. Chemically steric and spatial hindrances, such as those found in different sugar rings, are virtually nonexistent. Therefore, sorbitol is almost unbeatable in water binding. It also does not cause tooth decay and is less sweet than refined household sugar (sucrose). These properties are due to the molecular structure.

Sorbitol belongs, as already mentioned, to the osmotically and fermentable acting FODMAPs and thus causes flatulence, water inflow into the intestine, and increased peristalsis. This is known and not bad, unless too much of it is consumed. Then it can lead to diarrhea, as is known from the excessive consumption of some fruit varieties, such as plums [103, 104]. Plants and especially tree fruits develop more and more of it as they ripen, as a frost protection. The sugar alcohol binds a lot of water, which reduces frost damage at night.

3.6.3 Example of a Misunderstood Preservation—Rosemary Extract

Sometimes you read it on fish preparations, raw sausages, frying oils, or snacks: contains rosemary extract. This terpene mixture, obtained from the leaves of the plant, has strong antibacterial properties and is used to defend against herbivores. Rosemary, therefore, demonstrates an almost unbroken resistance and can thus, as a southern plant, cope well with adversities such as plant grazing, sun exposure, and the Mistral of the Rhone Valley. In domestic gardens, rosemary also fares well against everything except prolonged frost. The fact that this semi-shrub is so resistant is mainly due to its chemistry, which is also expressed in its excellent seasoning power. In the leaves and stems of the rosemary bush, in addition to herbaceous, resinous, and citrus-like fragrant aroma substances (which all connoisseurs appreciate in the "bouquet garni"), there are primarily tricky bitter taste substances that combine two properties: plant defense via terpenes and antioxidant properties via phenols. These helpful phytochemical constructs are called carnosol, carnosic acid, and rosmanol. Their chemical structure is very similar and is shown in Fig. 3.38.

Even for non-chemists, the chemical imagery is more than instructive at this point: These ingredients have two molecular parts. In the lower part (gray-green), there are typical diterpene structures that are highly effective against herbivores

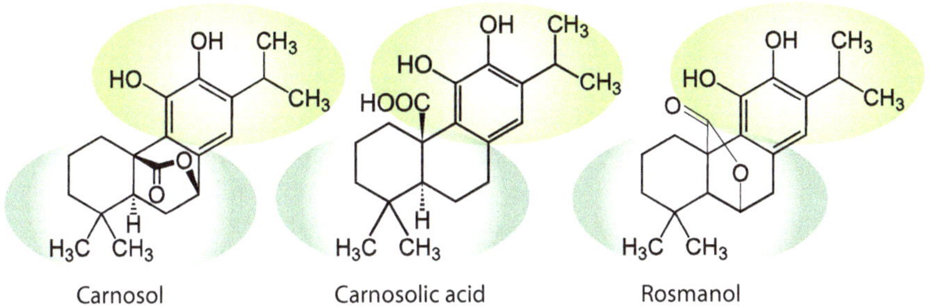

3.6 Misunderstood Preservation Methods

Fig. 3.38 The antibacterial effect of rosemary extract is hidden in the chemical structure of the basic components carnosol, carnosic acid, and rosmanol: the molecular synergy of properties

in plant physiology. In the upper part, however, a typical phenyl ring with two hydroxyl groups (OH) is shown, as is typical for most antioxidant-acting phenols in plants, such as in ellagic acids, quercetin, and other plant polyphenols. This molecule, therefore, works on several levels in rosemary.

It is therefore obvious to use this antibacterial and antioxidant-effective extract in well-dosed amounts as a natural preservative. Even the smallest amounts are sufficient to prevent oxidation processes or the growth of pathogenic germs. It immediately becomes clear why it is added to "healthy oils," such as fish or linseed oils with a high proportion of polyunsaturated fatty acids. Rosemary extract oxidizes first before fatty acids in the cells oxidize and can trigger cell damage. The sensitive double bonds are thus protected from their lipid oxidation—oils with polyunsaturated fatty acids, for example, become rancid less quickly. It is also used in the long maturation of raw sausages. Their fat oxidizes less, and rancid-tasting off-flavors are significantly reduced. This is crucial even in plant-based frying fats. During frying processes at these high temperatures, rapid oxidation of the fat must be prevented. Rosmanol, in particular, shows high stability [105]. The basis for the use of rosemary extracts or rosmanol is thus taken from nature. Often feared overdoses are self-prohibiting: The preserved products would quickly become too bitter and would not be acceptable in terms of taste. The effect of these bitter taste substances is known to everyone who adds rosemary sprigs in a braised roast for too long, turning sauces much too bitter.

3.7 Consumer Precariousness

The excessive availability of food is one of the negative developments of the strong industrialization of eating [106]. Food procurement is not a problem for the "*Homo industrialis*". It is constantly possible. Hunting, gathering, agriculture, livestock farming, and fishing are no longer necessary. These processes are delegated to farmers, producers, and the food industry. The advantages of industrial food production for people are obvious: The products become cheaper in relation to rising income. People have to work less and less for food. However, production becomes more opaque. A local baker, the butcher in the village, a local dairy, a cheese producer, or the vegetable farmer can usually still explain exactly how the food is made. This is no longer possible for a shelf-stocking assistant in a large supermarket. Even the artisanal methods of sausage production or bread baking were understandable to many people. The techniques were still very close to one's own cooking and baking at home. Even the animals that were processed into meat and sausage came from the immediate vicinity of the village or the surrounding area of the cities. In the modern industrial society, the relationships became more confusing.

The industrial production processes are complex, craftsmanship has been replaced by machines and production facilities, and food is transported over long distances during production, which in turn requires special efforts in preservation. Neither the paths nor the methods are immediately apparent. Uncertainty grows. The reasons for this uncertainty are as diverse as they are irrational [107]. It is primarily the many food scandals that contribute to people's uncertainty. Tangible crises like BSE (Sect. 3.3.3) have had an impact since the 1980s, even though, contrary to all predictions and projections, no significant measurable increase in the predicted brain diseases has occurred. Insufficient information ranks second among the reasons mentioned for uncertainty, closely followed by media reports. The latter is no surprise, as even the media considered serious do not always contribute to clarification. The reasons for this are often to be found in the pressure for ratings and the increase in circulation, according to the old news rule "*Good news are no news*". Scandals and exaggerations sell very well. But the ignorance of many media people reporting on food and nutrition is also alarming. Overly shortened presentations or simply non-existent knowledge of the physics and chemistry of food are indeed frightening and do not contribute to the enlightenment of readers and viewers. Also because physical-chemical facts are not translated correctly.

Fear of allergies, triggered by unknown additives and unclear origin of raw materials, also gives rise to uncertainty, as does the trend towards "free from" and "clean eating". Even high-quality natural ingredients such as gluten and whey proteins come under the scrutiny of the uncertain, FODMAPs such as lactose and fructose anyway.

Many of our foods, as well as our everyday and luxury items, have become black boxes. Most people use airplanes, cars, smartphones, computers, and televisions without having the faintest idea of how these devices actually work, or even what harmful, environmentally damaging, or exploitative materials are contained within them. These highly industrially produced black boxes are accepted and not questioned. However, when it comes to industrially produced food, many people draw the line, and black boxes are often not accepted. The blanket accusation that "the industry" wants to poison us [108] is, of course, unfounded. Any company that would sell unclean or even toxic food would undermine its own existence, not to mention the legal consequences.

Never before has our food been so well monitored, so safe, and so easily available as it is today, and analytical methods have never been as sophisticated and accurate as they are now. Consumers have the free choice between supermarkets and producer markets.

References

1. Simopoulos, A. P. (2011). Evolutionary aspects of diet: The omega-6/omega-3 ratio and the brain. *Molecular neurobiology, 44*(2), 203–215.
2. Foodwatch Futtermittel-Report. (April 2005). https://www.foodwatch.org/uploads/.../foodwatch_Futtermittelreport_komplett_0405.
3. Ziegler, J. (2011). *Hunde würden länger leben, wenn … Schwarzbuch Tierarzt.* MVG.
4. Vilgis, T. A. (2010). *Das Molekül-Menü: Molekulares Wissen für kreative Köche.* Hirzel.
5. Field, C. J., Blewett, H. H., Proctor, S., & Vine, D. (2009). Human health benefits of vaccenic acid. *Applied Physiology, Nutrition, and Metabolism, 34*(5), 979–991.
6. Jacome-Sosa, M., Vacca, C., Mangat, R., Diane, A., Nelson, R. C., Reaney, M. J., Igarashi, M., et al. (2016). Vaccenic acid suppresses intestinal inflammation by increasing anandamide and related N-acylethanolamines in the JCR: LA-cp rat. *Journal of lipid research, 57*(4), 638–649.
7. Phillips, M. C. (1972). The physical state of phospholipids and cholesterol in monolayers, bilayers and membranes. *Progress in Surface Membrane Science, 5*, 139–221.
8. Albrecht, O., Gruler, H., & Sackmann, E. (1981). Pressure-composition phase diagrams of cholesterol/lecithin, cholesterol/phosphatidic acid, and lecithin/phosphatidic acid mixed monolayers: A Langmuir film balance study. *Journal of Colloid and Interface Science, 79*(2), 319–338.
9. Demel, R. A., & De Kruyff, B. (1976). The function of sterols in membranes. *Biochimica et Biophysica Acta (BBA)-Reviews on Biomembranes, 457*(2), 109–132.
10. Albrecht, O., Gruler, H., & Sackmann, E. (1978). Polymorphism of phospholipid monolayers. *Journal de Physique, 39*(3), 301–313.
11. Segrest, J. P., Jones, M. K., De Loof, H., & Dashti, N. (2001). Structure of apolipoprotein B-100 in low density lipoproteins. *Journal of lipid research, 42*(9), 1346–1367.
12. Hevonoja, T., Pentikäinen, M. O., Hyvönen, M. T., Kovanen, P. T., & Ala-Korpela, M. (2000). Structure of low density lipoprotein (LDL) particles: Basis for understanding molecular changes in modified LDL. *Biochimica et Biophysica Acta (BBA)-Molecular and cell biology of lipids, 1488*(3), 189–210.
13. Huang, R., Silva, R. G. D., Jerome, W. G., Kontush, A., Chapman, M. J., Curtiss, L. K., et al. (2011). Apolipoprotein AI structural organization in high-density lipoproteins isolated from human plasma. *Nature Structural and Molecular Biology, 18*(4), 416.
14. Ivanova, E. A., Bobryshev, Y. V., & Orekhov, A. N. (2015). LDL electronegativity index: A potential novel index for predicting cardiovascular disease. *Vascular health and risk management, 11*, 525.
15. Sánchez-Quesada, J. L., Villegas, S., & Ordonez-Llanos, J. (2012). Electronegative low-density lipoprotein. A link between apolipoprotein B misfolding, lipoprotein aggregation and proteoglycan binding. *Current Opinion in Lipidology, 23*(5), 479–486.
16. Parasassi, T., De Spirito, M., Mei, G., Brunelli, R., Greco, G., Lenzi, L., Tosatto, S. C., et al. (2008). Low density lipoprotein misfolding and amyloidogenesis. *The FASEB Journal, 22*(7), 2350–2356.
17. Li, H., Dhanasekaran, P., Alexander, E. T., Rader, D. J., Phillips, M. C., & Lund-Katz, S. (2013). Molecular mechanisms responsible for the differential effects of ApoE3 and ApoE4

on plasma lipoprotein-cholesterol levels. *Arteriosclerosis, thrombosis, and vascular biology, 33*(4), 687–693.
18. Vigne, J. D., & Helmer, D. (2007). Was milk a "secondary product" in the old world neolithisation process? Its role in the domestication of cattle, sheep and goats. *Anthropozoologica, 42*(2), 9–40.
19. Evershed, R. P., Payne, S., Sherratt, A. G., Copley, M. S., Coolidge, J., Urem-Kotsu, D., Akkermans, P. M., et al. (2008). Earliest date for milk use in the Near East and southeastern Europe linked to cattle herding. *Nature, 455*(7212), 528.
20. Mlekuz, D. (2005). Meat or milk? Neolithic economies of Caput Adriae. In A. Pessina & P. Visentini (Eds.), *Preistoria dell'Italia settentrionale: Studi in ricordo di Bernardino Bagolini: Atti del Convegno* (pp. 23–24). Ed. del Museo fiulano di storia naturale.
21. Gillis, R., Carrère, I., Saña Seguí, M., Radi, G., & Vigne, J. D. (2016). Neonatal mortality, young calf slaughter and milk production during the Early Neolithic of north western Mediterranean. *International Journal of Osteoarchaeology, 26*(2), 303–313.
22. Greenfield, H. J., & Arnold, E. R. (2015). Go(a)t milk?' New perspectives on the zooarchaeological evidence for the earliest intensification of dairying in south eastern Europe. *World Archaeology, 47*(5), 792–818.
23. Papoli-Yazdi, M. H. (1991). *Le nomadisme dans le nord du Khorassan, Iran*. Iran: Institut français de recherche en Iran.
24. Prusiner, S. B. (1997). Prion diseases and the BSE crisis. *Science, 278*(5336), 245–251.
25. Prusiner, S. B. (1998). Prions. *Proceedings of the National Academy of Sciences, 95*(23), 13363–13383.
26. Elfrink, K., Ollesch, J., Stöhr, J., Willbold, D., Riesner, D., & Gerwert, K. (2008). Structural changes of membrane-anchored native PrPC. *Proceedings of the National Academy of Sciences, 105*(31), 10815–10819.
27. Boujon, C., Serra, F., & Seuberlich, T. (2016). Atypical variants of bovine spongiform encephalopathy: Rare diseases with consequences for BSE surveillance and control. *Schweizer Archiv für Tierheilkunde, 158*(3), 171–177.
28. Elmore, J. S., Cooper, S. L., Enser, M., Mottram, D. S., Sinclair, L. A., Wilkinson, R. G., & Wood, J. D. (2005). Dietary manipulation of fatty acid composition in lamb meat and its effect on the volatile aroma compounds of grilled lamb. *Meat science, 69*(2), 233–242.
29. Thanh, L. P., & Suksombat, W. (2015). Milk yield, composition, and fatty acid profile in dairy cows fed a high-concentrate diet blended with oil mixtures rich in polyunsaturated fatty acids. *Asian-Australasian journal of animal sciences, 28*(6), 796.
30. Daley, C. A., Abbott, A., Doyle, P. S., Nader, G. A., & Larson, S. (2010). A review of fatty acid profiles and antioxidant content in grass-fed and grain-fed beef. *Nutrition journal, 9*(1), 10.
31. Scheeder, M. R. L. (2007). *Untersuchungen der Fleischqualität von Bio Weide-Beef im Hinblick auf den Einfluss des Schlachtalters der Tiere und im Vergleich zu High-Quality Beef. Report*. ETH Zürich.
32. Dietrich, O., Heun, M., Notroff, J., Schmidt, K., & Zarnkow, M. (2012). The role of cult and feasting in the emergence of Neolithic communities. New evidence from Göbekli Tepe, south-eastern Turkey. *Antiquity, 86*(333), 674–695.
33. Haaland, R. (2007). Porridge and pot, bread and oven: Food ways and symbolism in Africa and the Near East from the Neolithic to the present. *Cambridge Archaeological Journal, 17*(2), 165–182.
34. Arranz-Oteagui, A., Gonzalez Carretero, L., Ramsey, M. N., Fuller, D. Q., & Richter, T. (2018). Archaeobotanical evidence reveals the origins of bread 14,400 years ago in north-eastern Jordan. *PNAS*, published ahead of print July 16, 2018. https://doi.org/10.1073/pnas.1801071115.
35. Greco, L. (1997). From the neolithic revolution to gluten intolerance: Benefits and problems associated with the cultivation of wheat. *Journal of pediatric gastroenterology and nutrition, 24,* 14–17.

References

36. Feldman, M., & Sears, E. R. (1981). The wild gene resources of wheat. *Scientific American, 244*(1), 102–113.
37. Nevo, E. (2001). Genetic resources of wild emmer, *Triticum dicoccoides*, for wheat improvement in the third millennium. *Israel Journal of Plant Sciences, 49*(sup1), 77–92.
38. Zhang, J., Liu, W., Han, H., Song, L., Bai, L., Gao, Z., et al. (2015). De novo transcriptome sequencing of *Agropyron cristatum* to identify available gene resources for the enhancement of wheat. *Genomics, 106*(2), 129–136.
39. Appels, R., et al. (2018). Shifting the limits in wheat research and breeding using a fully annotated reference genome. *Science, 361*(6403), eaar7191.
40. Busas, M. (2000). *Veränderungen bei der Brotherstellung*. Grin.
41. Grimm, H.-U. (2012). *Vom Verzehr wird abgeraten—Wie uns die Industrie mit Gesundheitsnahrung krank macht*. Droemer.
42. Vgl. Reichel, S. (2013). *So falsch ist unser billig Brot*. https://utopia.de/0/magazin/so-falsch-ist-unser-billig-brot.
43. Ternes, W. (2008). *Naturwissenschaftliche Grundlagen der Lebensmittelzubereitung*. Behr.
44. Schiedt, B., & Vilgis, T. (2013). Teig, Trieb, Textur—Proteine unter Stress. *Journal Culinaire, 15*, 47–59.
45. Belton, P. S. (1999). Mini review: On the elasticity of wheat gluten. *Journal of Cereal Science, 29*(2), 103–107.
46. Vilgis, T. A. (2011). *Das Molekül-Menü*. Hirzel.
47. Bucci, M. (2016). Protein folding: Minimizing frustration. *Nature Chemical Biology, 13*(1), 1.
48. Singh, H., & MacRitchie, F. (2001). Application of polymer science to properties of gluten. *Journal of Cereal Science, 33*(3), 231–243.
49. Kokawa, M., Sugiyama, J., Tsuta, M., Yoshimura, M., Fujita, K., Shibata, M., et al. (2013). Development of a quantitative visualization technique for gluten in dough using fluorescence fingerprint imaging. *Food and Bioprocess Technology, 6*(11), 3113–3123.
50. Visschers, R. W., & de Jongh, H. H. (2005). Disulphide bond formation in food protein aggregation and gelation. *Biotechnology advances, 23*(1), 75–80.
51. Vilgis, T. A. (2015). Soft matter food physics – The physics of food and cooking. *Reports on Progress in Physics, 78*(12), 124602.
52. Zielbauer, B. I., Schönmehl, N., Chatti, N., & Vilgis, T. A. (2016). Networks: From Rubbers to Food. In K. W. Stöckelhuber, M. Das, & M. Klüppel (Eds.), *Designing of elastomer nanocomposites: From theory to applications* (pp. 187–233). Springer.
53. Russ, N., Zielbauer, B. I., Ghebremedhin, M., & Vilgis, T. A. (2016). Pre-gelatinized tapioca starch and its mixtures with xanthan gum and ι-carrageenan. *Food Hydrocolloids, 56*, 180–188.
54. Woychik, J. H., Boundy, J. A., & Dimler, R. J. (1961). Wheat gluten proteins, amino acid composition of proteins in wheat gluten. *Journal of agricultural and food chemistry, 9*(4), 307–310.
55. Díaz-Gómez, J. L., Castorena-Torres, F., Preciado-Ortiz, R. E., & García-Lara, S. (2017). Anti-cancer activity of maize bioactive peptides. *Frontiers in chemistry, 5*, 44.
56. Cabra, V., Arreguin, R., Vazquez-Duhalt, R., & Farres, A. (2006). Effect of temperature and pH on the secondary structure and processes of oligomerization of 19 kDa alpha-zein. *Biochimica et Biophysica Acta (BBA)-Proteins and Proteomics, 1764*(6), 1110–1118.
57. Esen, A. (1986). Separation of alcohol-soluble proteins (zeins) from maize into three fractions by differential solubility. *Plant Physiology, 80*(3), 623–627.
58. Shukla, R., & Cheryan, M. (2001). Zein: The industrial protein from corn. *Industrial crops and products, 13*(3), 171–192.
59. Guzmán, A. Q., Flores, M. E. J., Feria, J. S., Montealvo, M. G. M., & Wang, Y. J. (2010). Effects of polymerization changes in maize proteins during nixtamalization on the thermal and viscoelastic properties of masa in model systems. *Journal of cereal science, 52*(2), 152–160.

60. https://www.zentrum-der-gesundheit.de/gluten.html.
61. Davis, W. (2013). *Weizenwampe: Warum Weizen dick und krank macht*. Goldmann.
62. Perlmutter, D., & Loberg, K. (2014). *Grain Brain: The Surprising Truth about Wheat, Carbs, and Sugar--Your Brain's Silent Killers*, Little, Brown Spark.
63. Uhlen, A. K., Dieseth, J. A., Koga, S., Böcker, U., Hoel, B., Anderson, J. A., & Moldestad, A. (2015). Variation in gluten quality parameters of spring wheat varieties of different origin grown in contrasting environments. *Journal of Cereal Science, 62*, 110–116.
64. https://www.food-detektiv.de/exklusiv.php?action=detail&id=87&volvox_locale=zh_CN.
65. https://bewusst-vegan-froh.de/schweineborsten-und-menschenhaar-im-brot-und-broetchen/.
66. Lauth, G. J., & Kowalczyk, J. (2016). Grenzflächenaktive Substanzen. *Einführung in die Physik und Chemie der Grenzflächen und Kolloide* (pp. 381–398). Springer Spektrum.
67. Shepherd, S. J., & Gibson, P. R. (2006). Fructose malabsorption and symptoms of irritable bowel syndrome: Guidelines for effective dietary management. *Journal of the American Dietetic Association, 106*(10), 1631–1639.
68. Hayes, P. A., Fraher, M. H., & Quigley, E. M. (2014). Irritable bowel syndrome: The role of food in pathogenesis and management. *Gastroenterology & hepatology, 10*(3), 164.
69. Brouns, F. J., van Buul, V. J., & Shewry, P. R. (2013). Does wheat make us fat and sick? *Journal of Cereal Science, 58*(2), 209–215.
70. https://www.zentrum-der-gesundheit.de/volksdrogen-milch-und-weizen-ia.html.
71. Venesson, J. (2013). *Gluten: Comment le blé moderne nous intoxique*. T. Souccar.
72. https://www.derwesten.de/panorama/wie-der-weizen-uns-vergiftet-wie-ungesund-ist-gluten-id10147460.html.
73. Shewry, P. R., & Hey, S. (2015). Do "ancient" wheat species differ from modern bread wheat in their contents of bioactive components? *Journal of Cereal Science, 65*, 236–243.
74. Schuppan, D., & Gisbert-Schuppan, K. (2019). *Schuppan, D., & Gisbert-Schuppan, K. (2019). Wheat syndromes*. Springer.
75. Von Koerber, K., & Leitzmann, C. (2012). *Vollwert-Ernährung: Konzeption einer zeitgemäßen und nachhaltigen Ernährung*. Georg Thieme Verlag.
76. https://www.zeit.de/zeit-magazin/mode-design/2016-01/kuechengeraete-design-fs.
77. Ziegler, J. U., Steiner, D., Longin, C. F. H., Würschum, T., Schweiggert, R. M., & Carle, R. (2016). Wheat and the irritable bowel syndrome—FODMAP levels of modern and ancient species and their retention during bread making. *Journal of Functional Foods, 25*, 257–266.
78. Buell, P. D. (2006). Steppe foodways and history. *Asian Medicine, 2*(2), 171–203.
79. Ballhausen, H., & Kleinelümern, U. (2008). *Die wichtigsten Erfindungen der Menschheit: Geniale Ideen, die die Welt veränderten*. Chronik.
80. Orthuber, H. (1968). *Handbuch für die Getränkeindustrie*. Gabler.
81. Pasteur, L. (1866). Pasteur, Louis. Études sur le vin: ses maladies, causes qui les provoquent, procédés nouveaux pour le conserver et pour le vieillir. Jeanne Laffitte (Editions) (Neuauflage: 7. Januar 1999).
82. Hartwig, G., von der Linden, H., Skrobisch, H. P. (2014). *Thermische Konservierung in der Lebensmittelindustrie*. (2. ed.). Behr's Verlag.
83. Star, B., Boessenkool, S., Gondek, A. T., Nikulina, E. A., Hufthammer, A. K., Pampoulie, C., et al. (2017). Ancient DNA reveals the arctic origin of viking age cod from Haithabu, Germany. *Proceedings of the National Academy of Sciences, 114*(34), 9152–9157.
84. Nedkvitne, A. (2016). The development of the Norwegian long-distance stockfish trade. In James Barrett (Editor); David Orton (Editor), Oxbow Books (Oxford) *Cod and Herring: The Archaeology and History of Medieval Sea Fishing*, 50–59.
85. Wicklund, T., Lekang, O. I. (2016). Dried norse fish. *Traditional foods* (pp. 259–264). Springer.
86. Malle, B., & Schmickl, H. (2015). *Essig herstellen als Hobby*. Die Werkstatt.
87. Cheng, L., Sun, D. W., Zhu, Z., & Zhang, Z. (2017). Emerging techniques for assisting and accelerating food freezing processes: A review of recent research progresses. *Critical reviews in food science and nutrition, 57*(4), 769–781.

References

88. Voigt, T. F. (2006). *Schädlinge und ihre Kontrolle nach HACCP-Richtlinien*. Behr.
89. Rodel, W., & Scheuer, R. (2007). Neuere Erkenntnisse zur Hürdentechnologie-Erfassung von kombinierten Hürden. *Mitteilungsblatt-Bundesanstalt für Fleischforschung Kulmbach, 1*(175), 3–10.
90. https://www.zeit.de/zeit-magazin/essen-trinken/2018-05/camembert-normandie-herstellung-pasteurisierter-milch-eu-recht.
91. Strahlmann, B. (1974). Entdeckungsgeschichte antimikrobieller Konservierungsstoffe für Lebensmittel. *Mitteilungen aus dem Gebiete der Lebensmitteluntersuchung und Hygiene, 65*, 96–130.
92. Heiss, R., & Eichner, K. (2013). *Haltbarmachen von Lebensmitteln: Chemische, physikalische und mikrobiologische Grundlagen der Verfahren*. Springer.
93. Barba, F. J., Terefe, N. S., Buckow, R., Knorr, D., & Orlien, V. (2015). New opportunities and perspectives of high pressure treatment to improve health and safety attributes of foods. A review. *Food Research International, 77*, 725–742.
94. Tappeser, B., Baier, A., Dette, B., & Tügel, H. (1999). Heute back' ich, morgen brau' ich. *Die blaue Paprika* (pp. 148–197). Basel.
95. Vilgis, T. (2011). Rauch und Raucharomen: Physik, Chemie, Struktur, Funktion. *Journal Culinaire, 13*, 42–59.
96. Grimm, H. U. (1999). *Aus Teufels Topf: Die neuen Risiken beim Essen*. Klett-Cotta.
97. Grimm, H. U. (2014). *Die Suppe lügt: Die schöne neue Welt des Essens*. Droemer eBook.
98. https://www.bioland.de/im-fokus/ihr-fokus/detail/article/tricks-an-der-kuehltheke.html.
99. https://slowfood.de/w/files/messepdf/slow_food_messe_qualitaetskriterien_2011.pdf.
100. Grimm, H. U. (2013). *Garantiert gesundheitsgefährdend. Wie uns die Zucker-Mafia krank macht*. Droemer.
101. Russ, N., Zielbauer, B. I., & Vilgis, T. A. (2014). Impact of sucrose and trehalose on different agarose-hydrocolloid systems. *Food Hydrocolloids, 41*, 44–52.
102. Vilgis, T. A. (2015). Gels: Model systems for soft matter food physics. *Current Opinion in Food Science, 3*, 71–84.
103. Jovanovic-Malinovska, R., Kuzmanova, S., & Winkelhausen, E. (2014). Oligosaccharide profile in fruits and vegetables as sources of prebiotics and functional foods. *International journal of food properties, 17*(5), 949–965.
104. Ma, C., Sun, Z., Chen, C., Zhang, L., & Zhu, S. (2014). Simultaneous separation and determination of fructose, sorbitol, glucose and sucrose in fruits by HPLC–ELSD. *Food Chemistry, 145*, 784–788.
105. Zhang, Y., Smuts, J. P., Dodbiba, E., Rangarajan, R., Lang, J. C., & Armstrong, D. W. (2012). Degradation study of carnosic acid, carnosol, rosmarinic acid, and rosemary extract (Rosmarinus officinalis L.) assessed using HPLC. *Journal of agricultural and food chemistry, 60*(36), 9305–9314.
106. Lang, T. (2003). Food industrialisation and food power: Implications for food governance. *Development Policy Review, 21*(5–6), 555–568.
107. Bergmann, K. (2013). *Der verunsicherte Verbraucher: Neue Ansätze zur unternehmerischen Informationsstrategie in der Lebensmittelbranche*. Springer.
108. Rickelmann, R. (2012). *Tödliche Ernte: Wie uns das Agrar- und Lebensmittelkartell vergiftet*. Ullstein eBooks.

Molecules Determine Our Food 4

Abstract

From the beginning of the universe, through the formation of atoms and molecules, from the first self-organized cells to today's complex biosystems, every process follows strict natural laws. Therefore, it is necessary to also view food and its nutritional value from this perspective. This opens up a different viewpoint and expands even the understanding of nutritional issues, clarifies some misunderstandings, and opens up new perspectives.

4.1 Where We Come From

For hunters and gatherers, food was probably not very complicated. It was all about survival. Before the use of fire, it was just about whether food passed the taste and smell test, and if it did, it was eaten. Before the use of fire—to return to Lévi-Strauss's ideas—there was no food culture, as this was only shaped by fire. People ate what nature provided and what sustained life. Hominids knew nothing about harmful ingredients, nothing about vitamins, secondary plant substances, or the like. They ate the available food and followed their senses and growing experience. The habitat of the very early hominids emerged without their intervention. The world and the universe are solely the consequence of a development since the Big Bang, which initially was based on pure physics, on not even a handful of fundamental interactions and a few fundamental particles, now called elementary particles. It banged about 13.8 billion years ago inaudibly, but this is still measurable today.

4.1.1 Genesis 1.0—In the Beginning was the Singularity

At the beginning of our current universe, there was neither place nor time. There was only an singularity, an incredible amount of energy [1]. The energy corresponded to an almost unimaginably high temperature. This concentrated energy was so high that 13.82 billion years ago, according to our chronology, the singularity burst and suddenly expanded. It inflated, step by step the universe was created up to the vastness we know today. In the process, the most important building blocks of matter were already formed in the first second (assuming today's definition of time). After only 0.000001 s, light particles (photons), quarks, electrons, and neutrinos emerged from the highly energetic radiation in an incredibly high number. Radiation and matter were now distinguishable (think of Einstein's famous formula of energy-mass equivalence, $E = mc^2$; the energy E results from the mass m and the square of the speed of light c). One second after the Big Bang, quarks merged to form atomic nuclei. Atoms could not yet form due to the high energy, as the kinetic energy was still so high that, for example, positively charged atomic nuclei could not yet capture negatively charged electrons. This required a further expansion of the universe and the associated temperature reduction. It took almost 400,000 years for this to become possible. Atoms gradually formed. Only about 100 million years after the Big Bang was the energy in the universe so low that stars, including suns and galaxies like our Milky Way, could form from atoms and growing meteorites through gravitation. Planets were captured by the gravity of suns as the energy decreased and the universe cooled, including our Earth, which has been orbiting the sun along with the other planets for 4.6 billion years. Quite apart from the fascinating processes, there is an important consequence in this context: Since the Big Bang, the universe and our Earth have emerged without any human intervention or control. Only in the most recent times did the *"Homo industrialis"* intervene a little in the events, and only since this epoch have we become very concerned about our nutrition. Before that, physics, chemistry, and biology followed the laws of nature in shaping the history of this Earth.

Even today, the echo of this Big Bang can be measured relatively easily. Everywhere, whether between the Moon and Earth or in galaxies light years away, the temperature in space is still 2.7 K, or $-276\ °C$. Just under three degrees above absolute zero. This cosmic background radiation is the remnant of the singularity, which, along with the stars and galaxies, still bears witness to the Big Bang [2, 3].

These are not imaginations of theoretical physicists, but can be precisely calculated, supported by models, and demonstrated on a small scale in experiments, for example at CERN in Geneva. Another key to these fundamentally simple models was provided by the discovery of the Higgs boson, that particle, which was disparagingly referred to as "divine," that assigns the exact mass to other elementary particles, such as quarks, leptons, etc. This discovery was indeed a direct proof of this fundamentally simple standard model, as the theoretical framework is called, even though there are still some gaps, the closure of which, however, will not overturn the current state of research [4], much like the extension of special to general relativity theory.

This model, together with the associated natural constants, determines [5] how molecules such as oxygen (O_2) or even water (H_2O) could form in the first place. Their existence is not at all trivial from a physical point of view. The fact that two hydrogen atoms combine with one oxygen atom to form the famous structure of the water molecule is subject to precise quantum physical and quantum chemical laws. For only this structure of the water molecule, known to us since chemistry lessons, results in the liquid that is the basis of life on Earth. If the water molecule were not slightly negatively charged at the oxygen and slightly positively charged at the hydrogen atoms, many things would be different from what we know. It would still take a long time before life was possible on a planet like our Earth. But the prerequisites were good, as there were carbon, nitrogen, hydrogen, and oxygen—the necessary (but not sufficient) building blocks of life.

4.1.2 Genesis 2.0—Elementary Life is Based on Self-organized Interfaces and Molecular Copy & Paste

The created by the physical conditions up to that point provided the best prerequisites for the emergence of life. Atoms could react to form molecules. The dead, but already self-organizing matter, merely had to learn a few tricks to start living, and molecules had to learn to reproduce themselves. This requires no magical creator, but merely the ingenious interplay of interactions on a molecular scale at moderate temperatures, the appropriate chemical environment for corresponding reactions of water, carbon, nitrogen, hydrogen, phosphorus, sulfur, and ionic compounds such as salts.

But there is something else to consider: Biomolecules, like all living things, must first be able to order and organize themselves. A disorganized heap of random biomolecules does not define life. For this, well-defined biochemical and biophysical processes must be able to take place reproducibly on a molecular scale. Thus, nature requires specific types of molecules that can perform these tasks. These conditions were ideal on Earth: oxygen and water were available in sufficient quantities. Likewise, matter that had accumulated over millions of years after the Big Bang and that—as has recently been discovered—was also enriched by regular impacts of meteorites from outer space. Even with molecules that had amino acid-like structures [6] and thus became the building blocks of proteins [7]. The life of the first single-celled organisms, the first elementary and organized nano-life forms, and their metabolism were, in a sense, coincidental. All physical-chemical parameters fit together until specifically shaped molecules were formed that were capable of organizing themselves—according to physical laws and through chemical reactions.

To understand what happened next, the cell membrane needs once again more attention. It consists largely of emulsifiers such as lecithin, as mentioned in Chaps. 2 and 3 several times. As the name suggests, lipids must have something to do with fat. In fact, a large part of lipids consists of two fatty acids and a head group. This head group is chemically composed of, for example, phosphorus, nitrogen, carbon, and hydrogen and is arranged so that this head group, like water, has a dipole moment, i.e., a rather positively and a rather negatively charged part. These phospholipids resemble, figuratively speaking, two-tailed "molecular tadpoles" (see Fig. 2.8). What sounds abstract is important for nature and the cell, because the polarity of the head group always speaks for water solubility, and the presence of water is the most important prerequisite for life—also in the cell. However, the two fatty acid tails are not water-soluble, but fat-soluble. One part of the molecule is therefore fat-soluble, the other water-soluble. The phospholipids, i.e., lecithin, are emulsifiers. So these molecules must organize themselves so that head and tail are physically and thermodynamically satisfied. This can only work if the molecules line up like soldiers: head to head and tail to tail. They stand in rank and file. Likewise, a second layer that joins them. The result is a lipid bilayer: The fatty acids repel each other, and the polar head groups form a boundary on both sides. If this structure closes into a flexible, movable ellipsoid, water can be enclosed in the resulting interior space and clearly separated from the surrounding water. The membrane and the lipid cell envelope are born, for every biomaterial, whether plant, animal, or human. Lipids, i.e., emulsifiers, are therefore among the basic building blocks of life along with proteins and RNAs [8]. This was the case in the primordial forms of life and has remained so until the *Homo sapiens*. What has changed and become more complex during evolution is the interaction of cells. They developed molecular communication mechanisms and joined together to form huge cell clusters, from which life developed. Even the complex biochemistry still works exclusively on these fundamental interactions dictated by physics, with which molecules recognize the finest differences of molecules when it comes to deciding what enters the cell and what does not, and how and when the cell fuel adenosine triphosphate is formed and degraded.

A simple cell is thus the smallest unit of life, the "elementary cell." Its structure, its emulsifier shell with embedded proteins and switching elements, whose stability with simultaneous flexibility, are already wondrous examples of the interplay between physical-chemical structures and vital functions. There are indications that the first functioning basic cell, known as "LUCA" (Last Universal Common/Cellular Ancestor) in evolutionary biology [9], emerged more than 3.5 million years ago. At that time, the universal functions were already established as we know them today [10]. A cell still has an exterior and an interior today, separated by a thin, flexible, and partially permeable membrane adjustable via the type of fatty acids (see Sect. 1.3) and proteins. Inside, there is the cell nucleus with RNA and DNA (the genetic information) and the mitochondria, which provide energy for metabolism and reproduction. The cell interior also contains water, which keeps the cell turgid due to its pressure (turgor). Everything is well regulated by electrical forces (salts, minerals, ions of different valence, the contrast between water and fat) and the reading and copying of four nucleotides. Biophysics and biochemistry in their purest form.

What is good for the cell and what is not is determined by its biological function and the elementary interplay of molecules, van der Waals interactions, electrical charges, and the contrast between fat and water. In essence, there would be no reason for an unnecessarily complicated relationship status with our food. This supports and helps understand the thesis of this book "before physiology comes physics, chemistry, and biology." Even we as *Homo sapiens* are based on those elementary biophysical and biochemical interactions of a handful of molecules. Nevertheless, food became complicated. But for entirely different reasons, namely when humans became sedentary and food procurement was significantly simplified compared to the Paleolithic period. People began to concern themselves with mystical things.

4.2 The Beginning of Ancestor Worship

4.2.1 From Ancestor Worship to Religion

Through sedentism, new and different social structures emerged among humans than was the case in the age of hunters and gatherers. Tasks could be divided, life structures were adapted to the circumstances. Rites and ceremonies were developed, and the first early religions formed [11, 12]. The starting point of these developments lies in an ancestor worship [13]. Hunters and gatherers developed a death cult that regulated dying, the transition from life to death. These quite rationally conditioned rites were entirely directed at the dying and the dead. With sedentism, the cult of death changed. In many tribes, ideas and beliefs developed that the deceased could positively and negatively influence the fate of their descendants. The belief developed that the ancestors were still part of the tribe and the people after death, even if not actually present. The ancestor worship thus guaranteed a continuous contact with the dead, and was therefore directed at the currently living. A spirituality seized humanity, which persists to this day in the most diverse religions. Humans understood death, but dealing with it is still difficult today. Death only becomes acceptable when a life after death is promised and the ancestors can be found there again. Humans created gods and believed from then on that these gods had created them. The reason for this has not changed to this day: people cannot accept their death, so they seek comfort in their grief in the mystical, unreal, and esoteric. Humans began to create more and more gods, who represented everything unexplainable and still find followers today, even if this takes irrational and absurd forms, as in creationism [14] and has led to wars for millennia.

With the beginning of cult, spirituality, and religion, even food took on new functions. They were chosen (or degraded, as they were no longer eaten) as sacrificial offerings. For the first time in their history, humans could afford not to slaughter animals for food. Rituals developed that were unthinkable for hunters and gatherers as well as for nomads. These sacrificial offerings have persisted in many major world religions and can be seen today as the earliest form of food waste. Although these offerings had a cultural meaning, they were no longer available as human food. Laboriously produced food was used in the service of religions.

4.2.2 Non-structuralist (Food) Cultures

At this point, it is worth revisiting Lévi-Strauss's culinary triangle, which was already introduced in Chapter 2 and further supplemented by a physical-chemical interpretation. This chapter focused primarily on the cooking achievements of early humans, who, through cooking and targeted fermentation, made the transition from nature to culture through a variety of cultural actions. This structuralism, which can even be clearly mapped onto molecular changes, discusses only the evolving cooking and food culture that serves to nourish humans over the millennia. These new spiritual forms do not appear in this transition from nature to culture. Nevertheless, they represent an essential form of food culture, especially with a current focus on Western culture, as food is seen neither as nourishment nor as a sacrifice, but is itself elevated to a remedy [15]. These ideas may seem new, but they are not, for the Christian Last Supper had long propagated the idea of "eating the body of Christ" [16]. To capture these new cultural approaches that have developed only since the advent of sedentism, the original culinary triangle needs an extension [17], which is shown in Fig. 4.1.

With sedentism, lifestyles changed significantly. The emerging prosperity was based solely on predictable food security. From the Neolithic Revolution onwards, people were no longer forced to search for food daily. Social structures formed, and societal differences became more pronounced. The models of energy input mentioned in Fig. 2.4, which secured both the lives of offspring and the elderly and regulated the available food supply, largely controlled population strength so that hunger was avoided. In the Neolithic, these models became obsolete. The sick and "unproductive" could be cared for and provided for on-site for the first time. The formation of new structures and superstructures became possible. Public health became an essential part of society, as did the ability to defend against dangers of any kind.

Art and language continued to develop. The use of grinding stones allowed for mush-like dietary forms, which initially changed human bite and tooth position, and only later enabled certain sounds like "F" [18]. Different talents and abilities could be tolerated and promoted. A differentiation of society became possible. Thus, hierarchical structures could emerge. Tribes and communities were organized. At the same time, ethical and moral standards could be developed, and coupled with ancestor worship, belief systems were born.

This is also shown in Fig. 4.2. While the cultural creation of cooking and eating humans, expressed through the culinary triangle of Lévi-Strauss (see Fig. 2.29 ff.), was predominantly driven by natural scientific, molecular facts *(hard sciences)*, they are characterized by structural (not structuralist) changes in the upper part of the triangle through *soft sciences* and no longer directly verifiable assumptions. The difference between culture and cult is illustrated.

Not coincidentally, the category "Physiology and Nutrition" is placed at the border between both triangles. A well-functioning physiology is, of course, the result of the edible food that provides people with energy, macro- and micronutrients on a daily basis, and thus the result of cooking and eating. And, of course,

Fig. 4.1 Sedentism brought forth new structures. Below is the (upside-down) culinary triangle of Lévi-Strauss, above are the resulting new structures as a consequence of the prosperity and security of the new lifestyles

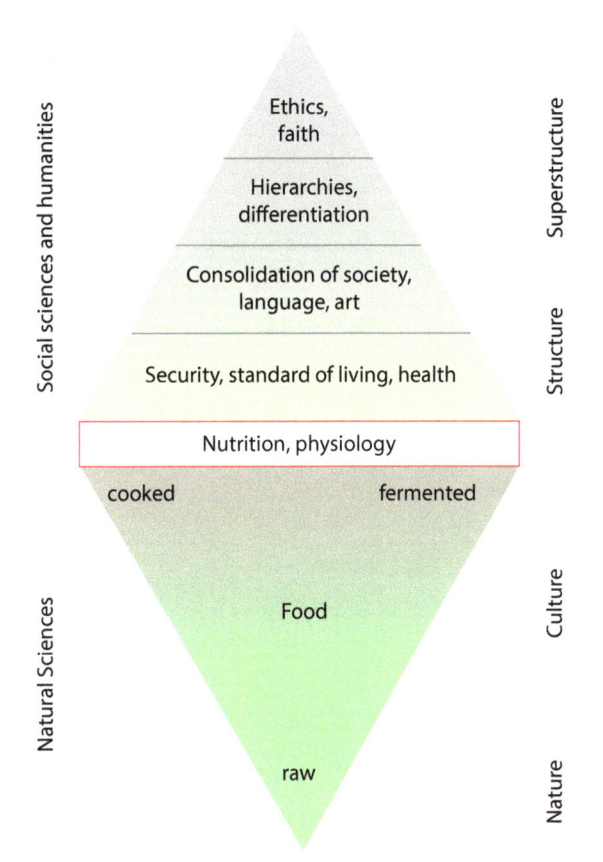

the performance gained from it is the result of a secured nutrition with high safety standards, as enabled by fire and fermentation. If we do not eat, we die. Even if we eat poorly or do not observe the safety standards of cooked and fermented food, we become ill, as recent archaeological finds and (fossilized) human feces in Northern Europe and the Middle East between 500 BC and 1700 AD have shown [19]. Chemical analyses show how parasites and germs took over, and the state of health deteriorated after inadequately cooked, half-raw, or even spoiled food was eaten.

4.3 Basic Food Meat: Physics, Chemistry, Taste

4.3.1 Proteins Everywhere

A sustainable food economy was unquestionable from the Neolithic Revolution until after the Second World War. Economic models, as discussed in Chap. 3, were maintained. Fields were cultivated according to specific rules, and when

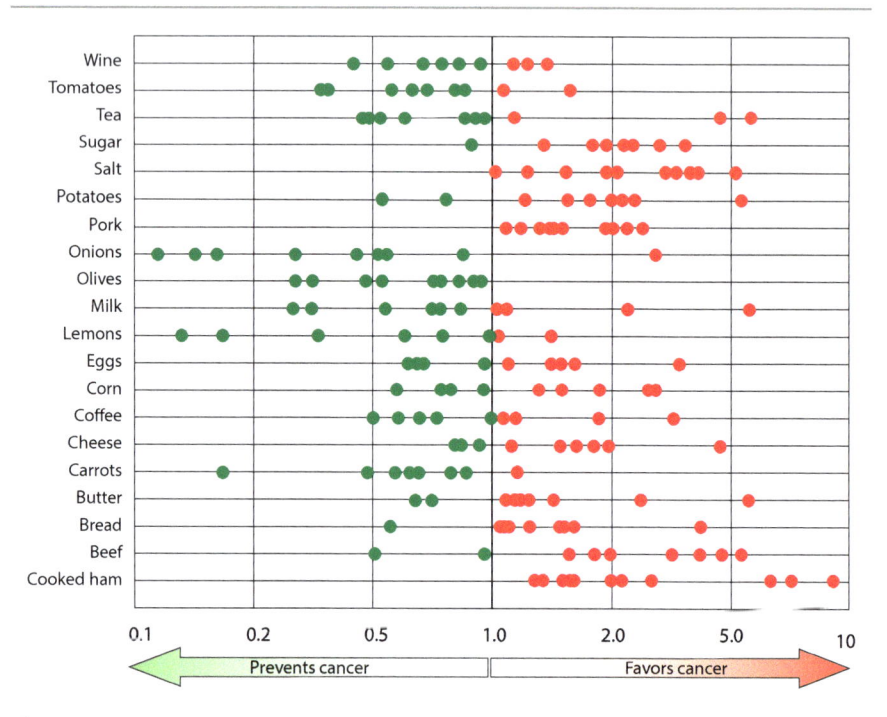

Fig. 4.2 Studies on food pro and contra their carcinogenicity. Each point corresponds to a published observational study (up to 2011). With few exceptions (salt, pork and cooked ham), there are fluctuations in both directions [25]. The interpretation is clear and hardly surprising: Nothing is known for certain whether food is generally harmful or generally healthy

animals were slaughtered for food, it was required to consume them completely. Non-edible parts could be used in various ways. Collagen-rich skin was made into leather and clothing, and bones were made into tools since the Paleolithic, later the first needles and awls.

Only in the years of prosperity after the Second World War did people develop fears of food. What created humans millions of years ago and nourished them best for millennia was declared poison, declared harmful, and denounced as a health risk. As explained in Chap. 1, valuable parts of animals have not been eaten for quite some time: offal, brain, collagen-rich meat, intestines, tripe, testicles, chicken feet and cockscombs, to name just a few. Even though in some food cultures these parts are considered delicacies: cockscombs in Spain, testicles in Mediterranean countries [20], intestines as sausages (Andouillettes, Andouille) in France [21] or chicken feet in Asia. If all parts of the animals were eaten, instead of focusing on the small percentage of so-called prime cuts such as loin, steak and short roast meat, a significant proportion of factory farming could be avoided. In China or Mediterranean markets, living animals are sold, which is not allowed in Germany for hygienic reasons. Of course, these animals, if they do not continue to live in their own yard/stable, are completely processed after slaughtering.

For many modern eaters, a chicken consists of two breasts, and possibly two thighs. However, a chicken consists of numerous edible parts, from which many dishes can be prepared. Offal such as liver, heart, lungs or kidneys can be used for an appetizer, such as salads, or made into a poultry cream. If these are combined with the two wings and the Sot-l'y-laisse (small muscles in the carcass), a complete meal is already created. The removed skin can be turned into crispy chips, which can be arranged on appetizer plates or with salads. The two thighs, which can be separated into upper and lower parts *(drumstick)*, are also sufficient for a ragout, Coq au Vin etc. for four people. Particularly, the glutamate bomb chicken broth can be cooked from the carcass with still adhering meat. The broth can be filled hot into sterile screw jars like jam, making it preserved and always available. And even the meat removed from the carcass after the broth cooking goes finely chopped together with herbs minced or mixed with cream cheese into meatballs, ravioli or Swabian pockets, which are enriched with vegetables, for example. Or directly into a chicken fricassee. Even the fat that settles on the broth and can be removed after cooling is suitable for stirring into vegetable ragouts, for lightly frying potatoes, just to name a few examples.

Far from the chicken breasts, the whole animal can satisfy several meals when the small parts are paired with vegetables. The *"Homo opulentus"*, the affluent person who developed after the war under the sign of the economic miracle, forgot how their immediate grandmothers and great-grandmothers dealt with food. Many of the mentioned parts of a chicken are considered waste even by some butchers today.

The factory farming is one of the major problems in the food industry. What happens in the barns and during animal transports is the result of thoughtless and boundless meat consumption. Sausage and meat counters at discounters and supermarkets reveal the misery. From a purely scientific and biological point of view, meat is far too cheap. Even if this statement sounds undemocratic and unsocial, the truth remains that the daily schnitzel or beef steak is not a fundamental right [22]. No one can seriously believe that humanity can be fed with affordable and good-quality meat as long as there are only prime cuts. Cheap meat can only be produced under adverse conditions and through disrespect for living creatures. These production methods have nothing to do with the shepherds and breeders of the past. The daily meat consumption does not either, even if it has recently been somewhat declining.

4.3.2 The Modern Western Individual and the Inclination Towards White Meat

The recent strong preference for white meat—turkey and chicken—does not come by chance. For years, red meat, such as beef, sheep, or pork, has been suspected of promoting colorectal cancer (see Ströhle et al. [23] for a comprehensive summary). As discussed in the previous chapters, meat was an essential building block in human evolution. It provided macro- and micronutrients in sufficient quantities, even if it was not consumed in large amounts and certainly not daily on the paleo menu. So how can meat suddenly become carcinogenic? From a naive perspective, this could only have two causes: We eat too much of it and we eat too few other foods. This discussion reveals one of the biggest problems of early nutritional science: The isolated focus on a product, in this case, red meat and its high content of heme iron; that is, the iron ion centered in the heme group of the protein myoglobin. Furthermore, viruses in undercooked meat are suspected of having effects that accumulate over time and could contribute to colorectal cancer in advanced age [24], as can be gleaned with difficulty from some associative observational studies. Other investigations go beyond the mere observation effect and provide more reliable data through *in vitro* studies on cells [26]. The transferability of these results to humans is unclear, but they clearly show ways in which future investigations *in vivo* on humans could be conducted.

Despite the studies, there is no clear statement [22]. It would be futile to quote the various possibilities read out in the subjunctive, only to discard them again. Even the much-cited polycyclic aromatic hydrocarbons do not have a significant effect with normal consumption of dark baked bread, fried and grilled meat, and sausages.

Heterocyclic aromatic amines—including many Maillard-reaction products from amino acids, such as those formed during frying and grilling—do not pose an immediate risk either. Although high-dose administration in animal experiments resulted in a risk of tumor formation, the cause is the breakdown products of these electrically charged amines (nitrogen compounds, nitrenium ions), which can interact with the nucleotides of DNA. On the other hand, there is a strong individual dependency in animal models themselves. Depending on the microbiome, tumors can form. However, these experiments also point to other, more fundamental problems. For one, the test animals, unlike humans, have not cooked for millennia, and their metabolism is not familiar with these amines; for another, the actual concentrations in real fried or grilled meat are 10,000 times lower.

As is well known from Chapters 1 and 3, even the much-cited saturated fatty acids in meat products are neither harmful nor healthy. They are necessary for membrane construction and are synthesized in the body as needed.

4.3.3 Many Studies, Little Clear Insight

It is best to stick to the only correct theory in principle: Everything we eat causes cancer and protects against cancer [25]. Two physicians from Harvard took the challange to pick out standard foods and compile the studies on each for their positive and negative statements. The results were astonishing, summarized in Fig. 4.2. We should internalize this.

As can be clearly seen from Fig. 4.2, the results from observational studies provide little causality. In fact, it is largely unclear whether a particular food, such as pork, actually has a "cancer-promoting" effect or not. To recognize this using the methods of observational studies, test subjects would have to be available who would eat only pork from now on. For years, for breakfast, lunch, and dinner. And what would happen if the control group ate only pork with onions and olives? For years, three times a day a handful of olives as an appetizer and then onion roast? Would the positive effects of the onions and olives outweigh the negative aspects of the pork? Or would one be immune to all diseases if only the onions and olives were eaten? Year in, year out? Hardly likely, this diet would be very unbalanced. The good news from Fig. 4.2 is: The scientific studies on individual foods should not only not be overrated, but they can be ignored for everyday life, as long as the personal menu is varied enough.

The bad news from Fig. 4.2 also shows precisely the situation we are in every day when such messages are spread from all kinds of media. There is always a study that fits the zeitgeist or a corresponding view. Whether it is books on strictly vegan or paleo nutrition: Several studies can usually be picked out that correspond to one's own ideas. In the end, all that remains is to critically question the results and check them for possible molecular connections, such as with nitrate and nitrite, which will be addressed in the next section (Sect. 4.3.3) on the topic of nitrite curing salt.

4.3.4 Of Heme Iron and Cancer

More specifically, the influence of haem iron and its role in treatment with nitrite curing salt is known, so that the red color is preserved during cooking or smoking (reddening). The heme group, which occurs in haemoglobin and the muscle pigment myoglobin, carries a nitrogen compound instead of a divalent iron ion after curing. Heme iron refers to the central iron ion Fe^{2+}, which is located in the haem group of myoglobin (a sarcoplasmic protein) and is responsible for the oxygen supply of the muscle and gives the meat its red color. In order for the ion not to oxidize during cooking and heating and for the meat not to turn gray (similar to chlorophyll, but with a magnesium ion there), it is reddened in sausage through curing processes with nitrite curing salt. A nitric oxide (NO) combines with the iron ion in the heme group, as shown in Fig. 4.3. Cured products, raw sausages, and hams thus remain reddish even during cooking, ripening, and fermenting.

This example is instructive because native and converted h aem iron can be directly compared. Heme iron, but more likely nitrated haem from cured and processed meat products, could under certain conditions form nitroso compounds that can trigger DNA damage, according to the hypothesis (Fig. 4.4). The second line of hypothetical approaches concerns saturated fatty acids, which are released

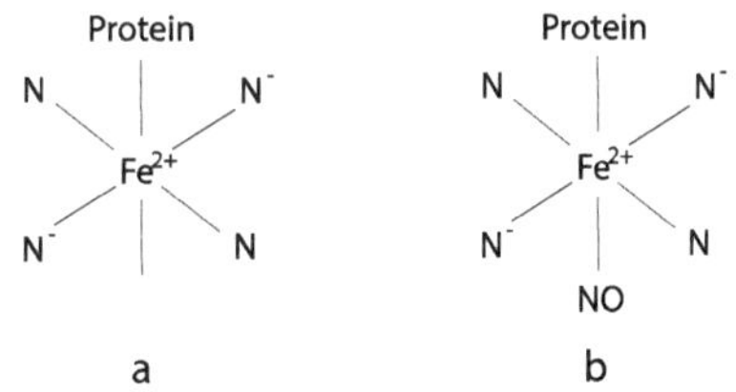

Fig. 4.3 Heme iron (**a**) occurs in myoglobin in meat (also in hemoglobin, blood pigment), during curing it becomes nitrosylated heme (**b**)

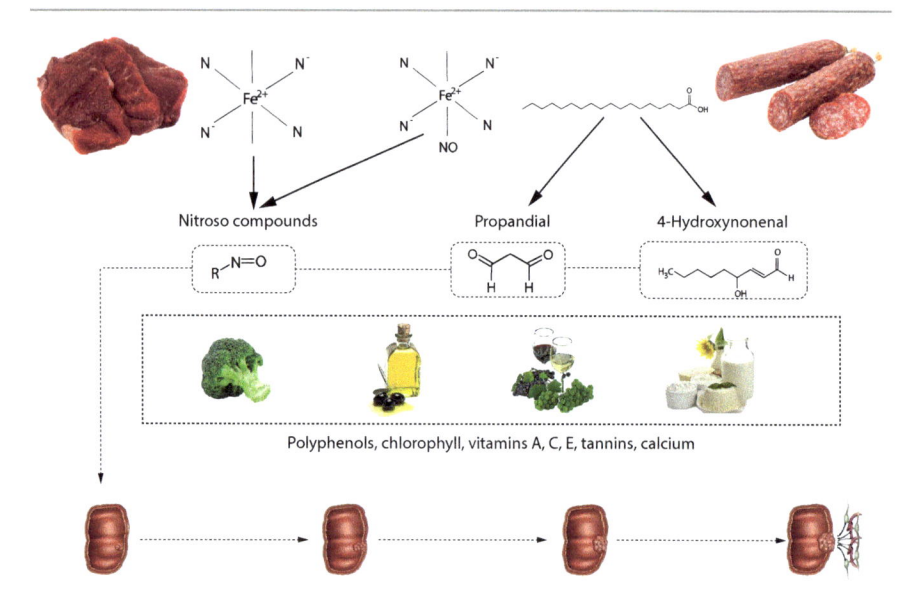

Fig. 4.4 Meat and processed meat in the diet. From haem, especially from nitrosylated heme, nitroso compounds with characteristic N = O group are formed, which together with the aldehydic degradation products of saturated fatty acids—propandial and 4-hydroxynonenal—can have carcinogenic effects through several pathways. Natural barriers to these processes are (poly-)phenols, tannins, vitamins and calcium

from fats under oxidation and break down into fragments such as propandial and 4-hydroxynonenal that can also damage cells, but this has not yet been conclusively proven [27]. This also applies to fats used for preparation and heated, for example, in pans or grills. On the other hand, fats and oils are carriers of the fat-soluble vitamins A and E and contain, if it is olive oil, a high amount of polyphenols (they taste bitter), which counteract these two processes. It is precisely the fat that shows its two sides. On the one hand, it can promote cancer, on the other hand, it even helps to prevent these conversions of heme and fat oxidation.

Also, tannins, as they are abundant in vegetables or grapes (and thus also in red wine), and chlorophylls form a strong barrier to this damaging process. They directly capture the haem, as chlorophyll can be seen as "plant heme," only the iron is replaced by magnesium. Even the doubly positively charged calcium ion counteracts these processes. Free calcium is already abundant in muscle meat, and for those who need more, they can enjoy some cheese from the cheese plate after the meat course or simply eat more vegetables with grilled meat or cured ham and drink an appropriate amount of red wine for the sake of (poly-)phenols and tannins. Even this naive enumeration points to an important point: The diversity of the ingested food is more decisive than the presence of a potentially harmful substance in one component of the meal.

So, if we eat a balanced diet, the required levels of micronutrients are already in balance. Therefore, there is not much need to worry. Similar findings are often expressed in studies. Participants who eat a lot of meat often consume too few other complementary foods such as fruits, vegetables, nuts, etc. The term "complementary" describes this situation much better than "healthy" and "unhealthy." Natural foods cannot be classified as healthy, unhealthy, or the like. Each natural food has its place in our diet, as the evolution teaches us. Moreover, each food has its own culinary value that should be promoted.

4.3.5 Once Guilty, Always Guilty

Unfortunately, hard scientific findings do not penetrate the general population, and the theory that red meat is carcinogenic while white meat is healthy persists. This is due solely to the high-quality chicken proteins, which are attributed a high biological value (see also Chap. 4). To this day, however, it is not at all clear whether white meat is actually "healthier" than red [28]. With regard to essential amino acids, as well as the fatty acid structure, an objective view of this claim does not emerge. In fact, red beef from pasture-raised animals on average contains significantly more long-chain polyunsaturated essential fatty acids than poultry meat and is also comparable in terms of purine content—the reasons for this will become clearer later. Such trends have absurd effects on animal husbandry: to increase the yield of white meat, breeders come up with the idea of overbreeding meat chickens and turkeys. In addition to factory farming, this also involves animal-contemptuous and breeding absurdities. The highly bred, mass-produced animals suffer so that consumers can eat the supposedly healthy poultry sausage in unconcerned quantities.

Wild game is, of course, also red meat. In terms of heme iron, it is even much "redder." Firstly, wild game performs significantly more muscle work, which is designed for both endurance and escape. Its meat is rich in the muscle pigment myoglobin due to the high oxygen demand and is therefore dark red. Secondly, wild game is usually shot, dying stress-free in the wild. Unlike stunned cattle in the slaughterhouse, whose carotid artery is opened while the heart is still beating, the wild game does not bleed out completely. Therefore, significantly more haem iron remains in the meat. As a reminder: In the Paleolithic era, wild game meat was a primary source of protein-rich food and helped *Homo sapiens* in their development.

4.4 The Chinese Restaurant Syndrome and the Chemistry of Umami Taste

The fear of glutamate as the trigger for the China Restaurant Syndrome persists stubbornly, despite a large number of scientific studies that come to the opposite conclusion. That this claim is hardly tenable was already hinted at in Chap. 1, at the latest, since the umami taste was recognized as the driving force behind the still valid food transformation. However, the "glutamate problem" only began in

1968 when author Kwok speculated in the *New England Journal of Medicine* [29] that his discomfort might be due to the glutamate-rich cuisine of Asian restaurants. In essence, only three severe diseases are known to be caused by proteinogenic amino acids. However, these are inherited and already manifest in infancy with very severe, life-threatening symptoms: maple syrup urine disease (leucinosis), tyrosinemia, and phenylketonuria. In the genetically determined and inherited disease phenylketonuria, the aromatic and essential amino acid phenylalanine cannot be broken down. Phenylketonuria leads to severe irreversible impairments and is neither treatable nor curable. Looking back at the evolution, people with this disease had low chances of survival. In maple syrup urine disease, the hydrophobic, essential amino acids leucine, isoleucine, and valine cannot be metabolized, while in tyrosinemia, the also hydrophobic, aromatic, but non-essential amino acid tyrosine cannot be metabolized. In a broader sense, homocystinuria can also be counted, in which the non-proteinogenic amino acid homocysteine cannot be metabolized.

With glutamic acid, the situation is quite different. Even the offspring of *Homo sapiens* received it in large quantities with mother's milk.

4.4.1 Mother's Milk Sets the Example: Umami and Sweet

Humans are already accustomed to the taste combination of umami and sweet as infants. No wonder, considering the evolution of hominids to *Homo sapiens,* which was only made possible by a high protein supply and thus a high density of essential amino acids. The offspring were conditioned to the essential taste directions sweet (lactose) and umami (glutamic acid) during the initial care after birth. In fact, human breast milk has the highest glutamate content compared to the milk of other mammals [30]. Breast milk essentially goes through three different stages during lactation for all mammals: colostrum (first milk) is produced during the first three days, followed by transitional milk during days four to 14, and mature breast milk is produced from day 15 onwards. During these phases, the ingredients are adapted to the needs of the infants, including the spectrum of amino acids [31]. It may initially seem surprising that the concentration of non-essential amino acids is significantly higher and increases much more rapidly than that of essential amino acids, with glutamic acid, i.e., glutamate, having the highest concentration of all amino acids. These relationships are exemplified in Fig. 4.5.

Why sweetness (lactose) and umami (free glutamic acid) dominate in the taste of mother's milk is ultimately unclear. However, it becomes plausible when we recall the results from Chap. 1: sweet and umami are innate taste preferences that, unlike salty, sour, and bitter, do not need to be learned. At the same time, milk sugar as a polar molecule and glutamic acid as a negatively charged ion can be transported as quickly directly through the medium blood (water) as all salts. Glutamic acid thus reaches the brain as an important neurotransmitter, where it plays a crucial role in neuronal function, unlike many other amino acids. Infants have not yet come into direct contact with adult food. It is clear that alarm signals

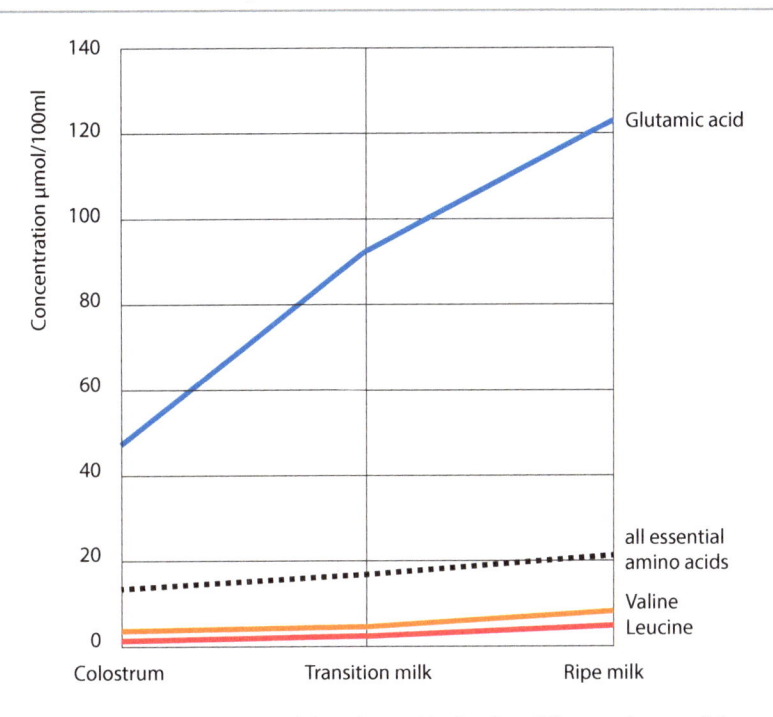

4.4 The Chinese Restaurant Syndrome and the Chemistry of Umami Taste

Fig. 4.5 Glutamic acid and essential amino acids in the different phases of breast milk. Glutamic acid is most strongly represented and gives breast milk a strong umami character in all phases. The sum of the concentrations of all essential amino acids (dashed line) is comparatively low

such as bitter, strong acid, or even salt levels above physiological concentrations cannot be perceived as food. The two most important remaining are umami and sweet for conditioning. A comparison of the various taste preferences of amino acids reveals very precisely how mother's milk tastes. This is shown in Fig. 4.6, in which the amino acid spectrum has been supplemented by the taste of amino acids.

When looking at Fig. 4.6, it immediately becomes apparent that the basic taste qualities of amino acids are dominated by umami and sweet. The main trigger of umami taste, the free glutamic acid, clearly dominates in concentration over all others. The bitter taste of essential amino acids is completely masked by the higher concentration of sweet-tasting amino acids. The conditioning for sweet and umami is also reflected in the receptors found on the taste of *Homo sapiens*. Both consist of a protein pair T1R2-T1R3 for sweet and T1R1-T1R3 for umami [32], with both sharing a receptor protein [33]. The abbreviation TR comes from English, standing for **T**aste **R**eceptor. In very elaborate studies and experiments, this close relationship between sweet and umami receptors was demonstrated, and it is greatly simplified in Fig. 4.7.

Amino acid	Taste
Alanine	sweet
Arginine	bitter
Asparagine	sweet
Aspartic acid	umami
Cysteine	bitter
Glutamic acid	umami
Glutamine	sweet
Glycine	sweet
Histidine	bitter
Isoleucine	bitter
Leucine	bitter
Lysine	sweet/bitter
Methionine	bitter
Phenylalanine	bitter
Proline	sweet
Serine	sweet
Threonine	sweet
Tryptophan	bitter
Tyrosine	bitter
Valine	bitter
Taurine	neutral

Concentration (relative units)

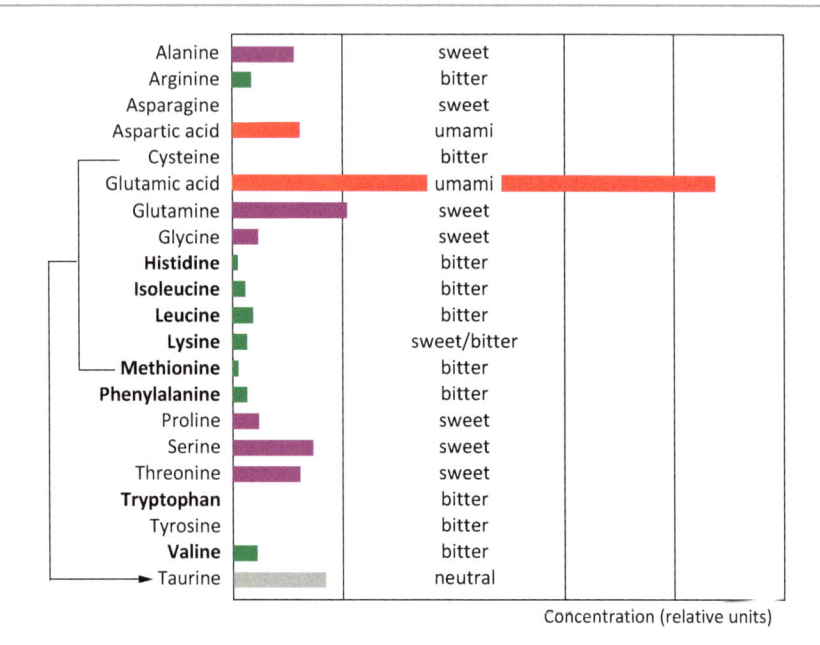

Fig. 4.6 The taste of mother's milk is also defined by the free amino acids. Apart from milk sugar, sweet-tasting amino acids (pink color) dominate alongside umami tasting representatives aspartic acid and glutamic acid (red bars). The bar length roughly reflects the average concentration in mature mother's milk. Essential amino acids (bold) are mostly bitter and occur only in comparatively low concentrations. The taste of amino acids is much more complex than shown here, but this has been omitted for the sake of clarity

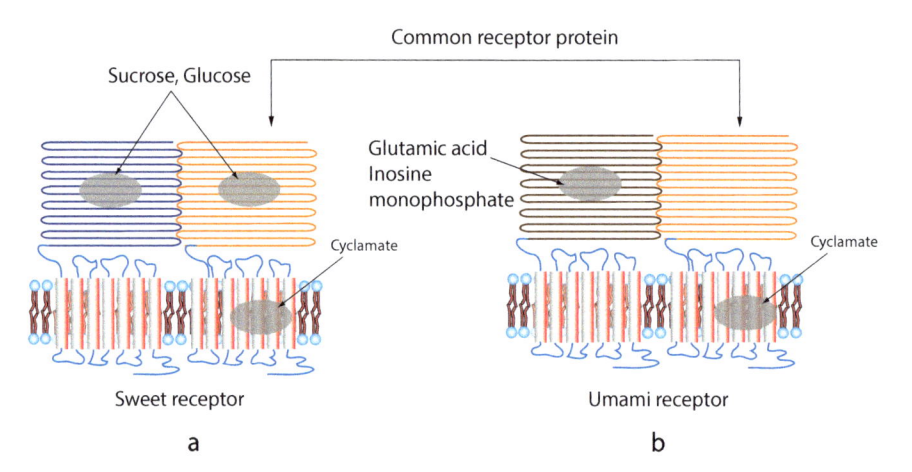

Fig. 4.7 The two receptor proteins for sweet, T1R2-T1R3 (**a**), and umami, T1R1-T1R3 (**b**), which are located in the taste receptors. The proteins, symbolically represented as wavy lines, are anchored in the cell membrane. The protein drawn on the right is identical for both receptors

4.4 The Chinese Restaurant Syndrome and the Chemistry of Umami Taste

Despite the greatly simplified representation, Fig. 4.7 shows the principle: The molecules of sucrose (table sugar) and glucose must encounter very specific sites (molecularly fitting pockets) on the receptors, as only there do structure and interactions match to trigger the stimulus current for sweet; similarly for umami. Glutamic acid or inosine- and guanosine monophosphate must encounter very specific sites on the protein of the partner receptor. At the same time, the shared protein, shown in orange in Fig. 4.7, remains active for the taste direction sweet, as demonstrated by the example of cyclamate—a synthetic sweetener.

This fact has several consequences. For one, it explains why even in old age and with dementia, sweet and savory dishes are still quite popular: Umami and sweet are the basic taste directions that are least likely to be forgotten. The coupling of receptors also contributes to this.

4.4.2 Comparison of Mother's Milk, Dashi, and Chicken Broth

What does "umami" mean in practice, where do we know the taste from in everyday life, and what molecular conclusions can we draw from it? Different cooking cultures often use different methods to achieve a culinary goal: When a European prepares a soup, they use meat, vegetables, and fire (or stove). However, when a Japanese person prepares soup, they add a Kombu seaweed and check their stock of bonito flakes (*Katsuobushi*) [34]. Both chefs have the same idea: to create a broth with a lot of flavor and depth. This can be achieved, as already mentioned in Chap. 2, by long cooking. But also by combining different ingredients. In Japan, Kombu seaweed provides plenty of flavor, as it contains not only sea salt and the important mineral iodine but also a lot of glutamate, i.e., free glutamic acid. All these ingredients can be best extracted cold to slightly warm, i.e., at temperatures of up to 50 °C (otherwise, the cell materials contained therein, such as alginate, carrageenan, or, depending on the type of algae, agarose, would dissolve from the cells). However, Dashi still needs a flavor enhancer, as the Kombu seaweed extract would so far be nothing more than water mixed with monosodium glutamate plus some iodized salt. Therefore, *Katsuobushi* is also added, which are thin, dried flakes of bonito fish (a small tuna), fermented and lightly smoked, thus providing smoke for the aromas and fermentation to expose important flavor carriers and flavor enhancers. Above all, the flavor enhancers play an important role, as the glutamate is no longer important, as it is already provided by the Kombu seaweed.

Central Europeans did not use Kombu seaweed for a long time, they fermented little besides vegetables, but they liked to cook and braise for several hours. For example, chicken or beef broth, in which bones, meat scraps, and tendons were cooked for hours, have long been part of the cultural standard repertoire of many communities and clans. They were cooked in advance and served as a basis for many dishes, even as a seasoning, as they rounded off and brought depth to every meal. So it had to be worthwhile to cook food for hours. The reward lies, as usual, in good taste. And this taste is close to mother's milk.

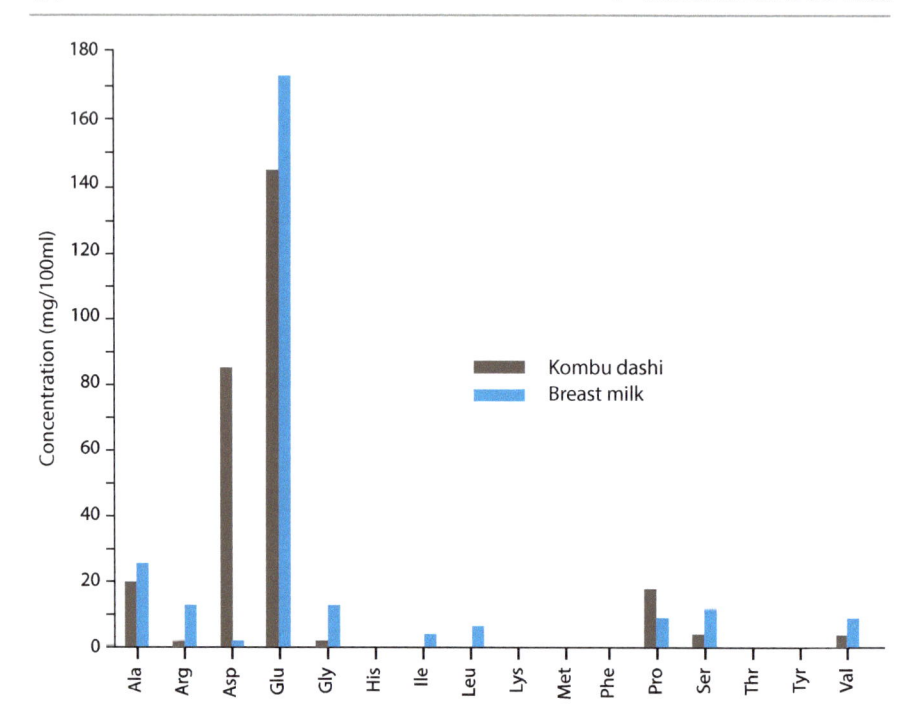

Fig. 4.8 The direct comparison of free amino acids in breast milk (blue bars) and in kombu seaweed-based dashi (brown bars) shows their close taste relationship despite different aromas

With about 20 mg per 100 ml breast milk, glutamic acid is by far the most concentrated of all amino acids, followed by taurine (5 mg per 100 ml) [35]. This amount of glutamate is roughly equivalent to that found in classic Japanese umami dishes as well as in many meat broths, as shown in Fig. 4.8. No wonder, then, that even children in Asia who have just been weaned and have not yet had the opportunity to get to know the other flavors, eat dashi without making a face.

A similar situation applies, for example, to chicken broth. Glutamic acid is present here in an even slightly higher concentration than in dashi and breast milk, but overall has a higher amino acid content due to the use of meat during cooking, with strong peaks in alanine, indicating a balanced sweet-umami interplay. More detailed analyses of all relevant taste enhancers show even more: glutamate is not everything [36]. Above all, two very important taste components also play a role: inosine- and guanosine monophosphate (IMP and GMP). They greatly enhance the taste impression umami. Thus, three essential components are responsible for the umami taste. They are shown in Fig. 4.9.

But where do these taste enhancers come from in the first place? Apparently, no one has added them extra in this case, leaving only the ingredients that can bring them along.

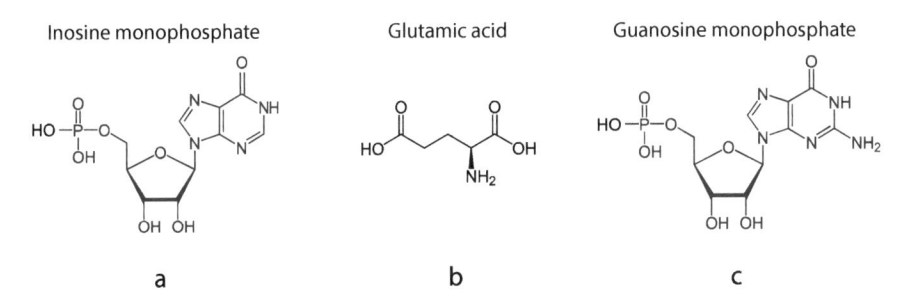

Fig. 4.9 The structure of the three main umami components: glutamic acid (**b**) and the taste enhancers inosine monophosphate(**a**) and guanosine monophosphate (**c**)

4.4.3 The Umami Taste as a Result of Purine Metabolism

In fact, it is about the actual flavor enhancers, specifically a degradation product of the cell fuel, energy supplier, and switch for muscle movements, ATP (Adenosine triphosphate). As it turns out, the two flavor enhancers are essential degradation products of the purine metabolism [37]. This is also often cited by nutrition experts and physicians when it comes to food recommendations for gout and similar diseases [38].

In fact, the breakdown of ATP occurs in steps, and at the end of this chemical reaction chain is the uric acid (purine), which—if found in high amounts in the body—becomes a control parameter for the diagnosis of gout. However, there is a long way between ATP and purine with two flavor-enhancing intermediate stations: IMP (Inosine monophosphate) and GMP (Guanosine monophosphate), which, paired with glutamic acid, multiply the umami taste. And this has been the case since humans started cooking and fermenting. This chemistry is not only relevant for cooking and combining, but also for slaughtering and maturing meat, maturing cheese, fermenting from sauerkraut to soy sauce and miso up to the Swedish Surströmming, the long-fermented fish from Northern Europe. Once again, it is demonstrated how a universally global chemistry determines the cultural techniques and the food culture.

In order to understand this, a brief excursion into chemistry is necessary. The most important stages of ATP degradation are summarized in Fig. 4.10. The cellular fuel, adenosine triphosphate (ATP; top left in Fig. 4.10) is a nucleotide, consisting of a four-fold electrically negative charged phosphate group (triphosphate) attached to a nucleoside (adenosine). This consists of a sugar (a ribose) and the base adenine. We know this base from DNA, in which the bases adenine (A), thymine (T), guanine (G), cytosine (C) (or uracil, U, instead of thymine in RNA) occur. Only adenine and guanine are so-called purine bases and therefore have taste potential. ATP gradually breaks down, for example during meat maturation or in vegetables, and initially forms di- and monophosphates (ADP and AMP) when phosphate groups are cleaved off. From AMP onwards, it becomes interesting for taste. Depending on the organism and enzyme status, inosine monophosphate

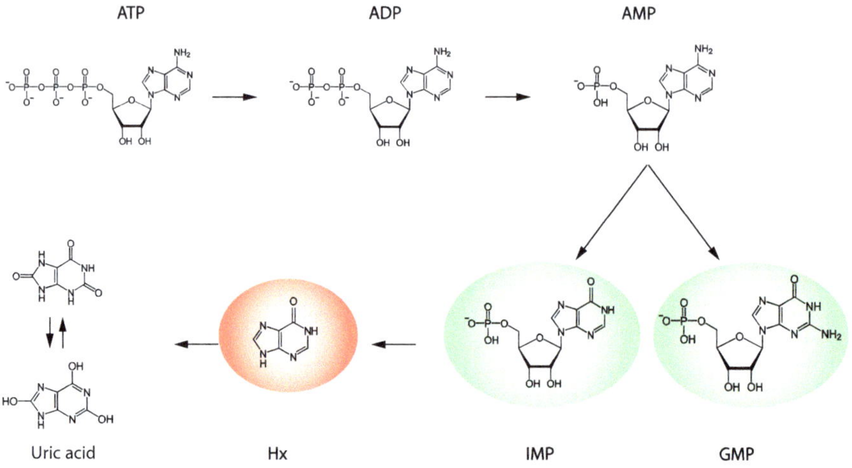

Fig. 4.10 Large parts of the taste formation umami are part of the purine metabolism and are due to the degradation of ATP (Adenosine triphosphate). From ATP, ADP is first formed (Adenosine diphosphate), then AMP (Adenosine monophosphate). The color-highlighted molecules guanosine monophosphate (GMP), inosine monophosphate (IMP) and hypoxanthine (Hx) are taste determining

(IMP) or guanosine monophosphate (GMP) are formed, both of which have an extremely high umami potential and support glutamic acid (Glu); IMP and GMP, by the way, to a similar extent [39]. The umami taste is intensified many times over. The synergy potential between glutamic acid and the nucleotides IMP and GMP shows: IMP and GMP are the actual flavor enhancers. As one of the next stages of ATP degradation, hypoxanthine is formed, which, with its bitter flavors from cooked meat, is often undesirable, but highly welcome in raw sausages and ham, for example. The enzyme xanthine oxidase is also involved in this purine metabolism, which further converts hypoxanthine towards uric acid. This enzyme, which is important for the purine metabolism of all living beings, is also said to have a high harmfulness, especially in connection with homogenized milk, as we will see later. That this is questionable from a molecular point of view will become clearer after further systemic preliminary work.

4.4.4 The Synergy Effect between Glutamic Acid and Nucleotides

The ability to enhance taste is evident even in simple experiments and has been known since 1967. In sensory experiments, test subjects were given aqueous solutions of certain taste substances for tasting, after which they had to record their taste impressions [40]. To do this, the two individual components glutamic acid and IMP were first dissolved in water and given to the test subjects for tasting.

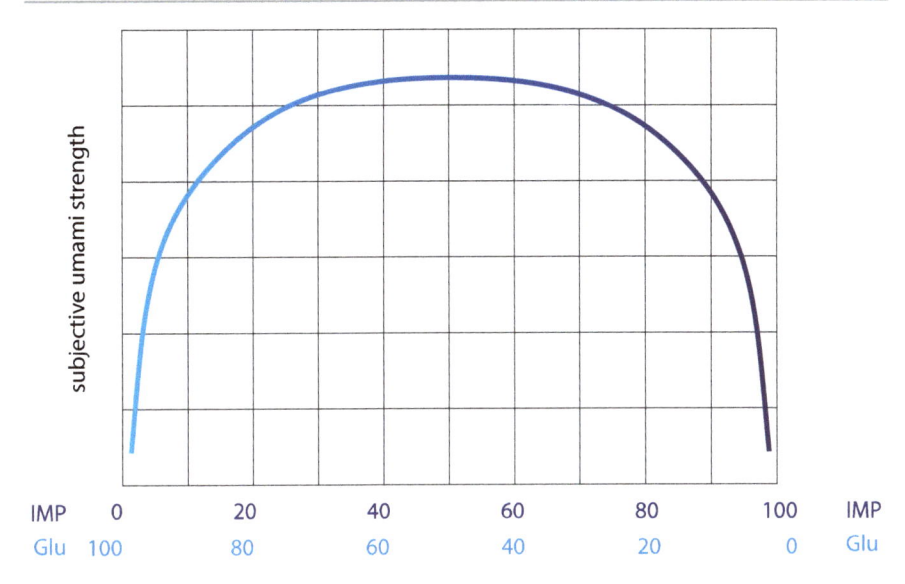

4.4 The Chinese Restaurant Syndrome and the Chemistry of Umami Taste

Fig. 4.11 Amplification and synergy effect of glutamic acid (Glu) and inosine monophosphate (IMP). Both taste umami; however, when they work together, the savory perception increases significantly. The subjective strength of the umami taste can even be mathematically modeled as an inverted parabola

With increasing concentration, the intensity of the umami taste (linear) increased, much more so for glutamic acid than for IMP. Of course, glutamic acid is the main trigger for this savory taste quality. However, when both glutamic acid and IMP were mixed in different concentrations, so that the sum of the concentrations was always the same, a very astonishing picture emerged. The two showed strong synergy effects, and the intensity of the umami taste was subjectively multiplied. When the concentration of glutamic acid was slightly reduced and replaced by IMP, the umami intensity increased significantly. A clear picture emerged, which became even more precise as the number of test subjects increased. This averaged result is summarized in Fig. 4.11. In the ratio between 10/90 and 90/10, the taste sensation is therefore strong, and at 50:50, it is even maximal.

Looking at the original publications, it is noticeable that they were conducted by scientists at the Central Research Laboratory, Ajinomoto Co. Inc. Kawasaki, in Japan. Now, Ajinomoto is precisely the company that manufactures and sells glutamate worldwide. Of course, one might think that these studies could be subject to a certain bias. However, this is not the case in this instance, as the subjective enhancement effects can be verified on one's own tongue. Moreover, the results can also be verified through objective experiments, without the involvement of humans in sensory panels, but only by measuring stimulus currents on biotechnologically expressed receptors [41]. Furthermore, when computer simulations are used, the binding sites and molecular changes of the receptor protein become

Receptor protein without stimulus	Receptor protein with Glu	Receptor protein with Glu + IMP
a	b	c

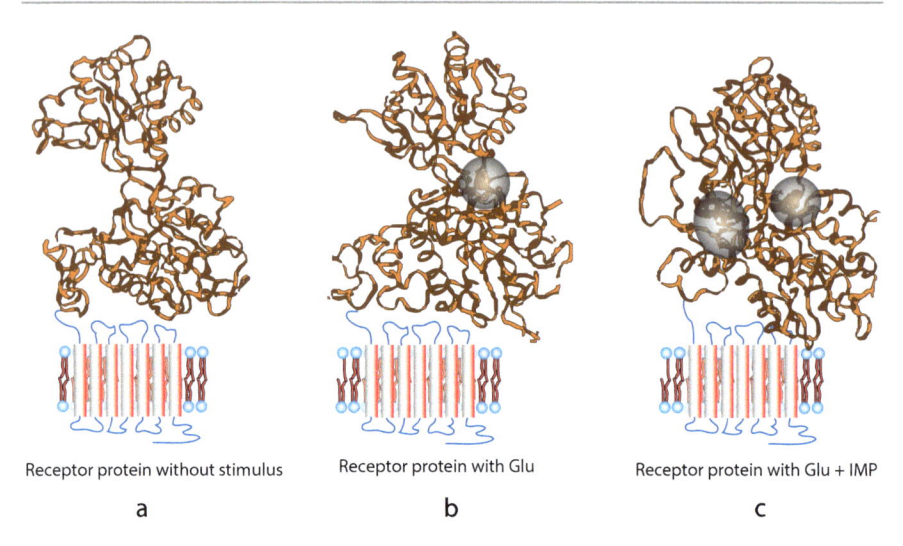

Fig. 4.12 The receptor protein T1R1 without stimulus (**a**), only with glutamic acid (**b**) and with glutamic acid and IMP (**c**), when the shape of the protein changes most significantly upon docking and thus the stimulus currents for the umami signal are highest

apparent [42]. The molecular effect of the substances can be precisely demonstrated on the receptor proteins, as shown in a simplified manner in Fig. 4.12.

Thus, the synergy of the two umami protagonists is objectively demonstrated through experiments and confirms the subjective perception in the sensory tests from Fig. 4.11.

4.4.5 Consequences for the Philosophy of Taste

To take a philosophical look at the topic of taste for a moment: Umami as the sum of stimuli from glutamic acid, aspartic acid, a whole series of peptides, and the nucleotides GMP and IMP is thus the "taste of life." The savory taste thus comes directly from the most important structural molecules, the proteins, as well as the derivatives of some nucleotides of DNA. This also includes the related taste sweet, because besides ATP is the glucose also an energy supplier for every single cell. The salty taste of the minerals supplied therein is the taste of the earth ("salt of the earth"), on which we live, because the soil contains hardly any macronutrients, but only provides minerals for the growth of plants that serve humans and animals as food. The basic taste sour is also hidden in cell function, all these acid-donating molecules live on protonation and deprotonation, which is regulated via the pH value, i.e., protons (H_3O^+). This is also visible in proteins: We always taste it just below the isoelectric point of most proteins, between pH 5 and pH 3. Below or far above (in the alkaline range), no cook can make a name for themselves. Bitter, on the other hand, is the taste of anti-life, of death. A strong bitter taste often means poison and warns of caution. But after these digressions, let us quickly return to the hard facts.

4.5 Glutamate and Nucleotides as Flavor Enhancers are the Cause of Global Cooking Cultures

4.5.1 How Taste Chemistry Determines Cooking Culture

The most important taste-causing agents for umami are thus glutamic acid, inosine monophosphate and guanosine monophosphate, which mutually reinforce each other. As already shown in Chap. 1, the umami taste was the driving force of evolution. This thesis can be further substantiated, because the two flavor enhancers IMP and GMP are also driving forces of universal cooking cultures and cultural techniques, regardless of where they were developed in this world. This is evident in many traditional recipes, whether they were developed in Europe, Asia, or Latin America.

As already noted with Dashi and chicken soup, there are two ways to create a tasty soup. Either the glutamate is added via Kombu seaweed and the flavor enhancers IMP and GMP through the bonito flakes, or it is cooked for a long time and, if necessary, various vegetables are added to the soup. What may seem banal on the one hand and from a cultural studies perspective can simply be the development of certain regional, local cooking cultures, follows the taste principle umami upon closer examination.

Let us take a moment to consider a classic European-style braised dish, regardless of the region. This always includes meat, acidity, usually in the form of vinegars or wines, specific braising vegetables such as celery, carrots, onions, and since tomatoes have been available in Europe, these as well. Above all, mushrooms are among the standard ingredients. When available, mushrooms or mushroom remnants are added to the braising pot. It is braised for hours at low heat. The country-specific differences are almost marginal: In some regions, anchovies are added to the braised dish (Spain, Italy, North Africa). And not so that the dish tastes like fish, but as a flavor enhancer, similar to the bonito flakes in Dashi broth. Therefore, it is not surprising if fish sauce is found in standardized industrial ready-made meals, even though the preparation does not go in the direction of seafood. This highly fermented fish sauce, which is mainly a seasoning standard (see Maggi) in Asian cuisine, does not taste like fish at all, but is a guarantee for glutamic acid Umami peptides and IMP—like the one or two anchovies in the pasta sauce of Italian cuisine. Or the "Maggi" of the ancient Romans and Greeks, the fish-based seasoning sauce Garum already mentioned in Chap. 1.

In some regions of Italy, cheese is added, even if the braising time of the pasta sauce is long. All of this has a deeper meaning: to achieve flavor depth, to hydrolyze proteins—hence the acidity, which supports this process. The combinations of

Fig. 4.13 Some examples of natural foods that participate in the interplay of glutamic acid, IMP and GMP. They belong to the basic ingredients of a universal cuisine, which in large parts is even independent of the respective cooking and eating culture. Meat provides plenty of IMP due to the purines in addition to glutamic acid, while mushroom cultures rather provide GMP

meat and root vegetables systematically show the selection of ingredients according to glutamic acid, IMP and GMP, as illustrated in Fig. 4.13.

Many traditional dishes and recipes can be found in this scheme, such as the beef goulash, which, in addition to meat, also contains tomato paste, stock, and plenty of onions as basic ingredients. Many inn cooks prefer to use the previously mentioned chicken broth instead of beef or veal stock, even though the goal is anything but a chicken fricassee. The chicken broth is merely a supplier of umami. Chicken and veal fricassees often include asparagus, peas, and root vegetables, as well as mushrooms and other fungi. Traditional recipes for chicken fricassee suggest using a soup chicken with soup greens (consisting of celery, leek, parsley, and carrot) along with mushrooms and peas (i.e., legumes). For Szegediner goulash, fermented cabbage, sauerkraut, is used in addition to broth, root vegetables, onions, and tomatoes to give the taste a strong umami component. Umami is therefore prominently placed alongside the sour taste. The Finnish classic Karjalanpaisti (Karelian meat stew, with meat, smoked and aged bacon, onions, carrots, and broth powder) follows exactly the same principle: umami enhancement on a broad front.

Similar combinations also occur on other continents. In Mexico, for example, it is the combination of meat and protein-rich legumes (chili con carne) that scores high in taste with plenty of glutamic acid, IMP and GMP. The traditional Peruvian dish Carapulcra combines pork fat, onions, chicken broth, and pureed peanuts, which naturally belong to the legume family—and, for example, are on the blacklist for gout patients because they contain too much uric acid—positively expressed (Fig. 4.10) a high umami potential.

European versions of the meat and legume combination are abundant: Be it the cassoulet in Perigord, the many recipes with Borlotti beans in Italy, or the Swabian lentils with smoked meat, which simmer for several hours. Soups and broths in China, Thailand, or Vietnam are prepared by cooking broth-typical umami ingredients, then enhancing them with miso pastes before being finished as noodle soups, fish soups, or poultry broths. Not to forget the countless recipes in many regions of China, where meat, cabbage, and broth are systematically combined.

The goal of many dishes in all cultures was thus to enhance the taste, optimize it, and develop as delicious dishes as possible, as shown in the combination scheme in Fig. 4.13. The goal is always a high level of umami taste. The molecular inputs came from the same categories in many places around the world.

Umami was apparently not only the driving force of the Evolution of cooking, but is also responsible for the development of similar culinary practices and food combinations in various cultures. Taste is therefore universal. Thus, the development of all cooking cultures is taste-driven. The *Homo sapiens* has been following its chemical senses for millions of years. It hasn't done much wrong in the process. However, modern humans are currently in the process of unlearning how to eat.

4.5.2 From Hidden Glutamate in Yeast Extract

An example of this is the discussions about yeast extract and the hidden glutamate [43]. The glutamate in yeast extract is not hidden, but lies like an open book for everyone to see (Fig. 4.14). In yeast extract, nothing more than hydrolysis takes place, i.e., a splitting of proteins. Only here, not soy protein (soy sauce), not wheat protein (Maggi) or not fish protein (fish sauce) is used, but the protein of yeasts. The process of production is simple biochemistry: Yeasts are allowed to multiply, as in bread or cake dough, in the wine cellar or brewhouse. They release a lot of enzymes that, for example, transform sugar into alcohol and flavors, but also proteases that can split proteins. If these proteases predominate, they begin their work. They break down the proteins of the yeast into their amino acids and peptides, until the yeast has largely digested itself. The amino acids are then dissolved in the nutrient solution, and the cell material remains as particles in the cells. This liquid is then centrifuged. A solid sediment, a paste, and the pure, glutamate-rich liquid with the other amino acids are formed—products that all have a high potential for umami.

Therefore, yeast extract belongs to the completely normal, even natural protein hydrolysates, like soy sauce, miso paste, Maggi seasoning, fish and oyster sauce. Nothing is hidden, nothing is unclear. Yeast extract is therefore no more harmful or even dangerous than the soy sauce with sushi or the grated Parmigiano over the long-simmered pasta sauce, which is already a glutamate bomb by itself.

How arguments are made ignoring facts is clearly demonstrated by this: Hardly anyone talked about the hidden glutamate in soy sauce or in Parmesan or

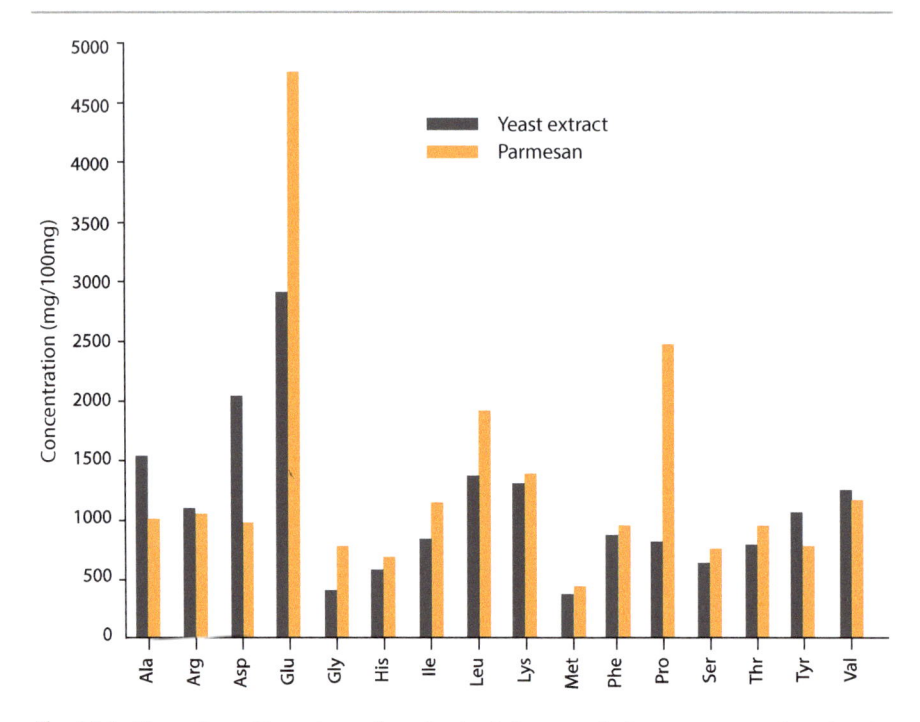

Fig. 4.14 The amino acid spectrum of yeast extract, for example from vegan spreads and pastes (dark bars), and Parmesan (yellow bars) is quite similar. Very ripe Parmesan provides, on average, much more free glutamate compared to yeast extract

other mature cheeses and demonized these foods. They are lobbied as "natural" in accordance with *slow-food* principles. Yeast extract sounds industrial, as does glutamate production, and these are demonized, even though they are chemically identical processes. It is merely about the taste of the amino acids. On the other hand, vegan yeast extract is considered "good," *"clean"* and politically correct. It is apparently an attempt to distinguish between good and evil. However, it is guaranteed: A cheese extract from long-aged Alpine cheese or Parmesan is at least as "bad" as yeast extract, as clearly shown in Fig. 4.14.

Anyone who dares to try sprinkling their pasta with yeast flakes from the health food store instead of grated Parmesan or Sbrinz will indeed experience a miracle: The taste differences are marginal, only the flavors are different.

4.5.3 Artificial and Natural Glutamate?

The difference between natural and artificial glutamate is said to be considerable, according to the most common objection. Similar statements can be found in the weekly magazine *Focus,* which states that glutamic acid also contains glutamate. This statement is as naive as someone claiming that water also contains H_2O.

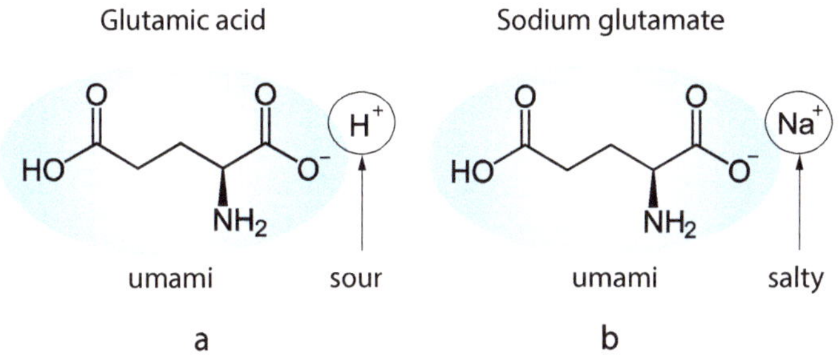

4.5 Glutamate and Nucleotides as Flavor Enhancers are the Cause ...

Fig. 4.15 The "natural" glutamic acid (**a**) and the "artificial" (mono-)sodiumglutamate (**b**). The two umami triggers (highlighted in blue) are identical in both cases. The only difference is the sodium ion in the glutamate. Glutamic acid tastes umami and sour (depending on concentration), glutamate tastes umami and a hint of salty

Such formulations hardly serve to educate, but rather stir up fears. The confusion is particularly great when it is claimed that only the artificially produced form of glutamate is of concern [44]. This claim is not substantiated. However, since it is labeled as "explained understandably," the unsuspecting reader is inclined to accept it as such. In cases of doubt, a look at the molecules, shown in Fig. 4.15, is useful and helpful.

As Fig. 4.15 clearly shows, the two umami triggering anions are identical. The fact that "artificial glutamate" releases a sodium ion is due to the nature of the substance. Deprotonated glutamic acid is electrically charged. It could not be stably incorporated into a solid. Only by neutralizing the negative charge, for example with sodium (or potassium), can it be filled as crystals in packaging for end consumers. "Nature-identical" glutamic acid would also exist: dissolved in water. Glutamate, i.e. sodium glutamate or its relatives, are thus the purest form of umami taste, even if it is hardly possible without a slight salt content via sodium (or potassium) in taste. With this knowledge, the pinch of glutamate available in the Asian restaurant is completely equivalent to a pinch of refined salt, a pinch of refined white table sugar, or even a pinch of citric acid (as crystals). When seasoning, only an independent taste direction is served. Advanced tasters therefore always have some caffeine for the final seasoning with fine bitter notes in the kitchen. A pinch of it adds a touch of clear bitter notes to any dish, if it fits. And with that, the taste quintet would be complete.

There is therefore no fundamental difference between natural and artificial glutamate—whether it is extracted, fermented, or even synthetic. None of it is harmful or even allergenic, as suspected by Kwok [29], because these molecules are identical. Our sensors, even complex molecules like taste receptors, only react to molecules. For a long time now, even the proteinogenic L-glutamic acid can be precisely characterized and produced, and even distinguished from the mirror-symmetric, non-proteinogenic D-glutamic acid. Nota bene: If living beings, including humans, were allergic to glutamic acid, neither proteins nor enzymes nor brain function could be built up. No embryo of *Homo sapiens* could develop and grow into a thinking human being.

4.6 Umami Never Comes Alone

So far, the umami taste has always been mentioned, but an important point has not yet been addressed: In contrast to sweet, salty, sour, and bitter, umami does not exist in its purest form. As already noted, the simplest form is glutamate, but it usually comes as a salt, so there is always a salt taste involved. Umami therefore never comes alone; it is always paired with other sensory impressions. This also includes the mouthfeel that we perceive in ripe cheeses, thick sauces, long-cooked broths, or long-fermented products, but not in pure glutamate. It is probably this that makes glutamate seem unnatural and disturbing to some people. Pure glutamate lacks mouthfeel.

This can be most easily seen in broths or braised dishes, which have been on the menu of *Homo sapiens* for a long time. Anyone who cooks their own broths knows the problem: A broth needs enough time to achieve this mouthfeel. After 20 minutes of cooking, it doesn't taste good, but after an hour and a half, you can guess what will be created after three or four hours: depth, mouthfeel, or Japanese "kokumi" (Fig. 4.16).

It becomes apparent how closely kokumi is linked to umami. During cooking, braising, and especially during the ripening of cheese, the protein hydrolysis already mentioned in Chap. 2 takes place with increasing cooking time. As long as the fragments of proteins (peptides) are large, their effect is only small (with a few exceptions). As they become smaller, di- and tripeptides are formed, which then—as soon as suitable peptides (γ-glutamyl peptides) such as glutathione are formed—provide the mouthfeel. However, cooking for too long is also of not

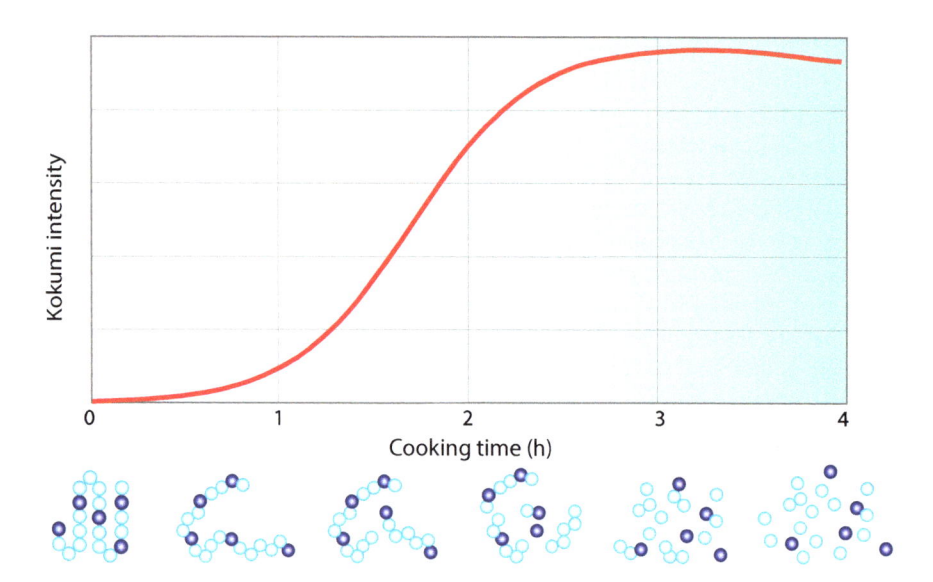

Fig. 4.16 The kokumi intensity as a function of cooking time and the progressive protein hydrolysis

4.6 Umami Never Comes Alone

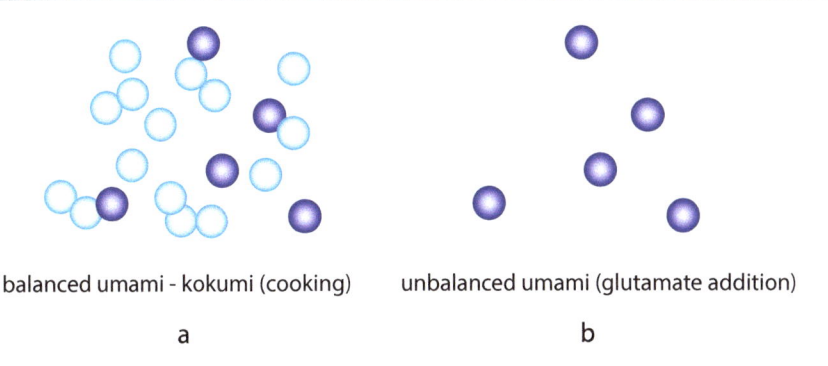

balanced umami - kokumi (cooking)　　unbalanced umami (glutamate addition)

　　　　　a　　　　　　　　　　　　　　　　b

Fig. 4.17 Natural umami taste is created through active cultural techniques and provides the entire accompanying program of protein hydrolysis, including mouthfeel kokumi (**a**). With the equivalent addition of glutamate, an identical level of glutamic acid is achieved, but no additional components for mouthfeel (**b**)

much use, as the mouthfeel can even decrease again if more free glutamic acid is formed as soon as the γ-glutamyl peptides continue to split under the supply of further thermal energy. This creates more glutamate, but less kokumi. The mouthfeel decreases, and the umami taste increases slightly.

This clearly shows: The taste quality umami never comes alone. It is always accompanied by the taste of other amino acids, even if—as in breast milk—the glutamic acid dominates. At the same time, the umami taste that arises naturally through cultural techniques such as cooking or fermentation is always accompanied by mouthfeel. In this respect, a "glutamized glass of water", i.e., water mixed with pure glutamate, has little to do with the "natural" umami of a broth or ferment. Not because the glutamate tastes different or worse, but because it lacks the entire molecular accompanying program of fine-tuning through the taste of the other amino acids of the peptides and the mouthfeel through the kokumi-relevant peptides. However, deriving a subjective harmfulness or unnaturalness from this remains a fallacy. What appears unnatural is—apart from flavors—above all the absence of the many sensory components on the tongue, as symbolically represented in Fig. 4.17.

In this context, the question remains: What about the Chinese restaurant syndrome? Is it merely imagination or real? The answer probably lies between these two extremes. Above all, however, it must be seen in connection with similar phenomena that have little to do with glutamate powder. Nevertheless, it is definitely not a result of glutamic acid.

4.6.1 The Chinese Restaurant Syndrome—Biogenic Amines and Secondary Products

Intolerances and reactions to fermented products are widely known. These include irritations when consuming red wine, matured cheese, very long riped ham or raw

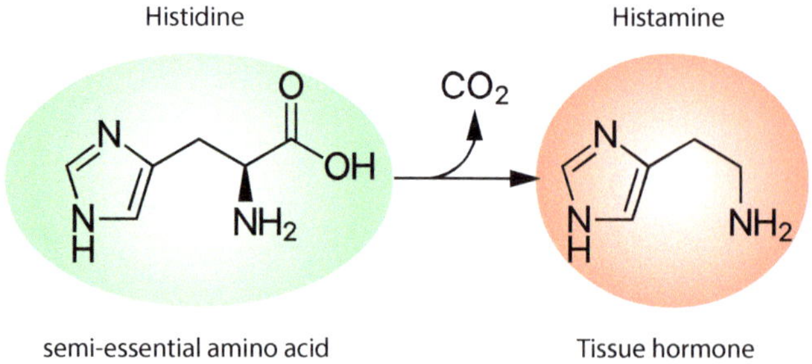

Fig. 4.18 From the (semi-essential) amino acid histidine, the tissue hormone histamine is formed, a biogenic amine

sausages, soy sauces, Maggi seasonings, and similar products. Naturally, higher concentrations of glutamic acid occur in the above-mentioned culinary items. However, the triggers for intolerance reactions in sensitive individuals are to be found in the other amino acids or in what results from them. During ripening or fermentation, some amino acids can chemically react to biogenic amines, which can trigger bodily reactions even in low concentrations. One of the best-known representatives of amines is histamine, formed from the amino acid histidine, which occurs in maturation and fermentation processes as well as in the body's own physiology. The relatively simple process via a decarboxylation is shown in Fig. 4.18. If there is too much free histamine in the body, reactions such as the narrowing of peripheral and central blood vessels, the narrowing of the airways, or other typical allergic reactions can be triggered.

The formation of histamine is just one example. Several amino acids are precursors of biogenic amines, which are responsible for the different reactions when consuming fermented and matured products. The reaction products that can arise from amino acids during ripening and fermentation have been systematically investigated some time ago [45], and the description is still valid. Further typical examples are listed in Table 4.1.

Considering the reactions and symptoms in Table 4.1, most of the symptoms associated with the China Restaurant Syndrome can be identified. In very high amounts, some biogenic amines actually have toxic effects [46]. Such high concentrations can hardly be reached in food, unless the fermentation process goes completely out of control [47]. In a scheme, amine-rich (fermented and matured) foods can be arranged according to their toxicity. Foods with a biogenic amine content between 0.1 and 1% are considered safe, while it becomes risky at 2% and above.

However, there is evidence that in people with histamine intolerance, there is a link between the enzymatic breakdown of histamines and other biogenic amines and free glutamic acid. Histamine from food is broken down by the body's own enzymes (diamine oxidase, DAO) in the small intestine. If this breakdown does not work and histamines enter the intestine, the symptoms described in Table 4.1

4.6 Umami Never Comes Alone

Table 4.1 Selection of typical amines formed during ripening and fermentation. The table is not complete, as more biogenic amines can form. The mentioned ones are sufficient for a basic understanding

Amine	Precursor	Extreme reactions and symptoms
Histamine	Histidine	Release of adrenaline, vasoconstriction, irregularities in the regulation of gastric acid, stimulation of smooth muscle in airways, uterus, and intestine
Tyramine	Tyrosine	Peripheral vasoconstriction, increased heart activity, increased tear and saliva production, increased respiratory rate, rise in blood sugar levels, can trigger migraines and headaches, releases noradrenaline
Cadaverine	Lysine	Lowering of blood pressure, slowing of heartbeat, lockjaw, relaxation and paralysis of limbs, enhances the effects of other biogenic amines
β-Phenylethylamine	Phenylalanine	Release of adrenaline and noradrenaline, high blood pressure, headache and migraine
Tryptamine	Tryptophan	High blood pressure

may occur. In people with histamine intolerance, the enzyme's effect is insufficient. Too much histamine (and other biogenic amines) enter the small intestine. Various foods can further block or even completely inhibit the activity of the DAO enzyme. These include some hazelnuts and walnuts, some mushrooms, fruit, and also fermented chocolate. Alcohol, tobacco, and gelatin are also suspected of inhibiting diamine oxidase. Free glutamic acid also shows an inhibitory effect on DAO in animal experiments [48]. This would mean that in people who are already sensitive to biogenic amines, the glutamate further intensifies the symptoms. They are, so to speak, doubly concerned, because all foods with high histamine content already contain relatively high amounts of glutamate due to the maturation and fermentation processes.

As can be seen from Fig. 4.19, most fermented foods are harmless, as the concentrations of biogenic amines remain low. High amine concentrations are mainly found in Asian fish and shrimp pastes, as they are part of the food culture in Malaysia and Indonesia, but are only used in small amounts, as their seasoning power is considerable.

As can also be seen, pure lacto-fermented foods such as sauerkraut, pickles, or yogurt appear to have relatively low amine concentrations. The reason is actually the fermentation process of lactic acid bacteria, which does not have strong protease activity and primarily responds to lactose, glucose, or other sugars. When working with yeasts, such as in beer and wine, protein-splitting enzymes can already release amino acids to a greater extent. The probability of them being converted to amines increases. When fermentation processes with a high proportion of proteases are carried out, such as with the koji mold (Aspergillus oryzae) often used in Asia, proteins are rapidly broken down and the many free amino acids consequently react to a greater extent and with a higher probability to form amines.

secure	tolerable	risky
0.1%	1%	>2%

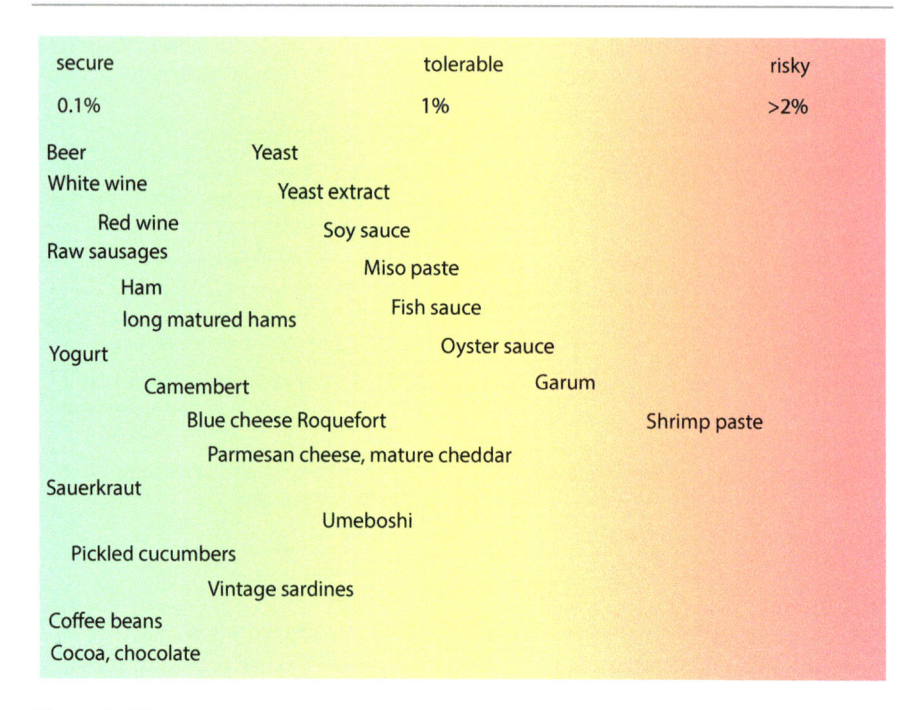

Beer
White wine
   Yeast
  Red wine
    Yeast extract
Raw sausages
     Soy sauce
  Ham
     Miso paste
   long matured hams
     Fish sauce
Yogurt
      Oyster sauce
   Camembert
       Garum
    Blue cheese Roquefort
         Shrimp paste
    Parmesan cheese, mature cheddar
Sauerkraut
     Umeboshi
  Pickled cucumbers
    Vintage sardines
Coffee beans
Cocoa, chocolate

Fig. 4.19 The scheme of fermented and matured foods. The green highlighted products are considered definitely safe, while amine concentrations that reach into the red area become more critical

This explains why yeast extracts, soy sauces, and miso pastes already have higher concentrations of biogenic amines.

If foods contain high amounts of protein (and water), such as cheese, more proteins are hydrolyzed with increasing ripening time or with the help of noble mold fungi (e.g., Camembert and blue cheese) through their proteases, which in turn suggests a higher amine concentration. Fish and seafood pastes use the muscle's own proteins to break down proteins almost completely with a high salt addition (up to 15% of the input weight). The consequence is that this is accompanied by a high content of biogenic amines, such as in the example of shrimp paste. Mediterranean anchovy pastes are in similar ranges.

There are indeed people who react more sensitively to biogenic amines and already react at lower concentrations (between the green and yellow areas in Fig. 4.19) with the symptoms exemplified in Table 4.1. Exact predictions are not easy, as some of the amines can mutually reinforce each other. Also, the distribution of amines in different foods is not identical. It is quite possible that, for example, soy sauces can be consumed with pleasure without problems, while very ripe cheese causes problems.

This all sounds worse than it is. It does not change the fact that the benefits of consuming fermented products outweigh the high concentrations of glutamate and corresponding concentrations of biogenic amines and persist [49]. FODMAPs are largely broken down, proteins become more available, acid-tolerant bacteria enter the microbiome, just to repeat some of the facts. Evolution also shows this. For millennia, fermented products have contributed to the health of tribes and peoples.

In spontaneous, wild fermentations, microorganisms are often at work that show a higher tendency to form biogenic amines. This is due to their enzyme equipment. That is precisely why today, especially with raw sausages or animal products such as fish pastes, the concentration of biogenic amines is controlled by the use of selected bacterial strains in the starter culture, even if this sometimes comes at the expense of flavor formation. There are always two sides to a coin: full flavor or total safety. Both are not possible at the same time without the help of (green) genetic engineering (e.g., the CRISPR/Cas9 method), compare raw milk cheese.

4.7 Glutamic Acid and Its Function Beyond Taste

In these medically proven and objectifiable reactions of predisposed people with reactions to biogenic amines, the free glutamic acid was no longer mentioned. This raises the question of what happens to glutamic acid during ripening. What amines and harmful secondary products are formed from it? The answer is simple: none, because that is its advantage. The glutamic acid remains and is responsible for umami taste! No matter how proteins are broken down: free glutamic acid remains largely stable. Therefore, it is present in all hydrolysates such as Maggi, soy, and fish sauces in considerable concentrations. That is why the glutamate concentration remains so high in fermented products, long-aged cheese and sausages, and long-cooked broths. It is due to the invariance of glutamic acid, which is resistant to reactions and degradation.

Thus, it becomes clear: the trigger for all discomforts associated with fermented products—whether soy sauces, miso pastes, mature red wine, or mature cheese—was never the glutamic acid, but rather the compounds formed from other amino acids were always to blame.

From the findings on biogenic amines, it can even be claimed that a pinch of glutamate in food is basically much healthier than the load of natural glutamate in soy sauce, matured Cheddar, or matured ham. Despite its bad reputation, pure glutamate does not contain a single biogenic amine that could cause irritation.

It is incredible and scientifically incomprehensible how the opinion that glutamate was the only evil became independent through copying and constant repetition. This misconception is completely incomprehensible from a scientific point of view. Systematically removing glutamate, including yeast extract, soy sauce, Parmesan, etc., from every meal would result in an unparalleled loss of taste. Chefs would be up in arms if stock bases were only allowed to be cooked briefly. Considering that glutamate and umami are the driving forces of evolution and food culture, these politically led discussions sound absurd and are completely detached from reality.

Overdosing Glutamic acid, would of course have no particular culinary appeal. However, this would be just as nonsensical as eating a pound of salt by the spoonful. No one would think of doing that. This is not uncommon for humans to do in the laboratory with animal models, such as rats and mice—to prove that glutamate, salt & Co are harmful. One can imagine how meaningful such studies are.

A new study shows the actual positive neurocognitive effects of glutamate [50]. Participants were given a glass of broth with glutamate to drink before making their meal choices at a buffet to see how this would determine their food selection. The result was astonishing: the participants subsequently chose healthier meals. However, monosodium glutamate did not drastically reduce the kilocalorie intake of the test subjects. The evaluation of eye-tracking also revealed that after consuming the soup with monosodium glutamate, the test subjects were much more focused on their chosen meal and did not pay further attention to the buffet. Increased activity was observed in the left prefrontal cortex, the part of the brain associated with self-control in food intake. Whether these new findings will actually be confirmed remains to be seen. However, they are definitely in line with the feature that the umami taste has been guiding *Homo sapiens* on the right path for a million years.

4.8 Physics of Sugar: Taste, Water Binding, Preservation

Sugar is unhealthy. This statement is true, but it is also fundamentally wrong. The claims go even further: sugar is poison [51], sugar is addictive [52]. Some internet pages elevate sugar to a drug [53]. Such statements naturally attract attention, which is exactly their goal. But what about the facts?

Time and again, sugar-free alternatives such as honey, maple- or rice syrup are referred to. This is, of course, scientifically untenable. The supposedly "sugar-free" alternatives contain a wide range of different sugars. Rice syrup consists practically of 100% glucose, meaning enzymatically split rice starch. This million-fold sugar is enzymatically split into glucoses, dextroses, and oligoglucoses. Sugar-free is often a deception. Clarification on many questions can be found again through more detailed physical-chemical considerations.

4.8.1 Sugar and Natural Sugar Alternatives

So what exactly is behind these "natural" sugar alternatives, especially honey, or the plant-based alternatives like maple syrup, agave nectar, or rice syrup? First, look at refined table sugar, as it is indeed an instructive basic system from a nutritional chemistry perspective. It consists of glucose and the recently much-maligned fructose; both are connected by a glycosidic bond, as shown in Fig. 4.20.

4.8 Physics of Sugar: Taste, Water Binding, Preservation

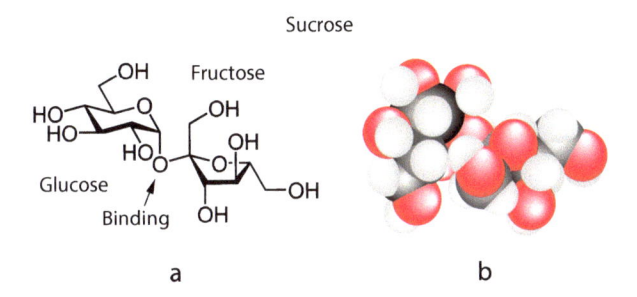

Fig. 4.20 Table sugar consists of one glucose and one fructose ring (**a**); in (**b**) the space-filling model is shown, carbon atoms are black, oxygen atoms are red, and hydrogen atoms are gray

Looking at the structural formula of sucrose, it immediately becomes apparent how glucose and fructose differ. The structure of glucose consists of a six-membered ring and therefore has six corners. Fructose is a five-membered ring and therefore has only five corners. They are different systems from the perspective of molecular structure, and for this reason alone, they must also be recognized and metabolized differently physiologically. As a reminder: Fructose belongs to the FODMAPs, which alone indicates that it must be metabolized differently from glucose. They also differ in terms of sweetness, which is, of course, due to the molecular structure.

Refined table sugar (sucrose) serves as the standard for the sweetness, against which all other sugars and sweeteners must be measured, as shown in Table 4.2 with just a few examples.

However, Table 4.2 should not be overrated from a physicochemical point of view, although it seems meaningful from a food technology perspective. The table compares apples to oranges, as the molecular structure of the various sweeteners is completely different, which is not only expressed in the sweetness, but also at the binding sites where the sugars and sweeteners must dock onto the receptor in order to trigger a sweet taste sensation at all. This cannot be explained in detail, but still, some universal and generally valid physicochemical rules can be

Table 4.2 The sweetness of some sugars and sweeteners in comparison

Sugar	Sweetness
Sucrose	1.0
Glucose	0.75
Fructose	1.7
Lactose	0.16
Maltose	0.35
Cyclamate	40
Steviol glycosides	300–400

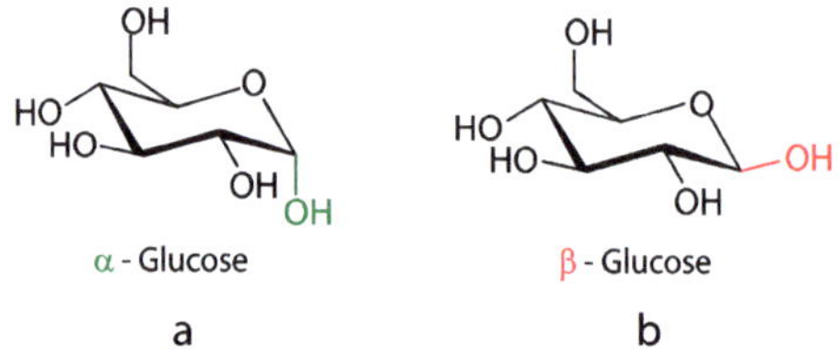

Fig. 4.21 The difference between α- and β-glucose appears marginal, but is fundamental for food: starch consists exclusively of α-glucose, cellulose of β-glucose

α - Glucose
a

β - Glucose
b

recognized that help with daily sweetening and, above all, shed some light on the darkness of sweet perception.

The chemistry of sugar is very complicated, as a large number of isomers can be formed from the respective chemical empirical formulas. There are also isomers of glucose, as shown by the difference between α- and β-glucose (Fig. 4.21).

As a small "extra sugar," let's briefly point out the relationship between molecular structure and sweetness and water binding, which play a major role in the search for sugar alternatives. The naturally occurring sugars we know so far are glucose, fructose, sucrose, galactose, lactose, and raffinose. They occur in both animal and plant foods. The sweetness of these sugars is related to the molecular structure and the strength of the hydrogen bond, which they can form. This binding strength can be imagined very simply: A polar water molecule likes to attach itself to the OH groups (hydroxyl groups) of the sugar; however, how strongly depends on the direction and position of the OH groups, as can be seen, for example, in 4.21. In the case of α-glucose, the green and black OH groups are very close together. Water is therefore bound far less, the water molecules would hinder each other and provoke a slight repulsion via the polarity. In the β-form, this effect is less relevant. The water binding is stronger, more water is bound in a hydration shell around the molecule.

And what does this have to do with the sweetness? The sweet-tasting molecules reach the receptors via saliva and are bound to the receptors via hydrogen bond interactions. However, this receptor binding is weaker when the water shell is larger, so the sweetness decreases, as can be seen in Fig. 4.22. It also decreases with the size of the sugar. Raffinose, a trisaccharide, already consists of three sugars: galactose, glucose, and fructose [54]. The sweetness and osmotic effect also decrease with the length of the sugar. These examples impressively demonstrate how molecular structures determine both physicochemical and sensory and physiological properties. Water binding, sweetness, and energy are significantly altered by slight changes in chemical structures.

4.8.2 The Sweet Receptor

As already noted in the case of umami taste, the perception of sweetness is also based on a pair of proteins, with one protein—the so-called T1R3—being identical in umami and sweet receptor. Given the variety of sweet-tasting substances, it is easy to imagine how complex sweet receptor proteins must be. The sweet receptor

4.8 Physics of Sugar: Taste, Water Binding, Preservation

Fig. 4.22 The sweetness decreases with increasing hydrogen bond binding, as can be seen in the sugars fructose (Fru), sucrose (Sucr), raffinose (Raf), glucose (Glc), galactose (Gal), mannose (Man), and the two isomers of lactose (Lac)

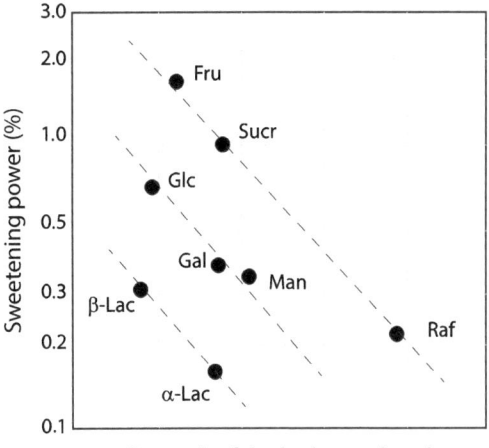

is equipped with many different binding sites [55] due to the highly diverse chemical structures [56], so that small molecules like glucose or fructose, as well as significantly larger ones like the various steviol glycosides can be detected [57] and lead to a strong sweet stimulus. Fig. 4.23 shows with just a few substances how different the sweet triggers are in size and shape. Small sugar-like molecules, such as sucrose and fructose, or the synthetic sweetener sucralose are equipped with the usual hydroxyl groups (OH). The chemical structure of other sweeteners such as aspartame, acesulfame, saccharin or the amino acid d-tryptophan is quite different. Nevertheless, they find suitable docking sites in the receptor protein T1R2. Sweet-tasting larger proteins like thaumatin, on the other hand, use the receptor protein T1R3, which is also present in the umami receptor. It is also worth mentioning that the classic table sugar, sucrose, as well as glucose (see Fig. 4.7) and steviol glycosides can trigger a sweet sensation at both receptor proteins.

In this context, sugar and sugar alcohols, which are known as FODMAPs, are not even mentioned in Fig. 4.23. The taste "sweet" is therefore far more complex, and a mere focus on sugar does not do justice to the basic taste. The complexity and diversity of sweet perception are, of course, due to evolution. Free sugars, which were eaten by hominids in fruits, fruit and sweet-tasting roots even before the use of fire, offered precisely this variety of complex and diverse compounds. These naturally included free amino acids, but also amino acid derivatives, i.e., sweeteners derived from amino acids, such as aspartame, a simple dipeptide of aspartic acid and phenylalanine, which is metabolized exactly into these two components. The circulating theories that these sweeteners would cause great damage in small amounts or even cancer have already been refuted [58], so this does not need to be discussed further here. Furthermore, related amino acids of phenylalanine and aspartic acid occur in many proteins that are consumed daily. Therefore, aspartame is also formed as an intermediate product during digestion in the intestine.

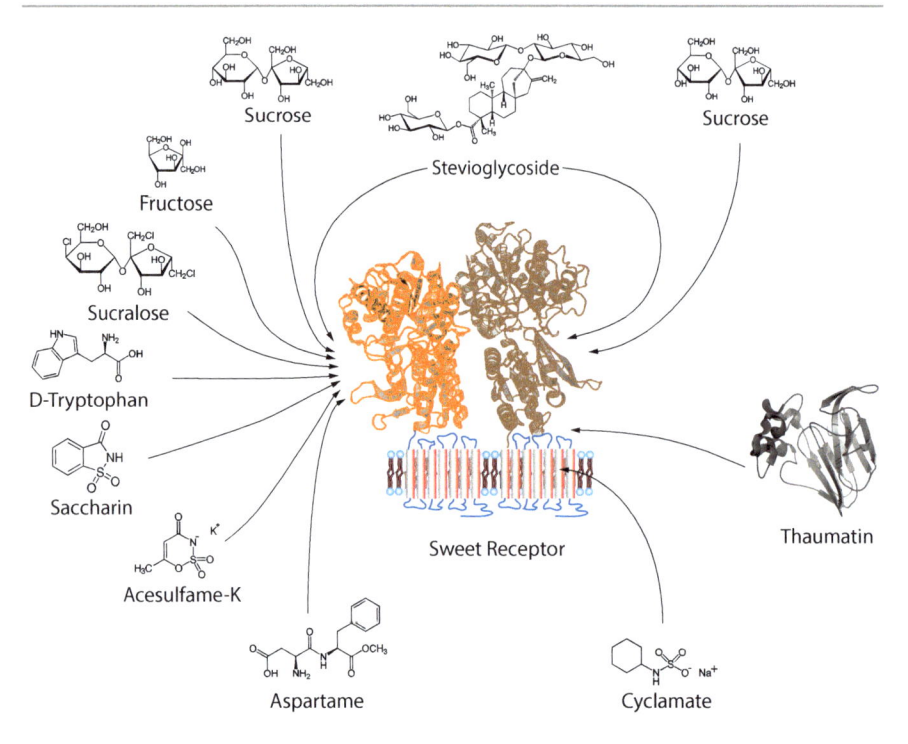

Fig. 4.23 The protein pair T1R2 and T1R3 can recognize a variety of water-soluble molecular structures of various sizes and shapes, which are responsible for the sweet stimuli of different strengths

4.8.3 Metabolism and Sugar

The problem with sugar, or more precisely with glucose, from a medical point of view, is actually its naturalness and function in the body. Glucose can, as already described several times, be absorbed directly, providing immediate energy for every cell. It does not need to be physiologically converted. Glucose is transported as a water-soluble molecule directly in the blood—also one of the reasons why blood tastes strongly sweet when licked. Therefore, clear physiological limits must be set for the concentration of glucose in the blood; also for physical reasons, since every dissolved substance in the serum changes the viscosity and thus the flow properties of the blood. For this purpose, there are two different systems: one for high, the other for low blood sugar levels.

If there is too much glucose circulating in the blood, the pancreas releases insulin. The task of this enzyme is to stimulate the fat cells (adipocytes) to take up glucose. In the best case, glucose is stored there temporarily. The glucose concentration in the blood serum decreases. These regulators perform their tasks as long as the blood sugar concentration does not fluctuate too much and, above all,

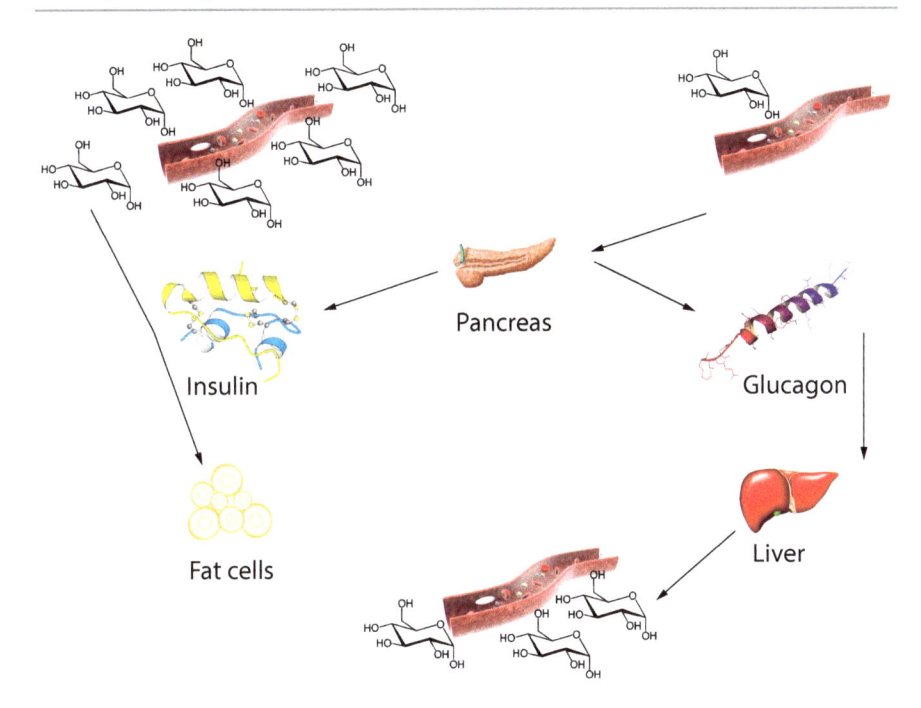

4.8 Physics of Sugar: Taste, Water Binding, Preservation

Fig. 4.24 Highly simplified representation of blood sugar level regulation

is not permanently too high. The lesser-known glucagon is released by the pancreas when the blood sugar level is low (right in Fig. 4.24). As a peptide hormone, it stimulates the liver to release some of the glucose stored there in the form of glycogen. This glucose then enters the blood, raising the blood sugar level. These processes strain the pancreas, which is constantly required to produce insulin or glucagon. It should be noted that even small differences in the chemical structure of sugar stimulate the pancreas differently, as can be measured in the case of α- and β-glucose [59]. Fig. 4.24 also schematically shows what happens to excess glucose: insulin stimulates fat cells to take up glucose, as the regulation of glucose concentration in the blood takes absolute precedence.

In simplified terms, the concentration of glucose molecules determines the blood sugar level. This level must, of course, only move within fixed limits and must therefore be regulated immediately by the enzyme insulin of the pancreas in case of fluctuations. The primary reason is, of course, that we do not get sick, do not develop diabetes or similar conditions.

The physical reason is the viscosity of the blood and its polarity as a solvent. On the one hand, the flow properties change significantly. On the other hand, the polarity and ionic strength of the blood, which is defined by the flowing ions and polar substances, must not change too much. Glucose binds water, which would be missing from the proteins dissolved in the blood. The molecular world in the blood serum would be thrown off balance.

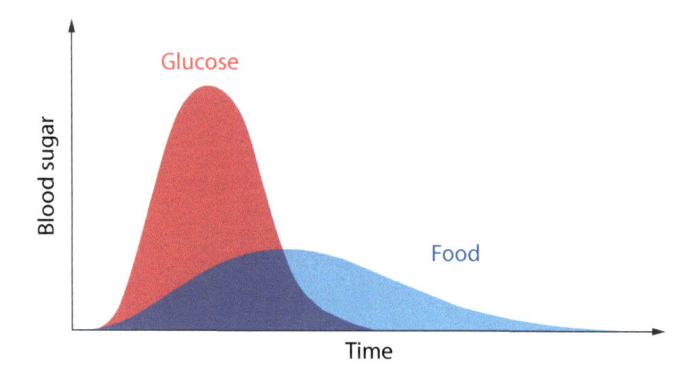

Fig. 4.25 The blood sugar level rises with food intake (food). With the reference substance, pure glucose (red), this increase occurs very quickly. However, the level also drops rapidly again

4.8.4 Glycemic Index

A permanently high, hardly controllable glucose level in the blood is a problem, as are the strong fluctuations. Phenomenologically, this fact is determined by the so-called glycemic index, which certain foods trigger. The temporal increase in blood sugar in the blood is measured after consumption. After eating, the blood sugar level rises, then insulin is released to balance it again and keep it constant for as long as possible.

For the determination of the glycemic index, glucose is used as a standard substance and other foods are compared with it. However, it is important how the increase in glucose in the blood with different foods compares to the administration of pure glucose. Glucose is suitable as a standard because this simple sugar is absorbed very quickly and without further metabolism. This is schematically shown in Fig. 4.25.

The glycemic index can therefore be defined by the increase and decrease of the blood sugar level. As a measure, the area under the curve of a food is set in relation to the area under the glucose curve. Therefore, the glycemic index (GI) of a food L is

$$GI = 100 \times \frac{\text{Area Food}}{\text{Area Glucose}}$$

defined.

How does table sugar behave in this context? A process is necessary to split this disaccharide (sucrose) into its components glucose and fructose. This is done by the active center of an enzyme. Subsequently, the glucose can be absorbed again by the corresponding receptors. The disaccharide molecule is cleaved with the help of the enzyme invertase, which breaks it down into its two monosaccharides. The process is relatively fast, i.e., sucrose has a high glycemic index accordingly. This is due to the very simple structure of sucrose.

4.8.5 Glucose versus Fructose

Is table sugar, sucrose, poison, as is often claimed? Now the prerequisites are in place to get to the bottom of this question. Household or table sugar is known to be a chemical compound of glucose and fructose. Consequently, the question arises, which of the two is the villain? Glucose cannot be it, because as mentioned several times, sugar is, next to ATP, adenosine triphosphate, the most important fuel for all cells. So it must be fructose, which has already been noticed as a FODMAP and can cause intolerances and irritations. The fructose molecule cannot be metabolized as such; it must be broken down via other enzyme pathways, especially when consumed in excess through fructose-sweetened beverages or foods. Then, fructose is processed differently than glucose in the chemical laboratory of the liver [60].

First, the question remains to be clarified what happens to glucose in the liver when too much of it is not needed as immediate energy due to too little physical exertion. The exact physiological processes are complex, but the final results are relatively easy to understand [61]. It is stored, but for osmotic reasons not as many individual glucose molecules, but as long-chain glycogen, a macromolecule that is branched similarly to amylopectin, but does not exist in a crystalline form. This stored supply can then be released when there is a higher energy demand. Therefore, sufficient physical activity is essential with a diet rich in sugar and carbohydrates.

The insulin concentration in the blood signals the brain via certain messenger substances, e.g., leptin, the degree of satiety. This protein leptin is expressed in the fat cells as soon as they are prompted by insulin to take up glucose. If the insulin concentration is high, there is hardly any feeling of hunger. At the latest, it is time to stop eating. All these regulations are lacking for fructose, which cannot be directly absorbed by the body cells but must be fermented or converted to fat in the liver via *de-novo*-lipogenesis. And this poses a serious problem with a constant excess of fructose, whether it comes from table sugar, fructose syrup, or even healthy fruit. Here, too, the principle applies: a molecule is a molecule, no matter what source it comes from.

Initially, sweetening with fructose appears attractive, as its sweetening power is significantly higher than that of glucose and sucrose. Also, compared to glucose, fructose puts considerably less strain on the pancreas [62], with insulin secretion remaining significantly below the level of glucose, as can be seen in Fig. 4.26. As a consequence, it can already be expected that, due to the absence of insulin, leptin as a satiety regulator will also be absent. But that's not all, because unlike glucose, fructose does not suppress the release of ghrelin (Growth Hormone Release Inducing), a peptide hormone produced in both the stomach and pancreas. In contrast to glucose, fructose, simply put, does not satisfy but rather stimulates further eating, which in turn increases fructose intake.

Furthermore, the burden on the liver is significantly higher, as fructose is taken up by the liver in large amounts and chemically processed. Two main effects are in the foreground: On the one hand, the chemical conversion requires energy, as fructose must first be phosphorylated to fructose-1,6-bisphosphate so that an enzyme,

Fig. 4.26 Immediate average increase in insulin in test subjects after glucose- (red) and fructose (blue) administration. However, the glucose and fructose doses administered to the subjects were extremely high at 75 g in order to measure the effects within the short observation period

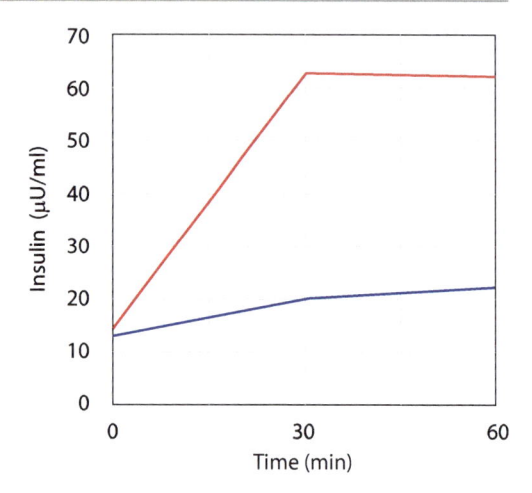

the so-called Aldolase B, can access it. The phosphorus comes from the cell fuel ATP, which gives off a phosphate group and, as with umami, becomes an adenosine diphosphate (ADP) and further adenosine monophosphate (AMP); its further conversion takes place via the purine metabolism (Fig. 4.10), among other things via the intermediate step inosine monophosphate (IMP). This produces uric acid, which can lead to gout and high blood pressure at persistently high concentrations (but not necessarily as the sole factor). This high concentration of uric acid does not occur during the synthesis of glycogen from glucose.

At this point, an incidental remark for context appears appropriate: This also shows why in Fig. 4.13 innards such as the liver are listed. Due to the purine metabolism, animal livers are rich in IMP and thus suppliers of umami. It is useful to see these processes in context.

On the other hand, fructose degradation products, so-called aldehydes, form fats and triacylglycerols (triglycerides) in liver cells. This *de-novo*-lipogenesis shows exactly the chemical pathways by which liver converts fructose into fats in the form of triacylglycerols. In the case of an excess of fructose and a permanently running *de-novo*-lipogenesis, the newly generated fat from fructose is stored in the fat cells of the liver, the long-term consequence is a non-alcoholic fatty liver. In Fig. 4.27, it is shown how, just a few hours after fructose intake in healthy individuals, fat synthesis begins and increases significantly. In this context, the strong similarity with alcoholic fatty liver is interesting, as too much ethanol also leads to the synthesis of fat in the liver.

By the way, little (or no) fat is synthesized from glucose. The direct comparison is shown in Fig. 4.27. So far, these facts seem to speak strongly against fructose. However, the theory is not that simple, as there is another point that has hardly been addressed so far. Although fructose does not trigger the release of the satiety hormone ghrelin, it suppresses appetite through a significantly higher release of

4.8 Physics of Sugar: Taste, Water Binding, Preservation

Fig. 4.27 *De-novo* fat synthesis in the liver from glucose (red) and fructose (blue) in direct comparison. Fructose is converted to fat, glucose to glycogen and not to fat

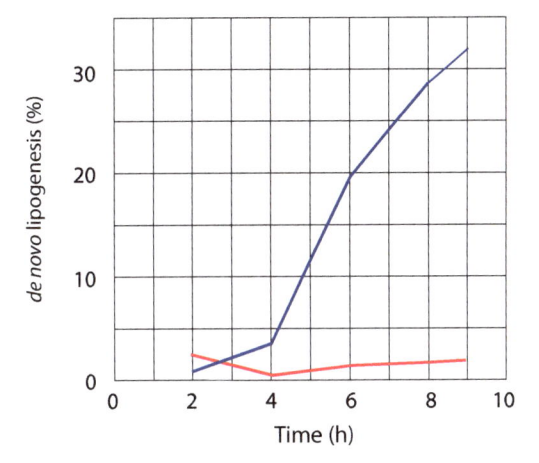

Fig. 4.28 Variation of the hormone and appetite suppressant PYY upon addition of high-dose glucose and fructose solution. In the first 30 minutes, the increase is very similar, then PYY decreases with glucose (red), while with fructose (blue) the concentration continues to increase (From the same study as the results of Fig. 4.26)

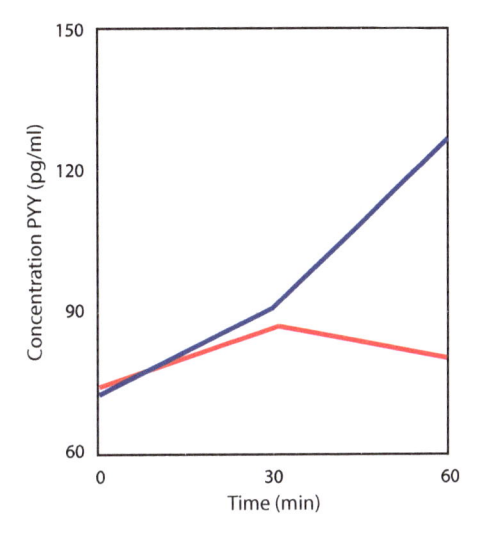

the peptide hormone and appetite suppressant PYY (Peptide YY), whose primary biological effect is the inhibition of gastric emptying. It delays the rapid emptying of the stomach of fatty food and thus allows for better digestion through a slower gastrointestinal passage. PYY has a strong influence on appetite and satiety and leads to reduced food intake. In fact, PYY from the small and large intestine causes satiety for a period of four to six hours. With a reduction in food intake of about 40%, this polypeptide has the strongest effect of all gastrointestinal hormones [63] (Fig. 4.28). This result is a plus for fructose and shows that all foods, even simple ones like fructose, have more than one side to the coin.

4.8.6 Honey, Rice, Maple, and Agave Syrup

Honey, although a natural product, whether unrefined, be it from an organic beekeeper or a traditional farm, naturally belongs to the same category. On average, honey consists of 80% carbohydrates, 17% water, and 3% proteins, enzymes, vitamins, and minerals. The sugar content of all honeys consists (Fig. 4.29) of over 90% glucose and fructose in varying proportions, the rest being three- to five-fold sugars plus a few enzymes that have remained in honey. Honey is, in good approximation, nothing more than a highly concentrated invert sugar syrup, made up of glucose and fructose, as well as a few tri- and oligosaccharides. At the molecular level, glucose and fructose dominate. The minerals supplied by the usual honey consumption, as well as the amino acids of the few enzymes, are largely negligible by the overall nutrition. Every bite of vegetables, meat, or sausage contains more minerals, protein, or vitamins than a spoonful of honey. It is therefore a fallacy to believe that one is fundamentally doing something good when sucrose in desserts,

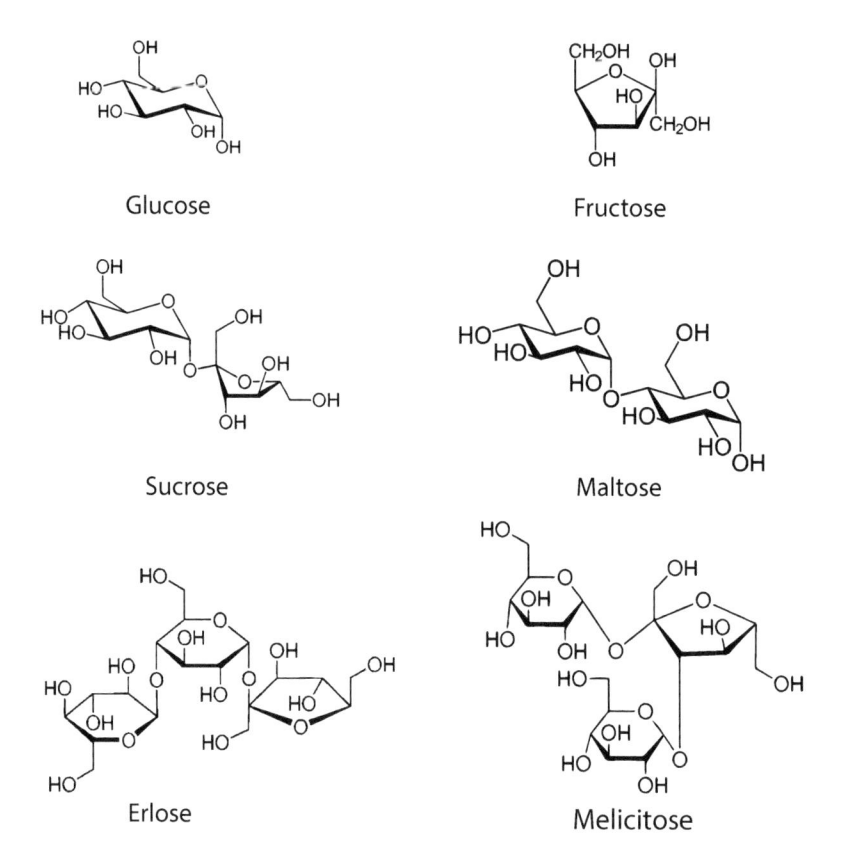

Fig. 4.29 Structure of the most common sugars in honey. The main portion consists of glucose and fructose

4.8 Physics of Sugar: Taste, Water Binding, Preservation

cakes, or other pastries is replaced by more or less the (almost) same amount of honey.

From this point of view, maple syrup is also no better for metabolism than table sugar. Apart from a few flavors and proteins, it remains a non-crystallizable sucrose-fructose mixture, which is at least as "un/healthy" as the strongly unhealthy classified fructose-glucose syrup made from cornstarch. However, high-fructose-glucose-corn syrup occupies a special position, as its fructose content is very high and thus serves as an intense sweetener. Consumed in excess, e.g., in sweet drinks, it stimulates *de-novo* fat synthesis particularly strongly.

Using such sugar alternatives is therefore not an effective method for targeted sugar reduction, but rather pure deception and certainly not a diet [64]. The glucose-fructose distribution in so-called "natural sugars" clearly demonstrates this, as shown in Fig. 4.30 visually and converted to an average teaspoon.

The direct comparison of various natural and industrial sweeteners illustrates their different composition of fructose and glucose. Other sugars (such as in honey) are not shown. The differences between natural sweeteners like honey, agave syrup, or maple syrup compared to industrial, refined sugar or fructose-glucose corn syrup are indeed marginal. Only rice syrup, hydrolyzed rice starch to glucose, naturally contains no fructose.

The sugar contents and the respective glucose-/fructose ratio from Fig. 4.30 are basically sobering. There are hardly any significant differences in terms of sugar content. There is nothing really healthy about the so-called natural sugar alternatives. Thus, the answers are simple: sugar, molasses, and other plant syrups are less refined, but they remain sugar, as do corn starch hydrolysates. And as such, they remain "poison" like sugar, if consumed in excess over long periods of time.

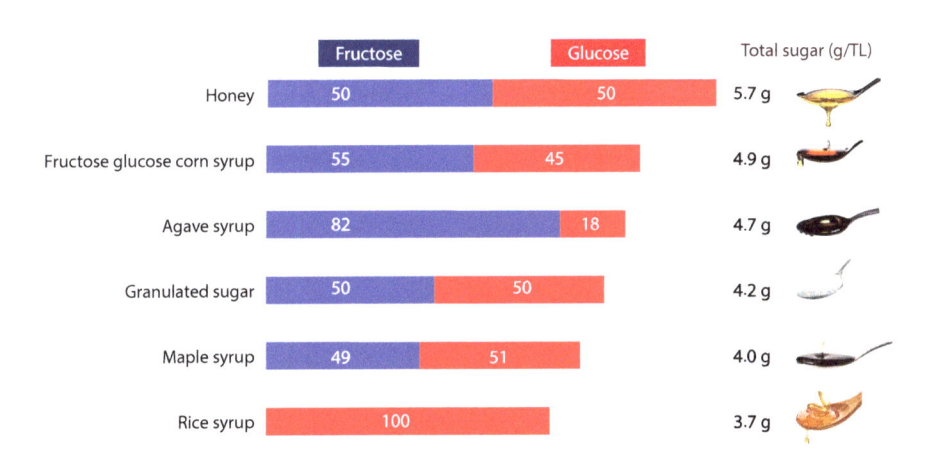

Sweetener	Fructose	Glucose	Total sugar (g/TL)
Honey	50	50	5.7 g
Fructose glucose corn syrup	55	45	4.9 g
Agave syrup	82	18	4.7 g
Granulated sugar	50	50	4.2 g
Maple syrup	49	51	4.0 g
Rice syrup	—	100	3.7 g

Fig. 4.30 Various sweeteners in comparison. The glucose-/fructose ratio is different in each case. The varying total sugar contents in grams strongly depend on the specific weight, which is visually expressed in the depicted teaspoon (TL)

This is not surprising, as sugar beet, sugarcane, maple tree, agave, and bee all rely on natural systems in which glucose is one of the most important cellular fuels.

The high fructose content in tree and plant juices (Fig. 4.30) is predetermined by nature. Fructose prevents the crystallization of plant juices at low temperatures, thus serving as frost protection. We can even see this fact in honey. If the fructose content in honey is higher than about 60–65%, it prevents the crystallization of honey, as in the case of acacia honey. This is precisely why bees ensure an adequate fructose content in honey: it must not crystallize in the beehive at around 37 °C. Otherwise, the bees and their offspring would have no food. Hard crystals cannot be used as food by bees.

The conclusion is therefore very simple: Those who have a high sugar consumption and replace sugar with natural alternatives such as honey, maple syrup, etc., change little in their actual sugar consumption. The usual arguments that honey is natural, contains enzymes and minerals, are irrelevant for the overall balance of nutrition. The micronutrients, minerals, and enzymes (from a nutritional point of view, nothing more than amino acids) contained in a tablespoon of honey, maple- or agave syrup are far surpassed by a bite of fruit, vegetables, or cheese. And the sugars consumed in comparison to honey & Cie are far undercut.

In Fig. 4.30, agave nectar and fructose-glucose-corn syrup stand out due to their particularly high fructose content, while honey (and invert sugar syrup from sucrose) appear balanced with a 50:50 ratio. With frequent and excessive use of these sugar alternatives, one is indeed doing oneself no favors, as they lead to an excessive intake of fructose. It is precisely this excess of fructose that contributes to the formation of non-alcoholic fatty liver.

4.8.7 Fruit Remains Valuable

Should fruit no longer be consumed now because of its high fructose content, as is often suggested? This is nonsense, as most of these research results were obtained with extremely high doses of fructose and glucose. Furthermore, when consuming raw fruit, the sugar must first be released from the hard plant cells, if this is even achieved 100%, which was already doubted in Chap. 1 (Fig. 1.18). It must not be forgotten that fruit, even those high in fructose, has far higher positive effects such as fiber, vitamins, polyphenols, secondary plant metabolites, etc., which outweigh the fructose in the overall balance. The daily apple, the daily banana, or the large amounts of sugar-sweet strawberries during their season are not the problem. Now it is also clear why: The excess fructose in fruits and fruit remains embedded in the natural food matrix and therefore never has the same impact as the fructose content in agave nectar or high-fructose syrup. Sweetened beverages, lemonades, and sweet soft drinks, on the other hand, are not the right choice in the long run. Apart from glucose, a lot of fructose, and a few flavors, there is little in them, a food matrix does not exist, all sugars are in aqueous solution. Therefore, an "overdose" is quickly reached. Here, too, a look at evolution is helpful: Fruits and fruit have been staple foods long before the use of fire. They are thus part of the diet of all primates, including representatives of the genus *Homo*.

Fructose and Glucose from fruits and vegetables are neither unhealthy nor toxic in typical consumption amounts. Nature sets the standard for the glucose/fructose ratio, as seen in honey, which has been part of the diet of *Homo sapiens* since the existence of wild bees, and thus before the Neolithic period. In this natural product, the glucose/fructose ratio is about 50:50. Similarly, in fruit and fruits, fructose and glucose each form in the range of around 50%. Naturally, there are fruits that contain more or less fructose, but this is irrelevant with a diverse diet, as *Homo sapiens* survived with all these products throughout time and developed splendidly. In the process, they learned that different sugars have significantly different sweetness levels. Table sugar, the standard sugar, is assumed to be 100% sweet, and it becomes apparent: Glucose, one of the cell fuels, is significantly less sweet, while fructose develops about 20% more sweetness. The 50:50 mixture of honey is therefore always sweeter. With less mass, more can be sweetened. This is even more true with fructose-rich corn syrups. However, these syrups do not form without human and enzymatic intervention. Therefore, restraint is advised with these ingredients.

What was absent from the diet of our ancestors, however, were sodas and other heavily sweetened drinks, as well as beverages with high alcohol content. Nor were fruit juices or highly processed smoothies. The problem becomes evident with smoothies, fruit juices, and energy drinks in their entirety: The matrix of these foods is largely destroyed. Smoothies are therefore highly processed foods, even if they are made at home, especially when using the intended high-performance blender. The goal of these blenders is indeed to extract every nutrient from fruit and vegetables. And thus, in addition to the desired vitamins and secondary plant compounds, also a high amount of fructose and glucose, which are immediately available and instantly take the metabolic pathways already shown—in contrast to consuming the same amount of raw fruit, which was only crushed with teeth. The "nutrition and body optimization" not only overshoots the mark at this point but also misses healthy eating by a wide margin.

One point must be explicitly emphasized: Despite all this general uncertainty, there is nothing wrong with a moderate and evolutionarily appropriate consumption of small amounts of sugar, fructose, or sucrose—on the contrary. At least for those whose diet consists of a variety of components, who are physically active, and who show no signs of obesity, hypertension, non-alcoholic fatty liver, or diabetes. Only those groups of people who suffer from these and similar pre-existing conditions are dependent on synthetic sweeteners and sugar alcohols, such as the already mentioned erythritol, sorbitol, xylitol, etc. And contrary to widespread opinions, these are not much harmful to health, as long as they are not consumed in large quantities.

4.8.8 Sugar is much more than just Sweet: OH loves H$_2$O

Sugar is far more than just a sweetener, as it has strong physicochemical properties that make it virtually irreplaceable for use in food. In Chap. 3, Fig. 3.33 this was already hinted at in a completely different context. Sugar has a very strong preservative effect, which is the simplest example of the molecular action of polar hydroxyl groups (OH) composed of an oxygen and a hydrogen atom. Due to the polarity of water, the water molecule H$_2$O is slightly negatively charged on the oxygen side and slightly positively charged on the hydrogen side. Therefore, the water molecules can directly interact with the OH groups, water is bound. Concentrated sugar solutions are highly viscous, their flow velocity decreases sharply with increasing sugar concentration, and the viscosity increases significantly. Water molecules are slowed down in their dynamics by the binding and their diffusion is greatly restricted. This is a prime example of strong water binding.

This high water-binding capacity of table sugar has long been used in cooking and patisserie. Since sugar has existed, it has been added to dishes and desserts in one form or another. Preparations of Blanc-mangers (a kind of almond pudding) are known from the Middle Ages, which combine sugar and (chicken) meat. Not only for the taste, but also to control the consistency and, above all, to extend the shelf life. The high water binding also means a reduction in water activity. Germs and fungal spores require free water for their metabolism and reproduction, i.e., water molecules that are outside the binding ellipses of the sucrose molecules in Fig. 4.31. Bound water is much too slow in terms of molecular movement, diffusion, to serve as a basis for life for germs. The best-known examples are jams, i.e., fruit preparations preserved in sugar.

Already in the Roman Empire, fruits were preserved with sugar by cooking them with sugarcane for a long time [65]. At the same time, a welcome and widely accepted sweetness developed, which cannot be attributed solely to sugar

Fig. 4.31 Sucrose binds water molecules via the OH groups. A hydrate shell is formed, represented by the blue ellipses. The water molecules in the blue ellipses are bound, the molecules outside are free

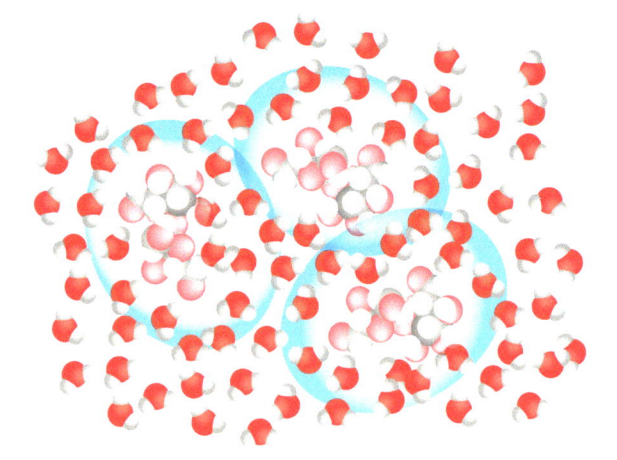

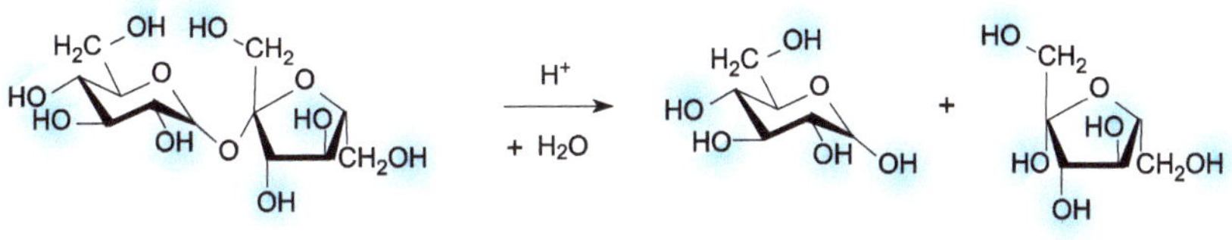

Fig. 4.32 The sugar invertase of sucrose under cooking with acid (protons, H⁺). Not only does the sweet taste increase due to the high fructose content, but also the number of OH groups (blue circles) as potential water-binding sites increases when sucrose is split into glucose and fructose from eight to ten per split sucrose molecule

and the sweetness of the fruits. The reason for this lies again in the physics of sucrose, which still makes table sugar unbeatable for preservation to this day: Due to the fruit's own acids, sucrose is split into its components glucose and fructose, resulting in invert sugar. This not only greatly increases the sweetness, but also the water-binding sites, as shown in Fig. 4.32.

The sugar invertase in the classic marmalade and jam cooking process thus increases sweetness and water binding at the same time and is therefore highly welcome for preservation. In addition, the two sugars fructose and glucose, as well as pectin and hemicelluloses from the fruit cells, are exposed during cooking. The result is a highly thickened, extremely long-lasting 50:50 fructose-glucose syrup that cannot crystallize further during storage—like acacia honey. These simple physical-chemical facts are often not mentioned or widely unknown in the discussion about the value of jams. No wonder, then, that a classic cultural technique becomes the perfect model for modern food technology. Invert sugar and honey are therefore the most common humectants in industrial food production as well as in home kitchens and patisseries. In every classic recipe, the sugar invertase is supported by the addition of lemon juice, while industrially produced jams often use the additive citric acid, which must then be declared.

4.8.9 The Appeal of Sucrose, and Why Stevia is not Always an Alternative

The high acceptance of sucrose is due not only to its sweetness but also to its water-binding capacity [66] and the resulting mouthfeel. Each molecule binds a precisely defined number of water molecules from saliva. The sweetness lingers sufficiently long in the mouth without being as sweet as fructose. These are rarely consciously noticeable side effects, but they are also responsible for the popularity of sucrose. Sucrose also lacks any aftertaste; with stevia, for example, there is a hint of bitterness that prevents stevia from being widely used despite its high sweetness. All these purely taste-related uncertainties can be resolved. However, stevia cannot be used in the excellent and long-established recipes of patisserie

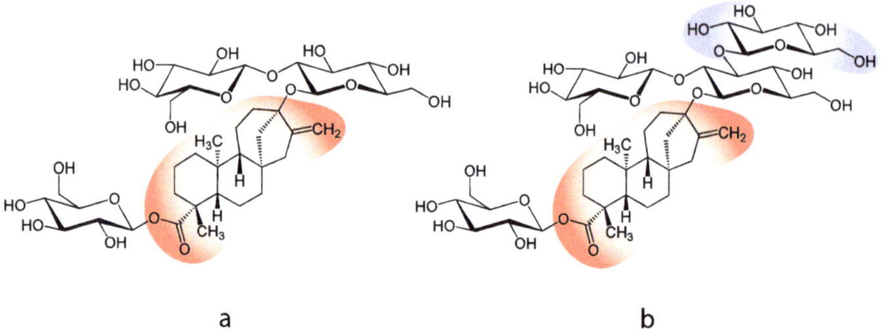

Fig. 4.33 Two typical representatives of steviol glycosides, stevioside (**a**), rebaudioside-A (**b**). They differ by an additional glucose (blue) in rebaudioside-A. The common center is a diterpene (steviol, red). The sugars are hydrophilic, the diterpene center is hydrophobic

and dessert culture. Stevia and the other low-energy alternatives do not have a comparable high water-binding capacity of sucrose, as can be seen from simple considerations below. In Fig. 4.33, two typical representatives of the sweet-tasting steviol glycosides found in the leaves of the plant *Stevia rebaudiana* (sweet herb) are shown.

In the stevia plant, up to ten different glycosides are present, whose structure is very similar. Glucose is connected to the diterpene steviol. Depending on the structure, these glycosides have an extremely high sweetness, which is up to 400 times higher than that of sucrose. Steviol glycosides are indigestible and do not contribute any energy, and the pancreas remains unburdened. Their sweet profile is very similar to that of table sugar, but the sweetness of the stevia plant has a bitter aftertaste [67]. Similar to saccharin and acesulfame K, some of these glycosides, depending on their molecular structure, stimulate both the sweet receptors and the bitter receptors more or less strongly [68]. Stevia is therefore not a universal plant-based solution for perfect sweetness due to taste-sensory reasons.

The other problem with Stevia is again part of physics. Precisely because of its high sweetness, much less of it is needed than table sugar. The physical consequence is clear: There are too few hydroxyl groups (OH) available for water binding. Steviol glycosides do have more OH groups than sucrose (Fig. 4.34), but due to the required low concentration, they never achieve the water binding of sugar.

A brief theoretical consideration illustrates this. The sweetness of a syrup made with 40 g sucrose per 100 ml water can be achieved with just 0.2 g rebaudioside-A per 100 ml water. However, the sucrose syrup has a 350-fold higher factor of OH groups (converted per mole). The viscosity of the syrup is therefore significantly higher than that of the corresponding stevia solution. That's not all, as the large diterpene center of the steviol glycosides is highly hydrophobic and strongly repels water molecules. The mouthfeel and textural properties of the stevia solution would thus be significantly worse. With Stevia, it is therefore not possible to

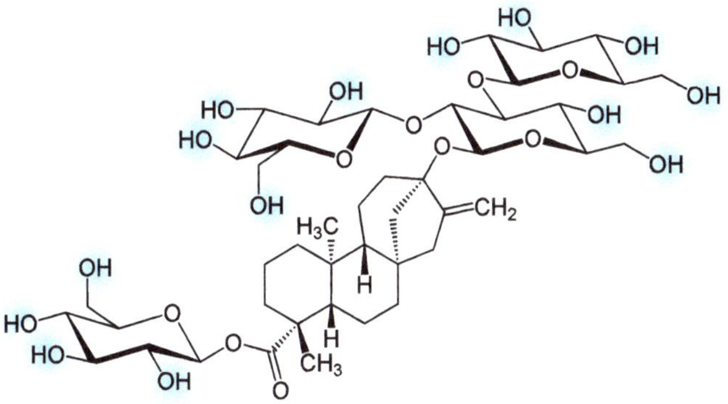

4.8 Physics of Sugar: Taste, Water Binding, Preservation

Fig. 4.34 Rebaudioside-A has six more OH groups than sucrose and the potential water binding is correspondingly high, but due to the 200-fold sweetness, far less of it is needed

produce comparable desserts or jams. Only for low-viscosity lemonades, teas, or coffee would stevia be suitable.

Sugar in the form of sucrose is thus much more than just sweet, and due to its water binding, its physical properties are in the foreground for many applications. They play a major role in most recipes, from pudding to crème anglaise to ice cream in patisserie. Developing recipes with sugar alternatives is therefore not just a matter of sweet taste, but primarily a physical challenge that must combine firmness, creaminess, mouthfeel, shelf life, and sweetness. A purely Stevia-based jam without further additives is therefore simply physically impossible. Therefore, in general, when reducing sugar, regardless of which sugar substitutes are used, considerable assistance or improvement must be provided on the physical side using additives. This can be most easily recognized in the case of heavily sugar-reduced jams.

4.8.10 What is the Purpose of Additives in Jam?

Without the water binding of sugar, the jam would be a tasty fruit preparation, but not as durable as desired. Therefore, water must be bound in another way, as in heavily sugar-reduced jams. The usual candidates are additives with E-numbers: hydrocolloids such as alginate, agarose, apple pectin, carrageenan, xanthan, etc. They act as indigestible additives (polysaccharides) to support the remaining sugar in the jams, bind water, and extend the shelf life after opening the jars. At the same time, they give the jam a similar gel-like mouthfeel that we know from classic jam.

Furthermore, in heavily sugar-reduced jams, preservation must be assisted, as an excessive addition of hydrocolloids tends to cause a slimy mouthfeel. No wonder, then, if mandatory (and sometimes very harmless) preservatives such as sodium sorbate, citrate, or sugar alcohols are added.

Since the sugar from grandmother's jams has been banned, additives with E-numbers must now be added so that the jam meets its requirements: The word jam means preserving in the best sense. It would probably be better to make jams the traditional way—only from fruits and enough sugar—and eat only a little of it, but with full and conscious enjoyment. Ultimately, the long-cultivated sugar fears and the associated diet craze have conjured up such inauthentic jams.

4.8.11 Conclusion: Is Sugar Poison or Not?

Sugar is, of course, no poison, as long as one does not lose control over how much one eats of it. Those who constantly consume lemonades, fruit juices, and cola, spread kilos of jam or honey on bread, or for whom the weight of a sticky-sweet ketchup exceeds that of the daily currywurst, are, of course, doing something massively wrong in their diet. But even then, sugar is the least to blame. It could be more about the attitude towards food and a loss of control that comes with it. Cooking for oneself still provides the best form of eating and is the best protection against all the alleged dangers.

Even the daily apple or the handful of seasonal fruit are not poison, contrary to some statements in relevant magazines. With our teeth, we do not expose all the available sugar, but we do receive many of the micronutrients that ensured the survival and further development of our ancestors before Wrangham's "cooking ape." More critical are even the high-performance mixer-produced fruit smoothies if consumed daily in larger quantities. In addition to the supposedly healthy release of all nutrients, as advertised for these devices, the full potential of sugars is also exposed. Then, the half-liter of "healthy" smoothie is quite comparable to a "quarter" of the unhealthy high-fructose-glucose syrup.

4.9 Lipid Digestion: Colloid Physics During the Gastrointestinal Passage

That fat is not unhealthy, but essential for life, has already been made clear in various places. In the digestion of fat, lipids, there are fundamental relevant questions, such as how it can succeed at all when the water-insoluble lipids from food enters the watery acidic stomach and intestine. Colloid physics is at the foreground here, as the fat absorbed from food is first emulsified and then repackaged and re-emulsified in a different way to prepare for digestion. This is done with the help of bile acids. These very special derivatives of cholesterol, such as bile salts or bile acids, are ideally suited for this purpose.

Exactly how the conversion of fats for digestion works is a highly complex process. Often we consume these fats in droplets, such as in sauces or as free spreadable fat on breads. During the formation of the food bolus (bolus) while chewing, the bolus formation, the fats are emulsified by the food components present in the bolus. Through oral processing, a swallowable, plastically deformable, semi-solid emulsion is formed. The fats enter the stomach as large droplets, emuslfied by the food bolus. There, a low pH value prevails, depending on the filling state of the stomach between pH 1 to about pH 3. This low pH value has little effect on the nonpolar fats, unlike other food components such as meat or vegetable pieces. These include proteins, but also cell material and especially phospholipids, which are used as auxiliary substances for fat digestion. The stomach contains acid-tolerant proteases, especially pepsin, which generate peptide pieces from long (denatured) proteins, which in turn have surface-active properties. Thus, they can support the phospholipids in the emulsion of free fats. Therefore, relatively large, complexly structured fat droplets form in the stomach and on the way to the small intestine, the surface of which uses all available emulsifiers to stabilize the fat droplets. What is a great advantage for transport turns out to be a disadvantage for digestion. Fat-digesting enzymes—pancreatic lipases—from the pancreas have no access to the fat enclosed in the droplet.

4.9.1 Stomach and Small Intestine—Mainly Colloid Physics

In fact, the immiscibility of fat / oil and water also plays a crucial role in digestion. Free fat or oil from meals must first be emulsified so that it can be transported from the stomach to the duodenum in a controlled manner. However, there are differences that manifest themselves through the chain length of fatty acids. This shows us new connections between nutritional issues and physical requirements. fat digestion is mainly determined by the comparison of length scales and volume-to-surface ratios. This occurs in the upper part of the digestive tract. It is hardly surprising that the oral processing, i.e., chewing, biting, insalivating, and swallowing, prepares the food for the stomach-intestinal passage. Less well known, however, is the fact that the predominantly occurring physicochemical changes in the stomach and small intestine prepare the chyme for physiological digestion. The physicochemically dominated area of the gastrointestinal digestive tract is schematically indicated in Fig. 4.35. The chyme encounters the watery environment of the stomach, which is characterized by a pH value of below 3 and thus mainly affects proteins—through changes in the charges of amino acids. In addition, the stomach is not a static structure: due to the constant movement of the muscles, the motility, which increases significantly when the stomach is filled, the bolus particles are constantly moved, sheared, deformed, mixed, and rearranged. The stomach contents are constantly exposed to shear forces. Since protein-splitting enzymes are already present in the stomach, protein connections are dissolved and broken down and changed by the shear forces.

After that, the particles enter the small intestine and are exposed to the enzymes of the pancreas, which can break down proteins, starches, and fats, as already discussed and illustrated in Chap. (Figs. 1.20 to 1.22). However, this only happens if the macronutrients are accessible to the enzymes. Tightly packed, emulsified fats and oils in huge droplets are not, nor are crystalline and thus degradation-resistant

Fig. 4.35 Schematic representation of the gastrointestinal tract. The circled area is responsible for the purely physicochemical effects of food processing. The same physical laws apply there as in the kitchen and laboratory

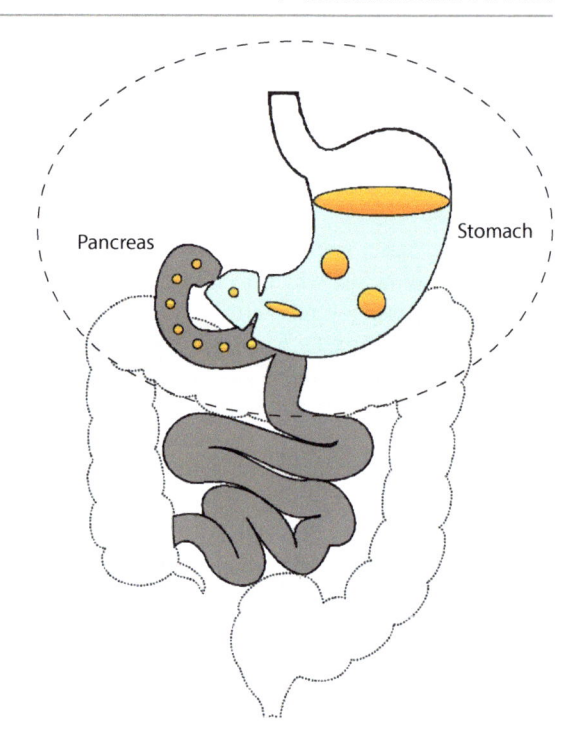

starches. The upper part of the stomach-intestinal passage serves the physicochemical provision of food for physiology. Thus, the first steps of digestion belong to colloid physics. For example, if we follow the path of fat from the food slurry in the stomach and small intestine, this becomes clearer.

4.9.2 The Path of Fat from the Mouth to the Intestine

Assuming one eats a buttered bread, a lard bread, a burger, a bread topped with high-fat salami, or a vegetable salad soaked in oil. In the mouth, the food is first broken down. The fat components that can melt at up to 37°C mouth temperature liquefy and are emulsified into the bolus. The bolus now has an inhomogeneous structure and is very complex in composition, depending on the food, as indicated in Fig. 4.36.

On the left in Fig. 4.36, the bolus is schematically indicated with an oil droplet (red). This droplet has a non-spherical shape and a highly heterogeneous surface, as shown on the right in the illustration. The fat/oil droplet is surrounded at its immediate surface by solid particles such as cellulose particles from vegetables or particles from muscle tendons. At the interface itself, there are solid particles, native and denatured proteins from raw and cooked components of the food, starches such as amylose or amylopectin, but also emulsifiers exposed by the

4.9 Lipid Digestion: Colloid Physics During the Gastrointestinal Passage

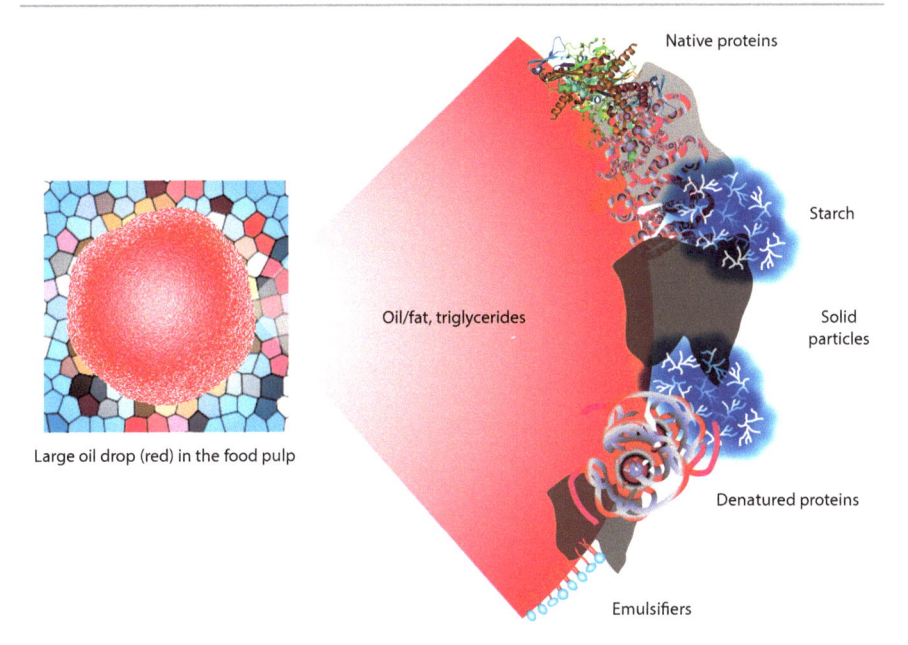

Fig. 4.36 Fats and oils in the bolus are emulsified in an irregular manner in the mouth during chewing. Most of the molecules that have already been discussed in earlier chapters are located at the interface of the fat and oil droplets

cooking or during oral processing. These include phospholipids, free fatty acids, or mono- and diglycerides from the food. The oil droplets in the bolus still have macroscopic dimensions at this stage, i.e., their diameter is in the range of millimeters to 100 μm. The surfaces are relatively flat and slightly curved compared to the molecular dimensions. From the enzymes' perspective, the interface is densely and thickly packed and flat. It would be an impossible task to digest these bolus pieces: the fat-cleaving enzymes—lipases—would have no physically feasible way to access the fat and oil packaged in the droplet. Therefore, already in the mouth, some of the starch is attacked and pre-cleaved with salivary amylases (see Chap. 3). Some of the starch surrounding the fat droplets is already loosened. This is a good thing, as no amylase is active in the acidic stomach that could accomplish starch degradation there.

Therefore, when fat arrives in the acidic stomach, it is still surrounded by complex and non-uniform natural emulsifiers and small bolus particles and must be further processed [69]. This occurs both physically and enzymatically. At the low pH value in the stomach, existing proteins are (partially) denatured. At the same time, acid-resistant pepsin begins to cut the proteins of the coating. The cohesion of the protein aggregates around the bolus particles decreases, and the heterogeneous layer around the oil droplets becomes increasingly loosened. Now the emulsifying layer around the fat droplets and fat particles becomes smaller, and in some cases even dissolved [70]. However, the droplets become unstable, fat is released and must be immediately re-emulsified. This is easily possible, as the stomach acid and the peptides have already done enough work and provided further emulsifiers: smaller denatured proteins, peptides with hydrophilic and hydrophobic parts, phospholipids from plant cells, as well as polar free fatty acids and mono- and diglycerides from food. Sufficient molecular material, therefore, to repackage the fats and oils in smaller portions.

The phospholipids released on a large scale from plant and animal cells through the digestive processes in the stomach are important for the further course of fat digestion. However, these phospholipids have a compatibility problem in the acidic and aqueous environment, as the fatty acids of the phospholipids lack fat and oil in which they could insert their fatty acids. The fat is still too tightly packed. Swimming around freely as individual phospholipids in the stomach is not an alternative, as the fatty acid tails are hydrophobic. Thus, at high concentrations, the phospholipids have no other thermodynamic way out but to put their own fatty acids together so that all their hydrophilic heads protrude into the water of the gastric juice. There are two possibilities for this: either they form micelles or liposomes, as shown in Fig. 4.37.

A liposome is a closed membrane around a cavity that encloses water (along with water-soluble substances). The minimum radius of a liposome depends on the saturation and the length of the fatty acids. Liposomes are therefore composed of lipid bilayers, as already known from cell membranes. Micelles, on the other hand, are significantly smaller. They consist of only a monolayer of phospholipids and can be compared to emulsion droplets that have a very small oil core, namely the fatty acids of the emulsifiers themselves. The phospholipids are another reservoir for the emulsification of dietary fats. From them, more and more packaging material and transport vehicles for fats and oils are built as the protein digestion progresses. This occurs after further protein digestion when the protein layer on the surface of the fat droplets becomes thinner and the droplets are just barely

Fig. 4.37 The self-organization of phospholipids allows, depending on the concentration, two spherical structures with minimal surface energy

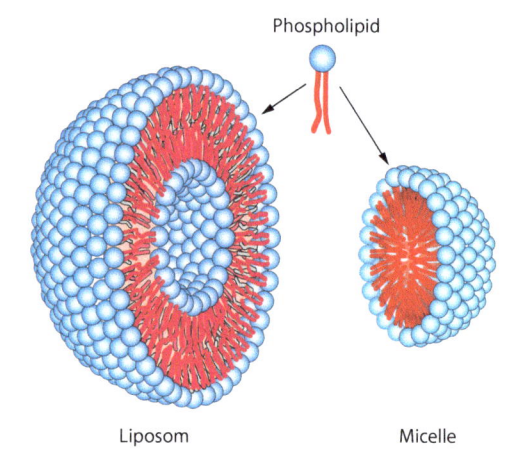

4.9 Lipid Digestion: Colloid Physics During the Gastrointestinal Passage

stabilized by heterogeneous emulsifiers or even merge into even larger droplets. Therefore, assistance from a universal emulsifier is necessary. This requires two tasks: First, it must be small and flexible enough to finally destabilize the large droplets, and second, it must repackage oils into smaller droplets and design their surfaces so that lipases have access to the fat, allowing its triacylglycerols or triglycerides to be cleaved into physiologically usable fatty acids. For this, bile acids are perfectly suited in size and structure.

4.9.3 Bile Acids, the Other Side of Cholesterol

Bile acids are based on the structure of cholesterol [71, 72]. This is intuitively obvious, as cholesterol has already been noticed as a control molecule in the cell membrane as a local solidifier and plasticizer (see Sect. 3.1.2). The molecule can thus deal with both membranes, i.e., lipid bilayers, and lipid monolayers, such as LDL and HDL particles. No wonder, then, that its basic structure is used for the remodeling of fat particles during fat digestion. Bile acids come in several structures, differing only in details that are not significant for the following. For the physical understanding, the limitation to cholic acid, a typical representative shown in Fig. 4.38, is sufficient.

Thus, bile acids in the small intestine with their very special structure become an emulsifier [73] and their effect on emulsions can be precisely understood [74]. The bile acids manage, on the one hand, to convert the large oil droplets into very small ones. Due to their broad hydrophilic structure, they can provide suitable docking sites for lipases and thus open the interface by taking fatty acids and phospholipids into the pliers and removing them from the interface of the fat droplets. The lipases now have access to the triglycerides and can cleave the fatty acids from them. This cascade of fascinating physical processes is shown in Fig. 4.39.

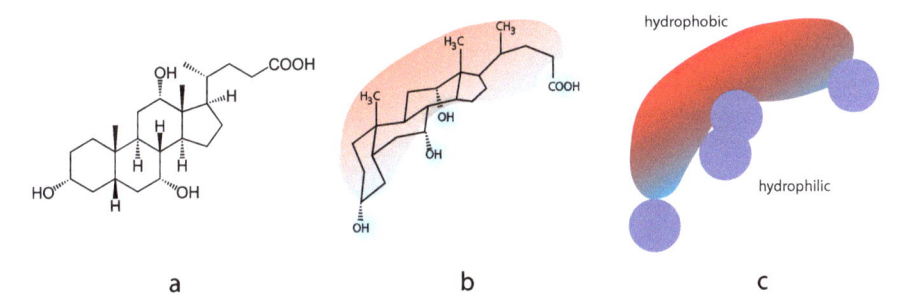

Fig. 4.38 Cholic acid, a typical representative of bile acids. **a)** The structural formula shows the great similarity to the triterpene backbone of cholesterol (compare Fig. 3.3). **b)** the spatial arrangement reveals the two sides of cholic acid: one side is hydrophobic, the other hydrophilic. **c)** a simplified physical model that will be used in the following to fundamentally understand the special emulsifying effect of bile acids

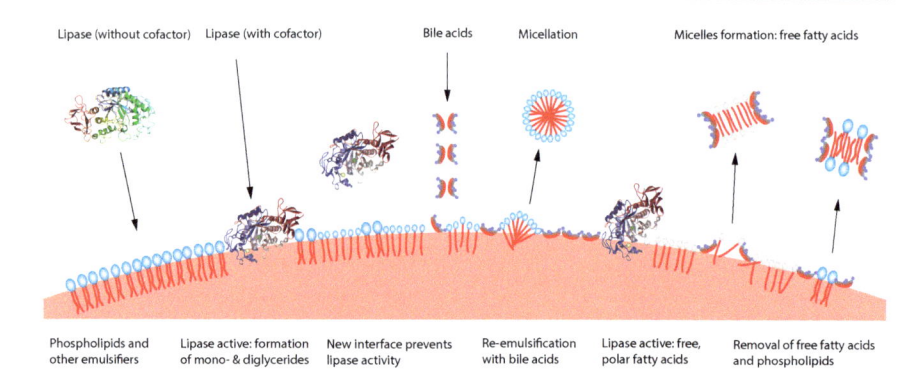

Fig. 4.39 The various steps that occur during the digestion of a fat droplet (red) after the digestion of proteins. Various micelles and other structures are formed, which are typical for the physics of soft matter and of high relevance

The bile acids first prepare the fat droplets to allow pancreatic lipases to access the fats at all. The lipases require a cofactor (e.g., a doubly positively charged calcium ion or a smaller protein, the colipase) to activate them. Without the cofactors, lipases cannot even access the fats trapped in the droplet, as shown on the left in Fig. 4.39.

Only when lipases are activated by the cofactors do they reach the surface with their active center and can cleave fatty acids from the triglycerides. This produces free fatty acids as well as mono- and diglycerides, which must also organize themselves at the fat–water interface. The interface is thereby reshaped and becomes highly heterogeneous with the emulsifiers of different polarity. This, in turn, makes it more difficult for lipases to adsorb on the surface of the fat droplet. Bile acids, on the other hand, manage to squeeze partially into the interface and push the existing emulsifiers together. The mono- and diglycerides, as well as fatty acids or phospholipids, have no choice but to evade. In doing so, the fatty acid tails will inevitably come into contact with water—an energetically very unfavorable situation. The only way to lower the energy again is therefore the formation of micelles to detach themselves from the droplet's interface in an orderly manner. This creates space for more lipases and their free access to the fat. Thus, more and more free polar fatty acids are formed, which settle at the interface and are transported away in small micelles by the bile acids in the last step (Fig. 4.39).

These steps follow the universal laws of colloid physics, regardless of animal or vegetable lipids. However, these purely physical steps are necessary to supply (long-chain) fats to the intestinal wall in order to make them available to physiology [75]. Only free polar fatty acids and monoglycerides can be absorbed by the intestinal wall and reassembled into triglycerides, which are packaged in large fat particles together with cholesterol and then supplied to the blood circulation via the lymphatic system, unpacked as LDL particles [71] (see Fig. 3.7).

4.9.4 Long-chain Fats (n > 12)

Long-chain fats—triacylglycerols or triglycerides—consist of fatty acids that are longer than twelve carbon atoms. These include most common food fats and cooking oils, whether it be lard, olive oil, or most vegetable oils. Some fats, such as butter (milk fat), have a very wide distribution of fatty acid chain lengths and thus have an exceptionally high proportion of medium-chain fatty acids. Long-chain fats are digested according to the processes just mentioned. These physical steps are not necessary for medium-chain fats.

4.9.5 Medium-chain Fats (12 > n > 6)

Triacylglycerols with medium-chain fatty acids can be absorbed quickly and predominantly without hydrolysis, i.e., cleavage of the fatty acids from glycerol by lipases, independently of the bile acids. They pass directly through the portal vein into the liver. Only there are they cleaved by lipases in the liver. The released fatty acids are thus available as an immediate energy source in the mitochondria of the cells. This can happen, for example, when consuming food based on coconut fat, as coconut fat consists mainly of medium-chain fatty acids.

In this context, the question of what happens with fats with a very wide fatty acid length distribution, such as milk fat (butter), is interesting. This question can be answered relatively easily. During the process described in Fig. 4.39, short- and medium-chain fats are selectively separated according to thermodynamic principles (phase separation). Due to the shortness of the fatty acids, the medium-chain fatty acids pack themselves into independent, very small droplets that are sufficiently stable in water and thus pass directly into the liver.

This is also physically based: medium-chain fats are liquid at body temperature of 37 °C, even if most (or all) fatty acids in coconut oil are saturated. The triacylglycerols can thus form extremely small droplets that are stable in an aqueous environment without emulsifiers. They therefore do not require the detours via the bile acids and are directly supplied to the energy system. Controversial statements, such as that the medium-chain, saturated fatty acids of coconut oil are pure poison, as they are popular on the internet [76], are therefore not generally tenable for physical reasons alone.

4.9.6 Plant Fats, Nuts, and Oilseeds in the Digestive Tract

Many digestive processes follow universal, food-independent laws of colloid physics and soft matter physics at the molecular level. They are also set in motion, for example, when we eat the nuts, oilseeds, or even legumes praised as healthy, which store relatively large amounts of oil, such as soybeans. Corn, soybeans, sesame, palms, sunflowers, olives, peanuts, and pumpkin—they are all oil fruits in agricultural parlance and have been sources of vegetable oil for centuries, which—depending on the composition of its fatty acids—serves various purposes in the kitchen. In botanical parlance, a distinction is still made between the fruit or pulp (olives, palms) and the seeds or seed, depending on which plant tissues the oil is isolated from. Little was known outside of botanical interest about where, why, and how plants store oil. The advantage for the seeds is evident: oil as a storage substance has the highest energy density of all macronutrients, and this is precisely what the plant uses during germination. The oil is packaged in the cells of the seeds, in small, very stable droplets ranging between nano- and micrometer size. These oil reservoirs are particularly well researched in oilseeds such as corn, sunflower seeds, or soybeans. They are referred to as oil bodies or oleosomes [77]. The typical size of the oleosomes is in the nanometer range—in soybeans, about 265 nm in diameter [78]—and is essential for the germination and growth of the seed. At the time of germination, when the plant cannot yet perform photosynthesis, the stored neutral fats are broken down into free fatty acids and finally monosaccharides, which are needed for the synthesis of carbohydrates [79].

The oil bodies of various seeds and legumes vary in size, depending on their habitat and the prevailing conditions. The most important sizes are summarized in Fig. 4.40.

Plant physiology and cell biology reveal how plants manage to package and distribute their water-insoluble oil in plant cells into these very small fat droplets. A natural emulsification process at the nano level takes place within the plant cell, which has been optimized by evolution over millions of years. At a specialized cell organelle (the endoplasmic reticulum), the finished fats produced by enzymes accumulate within its membrane (i.e., between the phospholipid bilayer). This leads to the membrane bulging into small droplets, which can pinch off from the membrane. The size of the droplets is also determined by additional proteins (oleosins), which simultaneously embed themselves into the emerging oleosome and prevent the individual lipid droplets from clumping together.

Apparently, soybean and rapeseed oil, as well as mustard seed, build oleosomes with diameters in the nanometer range. Their properties are therefore determined more by the surface than by the volume. Despite the size differences in the various

Soy	Rape seed	Mustard seed	Flax seed	Corn	Peanut	Sesame	Almond	Cocoa
0.26	0.65	0.73	1.34	1.45	1.95	2.00	2.60	7.0

Stability

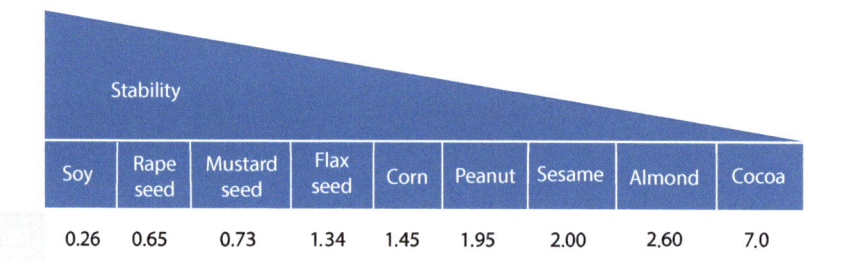

Fig. 4.40 Fat-storing oil bodies—oleosomes—have different diameters (D). Smaller particles are more stable

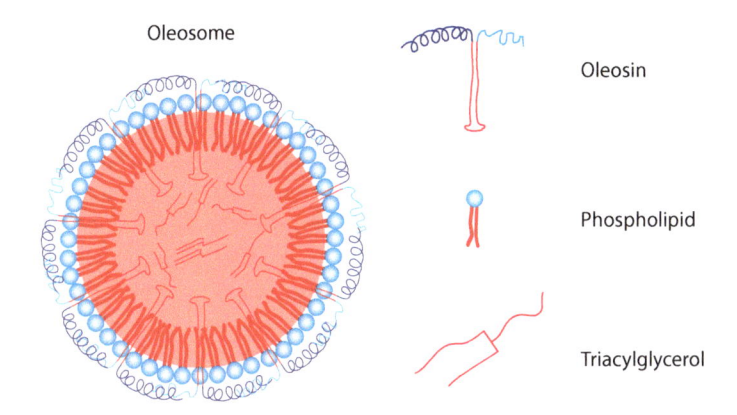

Fig. 4.41 Schematic representation of the oleosomes and their most important components. The oil of seeds and legumes is completely packaged in these oil droplets

species, oleosomes have a very similar structure. Since the oleosomes are incorporated into the aqueous environment of the cells of nuts and legumes, they must be stably emulsified. This emulsion formation in oil plants is a fundamentally universal process, as the requirements for stability are very high, and the nuts and seeds are exposed to high temperature fluctuations in the wild or take a long time to germinate. During this time, the fat stores must not be damaged. Oleosomes therefore receive reinforcement from very special proteins, in addition to the usual, tightly packed phospholipids around the oil, as schematically shown in Fig. 4.41.

As with animal HDL and LDL particles, a monolayer of phospholipids forms the basis of stability, albeit without a significant proportion of phytosterols. Additional stability is achieved through oleosins. These specially shaped proteins (oleosins) with a strongly lipophilic part, which is bent into a hairpin (top right in Fig. 4.41), are additionally incorporated and stabilize the oil particles. The hydrophilic parts (C- and N-terminal) of the protein consist of a helix and a highly unstructured polar part, and they carry a high proportion of polar and charged amino acids. Thus, oleosomes are electrically charged. Since all oleosomes carry the same charges, the droplets repel each other. They can therefore never agglomerate and combine into large oil droplets. The oil as stored energy is thus well protected from oxidation. These properties make oleosomes interesting systems for applications in biomedicine, cosmetics, and of course in the food sector.

4.9.7 Oleosins—Very Special Proteins

Oleosins are proteins that stabilize individual oleosomes. Their shape and function are unique [80]. Proteins or peptides are made up of amino acids. Nature incorporates a total of 22 different ones into proteins (proteinogenic amino acids).

Various properties can be attributed to the proteinogenic amino acids: Proteins are also amphiphilic, consisting of hydrophilic and hydrophobic amino acids, which are responsible for the formation of certain structural features of the proteins. One can imagine an oleosin like an umbrella. The outer umbrella part protrudes into the water (thus hydrophilic) and the handle is anchored in the oil (thus hydrophobic) (Fig. 4.41). A special primary structure is necessary for such stable anchoring. All oleosins therefore consist of an approximately 70 amino acid long sequence of hydrophobic amino acids, which forms a so-called hairpin structure. This sequence is one of the longest known hydrophobic amino acid sequence in nature. Another characteristic of oleosins, which are slightly modified depending on the plant species, is the proline knot. It is composed of, among other things, three amino acids proline and one serine and forms the 180° turn of the hairpin. This central hydrophobic sequence of oleosins is identical in all plant seeds. The only differences are found at two sequences at the ends sequences, the so-called N-terminal and C-terminal subareas, where the hydrophilic "umbrellas" are placed. They can consist of various hydrophilic domains, such as an amphipathic α-helix. The stable anchoring of oleosins in the oil phase is the main difference from the previously mentioned apolipoproteins in LDL and HDL particles. The charge present in the outer, hydrophilic umbrella part of the oleosins and the resulting repulsion of two oleosomes is mainly responsible for the stability of soy milk, but also the reason why some seeds have not lost their germination ability even after decades of storage.

This also explains why soy milk, tofu, or soy cream are very good emulsifiers. Soy milk contains the strongest emulsifiers of all: phospholipids, oleosins, storage proteins, i.e., proteins that contribute strongly to stability, and oil bodies, all of which contribute on different length scales. Oleosins are therefore the decisive stability enhancer in oil bodies [81]. They are extremely stable, far more stable than droplets in conventional emulsions, even cooking at 100 °C or stirring in strong mixers or technical homogenizers with high shear rates cannot destabilize them.

When eating nuts, crushed seeds from nuts, or soy products, such as tofu, most oleosomes still reach the stomach intact. However, the structure allows for rapid cutting of the oleosins by the acid-tolerant protease pepsin, as well as in the small intestine by pancreatic proteases such as trypsin. The oleosins of nuts and legumes are already pre-cut in the stomach by pepsin at some points, and the rest is taken care of by pancreatic enzymes. Bile acids slide into the simple phospholipid layer, breaking the already small micro- and nanodroplets into even smaller ones. The triacylglycerols of the oil become accessible to lipases, and the fatty acids are transferred for further utilization.

4.10 Basic Food Milk: Physics, Chemistry, Nutrition

So far, it has already been shown several times what useful insights the "physics and chemistry of digestion" preceding physiology allows. This idea also allows us to consider the structural aspects of milk and dairy products during their digestion from different perspectives. The food milk repeatedly leads to controversial discussions. In particular, the different structures of raw milk, pasteurized and homogenized milk, as well as extended-shelf-life (ESL) milk are of interest. The very long shelf life of ESL milk, which is offered as fresh milk, raises doubts about structure and digestibility. Homogenized milk also gives rise to speculation. The smaller fat particles of homogenized milk could penetrate the intestinal wall. The protein contained in homogenized milk is attributed a gluten-like structure, which is supposed to be responsible for problems in the intestine. Homogenized or pasteurized milk has also fallen into disrepute [83]. The enzymes present in raw milk, primarily xanthine oxidase, would be exposed and could damage the heart. In addition, "industrial cows" would produce too much of the unhealthy (a1-) casein, which is responsible for the formation of β-casomorphin-7, which in turn would harm humans [82]. These theses concern structural aspects of proteins and fats in milk and require a closer examination from the molecular perspective, as will be done in the following. However, milk must first be understood more precisely, not only its composition, but also the molecular structure, the organization of fats and proteins, and their colloidal structure. This alone will provide a new perspective on the food milk. Here too, physics and chemistry will show the fundamental way.

4.10.1 Raw Milk Proteins—Macro, Micro, Nano

In order to assess the properties of milk as well as the raised questions about health and non-health, the structure of the complex liquid must be considered. Let's start with raw milk and its structural elements, as provided by a cow. In milk, there are water, lactose, vitamins, water-soluble globular whey proteins, micelles packed caseins (total protein 2.5–6.0%) and, depending on the cow breed, 2–7% fat. In the milk that is usually offered, the fat content is adjusted to 3.5%, maximum 3.8% for whole milk. In raw milk from the farm or in preferred milk, protein and fat content are dependent on breed and feed (see, for example, the dissertation by Rätzer [84]). For the following, the detailed composition is not important, but the structure of the fat particles, the structure of the proteins, and their changes during pasteurization and homogenization.

The protein content of (raw) milk can be roughly divided into two groups from a structural point of view. On the one hand, there are the water-soluble, globular whey proteins, the lactoglobulins, and on the other hand, the caseins, which are structured in micelles (Fig. 4.42).

The fraction of water-soluble whey proteins consists essentially of α-lactalbumin and β-lactoglobulins. There is a whole range of immunoglobulins (antibodies), the structure and properties of which are not relevant in detail at this point. The secondary structure of the largest and most important whey protein is schematically shown in Fig. 4.43.

For the later question of the influence of homogenization and pasteurization, the thermal stability of whey proteins is important. α-Lactalbumins denature

Fig. 4.42 The most important (structural) proteins in milk

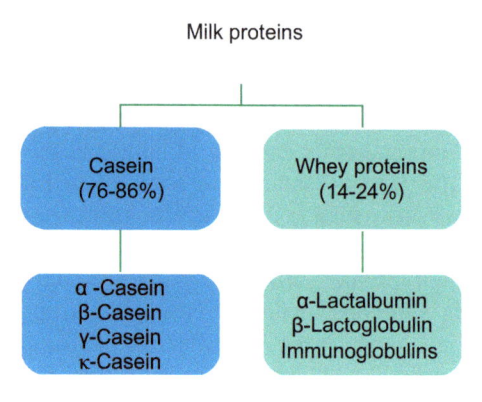

Fig. 4.43 The secondary structure of a monomer of β-lactoglobulin from (cow) milk. The yellow-colored area concerns the amino acid combination glutamic acid, glycine, and cysteine. (After uniprot.org)

already at 57.8 °C and β-lactoglobulins at 72.4 °C. This is known from the kitchen: When milk is boiled, a skin forms quickly. The whey proteins change their spherical shape, their water-repellent, hydrophobic sequences are exposed to water, the solubility decreases sharply, and any chance to find something hydrophobic is perceived—mainly through convection at the interface to the air. There, the whey proteins accumulate to form the nutritious milk skin. Mind you, this denaturation concerns the change in shape, but not the change in nutritional value. The amino acids do not change at these temperatures, of course. The higher denaturation temperature of β-lactoglobulin compared to α-lactalbumin is due to the structure, as the higher internal cohesion of native β-lactoglobulin is stronger through two disulfide bridges within the molecule. In addition, two β-lactoglobulins combine in native milk to form a dimer.

The two most important whey proteins, α-Lactalbumin and β-Lactoglobulin, are excellent amino acid suppliers. In terms of amino acid distribution, they are amongst the highest quality proteins available. They contain a large number of the essential amino acid Leucine, which is necessary for muscle building in humans. On average, 1 g of whey protein contains 105 mg of Leucine, 93 mg of Lysine, 69 mg of Threonine, 63 mg of Isoleucine, 58 mg of Valine, 32 mg of Tyrosine, 18 mg of Tryptophan, and 17 mg of Histidine. This combination in this concentration is rarely found in other proteins. Not to be forgotten are the 21 mg of sulfur-containing amino acids such as the essential Methionine and the semi-essential Cysteine. Although sulfur-containing Cysteine is not considered an essential amino acid for adults, as it can be enzymatically synthesized from Serine and Methionine. In infants, this is not possible due to the absence of the enzyme, so Cysteine is essential for the cross-linking of proteins (sulfur bridges, see Fig. 3.26) for example, for bone formation.

The short, maximum 170 amino acids long β-Lactoglobulins are more rapidly biologically available compared to packaged proteins, such as in muscle meat or even in legumes, but not immediately, as will be shown in more detail later. This relative availability of whey protein is, of course, desirable and highly welcome in offspring, the calves. Like all young children, they need to grow quickly and require the amino acid Leucine for muscle building. This is also why whey protein powder and whey shakes are highly popular among bodybuilders. No wonder that whey was considered a waste product of cheese production for a long time and was fed to omnivores such as pigs or chickens: The high Leucine content guaranteed rapid muscle mass build-up through natural feeding, and the high Cysteine, Calcium, and Phosphorus content guaranteed good cross-linking properties of hard structural proteins, tendons, and bones.

The Casein, the other protein fraction important for yogurt and cheese protein, has a far more complex structure. Casein is present in four forms and is packaged in Micelles [85], as shown in Fig. 4.44. In general, caseins are not, only slightly, or only partially (κ-Casein) water-soluble (Table 4.3). The proteins therefore form micelles (Fig. 4.44). The water-insoluble parts are located in the core of the micelles. The water solubility of the entire micelles is ensured by the κ-Casein molecules on the surface, which represent a strongly hydrophilic part of their structure.

αS1-Casein, αS2-Casein, and β-Casein form (via oppositely charged amino acids) strongly hydrophobic submicelles, which are connected to each other via calcium phosphate bridges [86]. Calcium phosphate itself can be present in larger (colloidal) complexes. However, the doubly positively charged calcium ion always binds ionically to singly negatively charged amino acids. Therefore, casein micelles are extremely stable and are not destroyed even when heated.

For the following, the average size of the casein micelles is crucial. Their radius (volume-averaged) has been determined to be 61 nm in recent detailed studies using careful neutron small-angle scattering experiments [86], so their diameter is approximately 120 nm. They are thus examples of natural nanoparticles and are significantly smaller than the fat particles, even after homogenization, as will

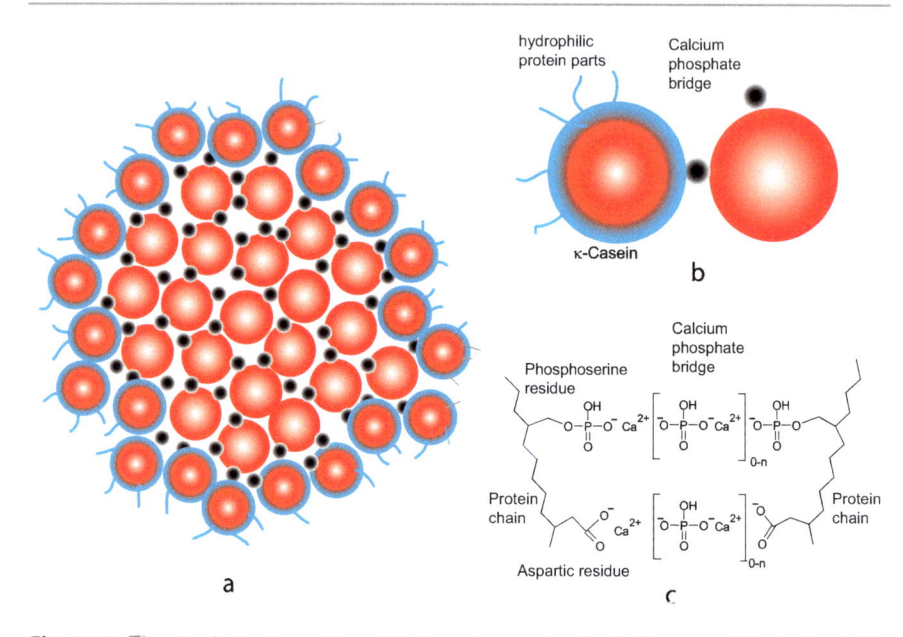

Fig. 4.44 The classic (and rough) model of casein micelles (left). The essentially hydrophobic core of α- and β-casein (red spheres) is coated with β-casein-rich submicelles (spheres with blue edges). The hydrophilic part of β-casein ensures stable dispersion of the micelles in the aqueous whey. All submicelles are connected via calcium phosphate bridges (colloidal calcium phosphate, nanoclusters), the more detailed chemistry of which is shown in the bottom right

Table 4.3 There are four different caseins in milk. Only κ-casein has a strongly hydrophilic part, which ensures the water solubility of the casein micelles

Type	Amount in g	Number of amino acids
αS1-Casein	10	199
αS2-Casein	2.6	207
β-Casein	9.3	209
κ-Casein	3.3	169

be shown shortly. Their specific weight is $\rho_C = 1.078$ g/ml, which is higher than that of water [87]. The noteworthy publication by de Kruif [86] reveals the length scales in the internal structure of casein micelles using neutron scattering. The radius of the calcium phosphate nanoclusters is between 1–2 nm, and their average distance is 18 nm. The average size of the submicelles is therefore in the range of about 15 to 20 nm.

The calcium phosphate bridges provide a high degree of cohesion for the micelles: The ionic binding energy is always greater than the thermal energy, even when fresh milk is heated significantly. Therefore, casein micelles remain intact during pasteurization and ultra-high temperature treatment. The ionic binding energy is about 20 times stronger than the thermal energy. The casein micelles will easily withstand denaturation during cooking and pasteurization, even if individual proteins, such as β-casein at 58.5 °C, change their shape within the micelle, which does not affect the cohesion of the micelles. This is consistent with the experience that casein in milk does not coagulate during cooking. The stability of casein micelles against aggregation (flocculation) is also very high: The hydrophilic parts of κ-casein protruding into the whey carry a negative net charge at a pH value of raw milk from 6.8 to 7, thus preventing the clumping of casein micelles. The casein micelles are therefore electrostatically stabilized nanoparticles that strongly repel each other as soon as they come close. Only by adding calf rennet, a protein-cleaving enzyme, are these charged parts removed and dissolved. The casein micelles aggregate, and curd forms. By adding acid (protons), the charges of the κ-casein arms are neutralized, and the casein micelles combine into large aggregates and precipitate. The understanding of the stability of milk is thus deeply rooted in the physics of colloids.

4.10.2 Structure and Composition of Milk Fats

In the fat particles, milk fats and oils accumulate and organize into small droplets with an average size of about 3 μm, but can vary between 1 μm and 20 μm. The size distribution of fat particles in raw milk is very broad, with both very small and very large particles present. The specific weight ρ_F is around 0.9 g/ml. These fat particles are in an aqueous environment, so they must be stabilized accordingly by emulsifiers to prevent or minimize coagulation, i.e., aggregation, and the formation of larger droplets. This is particularly a problem at higher temperatures, when cooking raw milk. Even then, fat does not separate from water, which speaks to a special structure of the native milk fat particles. The type of fats in milk plays a particularly important role in this regard.

In native cow's milk, there is a very broad distribution of fatty acids: Short-chain to long-chain, saturated and unsaturated fatty acids are present in milk fat. Apart from the high nutritional value of the fatty acid spectrum, the distribution and the associated molecular structure of the individual triacylglycerols (triglycerides) also determine a whole range of physical properties of the fat particles in raw milk (Table 4.4).

The chemical characterization of fatty acids of the general form C n:s (see Sect. 1.3) is the most important information. The number of carbon atoms n defines the length of the fatty acids, while the parameter s indicates the degree of saturation. Both together determine the melting and crystallization temperature of the fatty acids [88]. The strong variation in length and degree of saturation, and thus the melting temperature of milk fat, also shows how the triacylglycerols of milk must organize themselves to form droplets. However, since long and saturated fatty acids form crystals at room temperature, this can only happen if the liquid triacylglycerols, which consist of short-chain or unsaturated fatty acids, are surrounded by a solid, spherically crystallized core, as shown schematically in Fig. 4.45.

Table 4.4 The fatty acids of cow's milk; their average frequency and melting points determine the structure of the fat particles. The fatty acids come from different sources, the feed for the cows plays a decisive role in some fatty acids. Other fatty acids are produced by the enzymes Desaturase (desaturating) and Elongase (elongating)

Fatty Acid	Chemistry	Occurrence (%)	Melting Point (°C)	Origin
Saturated Fatty Acids				
Butyric Acid	C 4:0	3.0–4.5	− 7.9	Fermentation in the Rumen
Caproic Acid	C 6:0	1.3–2.2	− 1.5	Mammary Gland, Fermentation
Caprylic Acid	C 8:0	0.8–2.5	16.5	Mammary Gland
Capric Acid	C 10:0	1.8–3.8	31.4	Mammary Gland
Lauric Acid	C 12:0	2.0–5.0	43.6	Mammary Gland
Myristic acid	C 14:0	7.0–11.0	53.8	Mammary gland
Palmitic acid	C 16:0	25.0–29.0v	62.6	Mammary gland, adipose tissue
Stearic acid	C 18:0	7.0–3.0	69.3	Adipose tissue, feed
Unsaturated fatty acids				
Oleic acid	C 18:1	30.0–40.0	14.0	From C 18:0 Desaturase
Linoleic acid	C 18:2	2.0–3.0	− 5	Feed
Linolenic acid	C 18:3	0–1.0	− 11	Feed
Arachidonic acid	C 20:4	0–1.0	− 49.5	Desaturase, Elongase
trans-Fatty acids, conjugated fatty acids				
Vaccenic acid	C 18:1,11t	1	44.0	Rumen
CLA	C 18:2,11t	2	30–40	Rumen, adipose tissue

Fig. 4.45 Solid and liquid fats form a spherical fat droplet. For a simplified schematic representation, only one fat type is shown in the crystal phase and variations in the fatty acids are neglected. The solid fats organize themselves in a crystalline shell and enclose the liquid core of short-chain and unsaturated triacylglycerols (triglycerides)

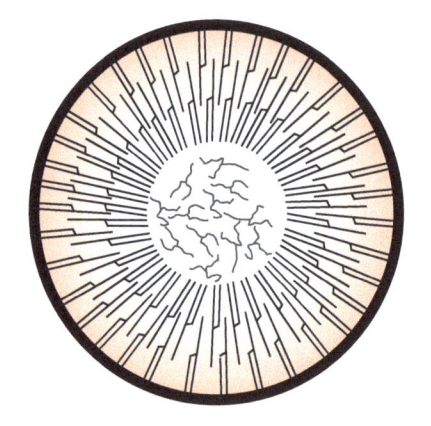

Table 4.5 Lecithin and emulsifiers of milk have various components that distribute themselves in membranes and around fats

Polar lipids	Percentage share
Phosphatidylcholine	36
Phosphatidylethanolamine	27
Sphingomyelin	22
Phosphatidylinositol	11

Since the fat globules are in an aqueous environment, the whey, additional stabilization mechanisms are necessary. Therefore, the surface of the fat globules is coated with a complex layer of emulsifiers [89]. These also ensure the stability of the fat globules under heating and the associated melting of the crystalline phase of the droplets (even if some of the droplets coagulate and the size distribution may shift slightly). Hence, the fat droplets are stabilized by a monolayer of phospholipids and by a surrounding membrane—a double layer of phospholipids. The phospholipids in turn are divided into four different forms in the milk in varying proportions, which are summarized in Table 4.5.

Phospholipids, as known from membranes, can adjust properties such as stiffness, bending energy, or local flexibility with different fatty acids. The membrane elements of raw milk take on almost all required biological properties. As with the oleosomes, the fat particles of oilseeds and legumes, the droplets would hardly be stable with a simple enveloping phospholipid layer, a monolayer. The fat droplets are further stabilized by enclosing them with a lipid double layer. This reinforcement strongly stabilazes the droplets physically, while at the same time many biofunctional molecules can be packed into the membranes that are important for the offspring, such as phospholipids, cholesterol, and especially enzymes.

The biological membrane around the fat droplets in raw milk does not have a comparable function as in cells, but it is an economical method to effectively enclose nutrients and functional components important for the calf. In the highly simplified representation of the surface of the fat droplet (Fig. 4.46), it is illustrated where, for example, the xanthine oxidases are located: between the lipid bilayer and the monolayer of phospholipids on the surface of the fat droplets.

4.10.3 Milk—Raw and Pasteurized

The aim of pasteurization is, of course, to inactivate the germs present in raw milk, which must not multiply beyond a critical number per volume in the milk. Pasteurization ensures food safety. As good as raw milk tastes and as high-quality as it is, it is also exposed to a whole range of possible contaminations with germs that can multiply even under refrigeration. Possible contaminations come from the udder skin, the bedding on the stable floor, residues of excrement in the stable, or hay and other (pre-contaminated) feedstuffs. Among them are also really dangerous germs like coli bacteria, which can cause diseases via enterohemorrhagic Escherichia coli (EHEC). There is hardly any difference whether the farm

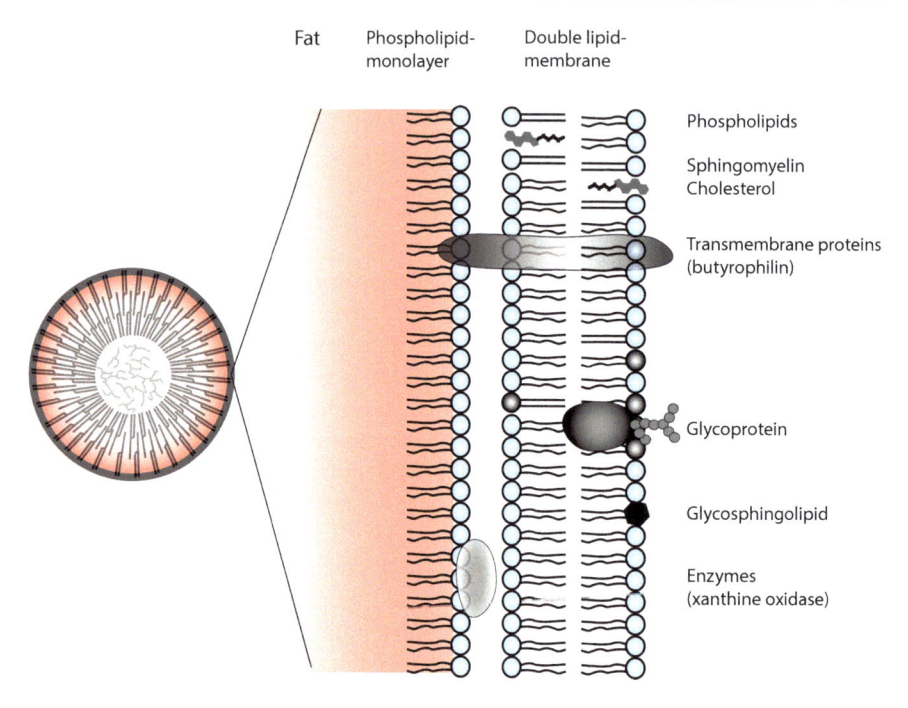

Fig. 4.46 The fat particles of raw milk are triple-packed. The fat is surrounded by a monolayer of phospholipids, around which a double-layer membrane of various phospholipids and cholesterol clings. In this case, the lipid bilayer is also capable of enclosing proteins, enzymes, and other functional molecules

is "organic" or "conventional". The aim of pasteurization and other preservation methods is to prevent the growth of such germs and to kill any germs that may be present. Depending on the method, thermal processes (pasteurization, ultrahigh temperature treatment) or microfiltration, in which germs can be filtered out, are suitable (ESL milk). In principle, these methods serve food safety, even if the opposite is repeatedly claimed and conclusions are drawn from scientific studies [90] or generally understandable publications [91, 92] that such treated milk would be the cause of intolerances and diseases.

As already noted, during thermal treatment, depending on temperature and duration, some whey proteins (partially) denature. It is only logical that this has no effect on their nutritional value in terms of macronutrients. Neither the number nor the biological value of the amino acids nor the structure of the fatty acids change during pasteurization and microfiltration, and thus the nutritional value of the macronutrients does not change either. Due to the denaturation of the whey proteins, they even become more easily digestible, as can be precisely tracked *in vitro* [93]. For this purpose, the native proteins are simulated under realistic conditions, i.e., at the prevailing temperatures of 37 °C and physiological pH values (around 2.0) in the stomach, and then the enzymes occurring in the stomach,

4.10 Basic Food Milk: Physics, Chemistry, Nutrition

mainly pepsin, are added. The formation of protein fragments and the speed of protein digestion can be precisely determined using suitable analytical methods. Afterward, these pre-digested and native proteins are passed on to the "*in-vitro*-small intestine." There, the temperature is also 37 °C, and the pH value is 8.0 when trypsin is added. Here, too, digestion can be precisely tracked and supported with analysis data. In this way, the whey proteins can be heated to different temperatures (and for different times) as would correspond to pasteurization, UHT-milk production, and ESL treatment.

In the case of β-lactoglobulins, digestion reveals quite astonishing information about the structure and cohesion of the molecules and, above all, where and under what circumstances they can be digested more quickly. This is exemplified for pasteurization temperatures and pH values using Fig. 4.47.

It is noticeable that native β-lactoglobulin is hardly digested by gastric pepsin in the stomach. Even with slight heating, the digestion rate remains comparatively low, even if the enzymes act on the proteins for more than 120 min. Only when heated to 70 °C and above does the digestion rate increase significantly, and after 60 min, most of what pepsin can cut is done. This is different under the conditions in the small intestine: There, trypsin is the decisive enzyme. Even the native β-lactoglobulins in raw milk are broken down there at significant rates. Preceding heat treatments of 80 °C and higher accelerate the digestion rate somewhat. If the disulfide bonds are separated beforehand, the digestion rate is significantly higher for both enzymes. In Fig. 4.47c, this is underlined once again. Special attention was paid to the separation of disulfide bonds in native and heat-treated

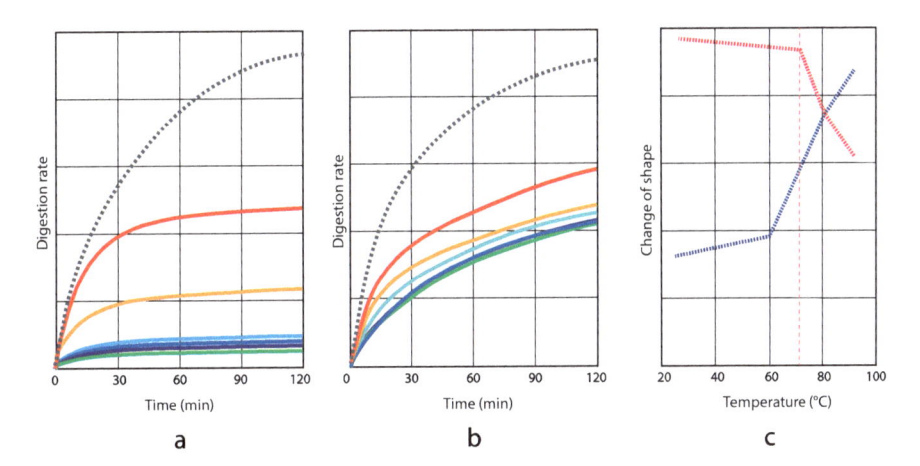

Fig. 4.47 Digestion (in relative units, measured by fluorescence spectroscopy) of β-lactoglobulin in the model *(in vitro)*. Data according to Reddy et al. [94]. (**a**) Digestion under pepsin. (**b**) Digestion under trypsin at different temperatures: green: native, dark blue 50 °C, green-blue 60 °C, light blue 70 °C, orange 80 °C, red 90 °C. For the values of the gray dotted curves (**a** and **b**), the disulfide bonds were previously enzymatically cleaved. **c**) The extent of the protein's conformational change in the stomach (red) at pH = 2 and in the small intestine (blue) at pH = 8

β-lactoglobulins. At the low pH values in the stomach, these are almost stable up to temperatures of 70 °C, even at higher pretreatment temperatures, no separation of the disulfide bridge occurs (red dashed curve). At higher pH values around 8, as they are present in the small intestine, the disulfide bonds break rapidly, and the proteins change their shape significantly from 60 °C. Whey proteins are therefore not digested as quickly as is often claimed elsewhere [94].

What can be deduced from such complicated but necessary experiments? Native β-lactoglobulins are resistant to pre-digestion in the stomach due to the intact disulfide bonds. Their digestion in the stomach becomes significantly easier with the thermal treatment of milk from 70 °C, and the whey proteins are more extensively pre-cleaved in the stomach. In the small intestine, however, the pretreatment of the milk plays hardly any role. Even native whey proteins are quickly broken down into peptides and amino acids. This observation is, of course, the consequence of the denaturation of β-lactoglobulin at about 72 °C. The denaturation temperature is shown on the right in Fig. 4.47 as a vertically dashed line. The shape changes of the protein triggered by this make it more accessible to the digestive enzymes, allowing the active centers of the enzymes to access better.

However, the experiments show even more: If the disulfide bridges are dissolved beforehand, the digestion of the protein begins almost unrestrictedly. Apparently, the internal cohesion via the cysteine-sulfur bridges is responsible for the high resistance of β-lactoglobulin to pepsin. This cohesion is very strong through the chemical bond, and the gastric acid does not weaken the bond, on the contrary. Only the high pH value in the small intestine allows the bonds of the disulfide bridges to be weakened. Disulfide bridges can dissolve more easily at pH values above 7. Therefore, in Fig. 4.47, it becomes clear why denaturation starts at lower temperatures at higher pH values. Due to the already dissolved disulfide bridges, the cohesion of the protein is weaker, and denaturation can occur at slightly lower temperatures, at about 60 °C. The enzymes then have full access to their interfaces. Pepsin can preferentially cut at the essential amino acids tryptophan, tyrosine, phenylalanine, leucine, and isoleucine. However, these interfaces are not accessible to the enzyme as long as the protein is stabilized by the stable sulfur bridges. Contrary to contrary assumptions, this does not result in an uncontrolled rapid uptake of amino acids from whey proteins [95]. The digestion experiments with the disulfide bridge-free β-lactoglobulin show this clearly. Only when sulfur bridges are broken can both pepsin and trypsin access their interfaces unrestrictedly.

These structural details, which are exemplified here in more detail for β-lactoglobulin, have nothing to do with tolerance or intolerance, or biological availability, nor with well-being or discomfort. However, they once again show how deeply one must look into the structure of proteins during digestive processes to capture physiological phenomena. Furthermore, something general about the digestion of proteins with disulfide bridges can be learned, which we have already encountered in bread (gluten). There, similar principles apply and show how the interplay of gastric and pancreatic enzymes systematically regulates protein digestion.

At the same time, the denaturation of proteins is apparently a matter of temperature and time. If the milk and the proteins are exposed to a short, intense heat pulse of 127 °C, they simply lack the time to denature extensively. The heat pulse distorts the membranes and some of the membrane proteins of the germs, which is enough to neutralize the germs, but not the β-lactoglobulins bound to themselves via some sulfur bridges. This is also defined by the physics of interaction and the resulting dynamics of the proteins according to fundamental physical laws. These physicochemical facts are not known to many nutrition populists. Therefore, there is also no danger from ESL milk. On the contrary: These processes guarantee a very high food safety, which a tasty raw milk does not have *per se*. This is in contrast to a small loss of micronutrients (some vitamins) that occurs during heat treatment. In pasteurized fresh milk, this affects up to a maximum of 10% of the heat-liable B-vitamins and vitamin C. In the case of UHT-milk heated briefly at significantly higher temperatures, the losses of the mentioned vitamins are up to 20%, while no significant losses are recorded in ESL milk due to filtration techniques. Therefore, it is a matter of weighing. Of course, high-fat raw milk, fresh from the farmer, stands for high enjoyment, but it can no longer be sold everywhere in supermarkets or organic stores (exception France). Discussing vitamins solely based on the food milk is futile for practice anyway. Milk is usually not the only food consumed, and the vitamin requirement can be easily met with a varied mixed diet without supplementation.

Regarding pasteurization and other thermal preservation methods and their effects on milk, a lot of nonsense is written, which can already be refuted with the knowledge gained so far [96]: The digestive aid lactase would be destroyed during pasteurization, which is why pasteurized milk is less well tolerated (taken from the internet [97]), which is unfortunately uncritically adopted and disseminated [98]. This is, of course, not correct, because there is no lactase in native milk, as the calves would lack an important nutrient, the carbohydrate lactose.

4.10.4 Whey Protein—A Glutathione Supplier

A frequently overlooked advantage of β-lactoglobulin is its potential to provide glutathione. Glutathione is a γ-glutamyl peptide, a term already mentioned in connection with kokumi. These glutamyl peptides are formed from normal peptides (α-bond) after the peptide bond has been altered by the enzyme γ-glutamyltransferase. In fact, glutathione is a kokumi-relevant molecule and also occurs more frequently during the ripening of animal products such as cheese, which then supports the mouthfeel.

Glutathione is a multitasker, as it acts highly antioxidative in all cells and is therefore part of the natural cell protection. Glutathione consists of the amino acids glutamic acid, glycine, and cysteine, and this combination occurs as a sequence in β-lactoglobulin, specifically at the yellow-marked site in Fig. 4.43. When the proteins are enzymatically broken down, there is a high probability that a direct precursor of glutathione is already present and can easily be converted from the normal α-bond to the γ-version via γ-glutamyltransferase. This can also be demonstrated *in vitro* [99] and *in vivo* [100].

This once again shows, as already explained in several places, how taste and function are combined in the human biosystem. Without the presence of glutamic acid, no cell or protein would function, even though it is not an essential amino acid. Nevertheless, it plays a significant role in umami taste, kokumi, mouthfeel, and the formation of the antioxidant glutathione.

4.10.5 Milk—Raw vs. Homogenized

As already mentioned, the size distribution of fat particles in raw milk is very broad, i.e., both small and very large fat particles are present. These large particles ensure that they quickly cream, i.e., in raw milk, large fat particles accumulate relatively quickly, within a day, on the surface and form a cream layer. Physically, this can be shown when the fat particles are described by macroscopic spheres with radius $R = D/2$, mass m, and a specific weight $G = mg$; g is the acceleration due to gravity. The mass m results from the density ρ_P of the particle and its volume V to $m = (4\pi/3)R^3\rho_P$. In the whey, the fat particles experience a buoyancy force F_A, which, according to the Archimedean principle, results from the weight of the displaced whey volume, $F_A = (4\pi/3)R^3\rho_W g$; ρ_W is the density of the whey. However, the particles are in motion as soon as the weight and buoyancy are different, i.e., when the particle and liquid density differ. Consequently, a frictional force acts, which in turn is defined by the particle size R, the viscosity η of the whey, and the particle velocity v and is described in the simplest case by the Stoke's force $F_R = 6\pi\eta R$. From the force balance, the time-independent creaming velocity of the particles, v, is obtained as

$$v = \frac{2}{9\eta}g(\rho_P - \rho_W)R^2$$

As expected, the creaming velocity v decreases with increasing viscosity and depends on the sign of the difference in densities. If the specific weight of the particles is lower than that of the surrounding liquid, the particles rise upwards (creaming), in the opposite case they sink (sedimentation). Of great importance, however, is the quadratic dependence of the velocity on the particle radius R. Large fat particles move disproportionately faster upwards. If the radius of the fat particle doubles, the creaming velocity increases by a factor of 4. The cream layer of raw milk thus consists of the large fat particles. This idea is correct, but not quite accurate, because during creaming, grape formation occurs. Individual fat particles join together and form larger aggregates. Their effective (hydrodynamic) radius is correspondingly larger, and these grapes rise faster. This results in an increase in the effective droplet radius over time, so creaming proceeds faster than one might expect and as predicted by the equation above. Storing milk at low temperatures (around 0°C in the refrigerator) also accelerates creaming. The thermal energy of the fat particles is lower, and the proteins in the membrane layer

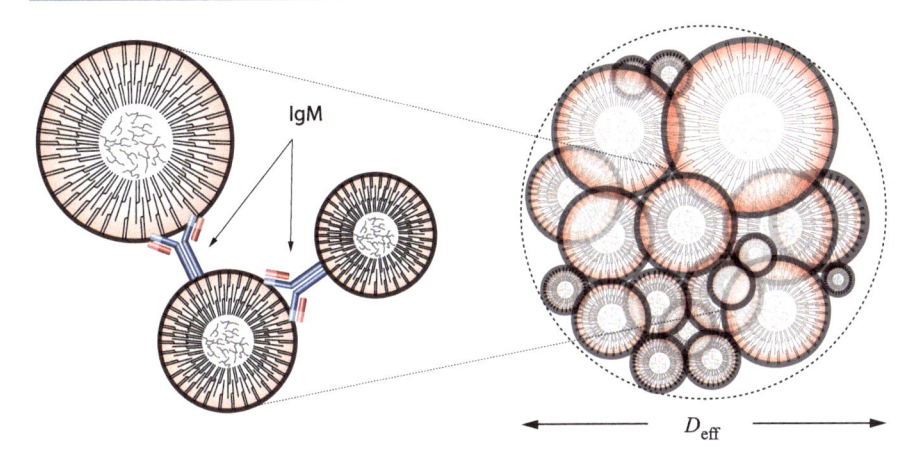

4.10 Basic Food Milk: Physics, Chemistry, Nutrition

Fig. 4.48 Aggregate formation in milk during creaming. Fat globules join together via the IgM proteins embedded in the lipid bilayer. The effective diameter, D_{eff}, steadily increases with increasing grape formation and is significantly larger than the average diameter of the individual fat particles. Consequently, the creaming velocity increases over time

of different fat particles connect via a protein, the immunoglobulin M (IgM) of the whey [101]. Grape formation is further accelerated at low temperatures. The creaming velocity is therefore no longer constant over time but increases dramatically with the size of the growing aggregates (Fig. 4.48).

At the normal body temperature of the cow, which lies between 38 and 39 °C, this grape formation does not occur, as the fat droplets move so quickly that they cannot connect via the IgM interaction due to their high energy and inherent velocity. Also, boiled and heated raw milk for an extended period will not cream as quickly. The proteins of the immunoglobulin IgM are denatured at 78 °C and no longer serve as links for aggregation. If the milk is only briefly pasteurized, the heat pulse is not sufficient to denature IgM, thus a thick cream layer still forms in pasteurized and non-homogenized milk after one to two days.

To avoid this, the milk can be homogenized (Fig. 4.49). The aim of high-pressure homogenization of milk is to reduce the size of the fat particles. To do this, the milk is forced through nozzles at high pressure, creating strong shear flows. This aims for maximum particle sizes of about 1 µm. The processes involved are considerable. If the radius of the particles subjected to the process is reduced by 1/10, this results in a thousand-fold increase in the number of particles due to volume constancy. Consequently, their surface area increases tenfold. This suggests a change in the microstructure of the surface and thus of the milk. In fact, fat droplets and casein micelles are significantly altered. The membrane bilayer is partially torn, and the casein micelles are reduced in size. Casein micelles attach themselves to the fat droplets, with hydrophilic casein submicelles near the monolayer close to the triacylglycerols. Whey proteins can also adsorb to the altered surface of the fat droplets [102].

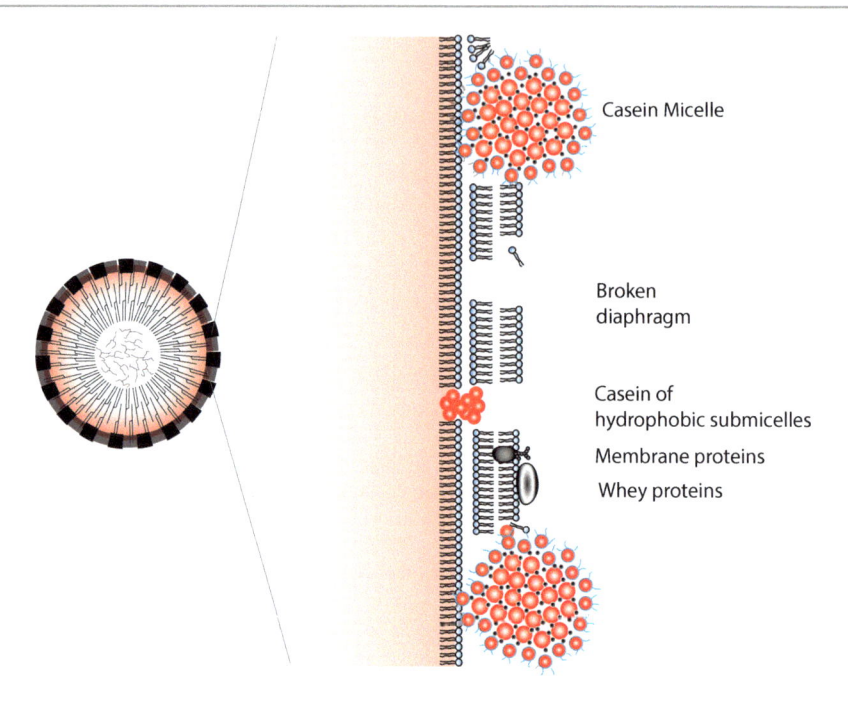

Fig. 4.49 Highly schematic (and not to scale) representation of the greatly altered surface of the fat particles after homogenization. The lipid bilayer (membrane) is partially destroyed. Casein submicelles and casein micelles themselves attach to the surface of the fat droplets

At the same time, enzymes are released from their natural sites, including lipases, whose catalytic effect separates fatty acids from the triacylglycerols; these are then degraded to aroma compounds that would give the milk a fatty-rancid odor. Therefore, it is actually imperative to heat the milk before homogenizing. This not only inactivates germs but also enzymes. The shelf life of homogenized milk is thus extended for two reasons: due to the low germ count and the reduction of enzymatically induced off-flavor formation.

The adsorption of proteins from the casein micelles onto the fat particles after homogenization can be determined using elaborate analytical methods [103]. It shows that, compared to raw milk, the relative proportion of proteins with respect to surface area and specific weight of the fat particles increases (Table 4.6).

Some parts of the detached membrane dissolve as lecithin. They help to foam homogenized milk better. An important, often overlooked point concerns the natural small fat particles of raw milk, whose size ranges between 0.5 μm and 50 nm. These particles, as well as many casein micelles, are not altered during homogenization. Raw milk also has a high number of nanoparticles that survive any homogenization process.

The surfaces of the fat particles are covered with small casein micelles after homogenization, thus reducing the density difference of the particles. The average

Table 4.6 Change in the protein content on the fat particles. Due to the reduction of the particle radius, the surface area increases significantly. At the same time, caseins are deposited on the fat particles. The proportion of bound protein increases, as does the proportion of protein bound to the fat

Milk (4% fat content)	Surface area of fat particles (m^2/ml fat)	Proportion of protein bound to fat (g/100 g protein)	Amount of protein bound to fat (mg/g fat)
Raw milk	1.2	3.8	28
Homogenized	4.8	39.8	345

density of the fat-casein particles in the homogenized milk is higher than that of the pure fat particles of raw milk and is barely different from the density of whey. The density difference is close to zero. The creaming rate becomes so low that no creaming can be detected within the usual storage and shelf life of the milk. At the same time, hardly any cluster formation occurs due to the changed surface. The thermal pretreatment of homogenized milk causes lactoglobulins and immunoglobulins to partially denature. IgM can therefore no longer act as a connecting bridge between still intact membranes of the fat-casein-particles in the homogenized milk. These effects, which accelerate creaming in raw milk, are no longer present after homogenization, and the density difference between fat particles and aqueous phase is smaller. Homogenized milk therefore does not cream in the refrigerator anymore.

4.10.6 Raw Versus Homogenized Milk in Digestion

The structural differences between raw milk and homogenized milk are often discussed in connection with health issues. Homogenized milk is held responsible for health problems, especially in pseudoscientific literature [104]. Nowadays, detailed *in-vitro* studies can show how raw milk and homogenized milk are digested on a molecular basis [105, 106]. When drinking fresh raw milk, it encounters the acidic environment in the stomach with pH values between 3 and 1. There, the milk "curdles," as when adding strong acid in the kitchen, the milk coagulates, which means nothing other than the casein micelles clump together into larger aggregates and thus show high resistance to pepsins in the stomach. When transported further into the small intestine, the pH value increases again, the aggregates can loosen, and are attacked by pancreatic proteases with suitable cleavage sites, such as trypsin and chymotrypsin. Thus, the different casein proteins of the casein micelles are gradually broken down.

In the case of fat digestion, the general aspects, as discussed in detail in Sect. 4.9 and in Fig. 4.39, can be referred to. In non-heated raw milk, the fat particles are surrounded by membrane proteins and the membrane bilayer. The fat particles also encounter stomach acid, where most of the membrane proteins are denatured, the membrane becomes porous and partially dissolves. Phospholipids

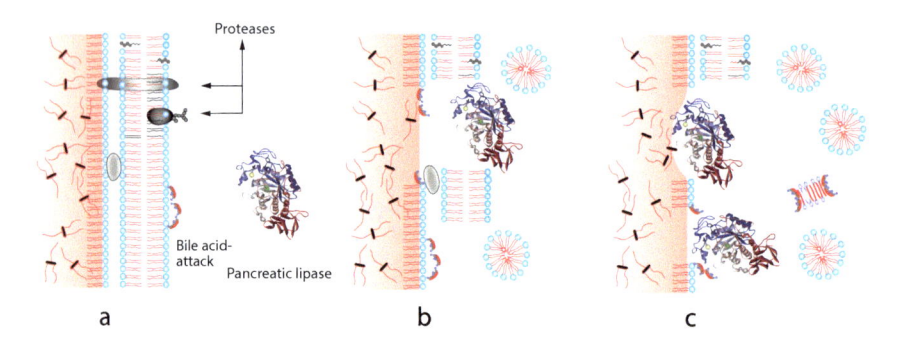

Fig. 4.50 Representation of various phases (**a–c**) of the digestion of fat droplets during raw milk consumption

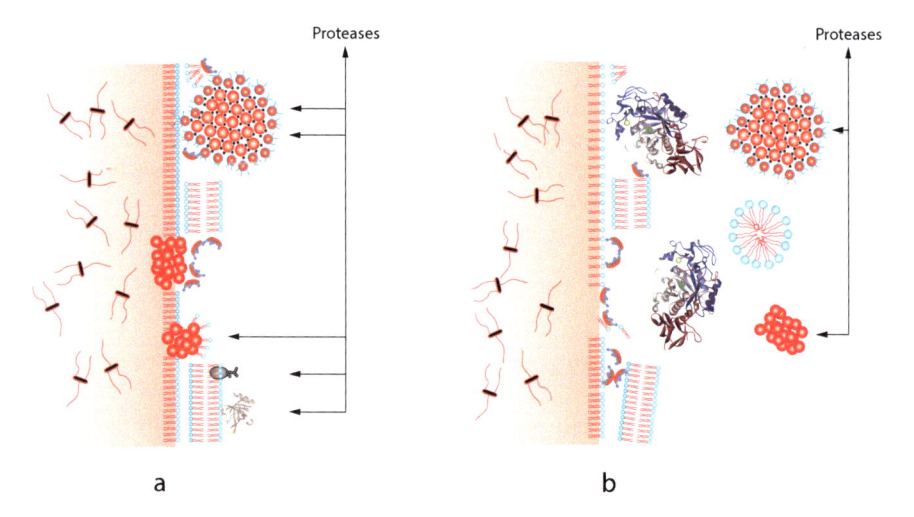

Fig. 4.51 The digestion of fat particles in homogenized milk. The altered surface of the fat particles (a) merely requires a different time management between proteases, bile acids, and lipases (b). The biophysical processes are nevertheless very similar to the digestive processes of all emulsions and dispersions and are not limited to milk

accumulate in the stomach to form micelles and are available for the emulsification of free fats. The membrane parts that have survived the passage through the stomach are dissolved by the bile acids. After that, the bile acids attack the monolayer of phospholipids and clear the way for the pancreatic lipase, as shown in Fig. 4.50.

In homogenized milk, the process is similar, but it occurs in a different time sequence. However, apart from the digestion of casein agglomerates via bile acids with the support of simultaneous protein digestion in the small intestine, the enzymatic processes are not particularly different (Fig. 4.51).

The milk fat particle membrane is already broken, instead the fat particles are covered with casein micelles. The structure of the casein micelles is "softened" at the low pH values in the stomach due to protein denaturation, and proteases begin protein digestion. While in the case of raw milk, the milk fat membrane must first be removed, this is only the case in some places for homogenized milk. The surface of the fat particles is mainly covered with casein micelles or casein submicelles, which must first be detached from the fat particles and proteases by means of bile acids before lipases can digest the fats. Here too, the high interfacial activity of the bile acids helps. They push themselves between the phospholipid monolayer and thereby reduce the adsorption energy of the casein micelles. These dissolve and move into the aqueous phase in the intestine. This clears the way for the pancreatic lipases, which then hydrolyze triacylglycerols. As in the case of raw milk, free fatty acids, bile acids, and phospholipids are transported in mixed micelles. The remaining casein micelles or parts thereof are then available to proteases for digestion. Gradually, the proteins of the micelles are cleaved into peptide fragments and amino acids, which can be recycled accordingly.

In this context, it is worth mentioning that the *in-vitro* simulation of such digestion processes can be carried out much more realistically than is generally assumed. There are now devices that very accurately reproduce stomach and intestinal movements. Even with these dynamic investigations, the results listed so far do not change [107].

So far, the molecular facts. The common myths still need to be briefly addressed in the following.

4.10.7 Homogenized Milk and Atherosclerosis

Since the 1970s, there has been the hypothesis [108] that the enzyme xanthine oxidase, exposed on the surface of fat particles during the homogenization process, is partly responsible for the development of atherosclerosis and coronary heart disease. The homogenization process could cause the fragmented milk fat globules to pass through the intestinal wall, allowing the enzyme xanthine oxidase to enter the organism. This, according to the hypothesis, could trigger a biochemical process in the heart, resulting in typical damage to the heart and arteries.

This is impossible for physical reasons alone. The fat particles are by no means so small that they could pass through the intestinal mucosa unhindered or undigested. They are still too large for that.

Moreover, heating, whether pasteurization or ultra-high temperature treatment, inactivates xanthine oxidase in homogenized milk. Denaturation, and thus inactivation of xanthine oxidase, occurs rapidly at 69 °C [109]. It is therefore neither active in pasteurized nor in ESL milk, and is present in denatured form, essentially as a protein. Since homogenized milk is always pasteurized, such claims are simply untrue. Furthermore, the enzyme xanthine oxidase is also present in the human organism. It does not differ in shape or function at the molecular level from the

4.10.8 a1- and a2-β-Casein

So far, the casein has not been examined in more detail, apart from the different hydrophilic and hydrophobic fractions. This was not absolutely necessary, as all phenomena could be understood without looking more closely at the details of the casein micelles. This will now be made up for, as behind it lies a controversy about β-casein, which sometimes shows a breed-specific change known as a1 and a2-β-casein. This provides plenty of discussion material, with the molecular causes and facts often not being known, but again being well understandable with the knowledge and methods discussed so far.

The β-casein of the first domesticated dairy cows in the Neolithic period consisted of pure a2-β-casein. It is known today [110] that there are a whole series of spontaneous, natural variants in cows at 14 positions of the protein chains for the β-casein protein, one of which occurs at amino acid number 67. The hydrophobic proline (in a2) is replaced by the hydrophilic and essential histidine in a1 β-casein. Through migration, offspring, and breeding, cow breeds have emerged since the Neolithic period in Europe and America that have both the a1 and the a2 variant, while, for example, only a2-β-casein is formed in Indian breeds, as shown in Fig. 4.52.

The difference in the variants of β-casein does not change the structure and size of the casein micelles, it only manifests itself in a point defect of the protein primary structure. However, this detail leads to a modification in the casomorphins

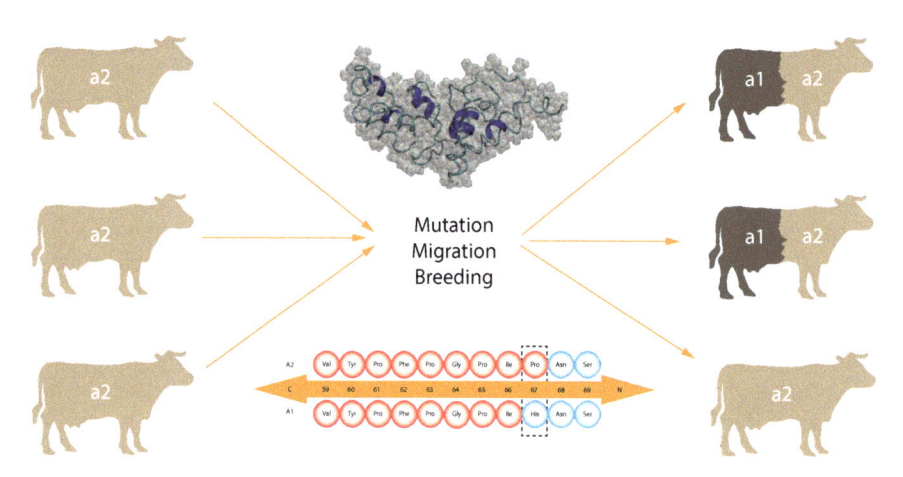

Fig. 4.52 Over time, through frequent mutation, migration, and breeding, cows have emerged whose β-casein (above) shows both the a1 and the a2 variant. The difference can be seen in the primary sequence (below)

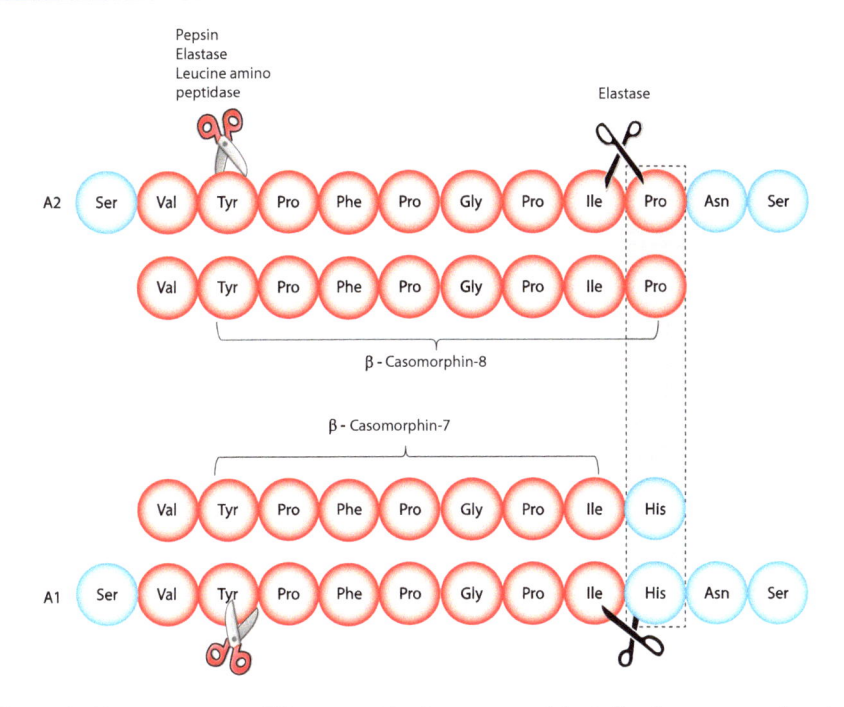

4.10 Basic Food Milk: Physics, Chemistry, Nutrition

Fig. 4.53 Enzymes cut two different peptides from a part of the β-Casein sequence. a1 and a2 form different precursors for β-Casomorphines, which are only formed after further enzymatically supported reactions. The fundamental difference: The hydrophilic Histamine is chemically responsible for this

during digestion, and many facts are based on this, but also fears, views, and opinions; even entire business models.

During the digestion of β-Casein, the interfaces of pepsin and the pancreatic enzymes Elastase and Leucine-Aminopeptidase play a role [111], which cut a bioactive peptide due to the altered amino acid from the a1 variant, precursor of β-Casomorphin-7 (Fig. 4.53).

Thus, the enzyme Elastase is not active at the indicated site in the a2 variant (above), the Proline is not cut at this stage, unlike the Histidine. Therefore, the a2 variation, which is also present in human Casein, is similar in enzymatic digestion processes. The release of Casomorphins can be enzymatically tracked [112] and systematically studied in *in-vitro* models [113]. In particular, β-Casomorphin-7 is more frequently detected in patients with neurological diseases [114]. This led to the conclusion that milk is associated with autism and sudden infant death. Due to the different peptide fragments during digestion, stronger reactions in lactose intolerance are also speculated [82]. There is no hard scientific data on this. There is therefore no reason to avoid milk and dairy products from a1 cows. In absurd-seeming animal experiments, mice were literally overfed with pure caseins, but hardly any significant results can be read from this [115]. Also, no β-Casomorphin-7 can be detected in the serum of humans after consuming milk. This can mean two things: Either the concentration is below the detection limit, or the molecule never arrives there because it was previously degraded or chemically altered in another way.

In these discussions, it must not be forgotten that a whole series of bioactive casomorphins are formed from both α- and β-casein. It is also forgotten that the a1-/a2-variants are only a tiny aspect. For example, β-casomorphin-7 is also formed from the so-called variants B, C, F, and G, in which an exchange of the amino acids proline and histidine also occurs at position 67 [116]. On the other hand, it is now also known that biogenic peptides actually perform functional tasks that have emerged during the course of evolution and the adaptation of human metabolism to available food. Even β-casomorphin-7 has positive effects. However, how relevant these are in interaction with all peptides that are formed during digestion remains to be seen. It is perfectly legitimate to focus on a specific molecule and research its function in all details, but it remains absurd to derive nutritional rules from it. This remark even applies to milk, as only the milk of a few breeds (Holstein) provides exclusively a1- (and B, C, F, G) casein. Many other breeds (Jersey, Brown Swiss, Guernsey) predominantly provide a2-casein, with a fluctuating a1 content. The high number of exclusively positively occupied bioactive β-casomorphins-5, -6, and -8 to -11, which are based on the alternating proline sequence from Fig. 4.53, show how unprofessional it is to attribute unproven negative properties of milk solely to number 7. This is not entirely true anyway, as these casomorphins show an agonistic opioid effect and, as mentioned in Fig. 4.53, are resistant to the cleavage sites of many digestive enzymes. Casomorphins therefore even prolong intestinal passage and can, for example, combat diarrhea [117]. Observations that they change stool frequency or stool consistency are therefore not surprising but are due to the physicochemical properties of these peptides.

Due to the special hydrophobic amino acid primary structure and the stepwise alternating proline sequence, these biogenic peptides also frequently occur in fermented and matured milk products, as summarized in Table 4.7.

Table 4.7 shows the classification of the most important casomorphins and their hierarchy of enzymatic protein degradation. The values in cheese and yogurt vary considerably. Occasionally, no casomorphins are found in Cheddar, blue cheese, Brie, and Limburger. The reasons lie in the enzyme status of the milk (e.g., raw versus pasteurized) as well as in the pH values and salt content of the cheese, which control enzyme activities [113]. The starter cultures used and their enzymes also play a role, for example, the peptidases of the lactic acid bacteria of *Lactococcus lactis* ssp. *cremoris* completely degrade casomorphins [118]. Other lactic acid bacteria have proline-specific peptidases that also cleave proline-rich sequences [119]. Therefore, generalized statements should be strictly avoided.

This table also reveals two nutritional relationships: Apparently, the even-numbered β-casomorphins, which have a proline at the end, only occur in relatively fresh products such as yogurt, buttermilk, or curdled milk when appropriate bacterial cultures are used. In most long-aged cheeses, these immune

4.10 Basic Food Milk: Physics, Chemistry, Nutrition

Table 4.7 Some of the detected β-casomorphins in milk products

BCM n	Sequential primary structure	Occurrence (only analytically proven examples)
3	Tyr–Pro–Phe	Edam, Yogurt
4	Tyr–Pro–Phe–Pro	Buttermilk, Sour Milk, Yogurt
5	Tyr–Pro–Phe–Pro–Gly	Caprino, Cheddar, Fontina, Grana Padano, Gorgonzola, Edam, Taleggio,
6	Tyr–Pro–Phe–Pro–Gly–Pro	–
7	Tyr–Pro–Phe–Pro–Gly–Pro–Ile	Brie, Cheddar, Gouda, Gorgonzola, Grana Padano, Fontina, Roquefort, Taleggio
8	Tyr–Pro–Phe–Pro–Gly–Pro–Ile–Pro	–
9	Tyr–Pro–Phe–Pro–Gly–Pro–Ile–Pro–Asn	Yogurt, Cheddar, Gouda
10	Tyr–Pro–Phe–Pro–Gly–Pro–Ile–Pro–Asn–Ser	Gouda
11	Tyr–Pro–Phe–Pro–Gly–Pro–Ile–Pro–Asn–Ser–Leu	Milk, Young Cheese

defense-strengthening casomorphins are apparently enzymatically degraded by specific enzymes. β-Casomorphin-10, discovered in medium-aged Gouda, plays a special role. It is taste-relevant, considered a cheese-specific bitter peptide [120] and is therefore indispensable for the *flavor* of some cheeses, like many α- and β-casein peptides, which are dominated by hydrophobic (and essential) amino acids. As a reminder: most essential amino acids taste bitter.

New studies using *ex-vivo* methods, combined with highly refined analysis methods, even show the occurrence of β-casomorphin-7 from both a1 and a2-caseins [121]. In contrast to test tube methods *(in vitro),* living animals have tissue and cells removed and added to the Casein. This also works with human cells, mucous membranes, and tissues. This allows the digestive processes to be *live* tracked and analyzed. If these already very conscientiously conducted experiments can be confirmed by other research groups, many concerns regarding BCM-7 and the seemingly significant differences between a1 and a2-caseins would be invalidated. Therefore, milk can be used as what it has represented since the Neolithic Revolution: a complex food from which enzymes, whether those of human digestion or those from the biotechnology of yogurt and cheese production, can make many positive things: They produce a whole range of bioactive substances from the proteins [122], which are good for people as a whole, as shown in Fig. 4.54.

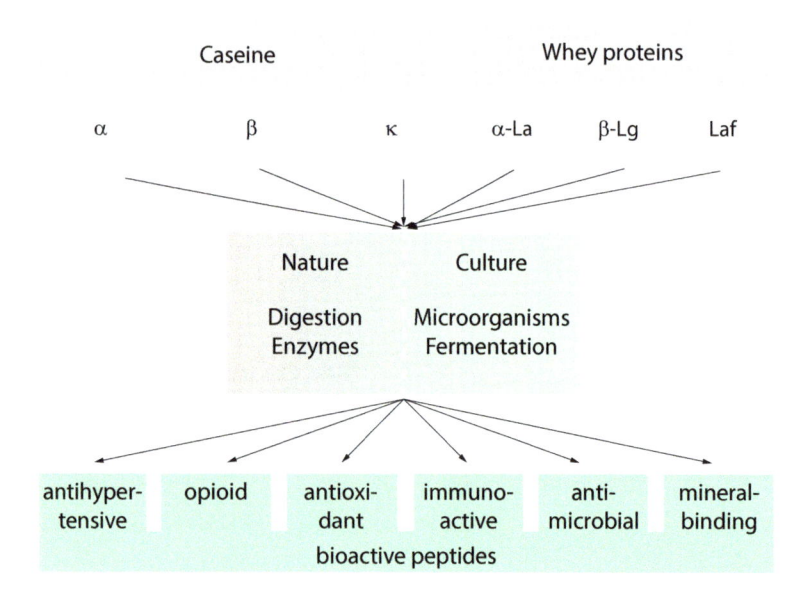

Fig. 4.54 What is produced from the proteins of milk, whether through the nature of the digestive tract or the enzymes of cultural techniques from biotechnology (yogurt, cheese), always results in bioactive peptides. The potential of all proteins, a1- and a2-caseins, as well as α-lactalbumin, β-lactoglobulin, or lactoferrin (Laf), is maximally utilized in humans or microcultures depending on the possibility

4.10.9 Micro-RNA in Milk

Milk contains even more, namely so-called growth factors, which enable the calf to have a good start in its new life. They are suspected of stimulating the cells of older children and adults to constant stress and are considered the main suspects in inflammatory processes up to rapid cell growth and tumor tendency [123, 124]. These growth factors are controlled, among other things, by so-called micro-RNA(miRNA; small ribonucleic acids). Such reports are quickly disseminated [125], without considering the backgrounds of these hypotheses or even delving deeper into the state of molecular research. Upon closer inspection, however, it becomes apparent how complex and, above all, unclear these questions are, but they can still be clarified when taking a closer look at the molecular functions.

So what do micro-RNAs do? In simplified terms, mi-RNAs can control or inhibit the reading and copying of information from DNA at specific sites, thus influencing the required protein production. More precisely: Micro-RNAs (miR-NAs) consist of 19 to 24 nucleotides, which are also known from DNA (see Chap. 2). They can block gene activity by attaching to specific regions of messenger or messenger-RNAs (mRNA). Thus, miRNAs control fundamental biological processes such as development, cell differentiation, cell growth and proliferation, or cell death. miRNAs can also accelerate the degradation of proteins, block protein synthesis, or promote the degradation of micro- and messenger-RNAs (mRNA). They are therefore useful regulators in the smallest units of the cell's biological life. Physicochemically, they are nothing more than small functional switches [126]. This is a good thing, as offspring must develop quickly, the immune defense must be strengthened, spontaneous cell changes, which also occur constantly in newborns (regardless of species), are quickly repaired [127]. Access to the brain must also be ensured, as some miRNAs apparently regulate neurological functions of fear coping and metabolic processes [126].

These miRNAs are found in the milk of all mammals, including humans. The diversity of these miRNAs depends on their biological functions and also on the stage of lactation [128], as is also the case with nutrient composition (see Fig. 4.5). The next question is where exactly these miRNAs are located in the milk and what happens to them during digestion. If they were digestible, they could not be absorbed and fulfill their functions. Therefore, miRNAs are packaged into nanometer-sized hollow bodies, so-called exosomes, vesicle-like structures between 30 and 90 nm in size. These exosomes are produced in the cells of the mammary glands, then transported through invagination mechanisms from the cell membranes, so that they are present in the milk in biological packaging. The packaging material consists solely of the membrane's own phospholipids, cholesterol, and membrane proteins to ensure a smooth invagination process and resembles a liposome, a closed nanosphere made of a phospholipid bilayer, as shown in Fig. 4.55.

It is by no means the case that miRNAs are exclusively ingested through food, as they are—as already mentioned—the basis of every organism, whether animal (and thus human) or plant. miRNAs are produced in cells, and thus also in

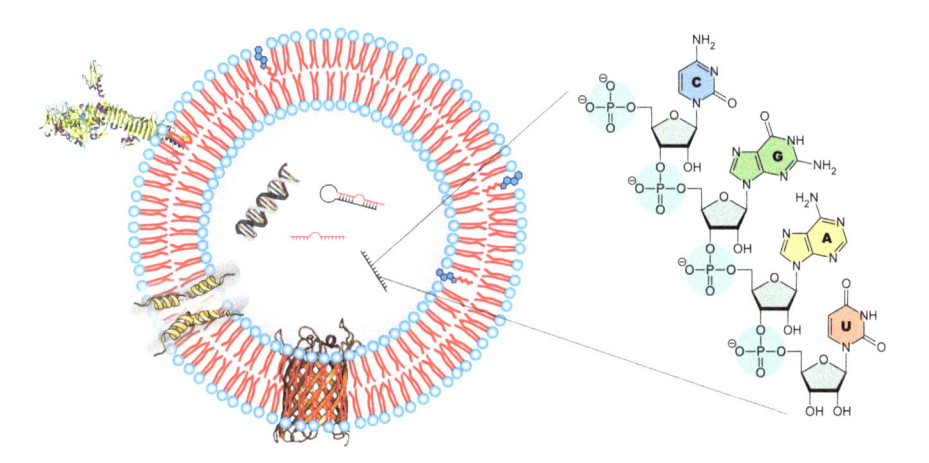

Fig. 4.55 Naive representation of an exosome: The shell consists of a double membrane layer, in which cholesterol and proteins are incorporated. Inside the aqueous interior are miRNAs in various states (precursors, argonauts, dicer, mature miRNA). Exosomes are prime examples of "soft nanomatter." The water-soluble miRNAs (symbolically represented on the right) consist of DNA nucleotides, with thymine not occurring, but instead uracil. The main chain of the miRNA consists of phosphoric acid and ribose, so all RNAs are highly water-soluble

every adult human. There are several hundred endogenous miRNAs in humans, which can be detected in the human body and there take on the tasks assigned to them exactly by the arrangement of the nucleotides of the DNA—U, G, A, C. These miRNAs do not diffuse freely in the watery body fluids (that would indeed be fatal), but are packaged by the cell through the already mentioned invagination processes into endogenous exosomes. For only in this way can they, enveloped in the membrane (Fig. 4.55), enter cells via invagination processes to empty themselves there and release the trapped content. They find their way to the relevant cells via very specific receptor proteins that exosomes carry in their envelope. These universal, cell-physical processes are also exploited in the administration of mRNA vaccines (Corona).

4.10.10 Cow's Milk Exosomes as Information and Drug Transporters

For some time now, miRNAs have been in the sights of research groups, for example, in the question of whether these nanoparticles from cow's milk, due to their nanosize and the cavity in the middle, could even serve as carriers for pharmacologically relevant active substances in order to transport them more specifically to the site of their action [129]. Milk critics, on the other hand, focus on the question of whether the exosomes spread gene-altering miRNAs from cow's milk in the bodies of milk drinkers and then act as switches or triggers for certain diseases, even tumors. Other research groups even find positive effects on tumors [130]. These questions stand or fall with the behavior of exosomes (and the active substances enclosed therein, be they miRNAs or active substances encapsulated in the laboratory) during the gastrointestinal passage.

4.10.10.1 Cow Milk Exosomes in the Digestive Tract

The list of miRNAs that are viewed critically is long [131], especially since the sequence of one of the cow milk miRNAs matches the human miRNA. Can it be concluded that drinking milk reprograms human genes? The question is open in detail, but there are good physicochemical reasons, which become understandable with the knowledge provided here, that this is unlikely.

In the cavity of the exosomes, there are also the miRNAs, of which there are a large number (several hundred) for the respective assigned task. The question of what happens to the miRNAs during digestion is therefore a question of the stability of the exosomes. Most studies concern *in-vitro* digestion. For this purpose, exosomes are extracted from the milk (pH-induced precipitation of caseins and subsequent centrifugation) and then subjected to simulated digestion with digestive enzymes (pepsin) from the stomach and pancreatic proteases. It turns out that the exosomes survive digestion [132]. This is not surprising, as the proteases only digest the (relatively few) proteins anchored in the exosomes. However, the lipid bilayer is densely covered with phospholipids. The little space that becomes free through protein digestion can be immediately filled with phospholipids, as is

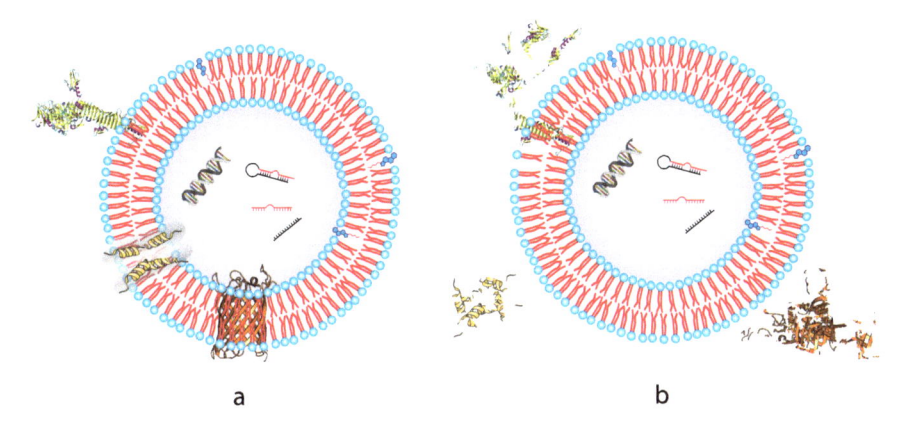

4.10 Basic Food Milk: Physics, Chemistry, Nutrition

Fig. 4.56 Schematic representation of exosomes during protein digestion. (**a**) An intact exosome with embedded membrane proteins. (**b**) After *in-vitro* addition of pepsin and pancreatic proteases, proteins are digested, and the phospholipids reorganize where gaps have formed

known from membrane physics, by moving closer together [133]. This is called the Marangoni effect. The phospholipids redistribute and fill the small gaps. In principle, these are no longer so closely packed, but they still keep the exosome stable, as schematically indicated in Fig. 4.56.

As also shown in *in-vitro* experiments [132], these intact exosomes can dock onto intestinal cells and be taken up by them. From this, it was concluded that exosomes would distribute the information contained in them, cow miRNAs, in the bodies of milk drinkers; human cells could thus be reprogrammed. However, this hypothesis contradicts observations, as a direct connection between the postulated diseases cannot be proven.

4.10.10.2 The Zurich Experiment—The Key Lies in Bile Acids

This contradiction can be illuminated with the *in-vivo* and *ex-vivo* experiments conducted in Zurich. There, the exosomes were not digested *in-vitro*, but *in-vivo* and *ex-vivo* [134], and thus analyzed more precisely [135]. It was shown that the components of the exosomes, including miRNA, are largely digested and the exosomes do not survive the gastrointestinal passage. Much of these observations can be attributed to the action of bile acids, as these amazing emulsion breakers can also break up lipid bilayers, even at the high curvature of the nanometer-sized exosomes. The mechanisms have already been defined in more detail in this chapter and are shown in Fig. 4.57.

The attack of bile acids initially occurs on the outer phospholipid layer when their hydrophilic side attaches to the exosomes (1). The changes in polar and electrostatic interactions allow individual phospholipids to be removed (2). However, the bile acids prevent the "sliding" and filling of the gaps by inserting their hydrophobic side between the fatty acid tails of the phospholipids (3, 4). This also allows the lower phospholipid layer to be attacked and phospholipids to be

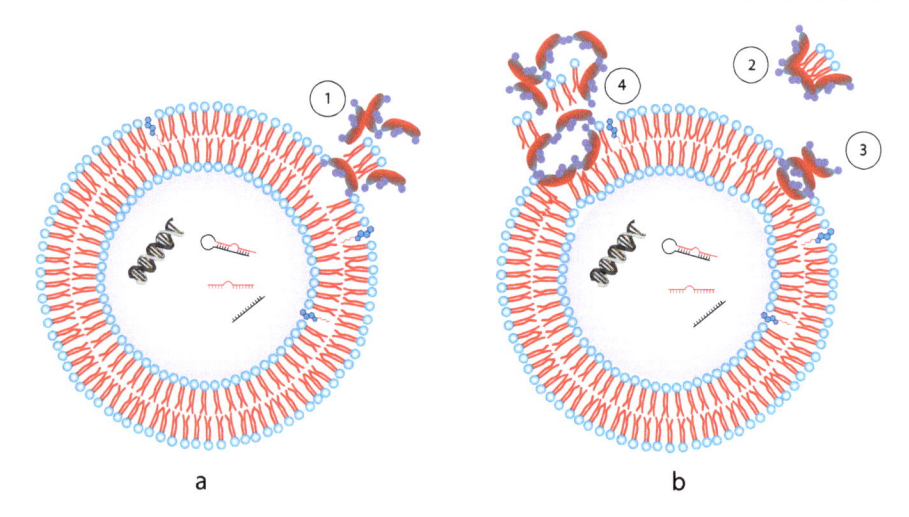

Fig. 4.57 Exosomes can become successively unstable during *in-vivo* digestion of the lipid bilayer (explanation see text)

removed in the next steps (4). The exosome becomes unstable and releases its content free. DNA, RNA and miRNAs can also be digested by the nucleotidases and RNases released from the pancreas.

These structural changes have not been pointed out anywhere in the scientific literature so far. This once again shows how much understanding becomes possible when the physicochemical view and the basic principles of the physics of soft matter are taken into account.

It must be expressly pointed out that these models apply to adults with a fully developed (healthy) intestine. In newborns, the microbiome is not yet fully developed and the paracellular uptake of exosomes and their contents is expressly desired [136]. Only with weaning is the offspring's intestine developed enough [137] to digest exosomes. From this point of view, milk consumption in adulthood does not seem to pose a problem either.

4.10.10.3 Endogenous and Exogenous miRNAs

On the other hand, using modern fluorescence methods in the laboratory on experimental animals or precise analyses of the blood of test subjects (in the study by Baier [138] there were only eight), miRNAs after milk consumption were detected. These are most likely endogenous [139, 140], because contrary to what was assumed, the encapsulation of miRNAs in exosomes only provides limited protection. On the one hand, bile acids cause significant instabilities in the double layer, as indicated in Fig. 4.57, on the other hand, the uptake and transport across and into the intestinal cells have so far only been demonstrated *in-vitro* with protease-digested exosomes [141] and thus contradict the *in-vivo* observed results. A new approach also promises further insights [142]. In this case, proteins and miRNAs can be stained with fluorescent dyes, packaged in exosomes, and thus tracked using fluorescence microscopy. Molecular research remains highly exciting, even if no dietary rules can be derived from it.

4.10.11 Milk Makes the Difference

Milk is initially nothing more than a well-structured, highly complex liquid. The structure formation results from the self-organization and the biomolecular processes of the glandular apparatus of the animals. One thing must not be forgotten: milk is a natural product. Therefore, the composition of fatty acids, protein content, and the forming aromas vary depending on the breed, season, feeding, and the stage of lactation of the animal. These subtle differences can be clearly tasted when enjoying raw milk. This is true whether it is from the same source at different times of the year or when comparing raw milk from different producers. In standardized, pasteurized, and homogenized products from large dairies, these subtle differences are hardly noticeable.

Furthermore, the different pretreatments have an impact on a whole range of culinary applications. For example, homemade yogurts made from ultra-high-temperature and pasteurized cow's milk have a significantly better water binding than yogurts made from untreated raw milk. Homemade and self-ripened cheeses made from raw milk or pasteurized milk have different aroma characteristics due to varying enzyme activity. Foam stability also depends on the fat content and pretreatment. Only with an understanding of the product's history can targeted enjoyment practices be carried out.

Also, as unfortunately often happens, the general demonization of cultural and industrial processes is unfounded. Such processes always serve food safety, such as pasteurization, ultra-high-temperature treatment, or even targeted shelf-life extension. The accompanying physical changes have no negative health effects on heart disease, allergies, or diabetes when objectively assessed, especially when considering molecular-physiological interactions (see, for example, the review by Michalski [143]). Even if the vitamins are partially oxidized during consistently short-term heating, milk remains a valuable food. A one-sided focus on vitamin C, the only truly heat-sensitive vitamin, does not reflect the nutritional value of milk. Milk is just a small part of a diverse diet. No reasonable person would drink milk solely for its vitamin C content. Rather, it is consumed for its fatty acid spectrum, the vitamin D precursors present, the high-quality protein content, or simply for enjoyment because it tastes good and feels good. If it doesn't taste good and doesn't feel good, one simply doesn't drink it. Or there may be genuine milk protein allergies. In case of doubt, it is always better to trust the microstructure and molecular composition, coupled with one's own body feeling, than to be unsettled by the countless pseudoscientific statements.

Another remark is necessary at this point: The occurrence of countless miR-NAs is not a singular problem of milk, as it is sometimes portrayed. They occur in all their diversity in all foods: in eggs, in meat, and of course in fruit and vegetables, in nuts, legumes, and seeds, where all information for plant growth is stored.

miRNAs are found in all cells of all biological systems. Natural, gene-free, and miRNA-free natural foods do not exist. In this respect, we consume countless mRNAs through our daily meals.

References

1. Bahr, B., Lemmer, B., & Piccolo, R. (2016). *Quirky quarks: A cartoon guide to the fascinating realm of physics.* Springer.
2. Gribbin, J., & Gribbin, J. (1995). *Am Anfang war...: Neues vom Urknall und der Evolution des Kosmos.* Birkhäuser.
3. Klein, S. (2014). *Die Tagebücher der Schöpfung: Vom Urknall zum geklonten Menschen.* Fischer.
4. Buchmüller, W. (2016). Das Higgs-Teilchen und der Ursprung der Materie. *Erkenntnis, Wissenschaft und Gesellschaft* (S. 209–223). Springer.
5. Hänsch, T. W. (2006). Sind die Naturkonstanten konstant? Zweiter Teil des Interviews mit Theodor W. Hänsch, Physiknobelpreisträger des Jahres 2005. *Physik in unserer Zeit, 37*(2), 62–63.
6. Granold, M., Hajieva, P., Toşa, M. I., Irimie, F. D., & Moosmann, B. (2017). Modern diversification of the amino acid repertoire driven by oxygen. *Proceedings of the National Academy of Sciences, 115*(1), 41–46.
7. Doig, A. J. (2017). Frozen, but no accident—Why the 20 standard amino acids were selected. *The FEBS journal, 284*(9), 1296–1305.
8. Mouritsen, G. (2014). *Life—As a matter of fat: The emerging science of lipidomics.* Springer.
9. Galtier, N., Tourasse, N., & Gouy, M. (1999). A nonhyperthermophilic common ancestor to extant life forms. *Science, 283*(5399), 220–221.
10. Reisinger, B., Sperl, J., Holinski, A., Schmid, V., Rajendran, C., Carstensen, L., & Sterner, R. (2013). Evidence for the existence of elaborate enzyme complexes in the Paleoarchean era. *Journal of the American Chemical Society, 136*(1), 122–129.
11. Wunn, I. (2005). *Die Religionen in vorgeschichtlicher Zeit* (2nd Edn.). Kohlhammer.
12. Drennan, R. D. (1976). Religion and social evolution in Formative Mesoamerica. In K. V. Flannery (ed.) *The early mesoamerican village.*
13. Pfälzner, P. (2001). *Auf den Spuren der Ahnen. Überlegungen zur Nachweisbarkeit der Ahnenverehrung in Vorderasien vom Neolithikum bis in die Bronzezeit.* Universitätsbibliothek Heidelberg.
14. Reimann, J. (2014). *Kreationismus vs. Evolution.* GRIN Verlag.
15. https://www.zeit.de/2016/06/ernaehrung-essen-palaeo-vegan; https://www1.wdr.de/fernsehen/quarks/gesunde-ernaehrung-essen-als-religion-100.html.
16. Schubert, A. (2018). *Gott essen: eine kulinarische Geschichte des Abendmahls.* Beck
17. Harris, M. (2001). *Cultural materialism: The struggle for a science of culture.* AltaMira Press.
18. Blasi, D. E., Moran, S., Moisik, S. R., Widmer, P., Dediu, D., & Bickel, B. (2019). Human sound systems are shaped by post-Neolithic changes in bite configuration. *Science, 363*(6432), eaav3218.
19. Søe, M. J., Nejsum, P., Seersholm, F. V., Fredensborg, B. L., Habraken, R., Haase, K., Hald, M. M., Simonsen, R., Højlund, F., Blanke, L., Merkyte, I., Willerslev, E., Moliin, C., & Kapel, O. (2018). Ancient DNA from latrines in Northern Europe and the Middle East (500 BC–1700 AD) reveals past parasites and diet. *PLoS ONE, 13*(4), e0195481.
20. Vié, B. (2011). *Testicles: Balls in cooking and culture.* Prospect Books.
21. Schönberger, M., & Zipprick, J. (2011). *100 Dinge, die Sie einmal im Leben gegessen haben sollten.* Ludwig.

22. https://www.zeit.de/wirtschaft/2018-03/tierschutz-fleischproduktion-tierschutzgesetz-verbraucher-fleischkonsum-5vor8.
23. Ströhle, A., Wolters, M., & Hahn, A. (2015). Rotes Fleisch—vom gehaltvollen Nährstofflieferanten zum kanzerogenen Agens. *Ernährung im Fokus, 15,*09.
24. Hausen, H. (2012). Red meat consumption and cancer: reasons to suspect involvement of bovine infectious factors in colorectal cancer. *International Journal of Cancer, 130*(11), 2475–2483
25. Schoenfeld, J. D., & Ioannidis, J. P. (2012). Is everything we eat associated with cancer? A systematic cookbook review. *The American journal of clinical nutrition, 97*(1), 127–134.
26. Seiwert, N., Wecklein, S., Demuth, P., Hasselwander, S., Kemper, T. A., Schwerdtle, T., Brunner, T., & Fahrer, J. (2020). Heme oxygenase 1 protects human colonocytes against ROS formation, oxidative DNA damage and cytotoxicity induced by heme iron, but not inorganic iron. *Cell death & disease, 11*(9), 1–16.
27. Corpet, D. E. (2011). Red meat and colon cancer: Should we become vegetarians, or can we make meat safer? *Meat Science, 89*(3), 310–316.
28. Bergeron, N., Chiu, S., Williams, P. T., King, M. S., & Krauss, R. M. (2019). Effects of red meat, white meat, and nonmeat protein sources on atherogenic lipoprotein measures in the context of low compared with high saturated fat intake: A randomized controlled trial. *The American journal of clinical nutrition, 110*(3), 24–33.
29. Kwok, R. H. (1968). Chinese-restaurant syndrome. *The New England Journal of Medicine, 278*(14), 796.
30. Rassin, D. K., Sturman, J. A., & Gaull, G. E. (1978). Taurine and other free amino acids in milk of man and other mammals. *Early Human Development, 2*(1), 1–13.
31. Baldeón, M. E., Mennella, J. A., Flores, N., Fornasini, M., & San Gabriel, A. (2014). Free amino acid content in breast milk of adolescent and adult mothers in Ecuador. *SpringerPlus, 3*(1), 104.
32. Li, X., Staszewski, L., Xu, H., Durick, K., Zoller, M., & Adler, E. (2002). Human receptors for sweet and umami taste. *Proceedings of the National Academy of Sciences, 99*(7), 4692–4696.
33. Behrens, M., Meyerhof, W., Hellfritsch, C., & Hofmann, T. (2011). Moleküle und biologische Mechanismen des Süß- und Umamigeschmacks. *Angewandte Chemie, 123*(10), 2268–2291.
34. Ninomiya, K. (2015). Science of umami taste: Adaptation to gastronomic culture. *Flavour, 4*(1), 13.
35. Kurihara, K. (2009). Glutamate: from discovery as a food flavor to role as a basic taste (umami). *The American Journal of Clinical Nutrition, 90*(3), 719S–722S
36. Dunkel, A., & Hofmann, T. (2009). Sensory-directed identification of β-alanyl dipeptides as contributors to the thick-sour and white-meaty orosensation induced by chicken broth. *Journal of Agricultural and Food Chemistry, 57*(21), 9867–9877.
37. Löffler, M. (2014). Pathobiochemie des Purin- und Pyrimidinstoffwechsels. *Löffler/Petrides Biochemie und Pathobiochemie* (S. 372–378). Springer.
38. Wolfram, G., & Colling, M. (1987). Gesamtpuringehalt in ausgewählten Lebensmitteln. *Zeitschrift für Ernährungswissenschaft, 26*(4), 205–213.
39. Wifall, T. C., Faes, T. M., Taylor-Burds, C. C., Mitzelfelt, J. D., & Delay, E. R. (2006). An analysis of 5′-inosine and 5′-guanosine monophosphate taste in rats. *Chemical Senses, 32*(2), 161–172.
40. Yamaguchi, S. (1967). The synergistic taste effect of monosodium glutamate and disodium 5′-inosinate. *Journal of Food Science, 32*(4), 473–478.
41. Mouritsen, O. G., Duelund, L., Bagatolli, L. A., & Khandelia, H. (2013). The name of deliciousness and the gastrophysics behind it. *Flavour, 2*(1), 9.
42. Mouritsen, O. G., & Khandelia, H. (2012). Molecular mechanism of the allosteric enhancement of the umami taste sensation. *The FEBS Journal, 279*(17), 3112–3120.

43. Grimm, H. U. (2011). *Die Ernährungslüge: Wie uns die Lebensmittelindustrie um den Verstand bringt*. EBook.
44. https://praxistipps.focus.de/ist-hefeextrakt-schaedlich-verstaendlich-erklaert_55558.
45. Nout, M. J. R. (1994). Fermented foods and food safety. *Food Research International, 27*(3), 291–298.
46. Linares, D. M., del Rio, B., Redruello, B., Ladero, V., Martin, M. C., Fernandez, M., & Alvarez, M. A. (2016). Comparative analysis of the in vitro cytotoxicity of the dietary biogenic amines tyramine and histamine. *Food Chemistry, 197*, 658–663.
47. Shalaby, A. R. (1996). Significance of biogenic amines to food safety and human health. *Food Research International, 29*(7), 675–690.
48. Brown, R. E., & Haas, H. L. (1999). On the mechanism of histaminergic inhibition of glutamate release in the rat dentate gyrus. *The Journal of Physiology, 515*(3), 777–786.
49. Marco, M. L., Heeney, D., Binda, S., Cifelli, C. J., Cotter, P. D., Foligné, B., Gänzle, M., Kort, R., Pasin, G., Pihlanto, A., Smid, E. J., & Hutkins, R. (2017). Health benefits of fermented foods: Microbiota and beyond. *Current Opinion in Biotechnology, 44*, 94–102.
50. Magerowski, G., Giacona, G., Patriarca, L., Papadopoulos, K., Garza-Naveda, P., Radziejowska, J., & Alonso-Alonso, M. (2018). Neurocognitive effects of umami: Association with eating behavior and food choice. *Neuropsychopharmacology, 43*(10), 2009–2016.
51. Lustig, R. H., Schmidt, L. A., & Brindis, C. D. (2012). Public health: The toxic truth about sugar. *Nature, 482*(7383), 27.
52. https://www.ugb.de/ernaehrungsberatung/zuckersucht/?zucker-sucht.
53. https://www.zentrum-der-gesundheit.de/zuckersucht-ausstieg-ia.html#toc-zucker-ist-eine-droge.
54. Shallenberger, R. S. (1963). Hydrogen bonding and the varying sweetness of the sugars. *Journal of Food Science, 28*(5), 584–589.
55. DuBois, G. E. (2016). Molecular mechanism of sweetness sensation. *Physiology & Behavior, 164*, 453–463.
56. Masuda, K., Koizumi, A., Nakajima, K. I., Tanaka, T., Abe, K., Misaka, T., & Ishiguro, M. (2012). Characterization of the modes of binding between human sweet taste receptor and low-molecular-weight sweet compounds. *PLoS ONE, 7*(4), e35380.
57. Jaitak, V. (2015). Interaction model of steviol glycosides from *Stevia rebaudiana* (Bertoni) with sweet taste receptors: A computational approach. *Phytochemistry, 116*, 12–20.
58. Leusmann, E. (2017). Stoff für Süßmäuler. *Nachrichten aus der Chemie, 65*(9), 887–893.
59. Yamazaki, M., & Sakaguchi, T. (1986). Effects of d-glucose anomers on sweetness taste and insulin release in man. *Brain Research Bulletin, 17*(2), 271–274.
60. Lim, J. S., Mietus-Snyder, M., Valente, A., Schwarz, J. M., & Lustig, R. H. (2010). The role of fructose in the pathogenesis of NAFLD and the metabolic syndrome. *Nature reviews gastroenterology and hepatology, 7*(5), 251.
61. Lustig, R. H. (2010). Fructose: Metabolic, hedonic, and societal parallels with ethanol. *Journal of the American Dietetic Association, 110*(9), 1307–1321.
62. Luo, S., Monterosso, J. R., Sarpelleh, K., & Page, K. A. (2015). Differential effects of fructose versus glucose on brain and appetitive responses to food cues and decisions for food rewards. *Proceedings of the National Academy of Sciences*, 201503358.
63. Rodeck, B., & Zimmer, K. P. (2008). *Pädiatrische Gastroenterologie, Hepatologie und Ernährung*. Springer Medizin.
64. McInnes, M. (2014). *The honey diet*. Hodder & Stoughton.
65. Woodroof, J. G. (1986). History and growth of fruit processing. In *Commercial fruit processing* (S. 1–24). Springer.
66. Kappes, S. M., Schmidt, S. J., & Lee, S. Y. (2006). Mouthfeel detection threshold and instrumental viscosity of sucrose and high fructose corn syrup solutions. *Journal of Food Science, 71*(9), S597–S602.

67. Goyal, S. K., Samsher, G. R., & Goyal, R. K. (2010). Stevia *(Stevia rebaudiana)* a bio-sweetener: A review. *International Journal of Food Sciences and Nutrition, 61*(1), 1–10.
68. Hellfritsch, C., Brockhoff, A., Stähler, F., Meyerhof, W., & Hofmann, T. (2012). Human psychometric and taste receptor responses to steviol glycosides. *Journal of Agricultural and Food Chemistry, 60*(27), 6782–6793.
69. Golding, M., & Wooster, T. J. (2010). The influence of emulsion structure and stability on lipid digestion. *Current Opinion in Colloid & Interface Science, 15*(1–2), 90–101.
70. Mackie, A., & Macierzanka, A. (2010). Colloidal aspects of protein digestion. *Current Opinion in Colloid & Interface Science, 15*(1–2), 102–108.
71. Löffler, M. (2014). *Pathobiochemie des Purin- und Pyrimidinstoffwechsels* (S. 372–378). Springer.
72. Latscha, H. P., Kazmaier, U., & Klein, H. A. (2016). *Steroide* (S. 481–487). Springer Spektrum.
73. Armstrong, M. J., & Carey, M. C. (1982). The hydrophobic-hydrophilic balance of bile salts. Inverse correlation between reverse-phase high performance liquid chromatographic mobilities and micellar cholesterol-solubilizing capacities. *Journal of lipid research, 23*(1), 70–80.
74. Garidel, P., Hildebrand, A., Knauf, K., & Blume, A. (2007). Membranolytic activity of bile salts: Influence of biological membrane properties and composition. *Molecules, 12*(10), 2292–2326.
75. Carey, M. C., Small, D. M., & Bliss, C. M. (1983). Lipid digestion and absorption. *Annual Review of Physiology, 45*(1), 651–677.
76. https://www.youtube.com/watch?v=Kkc-SQsaOTk, bzw. https://www.uniklinik-freiburg.de/index.php?id=17950. https://www.derstandard.de/story/2000086168642/expertin-entschuldigt-sich-fuer-kokosoel-ist-das-reine-gift.
77. Huang, A. H. (1996). Oleosins and oil bodies in seeds and other organs. *Plant Physiology, 110*(4), 1055.
78. Zielbauer, B. I., Jackson, A. J., Maurer, S., Waschatko, G., Ghebremedhin, M., Rogers, S. E., Heenan, R. K., Porcar, L., & Vilgis, T. A. (2018). Soybean Oleosomes studied by Small Angle Neutron Scattering (SANS). *Journal of Colloid and Interface Science.* https://doi.org/10.1016/j.jcis.2018.05.080.
79. Bresinsky, A., Körner, C., Kadereit, J. W., Neuhaus, G., & Sonnewald, U. (2008). *Lehrbuch der Botanik* (35th Edn.). Spektrum.
80. Hsieh, K., & Huang, A. H. (2004). Endoplasmic reticulum, oleosins, and oils in seeds and tapetum cells. *Plant Physiology, 136*(3), 3427–3434.
81. Maurer, S., Waschatko, G., Schach, D., Zielbauer, B. I., Dahl, J., Weidner, T., Bonn, M., & Vilgis, T. A. (2013). The role of intact oleosin for stabilization and function of oleosomes. *The Journal of Physical Chemistry B, 117*(44), 13872–13883.
82. Woodford, K. (2009). *Devil in the milk: Illness, health and the politics of A1 and A2 milk.* Chelsea Green Publishing.
83. Grimm, H. U. (2016). *Die Fleischlüge: Wie uns die Tierindustrie krank macht.* Droemer.
84. Rätzer, H. (1998). Wirtschaftlichkeit verschiedener Rindertypen: Vergleich von Milch- und Zweinutzungsrassen. Doctoral dissertation, ETH Zurich.
85. Ternes, W. (2008). *Naturwissenschaftliche Grundlagen der Lebensmittelzubereitung.* Behr.
86. De Kruif, C. G., Huppertz, T., Urban, V. S., & Petukhov, A. V. (2012). Casein micelles and their internal structure. *Advances in Colloid and Interface Science, 171,* 36–52.
87. Walstra, P. (1999). *Dairy technology: Principles of milk properties and processes.* CRC Press.
88. Vilgis, T. (2010). *Das Molekül-Menü: Molekulare Grundlagen für kreative Köche.* Hirzel Verlag.
89. Gallier, S., Ye, A., & Singh, H. (2012). Structural changes of bovine milk fat globules during in vitro digestion. *Journal of Dairy Science, 95*(7), 3579–3592.
90. Overbeck, P. (2014). Freispruch für Kuhmilch. *MMW-Fortschritte der Medizin, 156*(13), 29

91. http://www.kern.bayern.de/wissenschaft/107510/index.php.
92. Grimm, H. U. (2016). *Die Fleischlüge—wie uns die Tierindustrie krank macht.* Droemer.
93. Reddy, I. M., Kella, N. K., & Kinsella, J. E. (1988). Structural and conformational basis of the resistance of beta-lactoglobulin to peptic and chymotryptic digestion. *Journal of Agricultural and Food Chemistry, 36*(4), 737–741.
94. Melnik, B. C. (2012). Leucine signaling in the pathogenesis of type 2 diabetes and obesity. *World Journal of Diabetes, 3*(3), 38–53.
95. Melnik, B. C. (2009). Milk—the promoter of chronic Western diseases. *Medical Hypotheses, 72*(6), 631–639.
96. Bruker, M. O., Jung, M., Gutjahr, I., & Gutjahr, I. (2011). *Der Murks mit der Milch: [Gesundheitsgefährdung durch Industriemilch; Genmanipulation und Turbokuh; vom Lebensmittel zum Industrieprodukt].* emu-Verlag.
97. http://www.realmilk.com/safety/safety-of-raw-milk/.
98. Grimm, H. U. (2016). *Die Fleischlüge—wie uns die Tierindustrie krank macht.* Droemer.
99. Marshall, K. (2004). Therapeutic applications of whey protein. *Alternative medicine review, 9*(2), 136–157.
100. Bounous, G., Batist, G., & Gold, P. (1989). Mice: Role of glutathione. *Clinical and Investigative Medicine, 12*(3), 154–161.
101. Töpel, A. (2007). *Chemie und Physik der Milch: Naturstoff, Rohstoff, Lebensmittel* (S. 386). BehrE.
102. Lopez, C. (2005). Focus on the supramolecular structure of milk fat in dairy products. *Reproduction, Nutrition, Development, 45*(4), 497–511.
103. Kielczewska, K., Kruk, A., Czerniewicz, M., & Haponiuk, E. (2006). Effects of high-pressure-homogenization on the physicochemical properties of milk with various fat concentrations. *Polish Journal of Food and Nutrition Science, 15,* 91–94.
104. Dahlke, R. (2011). *Peace Food. Wie der Verzicht auf Fleisch und Milch Körper und Seele heilt.* Gräfe und Unzer.
105. Ye, A., Cui, J., & Singh, H. (2010). Effect of the fat globule membrane on in vitro digestion of milk fat globules with pancreatic lipase. *International Dairy Journal, 20*(12), 822–829.
106. Berton, A., Rouvellac, S., Robert, B., Rousseau, F., Lopez, C., & Crenon, I. (2012). Effect of the size and interface composition of milk fat globules on their in vitro digestion by the human pancreatic lipase: Native versus homogenized milk fat globules. *Food Hydrocolloids, 29*(1), 123–134.
107. Miralles, B., del Barrio, R., Cueva, C., Recio, I., & Amigo, L. (2018). Dynamic gastric digestion of a commercial whey protein concentrate. *Journal of the Science of Food and Agriculture, 98*(5), 1873–1879.
108. Oster, K. A. (1973). Evaluation of serum cholesterol reduction and xanthine oxidase inhibition in the treatment of atherosclerosis. *Recent advances in studies on cardiac structure and metabolism, 3,* 73.
109. Sharma, P., Oey, I., & Everett, D. W. (2016). Thermal properties of milk fat, xanthine oxidase, caseins and whey proteins in pulsed electric field-treated bovine whole milk. *Food Chemistry, 207,* 34–42.
110. https://www.uniprot.org/uniprot/P02666.
111. Kamiński, S., Cieślińska, A., & Kostyra, E. (2007). Polymorphism of bovine beta-casein and its potential effect on human health. *Journal of applied genetics, 48*(3), 189–198.
112. Jinsmaa, Y., & Yoshikawa, M. (1999). Enzymatic release of neocasomorphin and β-casomorphin from bovine β-casein. *Peptides, 20*(8), 957–962.
113. De Noni, I. (2008). Release of β-casomorphins 5 and 7 during simulated gastro-intestinal digestion of bovine β-casein variants and milk-based infant formulas. *Food Chemistry, 110*(4), 897–903.
114. Kalaydjian, A. E., Eaton, W., Cascella, N., & Fasano, A. (2006). The gluten connection: The association between schizophrenia and celiac disease. *Acta Psychiatrica Scandinavica, 113*(2), 82–90.

115. Haq, M. R. U., Kapila, R., Sharma, R., Saliganti, V., & Kapila, S. (2014). Comparative evaluation of cow β-casein variants (A1/A2) consumption on Th 2-mediated inflammatory response in mouse gut. *European Journal of Nutrition, 53*(4), 1039–1049.
116. Nguyen, D. D., Johnson, S. K., Busetti, F., & Solah, V. A. (2015). Formation and degradation of beta-casomorphins in dairy processing. *Critical Reviews in Food Science and Nutrition, 55*(14), 1955–1967.
117. Meisel, H., & Schlimme, E. (1996). Bioactive peptides derived from milk proteins: Ingredients for functional foods? *Kieler Milchwirtschaftliche Forschungsberichte, 48*(4), 343–357.
118. Muehlenkamp, & Warthesen, J. J. (1996). β-Casomorphins: Analysis in cheese and susceptibility to proteolytic enzymes from *Lactococcus lactis* ssp. *cremoris*. *Journal of dairy science, 79*(1), 20–26.
119. Matar, C., & Goulet, J. (1996). β-casomorphin 4 from milk fermented by a mutant of *Lactobacillus helveticus*. *International Dairy Journal, 6*(4), 383–397.
120. Toelstede, S., & Hofmann, T. (2008). Sensomics mapping and identification of the key bitter metabolites in Gouda cheese. *Journal of Agricultural and Food Chemistry, 56*(8), 2795–2804.
121. Asledottir, T., Le, T. T., Petrat-Melin, B., Devold, T. G., Larsen, L. B., & Vegarud, G. E. (2017). Identification of bioactive peptides and quantification of β-casomorphin-7 from bovine β-casein A1, A2 and I after ex vivo gastrointestinal digestion. *International Dairy Journal, 71*, 98–106.
122. Petrat-Melin, B., Andersen, P., Rasmussen, J. T., Poulsen, N. A., Larsen, L. B., & Young, J. F. (2015). In vitro digestion of purified β-casein variants A1, A2, B, and I: Effects on antioxidant and angiotensin-converting enzyme inhibitory capacity. *Journal of Dairy Science, 98*(1), 15–26.
123. Melnik, B. C., John, S. M., Carrera-Bastos, P., & Schmitz, G. (2016). Milk: A postnatal imprinting system stabilizing FoxP3 expression and regulatory T cell differentiation. *Clinical and translational allergy, 6*(1), 18.
124. Melnik, B. C., John, S. M., & Schmitz, G. (2013). Milk is not just food but most likely a genetic transfection system activating mTORC1 signaling for postnatal growth. *Nutrition journal, 12*(1), 103.
125. https://www.noz.de/deutschland-welt/gut-zu-wissen/artikel/949443/hat-milch-eine-krebsfoerdernde-wirkung.
126. Meydan, C., Shenhar-Tsarfaty, S., & Soreq, H. (2016). MicroRNA regulators of anxiety and metabolic disorders. *Trends in Molecular Medicine, 22*(9), 798–812.
127. Samuel, M., Chisanga, D., Liem, M., Keerthikumar, S., Anand, S., Ang, C. S., Adda, C. G., Versteegen, E., Markandeya, J., & Mathivanan, S. (2017). Bovine milk-derived exosomes from colostrum are enriched with proteins implicated in immune response and growth. *Scientific reports, 7*(1), 5933.
128. Alsaweed, M., Hepworth, A. R., Lefevre, C., Hartmann, P. E., Geddes, D. T., & Hassiotou, F. (2015). Human milk microRNA and total RNA differ depending on milk fractionation. *Journal of Cellular Biochemistry, 116*(10), 2397–2407.
129. Betker, J. L., Angle, B. M., Graner, M. W., & Anchordoquy, T. J. (2018). The potential of exosomes from cow milk for oral delivery. *Journal of Pharmaceutical Sciences*. https://doi.org/10.1016/j.xphs.2018.11.022.
130. Santiano, F. E., Zyla, L. E., Verde-Arboccó, F. C., Sasso, C. V., Bruna, F. A., Pistone-Creydt, V., Lopez-Fontana, C. M., & Carón, R. W. (2019). High maternal milk intake in the postnatal life reduces the incidence of breast cancer during adulthood in rats. *Journal of Developmental Origins of Health and Disease, 10*(4), 479–487.
131. Melnik, B. C., & Schmitz, G. (2019). Exosomes of pasteurized milk: Potential pathogens of Western diseases. *Journal of translational medicine, 17*(1), 3.
132. Liao, Y., Du, X., Li, J., & Lönnerdal, B. (2017). Human milk exosomes and their microRNAs survive digestion in vitro and are taken up by human intestinal cells. *Molecular Nutrition & Food Research, 61*(11), 1700082.

133. Panaiotov, I., Dimitrov, D. S., & Ter-Minassian-Saraga, L. (1979). Dynamics of insoluble monolayers: II Viscoelastic behavior and marangoni effect for mixed protein phospholipid films. *Journal of Colloid and Interface Science, 72*(1), 49–53.
134. Denzler, R., & Stoffel, M. (2015). Uptake and function studies of maternal milk-derived microRNAs. *Journal of Biological Chemistry, 290*(39), 23680–23691.
135. Title, A. C., Denzler, R., & Stoffel, M. (2015). Uptake and function studies of maternal milk-derived microRNAs. *The Journal of biological chemistry, 290*(39), 23680–23691.
136. Fujita, M., Baba, R., Shimamoto, M., Sakuma, Y., & Fujimoto, S. (2007). Molecular morphology of the digestive tract; macromolecules and food allergens are transferred intact across the intestinal absorptive cells during the neonatal-suckling period. *Medical molecular morphology, 40*(1), 1–7.
137. Patel, R. M., Myers, L. S., Kurundkar, A. R., Maheshwari, A., Nusrat, A., & Lin, P. W. (2012). Probiotic bacteria induce maturation of intestinal claudin 3 expression and barrier function. *The American Journal of Pathology, 180*(2), 626–635.
138. Baier, S. R., Nguyen, C., Xie, F., Wood, J. R., & Zempleni, J. (2014). MicroRNAs are absorbed in biologically meaningful amounts from nutritionally relevant doses of cow milk and affect gene expression in peripheral blood mononuclear cells, HEK-293 kidney cell cultures, and mouse livers. *The Journal of nutrition, 144*(10), 1495–1500.
139. Witwer, K. W. (2014). Diet-responsive mammalian miRNAs are likely endogenous. *The Journal of nutrition, 144*(11), 1880–1881.
140. Denzler, R., & Stoffel, M. (2015). Reply to diet-responsive microRNAs are likely exogenous. *Journal of Biological Chemistry, 290*(41), 25198.
141. Wolf, T., Baier, S. R., & Zempleni, J. (2015). The intestinal transport of bovine milk exosomes is mediated by endocytosis in human colon carcinoma caco-2 cells and rat small intestinal IEC-6 cells. *The Journal of Nutrition, 145*(10), 2201–2206.
142. Manca, S., Upadhyaya, B., Mutai, E., Desaulniers, A. T., Cederberg, R. A., White, B. R., & Zempleni, J. (2018). Milk exosomes are bioavailable and distinct microRNA cargos have unique tissue distribution patterns. *Scientific Reports, 8*(1), 11321.
143. Michalski, M. C. (2007). On the supposed influence of milk homogenization on the risk of CVD, diabetes and allergy. *British Journal of Nutrition, 97*(4), 598.

Physical Chemistry of Nutrition and Dietary Forms

5

Abstract

How harmful is food, which food components cause illness? And what is healthy? What should no longer be eaten? What must be eaten in order to make a long, healthy life possible? These questions can be heard and read almost daily. From a scientific point of view, there are usually no answers. In this chapter, these questions are investigated using typical examples from a physical-chemical and physiological perspective.

5.1 Healthy? Harmful? Where are the Dividing Lines

People's uncertainty about nutrition is great. Is the food healthy? Is the food harmful? If some foods makes sick, there must also be foods that heal, right? And what are the right superfoods for each indivual? To be cautious, someone prefers to eat free of gluten, lactose, and fructose. In the worst case, almost everything we eat is carcinogenic (see Fig. 4.2) [1].

Sometimes it seems that society, which has an abundance of food, is drifting unknowingly through a nutrition jungle that is becoming increasingly difficult to see through. Belief takes precedence over knowledge. Food religions have become the new faith communities, and their respective diets have become the sacred meal. Incorrect opinions persist for a long time—once they are out in the world, they remain there for a long time because the dynamics of their spread are uncontrolled. The best examples are cholesterol and eggs, the Chinese restaurant syndrome and glutamate, as well as the opinion that animal fats are "unhealthy" while plant-based ones are "super healthy."

These rather societal processes seem to be sociologically universal, as they are not limited to the area of food and nutrition. From scientific facts and their complex relationships, phrases emerge through excessive simplification, which, through uncontrolled dynamic communication systems, develop more or less into post-factual theories, conspiracy theories, or new ideologies. Nowadays, dynamic systems via social media like Twitter or Facebook are particularly fast. Within a very short time, unreflective opinions spread around the world. This is certainly one of the most global problems, as good and bad news spread much more slowly during the Neolithic period and while being conveyed on foot or with the first animal-drawn carts, people had time to think about their findings. Logical thinking does not seem to prevail on social media like Facebook & Co everywhere; it is merely about quick likes, attention, or self-affirmation, which usually only pleases the reward center for a short time, as the next posting follows just seconds later. Thus, it is inevitable that incorrect and erroneous opinions solidify, as shown in Fig. 5.1, and the filters, indicated by the dashed line, close so that peruations, post-factual views, and ideologies can no longer be compared with scientific facts.

Instead, it would be better to consult the natural sciences to form a well-founded opinion. Simplifications are necessary due to the high complexity of many scientific questions, but they must not be arbitrary, otherwise, the formation of opinions quickly ends up in the vicious circle of uncontrolled dynamics, from which there is hardly any escape.

Superficial associations promoted a strongly negative attitude towards natural sciences, resulting in a strong aversion to technology. Specifically, physics is more closely associated with strongly formative concepts such as nuclear fission and atomic physics, which have emerged as a result of quantum mechanics. This fundamental research, which has been awarded numerous Nobel Prizes, has only made possible what is now taken for granted. Despite daily interaction with the

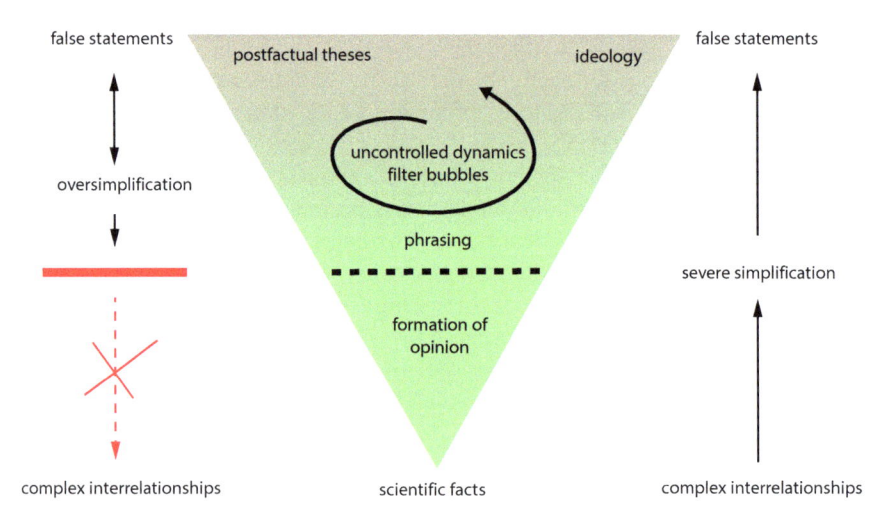

Fig. 5.1 The persuasion-forming triangle with impermeable filters and uncontrolled dynamics in opinion loops and filter bubbles. The red line (left) is no longer crossable to the complex relationships when simplification is too strong and phrase formation is too advanced

5.1 Healthy? Harmful? Where are the Dividing Lines

latest technology, such as the barely questioned smartphones, it is forgotten that all human progress from A for antibiotics to Z for Zener diode is based on solid foundations of natural sciences and mathematics and has not emerged from philosophical, historical, or cultural studies thoughts out of nowhere.

Above all, the subject of chemistry is held responsible for diseases and environmental damage. The standard example is "Seveso dioxin," an allegedly man-made industrial poison. This view is fundamentally wrong, as it has been on Earth longer than the chemical industry has existed. It often forms without human intervention when burning organic, therefore natural, material at high temperatures, such as forest fires or volcanic eruptions. Time and again, discoveries of dioxins in food, including organic eggs, make for fear-inducing headlines. Another synonym for the state of fear surrounding food [2], which results in an increasing hostility and inability to enjoy [3].

A deeper understanding of natural sciences is necessary to interpret such reports safely. However, how can this be achieved if scientists communicate with the public in a technical language that is incomprehensible to laypeople—and many communicators and influencers no longer present facts in the correct context for reasons of oversimplification? How omitting facts can distort meaning and lead to misinterpretations is symbolized in Fig. 5.2. If only the vowels are omitted from the real, detailed information "scientific facts" (scntfc fcts), the original information can be reconstructed practically without distortion despite the simplification. Omitting the consonants, on the other hand, is not possible. The remaining part (ieii a) leaves an infinitely large scope for completely arbitrary interpretation, such as "the fear in the boy." In the worst case, a mixture of misinterpretation and fear arises.

One of the essential problems is, therefore, finding a language that explains increasingly complex scientific relationships simply, without omitting essential information, so that connections can be recognized—and classified—[4].

When it comes to nutrition and healthy eating, it's even worse. Similar to watching football on Saturdays, when about 15 million men are the better coaches and experts (often overweight and visibly unathletic), there are also many, even more experts on the subject of nutrition. After all, everyone eats something, from

Fig. 5.2 Simplification is an art. Only if essential information is conveyed despite strong simplification can the original details be reconstructed from the passed-on information (right) (green circle). With too strong simplification, this is no longer possible (red circle)

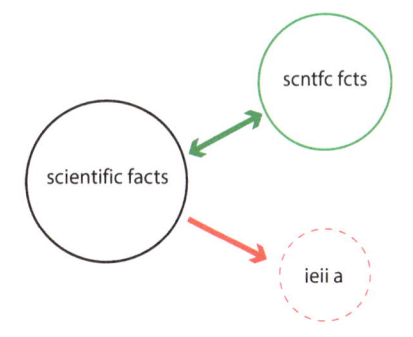

supermarket products to vegan raw food to Demeter pastries and Paleo smoothies. One has personal experiences that quickly make one a opinion leader. Paired with the latest news in the daily newspaper and articles in magazines, one's own eating biography ensures that there are about as many views on "healthy eating" as there are adults. If you follow all opinions, you really can't eat anything anymore: Everything seems "somehow unhealthy", unbalanced, and at least twice a year, either pesticide- or dioxin-contaminated. Too much sugar, salt, and glutamate—rapid death is imminent. But not everyone who vehemently represents an opinion also knows the facts and scientific backgrounds.

5.2　From Observational Studies and Popular Interpretations

Highly simplified nutrition books become bestsellers, not only the already mentioned faction with the tenor "all poison, evil industry", but also those that supposedly make everyone believe in X for U on a scientific basis. The best example is: *The Diet Compass: The 12-Step Guide to Science-Based Nutrition for a Healthier and Longer Life*[5]. A catchy title promises healthy nutrtion, even based on personal experiences at several points. The main recommendations of this book are quickly summarized according to Wenzel [6]:

- Eat real food—preferably unprocessed foods (vegetables, fruits, legumes, nuts, seeds, herbs, wheat germ, moderate amounts of fish and meat). Processed foods that are okay: whole grain products, oatmeal, yogurt, extra virgin olive oil, cold-pressed rapeseed oil, tea, coffee, small amounts of wine and beer.
- Make plants your main course—raw, cooked, steamed (just not too many potatoes and rice).
- Prefer fish over meat.
- Yogurt: yes; cheese: also okay; milk: so-so.
- Practice "time-window eating", e.g., from 8:00 a.m. to 8:00 p.m. Consume the majority of calories in the first half of the day, and eat nothing for at least two to four hours before going to bed.
- Enjoy!

This conclusion seems, on the one hand, like a single platitude, and on the other hand, it reflects exactly what became customary and practiced since the Neolithic period and was lived until the beginning of the prosperity phase of the Western world after World War II, as far as it was possible.

The book, like many others from lay literature, has a fundamentally questionable approach. It largely relies on associative observational studies, which even when combined into large meta-studies, do not correspond to the current state of research; therefore, the statements must be viewed critically. After all, observational studies fit into any worldview (see Chap. 4 and Fig. 4.2). Or not. Or they confirm one's own eating biography. The gain in knowledge is also low, even in some observation and some meta studies. Because this is exactly what happened, as shown in Fig. 5.1. The opinion "large meta-studies reflect the truth" has become self-sufficient and is no longer questioned in detail [7]. Ioannidis [8] discusses this problem of results from such associative studies very clearly and distinctly using the methods of mathematical statistics and calls for a reform of the publication method for observational and even some randomized meta-studies [9]. The reason is very simple: The quality of each meta-analysis depends sensitively on the quality and systematic nature of the statistics of the individual studies evaluated. Even small, statistically non-significant experimental studies are sometimes overrated in the meta-study. It is not uncommon for the data to be incorrectly extracted from the publications and weighted according to their significance. This also makes the statements of the meta-studies highly erroneous, the range of validity becomes very large, and it remains the classic "nothing is known for certain"—despite the enormous computing power required for such overarching evaluations of existing studies.

Sometimes it seems as if misunderstood studies are the best basis for major nutritional misconceptions. An enlightening example is the interpretation of the results concerning acrylamide.

5.3 The Inevitability of Acrylamide

5.3.1 Acrylamide—Brand New and Yet Ancient

The news "acrylamide causes cancer" caused quite a stir at the time, even reaching political levels. Long before the EU regulation came into force, the medical journal headlined: "Acrylamide more dangerous than nitrogen oxides" [10]. It was also pointed out that above 160–180 °C, the concentration of acrylamide in fried, baked, and deep-fried foods increases abruptly. Foods containing a lot of starch and protein are particularly affected. Therefore, according to the EU regulation of 2018, ready-to-eat French fries now have guideline values of 500 µg/kg, cookies 350 µg/kg, and roasted coffee 400 µg/kg. But that is even not all, because especially with homemade foods, the exposure is very high, as temperatures cannot be controlled very well at home. Thus, it is clear: even frying, deep-frying, or baking bread must be very dangerous. If the itake the saturated fat absorbed during deep-frying and the large amount of sugar in the form of carbohydrates (starch) in the potato, is added, one could, in an exaggerated way, commit slow suicide with fries. Indeed, one could if one were to eat only French fries and nothing else for years. But then it is still far from certain whether death was caused by acrylamide, fat, or the large amount of starch—or perhaps rather by the somewhat one-sided, micronutrient-poor diet.

Moreover, at this point it must be questioned how humanity survived before the years 2000 [11] and 2002 [12], when the first study was published twice. Since the Paleolithic era, people have been roasting, frying, and grilling over fire whatever they define as edible. Be it starchy tubers and roots, or meat. No one, except a few polymer scientists, knew the term acrylamide; it is simply a monomer unit of the material polyacrylamide. Suddenly, it became the "fries, toast, cookie, coffee, gingerbread, and beer poison." To really understand this, one has to leaf back a few decades in the original literature.

First, let's look at the experiments of the Swedish working group, which published the first indications of the acrylamide problem. In two different experiments, six and eight (four male and four female) laboratory rats were fed with standard rat food. Standardized food is important for the comparability of laboratory results. The composition is based on the typical nutritional profile of the animals and is precisely balanced in terms of carbohydrates, sugars, amino acids, fatty acids, and micronutrients. This food is available in pellets. From these pellets, a dough-like paste was first made with water, which was simply dried for the control group and fried golden brown for the other group. The rats were then fed this for 102 days, and afterwards, their blood values were measured. In the control group, about 20 pg of acrylamide metabolites were detected in the serum, while in the group fed with fried standard food, 160 pg were found. The unit pg stands for picogram, which is one trillionth of a gram. In another experiment, the researchers changed the frying temperature. One part of the food was fried at 180–200 °C, the other at 200–220 °C, and one group of rats was fed this for 30 days and compared again with the control group. In the control group, 5 pg of metabolites were detected, while in the "fries group," 65 pg ($\pm$15 pg) were found. There were no significant differences between male and female animals. The animals did not show any carcinomas, but only the increased values of the degradation products of acrylamide. The animal experiments clarified how acrylamide is metabolized in the living organism of rats.

What can we really deduce from this for the effect of acrylamide on humans? This is initially completely unclear, as the following thought experiment shows. To establish comparability, a comparable number of humans, quasi *"Homo laborrattis"*, would have to be locked in sufficiently large laboratory cages. First, a standard human diet would have to be defined, which corresponds approximately to geriatric puree or liquid food. This standard diet—a mixture of proteins, fats, carbohydrates, vitamins, minerals, and trace elements—would have to be stirred into a puree. The control group in cage 1 would only get the cooked puree to eat, the other one the pan-fried version. No apple in between, no salad, no meat, exclusively this standard. They would only drink water. The movement patterns of the laboratory humans would also have to be standardized. This inhumane human experiment would last about ten to 15 years, converted to the lifespan of real laboratory rats. In addition, the test subjects would have to undergo regular, sufficient controls. And yet the informative value would be low, because the acrylamide group would have to be defined simultaneously supplemented. By frying the standard food, macro and micronutrients are naturally lost, because some vitamins are heat-sensitive, free sugars caramelize when heated and are therefore no longer available as energy-providing macronutrients, as well as amino acids that react

5.3 The Inevitability of Acrylamide

with sugars to form flavors and thus have no nutritional value and are missing in the overall diet plan. Even if it is only small amounts, they add up significantly over the course of the experiment. This thought experiment quickly shows how little such animal experiments can be transferred, despite all their usefulness for basic research.

Certainly, the acrylamide-free control group would not be healthy after this experiment either, because something crucial is missing in human nutrition: variety, joie de vivre, enjoyment. So everything that has characterized the biography of the nutrition of *Homo sapiens* for over a million years, from raw to cooked to fermented. Food diversity and dietary forms that cannot be covered by standardized dietary forms. The *Homo sapiens* consumes a mixed diet, unlike experimental animals, which contains a whole range of macro and micronutrients. In the spirit of raw, cooked, and fermented, the basic pillars of nutrition since time immemorial. Therefore, deriving direct comparability from standardized animal experiments is simply illogical and not permissible.

No animal experiment and no meta-study provide absolute certainty about the effect of individual substances or the meaningfulness and quality of certain dietary forms. However, in the case of acrylamide since 2002, a political machinery has been set in motion like rarely before. Quite quickly, the original study's results were abandoned, and an uncontrolled dynamic was entered, as shown in Fig. 5.1. The results of the publications remain true in essence and are not diminished in the slightest, but the reactions to them served politics.

Also by scientists, having a look at the original literature today, over 3000 scientific papers on this topic on the scientific data server (until the beginning of 2020) *Web of Science* will be found. The topic is booming in science, and it is almost impossible to draw a conclusion of all studies on this topic alone, similar to the diet compass, especially since not all publications can be checked in detail for the actual analytical methods. Therefore, as always, it is necessary to focus on the elementary molecular facts that allow the problem to be assessed from other perspectives. Then, relevant conclusions can indeed be drawn. Especially for the preparation techniques and the culinary context.

Precisely for this reason, experiments with laboratory rats like these are useful, as they provide a small part of a much more complex puzzle. In contrast to pure observational studies, which cannot provide any information about causal relationships, such *in-vivo* animal experiments allow cause and effect to be observed very precisely. When genetically modified experimental animals are used, the results can be even more clearly substantiated. Nevertheless, conclusions about humans are hardly possible. Animal experiments can therefore only provide very weak indications of possible relationships. Quite apart from the different physiology of the experimental animals compared to humans, the living conditions of the experimental animals are never comparable to those of humans in everyday life. In the last part of the book (Chap. 7), these ideas are once again embedded in a larger, more scientific context.

5.3.2 Acrylamide and Non-Enzymatic Browning Reaction

Acrylamide is always formed during non-enzymatic browning, i.e., the Maillard reaction [13], and is therefore practically unavoidable, although the process-related concentration can be controlled. In principle, there are two universal pathways for the formation of acrylamide: the "carbonyl pathway" and the "fat pathway" [14]: Fats must react with amino acids or sugars with amino acids. For the most important reaction pathway, a carbonyl compound, usually glucose in food, and the amino acid asparagine must come together and be exposed to higher temperatures. Inevitably, both react to form acrylamide. With the involvement of fat, fatty acids are first cleaved from the triacylglycerols (triglycerides) at high temperatures, resulting in the formation of free fatty acids, glycerol, and the compound acrolein, which in turn reacts with other components to form acrylamide (Fig. 5.3). Acrolein is also known as 2-propenal, a fragrance found in candles, the smoke point of oils, or as a typical frying smell, and is thus attributed to both fats and paraffins.

Without having to have comprehensive knowledge of the individual reaction processes, it can be seen from Fig. 5.3 where acrylamide can be formed. Not only from the amino acid asparagine, which reacts with sugar (reactive carbonyl compound from the Maillard reaction), but also from fatty acids and glycerol from fats, provided that the amino acid degradation products β-alanine and carnosine are already present during the frying process. We already know β-alanine from

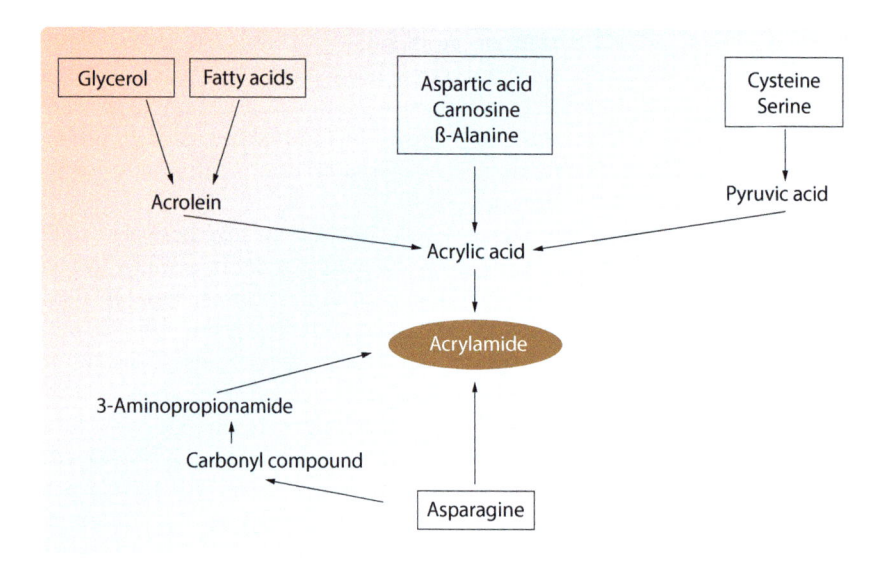

Fig. 5.3 Various pathways from some components of macronutrients (in small boxes) lead to acrylamide. Carnosine and β-alanine are already formed during cooking at 100 °C and are perceived as flavor enhancers

biogenic amines, it is formed from the amino acid alanine. Carnosine, a compound of the amino acids alanine and histidine, is a typical component of meat broths. It is formed at low temperatures during cooking and has a sweetish and sourish enhancing taste. Moreover, it is even formed during the ripening of meat at low temperatures. Certain precursors and precursors of acrylamide are therefore always formed. The unavoidability of acrylamide is thus due to the foods themselves, as they all consist of proteins, sugars, and fats. Even if sugar and fat are not obviously present, they are abundant in all natural foods: glucose from starch, sugar from the glycoproteins of the cell membrane; fatty acids are also abundant in the phospholipids of the cell membrane. The acrylamide problem has thus been in the world longer than humans have controlled fire. It is always formed when biological material is treated with temperatures significantly above 100 °C, be it in fires, be it in the food over the fires of hominids, or be it on the grill at a beer festival.

5.3.3 Acrylamide—Amino Acids and Sugar

However, if acrylamide is fundamentally unavoidable at higher cooking temperatures, it is desirable to learn something about the reaction pathways. The simplest experiment is to heat glucose and asparagine at different temperatures and measure the resulting concentration of acrylamide, as shown in Fig. 5.4.

In this model experiment [13], asparagine (0.1 mmol) and glucose (0.1 mmol) were heated in 0.5 molar phosphate buffer (100 ml, pH 5.5) in a sealed glass tube for 20 min. Afterwards, the acrylamide concentration was determined using various analytical methods. A clear increase in concentration was already observed from 130 °C, reaching a maximum at 170 °C, before this reaction pathway in the aqueous solution produced less acrylamide at higher temperatures.

Fig. 5.4 Concentration of acrylamide formed from asparagine and glucose as a function of temperature. In this model experiment, the concentration is highest at 170 °C [13]

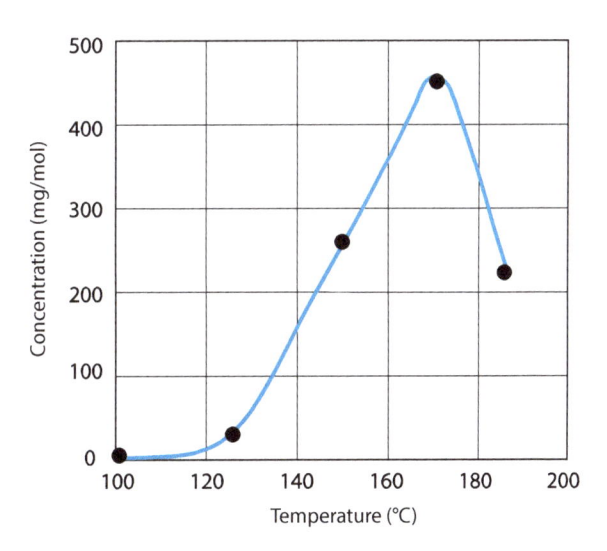

This reaction pathway indicates that starchy foods, which provide a high proportion of glucose through hydrolysis during heating, release significantly measurable concentrations of acrylamide. This becomes clear when real foods are used. The example of oven-baked French fries illustrates this. This model experiment is informative because it excludes the fat component of Fig. 5.3 and primarily shows the influence of amino acids and carbonyl compounds [15]. From a chemical point of view, a systematic correlation between time and temperature for acrylamide formation is shown, as illustrated in Fig. 5.5.

It becomes clear how baking time and temperature affect the process. At 195 °C, the formation of acrylamide even stagnates with increasing time in the range of 24 min of baking time at about 300 µg/kg French fries. The acrylamide concentration thus remains very low overall. At higher temperatures above 200 °C, it increases disproportionately with baking time.

The explanation is relatively simple, because: The higher the temperature, the faster water evaporates from the prefabricated (and partially pre-baked) potato sticks. The Maillard reaction is directly correlated with the water content or water activity. The faster the water evaporates, the faster the Maillard reaction begins at the less moist spots on the surface. Where browned spots with their deliciously fragrant roasted aromas appear, acrylamide inevitably forms as well. At low temperatures, the evaporation rate is lower, consequently the slowly evaporating water cools the process through its (latent) heat of evaporation over the entire duration, and less acrylamide forms.

This is also clearly demonstrated when using different cooking methods [15]. The acrylamide content of prefabricated potato pancakes from the convenience sector, which were finished using various methods, was compared. These included

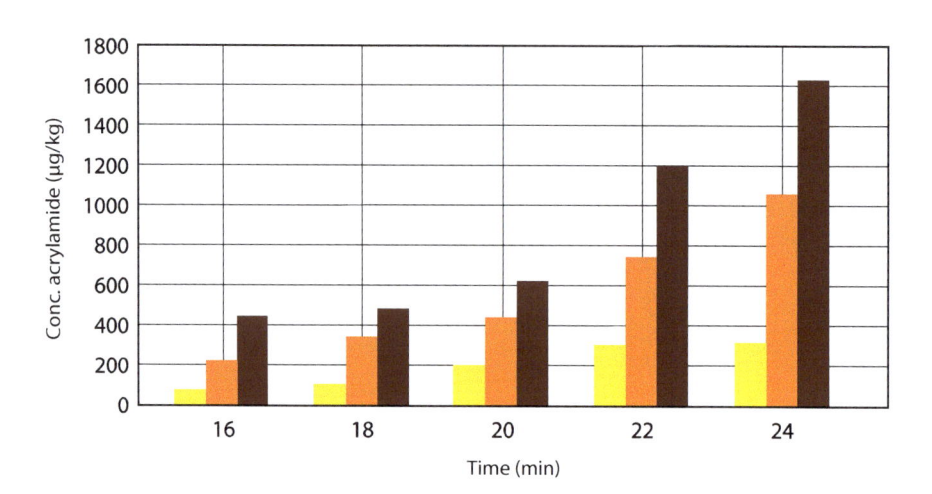

Fig. 5.5 Acrylamide formation in oven-baked French fries. At an oven temperature of 195 °C (yellow), the increase is very moderate, at 210 °C (orange) significantly higher, at 225 °C (brown) it increases disproportionately from the beginning

5.3 The Inevitability of Acrylamide

pan cooking at various temperatures, as well as oven baking and a frying process. Although the measurement results are not directly comparable, as too many parameters were changed simultaneously in this study, trends for the corresponding household cooking techniques can still be observed (Fig. 5.6).

It is shown that low temperatures in the pan and even 220 °C in the convection oven hardly increase the acrylamide content. High temperatures in the pan yield values that are even higher than the concentrations that occur during short-term frying. Although the color of the potato galette appears significantly darker after frying. The degree of browning is therefore not always a visible measure of acrylamide concentration. This is also clear, because the hot oil surrounds all areas of the potato pancake, while pan frying only browns the direct contact points more strongly. Consequently, during frying, the entire, significantly larger surface is exposed to the high temperature, while pan frying only browns the direct contact points more.

5.3.4 Acrylamide—Amino Acids and Fats

The comparison between pan-frying (180 °C/12 min) and deep-frying (180 °C/3.5 min) shows a clear influence of the oil on the formation of acrylamide, as can already be seen from Fig. 5.3. Fats become unstable under high heat, forming free glycerol and degradation products from the free fatty acids, which in turn can serve as precursors for the formation of acrylamide [14], as indicated in Fig. 5.3. This is not only relevant for deep-frying, but also for high-fat baked goods such as cookies or shortcrust pastry products, which are baked at high

Fig. 5.6 The acrylamide formation in direct comparison. Due to the pre-baking, a low concentration is already present. Pan frying and oven with convection in the oven only lead to a slight increase

temperatures in a short time [15]. When oil is heated to temperatures above the smoke point, the fatty acids split from the glycerol. A portion of the free glycerol is then dehydrated to acrolein. However, fragments of fatty acids also affect the formation of acrylamide, as soon as asparagine is present (Fig. 5.3). Thus, it is clear that more acrylamide is formed the more unsaturated fatty acids are present in the frying oil. Conversely, it is also clear that stable, long-chain saturated fats (dominant in the fatty acid C 18:0), such as those of animal origin, support the formation of acrylamide much less.

This fact once again shows that without recognizing the overarching relationships, valid recommendations cannot always be derived. Traditional Belgian fries (they are still considered the best in the world) were prepared in beef tallow for centuries. Then, nutrition science concluded that animal fats were largely saturated and therefore unhealthy. Fries should no longer be fried in beef tallow. Instead, vegetable fats were hardened, resulting in truly unhealthy *trans* fats. The taste of the fries suffered, so frying fat mixtures were developed to circumvent the *trans* fatty acid problem, with a lower smoke point than traditional beef tallow. The result: acrylamide concentration increased, leading to the recommendation to no longer bake fries, fried potatoes, and toast crispy, but, at the expense of aroma and taste, only to "golden." Politics and consumers are satisfied with simple measures, but have forgotten that traditional frying in beef tallow not only achieves better flavour but also keeps the formation of the dreaded acrylamide in check. Without scientific reason, a traditional method has thus disappeared from food culture. As often happens when molecular relationships are disregarded.

5.3.5 Glycidamide—Fatty Partner of Acrylamide

In this context, the secondary effects of acrylamide are also worth mentioning, especially when it encounters unsaturated fatty acids during heating [16, 17]. Here, it can react with oxygen to form glycidamide, a compound that is considered highly harmful to health. Model experiments can demonstrate how acrylamide forms glycidamide in a fat matrix of unsaturated fatty acids. In one case, the acrylamide concentration was kept constant and the temperature was varied, while in the other case, the temperature was kept constant and the acrylamide input was varied. In both cases, the amount of glycidamide formed was measured. The results are summarized in Fig. 5.7.

As this model experiment shows, a secondary substance is formed from acrylamide as soon as unsaturated fats are involved. Another reason why the fats used for frying, baking, or roasting should be saturated. This was also demonstrated in this study, as the glycidamide levels of French fries prepared in different fats were determined. Prefabricated, commercially available potato sticks were prepared in sunflower oil (rich in C 18:2 fatty acids) and coconut oil (predominantly saturated medium-chain fatty acids). And again, the formation of glycidamide correlated with the amount of acrylamide also formed.

5.3 The Inevitability of Acrylamide

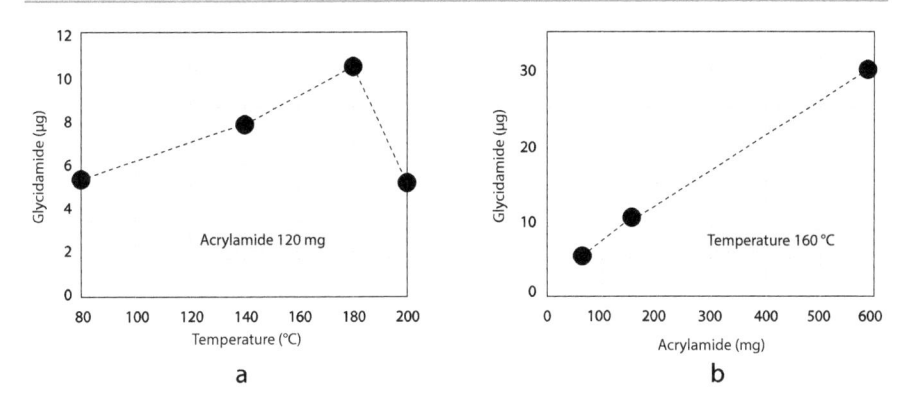

Fig. 5.7 The amount of glycidamide formed in the presence of unsaturated fats: **a** with a constant input of 120 mg acrylamide; **b** at constant temperature with increasing acrylamide concentration [17]

Thus, it seems clear: French fries are poison, because acrylamide is absorbed during the gastrointestinal phase and distributed throughout the body. In the liver and cells, some of it is converted into glycidamide with the involvement of glutathione during metabolism. In this respect, it is quite reasonable to permanently avoid bread, French fries, and chips. True or not?

5.3.6 How relevant is glycidamide?

Answering this question is not easy, as the chemical reaction from acrylamide to glycidamide is not only found in potato products, but also in bread and baked goods [18], especially in cake crusts and cookies when prepared with plenty of fat; but also in coffee, breakfast cereals, roasted nuts, or toast based on classic butter recipes. Nevertheless, the actual danger in all areas is unclear [15, 19], so that certain diseases cannot be traced back to acrylamide and glycidamide. Human nutrition is too diverse for that.

The EFSA (European Food Safety Authority) concluded [15] that the current level of exposure to acrylamide is not of concern with regard to cancer. Although epidemiological studies have not yet proven that acrylamide acts as a carcinogen in humans, the margins of exposure indicate concern about neoplastic effects based on animal experiments. In other words: Nothing is certain, especially when standardized animal experiments are directly compared with human nutrition (see Sect. 5.3.1). However, it is also certain that *Homo sapiens* developed splendidly in the past despite much cruder preparation methods than are common today.

But the acrylamide result and the political activism it triggered had one major advantage: research benefited. Suddenly, methods were developed that allowed for a closer examination of Maillard processes and the influence of mechanisms on the formation of compounds such as acrylamide. This is indeed a great advance for research, as much new and relevant physical-chemical knowledge was generated. This includes, among other things, how the formation of acrylamide can be somewhat contained (although never completely avoided).

5.3.7 The Influence of pH Value, Water Activity, and Fermentation on Acrylamide Formation

It is known that the browning reaction strongly depends on the pH value [14, 15] and how it can be controlled via the pH value. Foods brown faster when the pH value is increased, i.e., shifted towards alkaline [20]. This can be easily verified in one's own kitchen, as, for example, when roasted onions are sprinkled with some baking soda, they darken more quickly. The roasting process starts faster, and desirable roasting flavors as well as melanoids, which are responsible for the brownish shimmer of the surfaces of bread, French fries, or meat, form more quickly. The physical chemistry behind it is clear: the higher the pH value, the faster the deprotonation (the detachment of protons from a compound) of the amino acids released under heat occurs: they release their hydrogen much more easily in an alkaline environment and are available for chemical reactions of aroma formation.

But what about the formation of acrylamide? This mechanism is also relevant for the reaction of asparagine with sugars. Therefore, the formation of acrylamide should also accelerate at higher pH values. Conversely, less acrylamide should form in acidic foods. To establish a clear relationship, clearly controlled model experiments must be carried out again, causing as little measurement error as possible. For this purpose, potato preparations made from potato powder with different pH values were roasted and the resulting acrylamide was measured [21].

The "potato model" consisted of a dried and sieved potato powder with 0.03 g fructose/100 g, 0.03 g glucose/100 g, and 0.89 g asparagine/100 g. The latter was added to be able to measure a controllable and significant acrylamide concentration. The potato powder was mixed with water and standardized to a homogeneous mass of 41% potato powder, 38% water, and 21% oil. The pH value was adjusted before heating. This can be precisely adjusted with various acids, such as ascorbic, citric, acetic, and lactic acid, as well as hydrochloric acid in appropriate concentrations. Various mineral salts from calcium and magnesium can also be used to obtain a wide variation of pH values using different methods. Sodium hydroxide (alkalis) or various potassium and phosphorus compounds, as they also occur in foods, are also suitable for obtaining a good overview with statistical relevance. Regardless of the method used to adjust the pH values, a product-independent and thus universal picture emerges. The results are summarized in Fig. 5.8.

A measure of the Maillard reaction and thus the aroma formation and browning of foods are the reactive amino acids. The concentration of the resulting acrylamide increases, as expected, rapidly with the pH value. At pH values around 4, browning occurs rather cautiously, while it increases strongly at pH

5.3 The Inevitability of Acrylamide

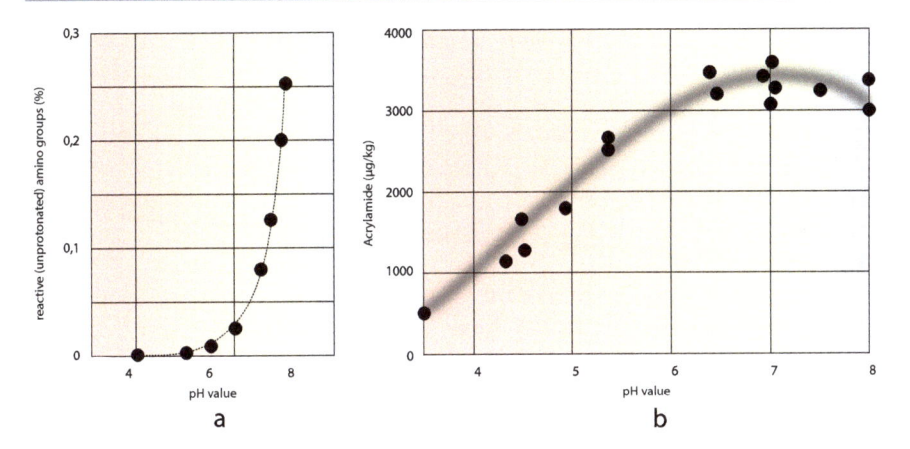

Fig. 5.8 Reactive amino acids (**a**) and the formation of acrylamide (**b**) as a function of the pH value

values around 8, i.e., slightly alkaline (Fig. 5.8a). The formation of acrylamide is similar (Fig. 5.8b). At pH values around 4, the concentration of acrylamide in the potato model is rather restrained, but it increases almost linearly with the pH value between 4 and 6. At pH values around 7 (neutral), saturation seems to set in, while it decreases slightly at even higher pH values. The basic mechanism for this is also, in a highly simplified way, the blocking of chemical intermediates (Schiff bases) during the Maillard reaction.

Ascorbic and citric acid, as well as phosphates as aids in industrially produced baked goods, thus have more functions than is commonly thought. They also help to curb the formation of acrylamide. It becomes clear that, for example, less acrylamide is present in sourdough preparations than in comparable non-sourdough breads. Sourdough fermentation lowers the pH value more strongly and also reduces free asparagine through the microorganisms. Thus, the formation of acrylamide in sourdough breads is significantly lower [22].

The low acrylamide concentration in sourdough fermentation also provides a clue in a completely different direction, namely towards enzymes, asparaginases, which can break down free asparagine. In some yeasts and other microorganisms involved in sourdough fermentation, asparaginases are produced in small amounts [23].

5.3.8 Asparaginase—A New Enzymatic Tool for Acrylamide Reduction?

In medicine, the enzyme asparaginase has been used for some time to treat leukemia and other types of cancer. Asparaginase acts catalytically on free asparagine and causes it to react to aspartic acid. If free asparagine is involved in pathological processes in tumors, treatment with asparaginase is initiated [24]. It was therefore obvious to try using this class of enzymes to eliminate one of the components involved in acrylamide production, namely the amino acid asparagine.

Therefore, this enzyme is also used in the food industry [25], for example in cookies, biscuits, breakfast cereals, or potato chips, to significantly limit the formation of acrylamide [26]. Since aspartic acid is only involved in secondary reaction pathways according to Fig. 5.3, the hope for reduced acrylamide formation is justified.

For the reaction of the amino acid asparagine with sugars, it is essential that the amino acid is present in its free form. Therefore, before baking, at least free, non-protein-bound asparagine can react to aspartic acid via the catalytic action of asparaginase. Bound asparagine cannot react to acrylamide as long as it remains bound during baking, frying, or grilling. If asparaginase is added to the dough, the formation of acrylamide is reduced. Sufficient water activity is important in this process. Dough and potato preparations must therefore contain enough water to keep the reaction of asparagine to aspartic acid going.

For medical application, asparaginase is fermented from microorganisms. The biotechnical yield of the enzyme from fungi is low for the food industry, and the biotechnological effort is too high. Thus, the question remains which more readily available plant sources of asparaginase can be identified for acrylamide reduction. There are indications that some seeds of legumes, such as lupins or peas, contain significant amounts of asparaginase, the activity of which can be increased during germination. If finely ground pea flour is added to wheat flour in three different proportions (1%, 3%, and 5%), bread can be baked from it and the acrylamide concentration determined. In these experiments, bread was baked at 220 °C for 22–25 min. The acrylamide reduction depends strongly on the flour used. In a wheat bread with light flour, the addition of pea flour at all concentrations showed only an acrylamide reduction below 10%, while in bran and whole grain breads, a 5% addition of pea flour achieved a reduction of acrylamide levels in the crust by 57% and 68%, respectively. This alone shows the interplay of various physical parameters. Bran and whole grain flours with higher water binding through soluble fiber and minerals cool the bread during baking through latent heat, the asparaginase remains active longer and can convert more asparagine into aspartic acid.

A targeted application of asparaginase even has positive sensory effects, namely a contribution to aroma formation. On the one hand, not every free asparagine reacts to acrylamide, but forms a small part of furans [27] and pyrazines [28], specific aroma types that have a roasty smell. When free asparagine is converted to aspartic acid, there is also a small proportion of aromas with aspartic acid, which points in the fruity, sweet, and caramel-like odor direction [29]. An important aspect is the inertia of aspartic acid in the Maillard reaction, as already mentioned for glutamic acid. This means that a significant part of the aspartic acid remains after baking, frying, or roasting and contributes, together with the free glutamic acids, to the umami taste. These ideas are summarized in Fig. 5.9.

5.3 The Inevitability of Acrylamide

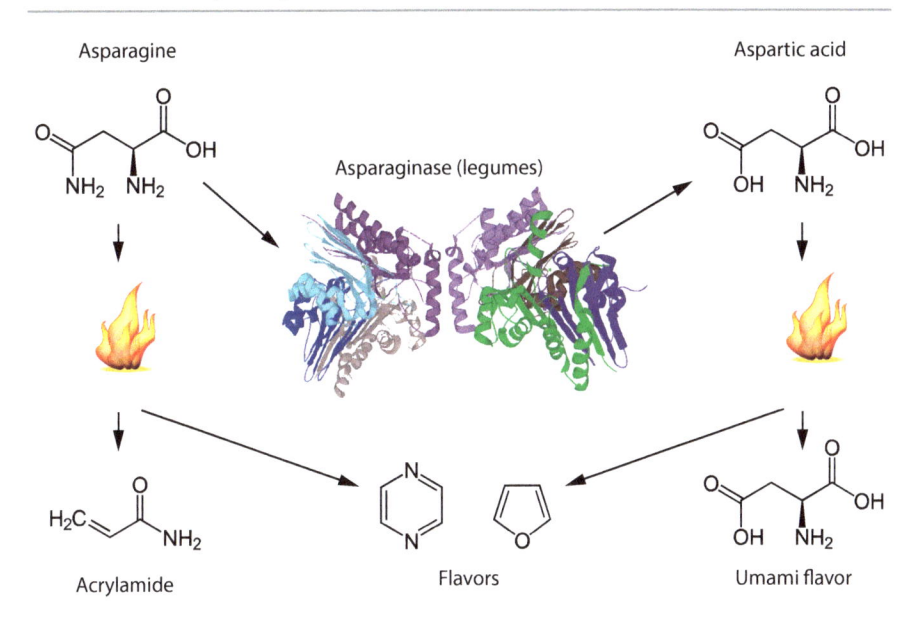

Fig. 5.9 The asparaginase (center) converts asparagine (top left) into aspartic acid (top right). Under the influence of heat, asparagine ultimately forms acrylamide (reaction pathway on the left), while various aroma compounds, such as pyrazines and furans, are formed from aspartic acid (reaction pathway on the right)

5.3.9 The Other, Good Side of the Maillard Reaction

Contrary to the general belief that the Maillard reaction only produces harmful and carcinogenic substances, there is another side to this coin, which is less well known. In addition to compounds suspected of being carcinogenic, the Maillard reaction also produces substances with antioxidant effects [30]. The best-known representative of these is the compound Pronyl-Lysine, whose strong positive effect is well known [31]. It is formed in bread crusts [32], in coffee during roasting, or during the malting of cereals [33] and has long been the focus of research as a positive Maillard product [34]. Pronyl-Lysine is formed from the amino acid lysine upon heating, just as inevitably as acrylamide.

Furthermore, especially water-soluble and weakly water-soluble products formed during the Maillard reaction show a significant antioxidant and anti-inflammatory potential in laboratory experiments [35]. This has also been found in a series of experiments using model systems [36], as well as in substances formed in dairy products (such as cream and cheese topped gratins) from milk casein and sugars [37]. There are many more examples in the current original literature pointing in this direction, and it has also been shown to what extent these products act during the gastrointestinal passage [38].

Not everything that is baked, fried, or deep-fried should be classified exclusively as harmful to health. The different eating biographies of the representatives of *Homo sapiens* in various cultural circles, which have formed since the Neolithic Revolution, bear witness to this in a very special way. If these methods had proven to be unhealthy, it would have been reason enough to extinguish fire and abandon high-temperature cooking techniques. It was only thanks to high-resolution analytical methods that acrylamide became known. In fact, this alarmed the food industry, which began to investigate and change production methods with targeted measures [39]. The extent of a real danger to the diverse eating human remains unclear to this day.

5.4 Biological Value and Food Proteins

5.4.1 Biological Value

As already clearly addressed in Chap. 1, a sufficient amount of protein in the diet is of crucial importance. Only with a sufficient and balanced intake of protein, or better amino acids, can the physiochemistry produce proteins for physiological needs. This makes it clear what balanced means: Above all, the essential amino acids must be supplied through the diet so that functional proteins such as muscle proteins can be constantly repaired, built up, and renewed. Because these functional proteins also require these essential amino acids. It is clear that animal proteins are most similar to the proteins of *Homo sapiens*. Humans are physiologically and biologically animals; they must build muscles, and our brain must be supplied with enough energy so that it can function at all. Therefore, animal proteins are attributed a high biological value. Even without going into detail, it is clear upon some reflection that meat is indeed a piece of vitality, as an all-too-short advertising slogan once suggested. Because when consuming meat, without side dishes and without wine, we receive a high amount of protein that humans can utilize up to 95%, because nothing interferes with metabolism. No tannin, no polyphenol binds protein and slows down or prevents the almost complete utilization.

Of course, it is clear at the same time that this view, as biomolecular as it may be, is sensible for pure carnivores such as cats, dogs, etc., but only partially for the complex system of humans, whose nutrient requirements have been able to adapt incredibly to the respective local conditions and the available nutrient supply since evolution—unlike other living beings. Well-known examples are pandas, which, despite having the gastrointestinal tract of a carnivore, feed exclusively on low-protein bamboo. Long gastrointestinal passages are necessary for this in order to utilize every essential amino acid or to produce them from other food components. Due to the lack of sources and environmental changes, the panda can only survive with human help.

How biologically valuable a protein is can be determined using (hypothetical) reference proteins [40]. The essential amino acids have already been mentioned in the previous chapters, and their taste and predominant hydrophobicity (water

5.4 Biological Value and Food Proteins

insolubility) have been described. Essential amino acids can be remembered using a mnemonic: The initial letters of the individual words of the true sentence "**L**uckily, **F**ew **W**orthy **M**olecules **I**n **K**ids' **V**arious **T**issues" (L, F, W, M, I, K, V, T) result in the single-letter code for the essential amino acids: Leucine, Phenylalanine, Tryptophan, Methionine, Isoleucine, Lysine, Valine, and Threonine (also common is the mnemonic PVT TIM HaLL ("private Tim Hall") for all first letters in the three letter code). If these amino acids are present in a protein in sufficient quantities, it is a high-quality protein.

What happens to proteins when they are digested? In Chap. 4, the effect of enzymes has already been discussed, and protein digestion was addressed. Protein digestion takes place successively.

Already in the stomach, protein chains are denatured by acid while falling below the isoelectric point, i.e., brought out of shape, and cleaved into peptides by the acid-resistant enzyme pepsin. Further cutting of these peptides is carried out by pancreatic enzymes in the small intestine. They break them down—ideally—into individual amino acids. These can be absorbed in the intestine and assigned to their various functions, as shown schematically in Fig. 5.10. After that,

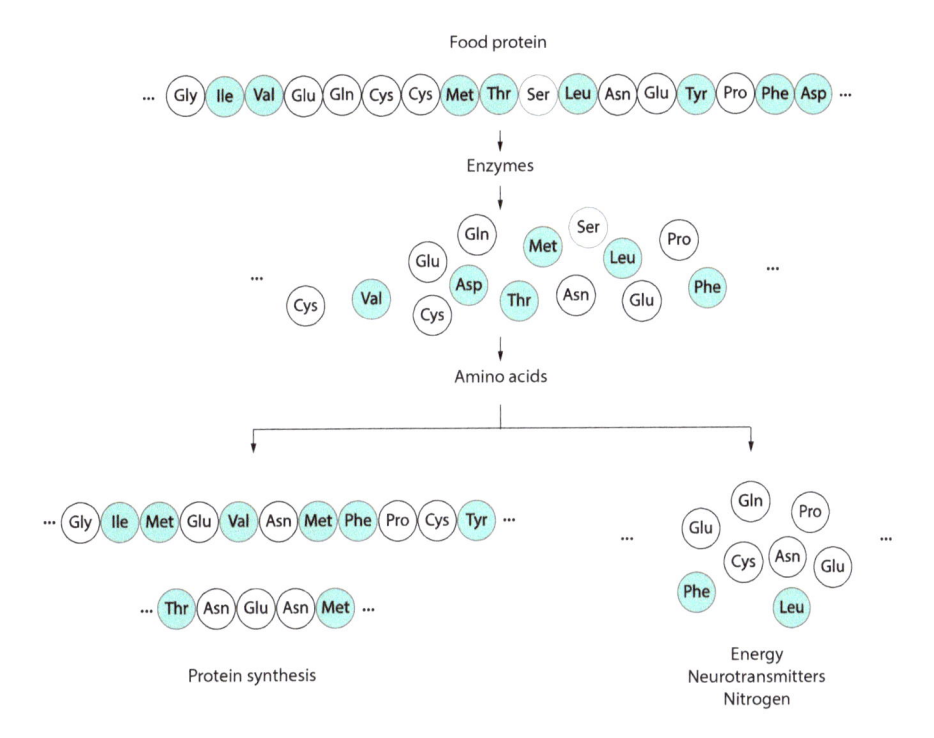

Fig. 5.10 Complete digestion and function of protein chains in the body. Protein chains (top) are broken down into their amino acids by enzymes. Essential amino acids (highlighted in color) are needed for both the construction of vital proteins and the biochemical synthesis of functional substances (messenger substances). (Non-essential) amino acids also provide energy (saturation)

biochemistry decides what happens based on specific signal carriers. Where and when do proteins need to be repaired, replaced, or renewed? Which messenger substances need to be synthesized? Then new proteins are assembled and folded using enzymes; primarily, essential amino acids from food are used. Of course, the assumptions that muscle protein from a beef steak would become a muscle protein of the biceps or even that braised shank of an osso buco would become an endogenous muscle protein are not correct. Instead, where there is a need for protein, the required protein is assembled from the available stock of amino acids. For example, some leucine from a whey or oat protein also migrates as an amino acid into a membrane protein or into hemoglobin.

Therefore, it is necessary to consume a broad spectrum of different high-quality proteins through the diet so that protein synthesis can take place to the necessary extent. To this end, nutritional science has long tried to define protein values.

5.4.2 Liebig's Minimum Theory

The simplest definition is derived from Liebig's minimum hypothesis, which was actually established for plant growth [11]. A plant can only grow as long as the limiting substance (e.g., minerals in the soil) is not depleted. This approach was roughly transferred to the body's own protein synthesis. Proteins can only be fully synthesized until the essential amino acid with the lowest concentration in the body is already incorporated. This can be illustrated by an "unfinished" barrel made of different staves, as shown in Fig. 5.11. The filling level is determined by the lowest stave.

This barrel model applies exclusively to essential amino acids, as non-essential ones can be synthesized in the corresponding cells. However, if essential amino acids are missing, proteins cannot be fully synthesized according to the required blueprint, but remain incomplete and thus without biological function. For example, if methionine were missing, the two proteins in Fig. 5.10 could not be fully

Fig. 5.11 Limiting amino acids in a barrel model. In this example, the amino acid histidine is the limiting amino acid. Even if the other amino acids are present in high concentrations, no further proteins requiring histidine can be synthesized

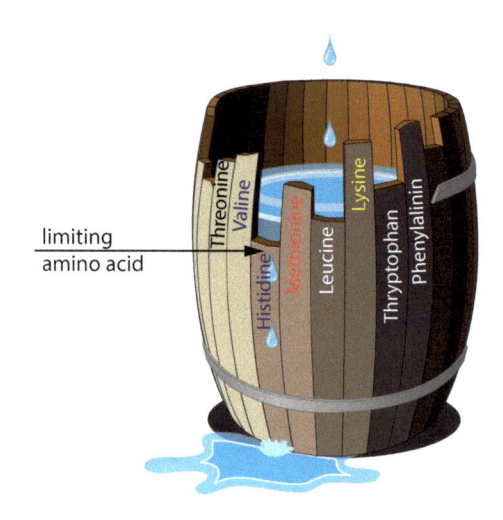

5.4 Biological Value and Food Proteins

assembled. There would remain peptide fragments or individual amino acids that would then provide energy or be used as messenger substances.

Recently, there have been increasing indications in the scientific literature of molecular details of the high biofunctionality of peptides [42], for example when consuming meat [43], dairy products [44], many different animal products [45] and also plants [46]. These functional peptides have a length of about three to 30 amino acids and have a variety of effects on cell physiology. They prevent cardiovascular diseases, have an antioxidative effect, show positive properties for the immune system, and much more. Although most scientific studies focus on peptides that are formed during digestion, it would only be logical if peptides formed in cells already have corresponding functions there if the prolongation phase cannot always be completed. In nature and evolution, nothing happens in vain, not even on the smallest length scales; *Homo sapiens* still bears witness to this today. Malnutrition has been almost effortlessly overcome by humans during natural disasters and long wars. Our grandfathers, who survived two world wars in famines, still bear witness to this today.

The minimum hypothesis can therefore only be a rough guideline and only in cases of real deficiency in the sense of malnutrition, such as in old age [47], extremely unbalanced diets, or self-imposed, permanent special diets like vegan raw food [48], which must lead to a deficiency supply for physical-chemical reasons alone. But it is not the case that the other amino acids do not make a meaningful contribution. At least they contribute to energy intake.

5.4.3 Classic Definitions of Biological Valency

In principle, textbook sentences on the definition of the concept of biological valency read as follows: The biological value (*BV*) is a measure for the assessment of dietary protein. The higher the biological value of a dietary protein, the more endogenous protein can be built from it. For example, the protein mixture of chicken egg, which serves as the standard and has the highest biological value, is rated at 100%, while proteins from cereals can only achieve 50–70%. Therefore, nutritional science has tried to define the concept of biological value more precisely, and there are various ways to quantify it. The biological value of a protein was defined as the product of the ratios of the concentrations of the essential amino acids relative to a reference protein, in the classical definition, the ovalbumin in chicken egg white:

$$BV = \sqrt[n]{\prod_{i=1}^{n} \frac{i - \text{th essential amino acid (food protein)}}{i - \text{th essential amino acid (reference protein)}}}$$

In this definition, for example, it becomes clear how serious a very small proportion of an amino acid affects the product: If an amino acid were not present in a protein at all, one of the factors would be zero. Consequently, the whole product would be zero. The biological value would be $BV = 0$, as in the case of collagen, which, for example, lacks the essential amino acid tryptophan.

An alternative definition of biological value is determined by the nitrogen balance.

$$BV = 100 \times \frac{\text{retained nitrogen concentration}}{\text{consumed nitrogen concentration}}$$

It seems plausible that any protein that is closest to egg white in its amino acid pattern and sequences is characterized by its biological value. On the other hand, this also means that many of our body's own proteins, which are similar to albumin, can simply be recognized particularly well by enzymes. For this reason, foods such as egg, milk, and meat are attributed a high biological value. In general, it follows that proteins of animal origin are biologically more valuable than proteins from plant-based foods because they usually contain more essential amino acids.

If certain essential amino acids are not present or only present in small amounts in certain proteins, this limits the biological value according to the cited formulas. For example, a lower protein value in rice, wheat, and rye is due to a low proportion of lysine in the cereal proteins; compared to whole egg, it is only 40%. In corn, it is the low proportion of tryptophan and in legumes, the amino acid methionine. For corn, it has already been pointed out elsewhere which cultivation technique, nixtamalization, the Incas developed to make corn a valuable food at all (see Sect. 3.4.10).

Such limiting amino acids determine the biological value BA sometimes dramatically, in the case of collagen, for example, the tryptophan is not present at all. The product in the above equation is thus equal to zero, consequently, collagen or gelatin have a biological valcency of zero.

5.4.4 Collagen Drinks—Collagen Against ellulite?

At this point, a small digression is appropriate, which could not better express common misinterpretations. Collagen-/gelatin-containing drinks are offered, and their skin-tightening effect is emphasized as well as a scientifically proven "anti-aging effect". From a scientific point of view, this idea falls short: This is more wishful thinking, as can be seen from Fig. 5.10 alone. Protein synthesis is not regulated according to current beauty requirements, but according to biological criteria. The life-sustaining function and the respective blueprints in the cells take precedence.

Orally administered native collagen is present in a triple helix and, if not thermally treated and denatured to gelatin, is difficult to digest. Native connective tissue is only usable to a small percentage, with 40% passing through the gastrointestinal phase undigested [49]. No wonder, as this triple helix, three gelatin chains, natively in helix form, are braided like a wire rope [50, 51] and is thus only slightly accessible to proteases. Together with the protein elastin, collagen is responsible for the high stability, tensile strength, and elasticity of connective tissue and thus also for smooth skin on the face or other body parts. Whether in animals or

5.4 Biological Value and Food Proteins

humans, it is almost the identical protein. The idea is therefore obvious to administer animal collagen to humans so that they can constantly build connective tissue. That this simple "equation" is hardly scientifically justifiable can be clarified in several steps.

In order for collagen to be braided into the physiologically required stable triple helix, a specific amino acid pattern must be repeated over long stretches along the chains, with the winding framework being carried out via the polar amino acid glycine (Gly), which occurs exactly (Gly) at every third position of the protein chain (Fig. 5.12).

The collagen-typical patterns are Gly-Pro-Hyp (10.5%), with the amino acids proline and hydroxyproline (Hyp) contributing to the stability of the triple helix. Other common tripeptide sequences are: Gly-Leu-Hyp (5.5%), Gly-Pro-Ala (3.4%), Gly-Ala-Hyp (3.4%), Gly-Glu-Hyp (2.8%), Gly-Pro-Lys (2.7%), Gly-Glu-Arg (2.7%), Gly-Pro-Arg (2.6%), Gly-Glu-Lys (2.5%), Gly-Phe-Hyp (2.5%), Gly-Pro-Gln (2.5%), and Gly-Ser-Hyp (2.3%). This sequence formation is therefore mandatory to ensure the function and properties of collagen.

So, when consuming collagen in its native form and its enzyme-accessible form, gelatin, a cleavage into amino acids and corresponding peptides occurs. The likelihood of these being reassembled in the dermis in the sequence required for collagen is very low. However, there is evidence that some of these peptides from gelatin are biofunctional and stimulate the synthesis of connective tissue. Therefore, attempts are being made to trigger this stimulation through orally administered peptides with amino acids of the collagen pattern, especially Gly-Pro-Hyp, Gly-X-Hyp, and Gly-Pro. In fact, these peptides can be detected in the blood. Small effects on binding to corresponding ligands can be observed, but they are very low in clinical tests. The factor of *ex vivo* built collagen when taking such special preparations compared to placebos is only 1.1. The effect is therefore statistically not very relevant [52]. Skin tightening through collagen drinks remains a very dubious business from a scientific point of view. Factors other than the non-essential amino acids glycine, proline, and hydroxyproline seem to stimulate

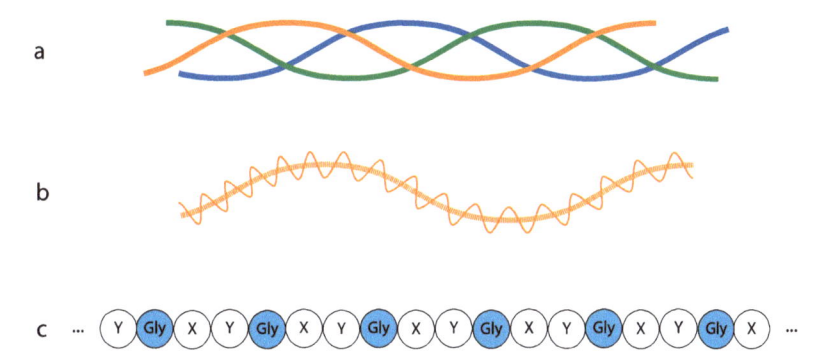

Fig. 5.12 The universal basic pattern of all collagens (connective tissue). **a** The basic unit of collagen fibers is a triple helix. **b** Each protein strand of the triple helix is itself a helix. **c** The amino acid structure of each individual chain follows a pattern with representative repetitive sequences, in which glycine (Gly) occurs at every third position

the cell's collagen synthesis more importantly. These include vitamins, omega-3 fatty acids, minerals [53] and phytochemicals [54] such as polyphenols. All of this sounds more like a diverse and enjoyable diet from the perspective of micro- and macronutrients than pills and drinks with unclear results.

5.4.5 Biological Valency and Bioavailability are Different

The biological value is only one criterion, as it expresses the presence of essential amino acids. Food proteins and their amino acids not only have the function of being reassembled into the body's proteins but also provide energy and thus lead to long-lasting satiety [55], as can be seen on the protein lever (see Fig. 1.11).

However, this information is not enough, because whether and to what extent these proteins are available to humans during gastrointestinal passage is a completely different matter. Corn is a perfect example of the difference between value and availability. As already mentioned in Chap. 2, one of the main proteins in corn is zein, which is difficult to denature due to its strong hydrophobicity and resistance to thermal denaturation. The pH values during the gastrointestinal passage never rise high enough for Zea mays proteins to unfold. Enzymes have no chance of bringing their active centers to the corresponding interfaces. This was ultimately the main reason why the Incas, whose main grain was corn, recognized how corn changed through treatment with ash, became workable, and more digestible. The macronutrients of this protein only became available after this physical-chemical trick, nixtamalization (see Sect. 3.4.10).

We also explained in detail in Sect. 4.10.4 how resistant whey proteins can be to digestive processes. Only cutting the cysteine double bonds allows for much faster digestion and absorption of the essential amino acids contained therein, such as leucine. Therefore, if a food contains proteins and thus amino acids, this is not necessarily a criterion for their availability to human physiology.

Now, in classical nutritional science, it is quite common to argue with the different biological values and to make suggestions for certain diets, such as vegetarianism or veganism, on how to supplement the missing essential amino acids with certain combinations. The best-known example is potatoes with cottage cheese, as the missing or low-concentrated essential amino acids in the lower-value potato protein are supplemented by the high-quality spectrum of proteins in dairy products. Especially in veganism, an adequate intake of essential amino acids from vegetables and fruits is not trivial, so the obvious suggestions to combine legumes, nuts, and cereals accordingly are quite understandable. Therefore, suggestions from medical research are adopted and "health-relevant" combinations of animal and plant protein sources are proposed, which, according to observational studies, should prevent cardiovascular diseases [56]. Incidentally, these—as expected in

5.4 Biological Value and Food Proteins

such observational studies—do not show clear results. A positive effect of a purely plant-based diet could not be proven on a scientific basis (as in many other intervention studies), contrary to the many theses.

In this study [56], the deficits of various exemplary plant proteins were taken into account based on the limitingamino acids: In legumes, these are often the sulfur-containing amino acids cysteine and methionine, in cereals they are lysine, isoleucine, threonine, leucine and histidine, in nuts and other oilseeds they are tryptophan, threonine, as well as isoleucine, lysine, and again the sulfur-containing amino acids. Fruits and vegetables are relatively low in protein anyway and are not suitable on their own for adequate protein supply. Even if huge amounts were eaten, methionine, lysine, leucine, and threonine would limit sufficient protein synthesis in the cells. Only the storage proteins of the soybean have a broad distribution of essential amino acids. This is also a reason why tofu has a large share in many dishes in many parts of Asian food culture and is recommended for vegans together with lupine protein. In industrial processes, highly processed combination products made from tofu, pea, lupine, and soy proteins are produced as vegan convenience products, which can be found in supermarkets and serve as a substitute for animal foods.

From these complementary combinations to increase the biological value of a meal, it is then often concluded that the intake and supply of essential amino acids are ensured. The intake is, but the supply is not, because these approaches assume that the concentrations of amino acids during gastrointestinal passage have an additive effect, as symbolically illustrated in Fig. 5.13.

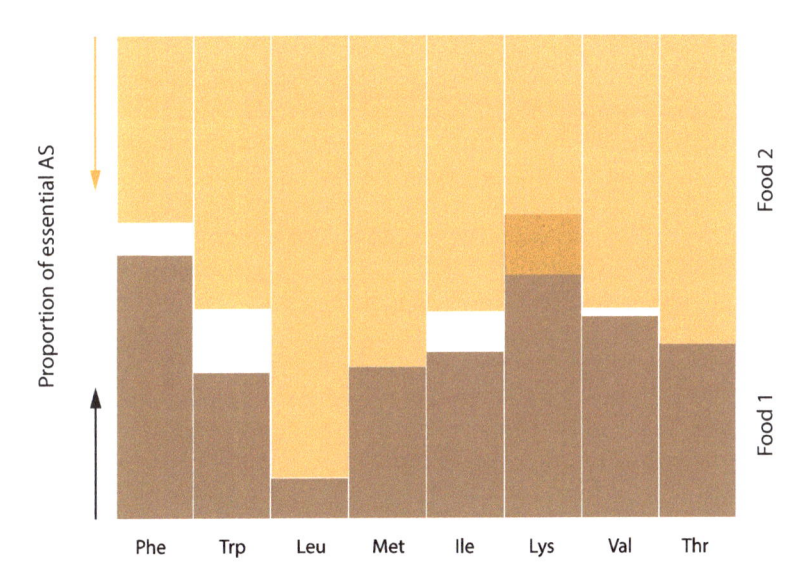

Fig. 5.13 Two (hypothetical) foods with their (hypothetical) essential amino acids. Food 1 is low in leucine (Leu), Food 2 is rich in leucine. With the combination of LM1 and LM2, it is hoped to close the gap in leucine in LM1

In this, the concentrations of essential amino acids are represented by the length of the bars of two foods, LM1 and LM2. LM1 is low in leucine, while LM2 has a more balanced ratio of amino acids. The hope is, according to the approach of biological value, to balance the deficits of LM1 in its limiting amino acid by combining LM1 with LM2, thus ensuring a balanced supply of amino acids. Although the concentrations of amino acids on a plate of LM1 and LM2 are formally as shown in Fig. 5.13, this does not guarantee the additivity in biological availability. The amino acids are bound in proteins, they are embedded in the swallowed food pulp. The structure of the proteins depends on the preparation. All these are factors that determine the enzymatic degradation, its kinetics, and the temporal course of digestion.

In fact, a simple additivity of essential amino acid concentrations after digestion is not detectable even in simple protein combinations in the blood, as can be impressively demonstrated in simple model experiments *in vivo* [57]. In the study by Revel et al., minipigs were fed with the typical milk proteins whey protein and casein in an animal model, and in comparison, others with a mixture of whey protein with wheat and pea proteins adjusted to the leucine content. Subsequently, the blood-measurable content of the specific essential amino acids leucine and phenylalanine, as well as the total concentration of all essential amino acids, were determined, as well as the temporal course of digestion. The resulting anabolic reaction of the muscles via leucine was also investigated. It turned out that, although a muscle reaction was triggered independently of the protein source, there were observable differences. Even when the protein mixture of plant and milk proteins was adjusted to the same leucine intake as with the whey protein source, the concentration of leucine in the plasma was lower. The protein mixture thus does not have the same effect as the consumption of pure whey protein. This shows that apparently various components of the food matrix can influence and change the provision of essential amino acids. In minipigs suffering from muscle wasting, only whey protein intake showed a sufficient effect, while all other protein mixtures remained inefficient. Apparently, the matrix already plays a major role in digestion. This was already intuitively recognizable from Fig. 4.36. Especially for older people, geriatrics should also pay attention to an effective protein intake, which must take place apart from vegan belief rules if the existing muscle mass is to be preserved as well as possible with increasing age.

To emphasize this point explicitly: This discussion is not a plea for excessive intake of animal proteins. It is physiologically not necessary to eat large portions of meat daily. However, the fact remains that animal protein is closest to the protein of *Homo sapiens*. Therefore, it is not surprising that a) *homo sapiens* utilize these proteins most effectively and that these proteins b) provide everything humans need.

The results of this just-discussed model experiment on minipigs have direct and far-reaching consequences for a new sustainable food production on a large scale when animal proteins are to be replaced by plant proteins, even in smaller quantities [58]. With alternative "future proteins" from bioreactors, future foods, and sustainable replacement products, more must be taken into account than appears to be the case in many places. The main problems lie neither in good intentions nor in feasibility, but as always deep in the nanoscale and hidden in molecular interactions.

5.5 Intolerances and Gluten-free Baked Goods

For some time now, many "free-from" foods can be found on supermarket shelves, which are attributed to a certain health potential. The reason for this increasing offer lies in various theses such as, gluten makes you sick, it clogs the intestine. The "free from" list can be extended: free from histamine, free from glutamate, free from refined sugar, free from refined salt, free from fat, etc. Time and again, a common and fundamental problem emerges. When a substance is removed from food, e.g. fat from cheese, then alternatives must be added due to the lack of enjoyment. Fat is replaced, for example, by means of clever emulsions, by dietary fibers such as cellulose fibers, oligofructoses, and by binding agents such as guar gum, xanthan or indigestible waxes. The principle is clear: in most cases, a macronutrient is replaced by a non-nutrient. Inevitably, the nutritional value and nutrient content of the modified foods decrease with the reduced energy.

5.5.1 Gluten-Free

One of the most scientifically questionable hypes are gluten-free baked goods and foods. At least for people who do not suffer from celiac disease and from gluten-induced and recurring inflammation of the intestine. In many cases, the trigger for these ailments is the gliadin (Chap. 3), that self-crosslinked part of gluten. Above all, its amino acid sequences at some points of the gliadin molecule chain lead to misinformation when antibodies (also proteins) dock onto them and recognize them [59]. The resulting and misguided autoimmune reaction causes massive inflammation of the small intestinal mucosa, as the antibodies produced by the immune system in turn attack the body's own proteins of the intestinal mucosa. As recently shown by modern sequence analyses (proteomics), celiac disease-relevant sequences are mainly found in α-gliadin [60]. However, the most common cereals (Fig. 5.14) have precisely these amino acid sequences built in and therefore cannot be consumed by those affected either.

The proteins of the durum wheat precursor emmer also use gluten as a storage protein. In this respect, emmer is not an alternative to wheat. Oats could be considered as a substitute at most, but they do not convince with their water-binding capacity and do not allow for stretchable doughs. We will have to come back to such physical problems later when it comes to replicating the textural properties of gluten-free baked goods, for example.

In addition to the FODMAPs mentioned in Chap. 3 Fermentable Oligo-, Di-, Monosaccharides And Polyols (FODMAPs), specific proteins play a significant role in non-celiac wheat sensitivity. This issue does not seem to have been fully

Fig. 5.14 The common cereals are not gluten-free, neither are rye and barley

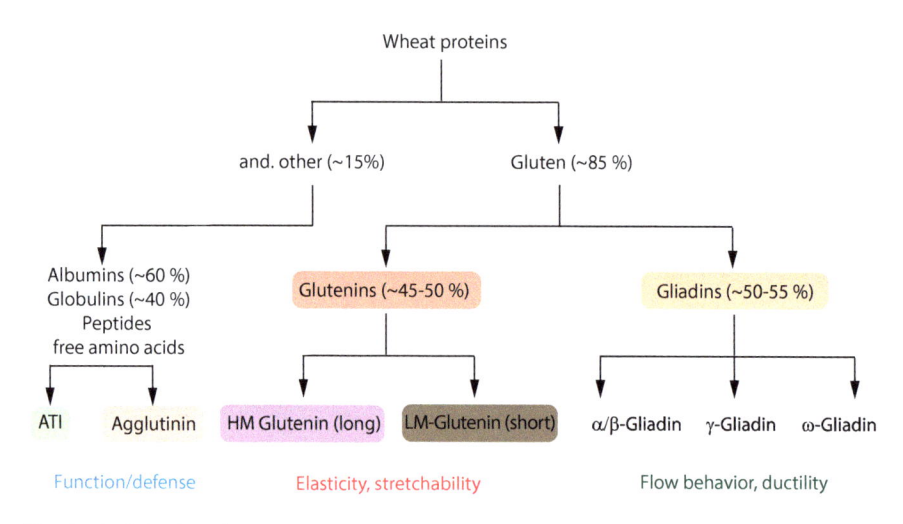

Fig. 5.15 Selection of the most important proteins in wheat and their function

researched in all its details. However, it is clear that apart from FODMAPs, the intolerances can be traced back to protein structures. These can be roughly summarized for wheat as shown in Fig. 5.15.

The main component of wheat protein, accounting for an average of 85%, is the gluten, which can be further divided into glutenins and gliadins. Amino acid patterns found in these proteins appear to be responsible for triggering inflammation in the presence of celiac disease.

5.5.2 Wheat Germ Lectins

For further protein-related intolerances, lectins, such as wheat germ agglutinins (ommonly abbreviated as WGA*s*), have been held responsible. Lectins are found in the germ of wheat, but also in the outer layers of the grain, i.e., in those parts that serve as the first attack surface for insects and microorganisms such as bacteria or wild yeasts. WGAs serve as a defense mechanism for the grain against predators. The structure and function of WGAs have been largely elucidated [61]. In wheat, three variations are relevant, agglutinin-isolectin 1 to 3, which have similar structures. WGAs, the three mentioned lectins, consist of 186, 212, and 213 amino acids [62] and always occur as dimers. This means that two of these lectins—not necessarily the same ones—aggregate.

Their task is, for example, to bind to the chitin of insects and thus reduce their appetite for the grain. In microorganisms, lectins bind to specific cell membranes, thereby preventing their proliferation on the plant. In mammals, WGAs bind to N-acetyl-D-glucosamine, for example in the conjunctiva of the eye, or in the matrix of connective tissue when they bind to sialic acid in the mucous membranes.

The lectins common to all plants, for example in beans, potatoes, soybeans, legumes, etc., are therefore considered "toxic." Meanwhile, books written by popular medicine authors exist [63] about how harmful this class of molecules is and that humans should avoid this class of substances as much as possible. Of course, imitators continue to spread scientifically unclear stories [64].

But what are the facts about WGAs? The fact is: The proteins must be extremely stable in order to fulfill their function, so that they do not change their shape in the face of pH changes, temperature fluctuations, or different aqueous environments. In nature, this can only be ensured by incorporating a sufficiently large number of sulfur bridges (via the amino acid cysteine). Many lectins are therefore considered difficult to destroy. Does this mean that whole grain bread or the highly regarded wheat germ oil [65] (for which, by the way, over 1 ton of wheat is needed to obtain 1 liter of oil) is unhealthy?

One could indeed hastily conclude this, because the high number of disulfide bridges, 16 in agglutinin lectins, suggests a high resistance to gastric and pancreatic enzymes, even though the pH value in the small intestine rises to about 8 and disulfide bridges can be dissolved above this value. Despite intensive research and investigation, nothing can be found in the specialist literature about negative effects. In a recent, comprehensive review article, the state of research is summarized [66]. It turns out that there are no data from reputable studies that support the suspicion that WGAs lead to health effects after consuming whole grain products. Despite numerous assumptions that WGAs cause intestinal damage and diseases, there is to date neither evidence for this nor a reason to recommend the normal healthy population to avoid whole grain products. In contrast to the amino acids of the gluten protein sequences, WGAs are also not responsible for inflammation related to celiac disease. In short, the amount of ingested WGAs when consuming whole grain bread is simply too low for a health problem to manifest itself.

5.5.3 Amylase-Trypsin Inhibitors

So-called ATI, Amylase-Trypsin Inhibitors, also proteins, seem to play a role in many complaints of sensitive people. This is explained in detail and yet understandably in a highly recommended book [67]. The authors of this book present the current state of knowledge on a scientific basis, as they actively research these questions and use the natural scientific methods of molecular medicine. This is in stark contrast to the widespread popular literature [68–70] on this topic, of which hardly any original literature can be found in refereed journals. For this reason alone, there are few reasons to believe anything from these books, let alone take the unfounded content seriously.

The name of the ATI proteins stands for their molecular program: They prevent the activity of the enzymes amylase and trypsin. The wheat grain incorporates these inhibitors to regulate the activity of digestive enzymes. During germination, the storage proteins (gluten) are digested by trypsin (and other proteases) to obtain amino acids and energy. Starch, i.e., amylose and amylopectin, are also digested by amylases to obtain the cellular fuel glucose. However, the energy and amino acid stores must not be consumed too quickly due to unpredictable germination times (unclear humidity, temperature, climate). Therefore, ATI regulates the activity of the enzymes. The primary goal is always the survival of the grain so that new wheat can emerge from it, which is why inhibitors (ATI) are not a product of industry or breeding, but are already present in wild wheat, grass seeds, for good biochemical reasons. They prevent premature depletion of energy stores under unfavorable germination conditions through their molecular management. The ATI are mainly found in the outer layers of the grain, i.e., outside the flour body, to be used where it is important. There are up to eight variations of these ATI in wheat, containing between 124 and 186 amino acids, and another two containing only 27 and 44 amino acids. The structures are extremely compact and cannot be altered by heating or changes in pH value, which is also due to the four disulfide bridges in the protein. Also, due to their biophysical and biochemical requirements, these inhibitor proteins prove to be extremely stable through appropriate secondary structures, as the structure in Fig. 5.16 already illustrates.

The high stability shows, that ATI is not altered or enzymatically degraded by the usual proteases during the strong changes in pH value during the gastrointestinal passage. Another point to consider is that human pancreatic amylases and trypsin must meet similar requirements as the amylases and proteases of the wheat grain during germination, cutting proteins and starch as much as possible.

Therefore, the ATI can flow into the small and large intestine in their native form. They reach the mucous membranes via wheat products (mainly from whole grain) and trigger nonspecific inflammations that are neither clearly detectable by endoscopy nor by tissue samples [66] and disappear after a few days when wheat is avoided. However, this does not happen in all people, but only in those with a certain predisposition. *Ex-vivo* tissue examinations indicate a reaction of the innate immune system. Non-celiac wheat sensitivity is therefore not related to allergies

Fig. 5.16 The structure of a representative of the family of Amylase-Trypsin Inhibitors (ATI). The high stability is visible in the yellow-colored areas (disulphide bonds). The helices are also stabilized by disulfide bridges

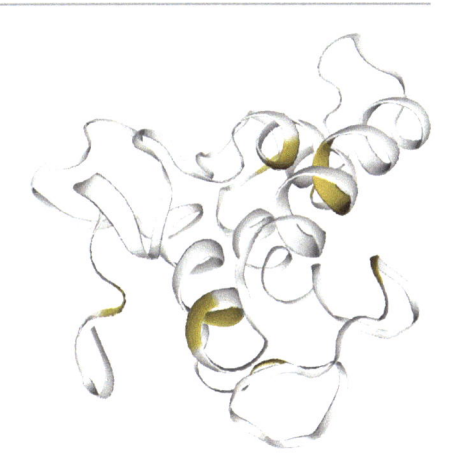

or celiac disease [71]. The proteins of the ATI class bind to specific receptors anchored in the membranes of immune cells. The reactions to this are the same as those that occur in bacteria or other germs. Apparently, misinformation is spread in the immune system, which is responsible for the intolerance reactions.

In this context, another point is important from a physicochemical point of view: If ATI reaches all areas of the gastrointestinal tract unhindered, what happens to the released pancreatic enzymes amylase and trypsin? In principle, the ATI ingested via wheat and other cereals should also inhibit the pancreatic enzymes released in the small intestine and prevent them from doing their work. There is an impressive experiment that includes detailed computer simulations [72]. It turned out how wheat ATI binds trypsin and amylase to inactive complexes. An example is shown in Fig. 5.17. This complex formation is a consequence of the physical interaction of certain amino acid sequences that are attractive. However, if the enzymes are (weakly) bound in such complexes, they remain inactive. When these complexes are formed, the active centers of the digestive enzymes can no longer cleave proteins or starch. A portion of the macronutrients remains undigested, continues in the intestine, and must be partially fermented in the large intestine, whether we like it or not.

Such processes are fundamentally important insofar as the effect of the ATI not only physiologically triggers reactions via docking to certain receptors (Toll-like receptor-4, TLR4), but can also inactivate pancreatic enzymes in the intestine beforehand. Much of what happens in non-celiac wheat intolerance can thus be traced back to the ATI. This also applies to this phenomenon: The molecular interactions that control processes are always fundamental.

Another aspect is also revealed through this consideration: Unlike small oligosaccharides (FODMAPs), ATIs cannot be easily pre-digested by microorganisms, such as those used in previous dough management through yeasts and sourdough cultures. The enzymes released by the microorganisms during dough management also do not break down ATIs. Incidentally, it would be a challenge for research

human trypsin   Wheat ATI   human amylase

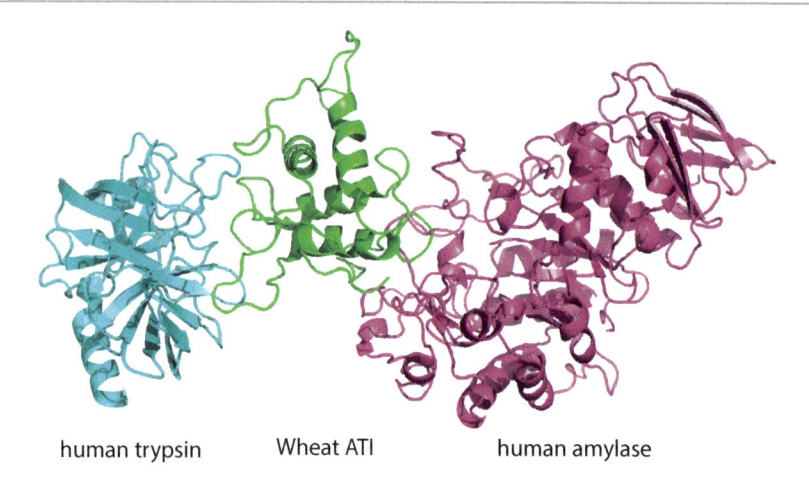

Fig. 5.17 Complex formation between the human enzymes trypsin and amylase in the digestive tract and in the ATI ingested via wheat. The enzymes are inactivated by the complex formation

to contain ATI, i.e. to biotechnologically develop enzymes that partially decompose and inactive the ATIs. However, it would not be a good idea to eliminate ATIs through genetic modifications or other breeding methods, as they are the most natural insecticide that plants have in their program. Wheat without ATIs would yield much less.

5.5.4 Why is Wheat Intolerance Increasing?

In this context, an essential point must be made: These intolerances that have occurred in recent decades, apart from the genetically determined disease of celiac disease, are apparently not the result of "overbreeding" or even "industrial manipulation," but also partly due to the increased consumption of whole grain products. Whole grain breads certainly provide significantly more minerals and fiber than white flour, and this is a hearty, grainy pleasure for a large part of the population. Hearty breads made from whole grain, produced under long sourdough management and precisely baked, taste excellent. The downside, however, is a permanent increase in the intake of non-nutritive and indigestible substances such as FODMAPs, wheat germ agglutinins, or ATIs. In the early years of the Neolithic, people did not eat too much of the whole grain. Over millennia of grain cultivation, techniques and methods were wisely developed not only to separate the wheat from the chaff but also to process the grain so that only protein and starch remained. The long bread history of France, Italy, and Germany shows this. It was only in the 1970s when whole grain utilization, including fresh grain porridge, entered the "whole food culture" of the broader population that dark whole meal breads with bran or raw crushed whole grain cereals suddenly appeared on breakfast buffets. Soon after, the stomachs and intestines of some sensitive *Homo sapiens* began to grumble.

The solution would be straightforward: All people who cannot tolerate bread and baked goods made from whole grains simply do not eat them. All without spreading theses without a scientific background.

5.5.5 Is Gluten-Free Healthy?

Gluten-free is not healthy for healthy people. Many benefits of wheat have already been discussed in Chaps. 2 and 4. Since the Neolithic, people have been breeding grains that are easy to process and nourish humans. Both aspects are inextricably linked from a molecular point of view. If this were not the case, the breeding efforts 8000 to 10,000 years ago would not have been continued. The advantages achieved through these efforts are quite comparable to the efforts of hunting and controlling fire.

Wheat is the only grain that combines both nutritional and baking technological effects. It forms high molecular weight glutenin (see Sect. 3.4.5), which ensures leavening and dough loosening, and furthermore, gluten as a storage protein has all essential amino acids available. That this is an exceptional situation becomes clear when comparing the storage proteins of different cereal species that have developed from grasses since the Neolithic period in various climatic regions. This is summarized in Table 5.1.

In this context, the strength of wheat compared to other cereals and pseudocereals becomes apparent. Their storage proteins have fundamentally different properties. As already known from Sect. 3.4.9, corn must be pretreated in order to be baked at all. Wheat is the only species that produces enough high-molecular glutenin to ensure excellent baking properties (Sect. 3.4.11.3). The storage proteins of rice reach slightly more than half the length of high-molecular (HM) and low-molecular (LM) glutenins, but have far less cross-linkable cysteine. Teff flour (teff is a grain known in Africa since the Neolithic period) also has about half the length of high-molecular glutenin, but its cross-linking properties are significantly higher than those of rice storage proteins. However, the cysteine groups, unlike glutenin, are not predominantly located at the end of the protein chains, which also severely limits the baking properties. This is also known from rye, which contains gliadin but no glutenin, and often shows a "spreading" during baking due to its poorer cross-linking properties. Hobby bakers are familiar with the phenomenon: rye bread often spreads during baking, becomes flatter, and shows a tendency towards flatbread, as was standard at the beginning of the Neolithic period before the properties of wheat were recognized. The shape of millet breads in parts of Africa and rice breads in India (which only exist as thin flatbreads or papadams) also has its deeper cause in the molecular properties of the storage proteins, which ultimately dictate the shape of the breads through the baking properties. Once again, physics and chemistry define culture—and nutrition. The value of the storage proteins also determines the supply of the ethnic groups that consume this particular grain. At this point, it is worth recalling the problem of the poor biological availability of corn (Sects. 3.4.9 and 3.4.10), which could only be solved after the corresponding cultural techniques, the alkalinization through nixtamalization, were developed.

5.5.6 Gluten-Free—Often Nutrient-Poor

Producing gluten-free baked goods and pasta poses a physical-technical challenge—in several respects. The physical consequence is: if you leave out gluten, it must be replaced by something else. A gluten-free country bread, for example, consists of [73]:

Water, cornstarch, sourdough 16% (rice flour, water), buckwheat flour 6.5%, rice flour, sorghum flour, vegetable fiber (psyllium), rice starch, rice syrup, sunflower oil, soy protein, yeast, thickener hydroxypropyl methylcellulose, salt, natural flavor. For a gluten-containing artisanal country bread, however, a simple basic recipe is sufficient: Water, wheat flour, sourdough, salt.

A direct comparison of the ingredients objectively reveals a large part of the physicochemical misery that one is forced into with gluten-free products. To achieve the physical properties of wheat gluten, various ingredients must be combined. From Table 5.1, it is already apparent that this can only be achieved to a limited extent. No other storage protein has a molecular weight as high as glutenin. The cross-linking property via cysteine, which is located at the chain ends of high-molecular-weight glutenin (see Sect. 3.4.6), cannot be compensated for by any of the other proteins. High elongation and equally high tensile strength can only be achieved through the combination of many gluten-free ingredients. The high water binding capacity that gluten provides in the cold state during kneading is not guaranteed with alternative proteins. Therefore, hydrocolloids, binders (usually xanthan, guar gum, or similar) or even, as in the mentioned example, highly viscous syrups are added (which also serve as browning and flavor aids). Also, methylcelluloseof different methylation, carboxymethylcelluloses, or hydroxypropylmethylcelluloses are used to prevent the "spreading" of bread during baking and to support bread in leavening, crumb and bubble structure, and the volume of baked goods [74]. While gluten forms sulfur bridge bonds relatively quickly at 60–70 °C, this does not happen or happens much worse with gluten alternatives, as can be seen from Table 5.1. On the other hand, the molecular movement, the average speed of the molecules, increases with temperature in the oven. The result is that the dough pieces spread apart in the oven before they can solidify. This can be prevented with methylcellulose, as they "gel" under heating, depending on the number and distribution of the (water-insoluble) methyl groups between 45°C and 55°C, thus preventing spreading. To support water binding, psyllium fibers are also added. Behind this term are psyllium husks, which have an extremely high water-binding capacity due to the non-starch carbohydrates present [75] and can also form gels under certain conditions [76]. Responsible for this are the cell materials known as mucilages, which ensure the stability and flexibility of plant cells.

5.5 Intolerances and Gluten-free Baked Goods

Table 5.1 Common cereals from the family of grasses. The main storage proteins (prolamins), the maximum molecular weight of the proteins, as well as the strength of the cross-linking property through the available cysteine are listed (+++ very strong, – not present). AS = amino acid

Family	Grasses							
Subfamily	Bambusoideae	Pooideae				Panicoideae		Chlorodoideae
Tribes	Oryzeae	Triticeae			Avenae	Andropogoneae		Chlorideae
Species	Rice	Wheat	Barley	Rye	Oats	Corn	Sorghum	Millet	Teff
Prolamin	Oryzein Cupincin	Gliadin HM-Glutenin LM-Glutenin	Hordein	Secalin	Avenin	Zein	Kafirin	Zein Kafirin	Eragrostin
Max. Number of AS	470	800	305	477	242	240	269	270	400
Cys	+	+++	–	++	–	+	–	–/+	++

Fig. 5.18 A whole range of different ingredients is necessary to compensate for the physical properties of wheat. In all aspects, this is not entirely successful, so the molecular balance must be somewhat restored with various technical tricks

Figure 5.18 thus vividly demonstrates the strength of gluten. Water, sourdough, and salt are present in both country breads, so they are equivalent in both preparations. However, to compensate for the physical properties of wheat flour, which become apparent during dough preparation and baking, a whole range of ingredients is required.

The technical side of gluten-free baked goods, bread, and pasta also reveals the weakness in terms of nutrition. When the concentrations of essential amino acids (including the cross-linker) are compared in Table 5.2, it becomes clear that the gluten-free alternatives are inferior in this respect.) for the example of gluten-free bread, the strength of gluten is also evident in this respect. With the exception of lysine and threonine, it is superior to even soy protein. Gluten leads in cross-linking properties, as the cysteine is located in the right places within the protein.

From a scientific point of view, it is not advisable for adults without gluten intolerance to unnecessarily avoid this high-quality protein. Furthermore, it should not be forgotten that all other ingredients such as psyllium fibers, xanthan, methylcellulose are completely nutrient-free. They only contribute as soluble dietary fibers.

These considerations once again show how closely the development of cultural techniques, the respective molecular composition of food, and human nutrition are interlinked. This strong coupling is completely ignored in most discussions on many issues. Therefore, opinion loops that end in ideologies, as shown in Fig. 5.1, are unfortunately inevitable.

Of course, it is easy to bake gluten-free, even without any grains: with starch, regardless of the food source, with protein, whether from eggs, peas, lupines, soy, or nut flours, or potatoes. Some leavening agents and, of course, hydrocolloids

5.5 Intolerances and Gluten-free Baked Goods

Table 5.2 Essentialand semi-essential amino acids from cereals, pseudocereals, and soy compared (values from nährwertrechner.de, except for teff). Gluten performs best with essential amino acids, rice and the knotweed plant buckwheat perform worst

	Gluten	Soy protein	Sorghum	Rice protein	Buckwheat	Teff
Amino acid	Proportion (mg)					
Isoleucine	**3240**	3105	462	**321**	336	500
Leucine	**5508**	4968	1151	587	**535**	1007
Lysine	1134	**4278**	226	260	499	380
Methionine	**1296**	897	207	**123**	154	430
Cysteine	**1539**	1104	128	**82**	181	240
Phenylalanine	**4050**	3105	384	342	345	700
Tyrosine	**2835**	2415	226	307	**181**	460
Threonine	1944	**2691**	344	266	354	510
Tryptophan	**810**	759	148	**68**	127	140
Valine	**3321**	3243	512	458	499	690
Arginine	2349	**5244**	305	396	717	520
Histidine	**1620**	1587	167	**109**	190	300

such as xanthan, methylcelluloses, guar gum for the respective water management and the crumb stability in the dough as well as during baking. Kneading, empiricism, and hoping. This works technically and technologically sometimes better, sometimes worse. However, the results have little to do with the cultural asset of bread. It is basically analog bread. For healthy people, consumption for prevention is unnecessary. Gluten-free bread offers a real advantage exclusively for the few truly sick people [77].

5.5.7 Super Grain Teff?

Teff, the dwarf millet from Ethiopia and other regions of Africa, also gained significant momentum due to the gluten-free wave. It originated (Table 5.1) like millet from the sweet grass tribe Chlorideae. Predestined for hot, dry regions that do not allow wheat and hardly any other genera, the Abyssinian lovegrass proves to be extremely adaptable and resistant and can be easily cultivated even in the mountains of tropical Africa, as well as in the valleys. Flatbreads can be baked and beers brewed from the gluten-free seeds. Furthermore, the straw is a welcome feed for the livestock of the local agriculture.

In fact, the Abyssinian lovegrass has a very broad spectrum of storage proteins [78], which allow it this flexibility in cultivation and growth. The processing of teff is simple, but falls into the category already addressed in Chap. 2 and 3: It is hardly possible to bake breads with a high rise, high water binding, or stable crumb, unless teff is mixed with other proteins. Teff on its own is therefore suitable for flatbreads, as is still practiced in the food culture of its cultivation areas today. This has also been demonstrated on a scientific basis in a series of systematic investigations [79]. In fact, its adhesive property is not very pronounced, as shown in Fig. 5.19 in comparison with other gluten-free breads and wheat. The leavening agent used in these model experiments was yeast.

Visually, this shows exactly what could already be guessed from Table 5.1. Unchallenged at the top is white wheat flour, followed by whole wheat, buckwheat, and barley. Their crumb structure is stable, and they withstand the baking process very well for various reasons (starch-protein ratio). Barley already shows a poorer binding on the surface, where the bread appears brittle. Because the temperature during baking (in the mold) is very high, the bread dries out and becomes brittle. Corn, for well-known reasons, hardly shows a stable volume, and sorghum quickly becomes brittle even inside the bread. The proteins cannot withstand the dough rise and high baking temperatures. The situation is similar with rice. Quinoa also shows significant baking errors due to the lack of highly elastic proteins. Teff breads are more towards the lower end. The crumb is uneven according to the distribution of baking temperature. The bubbles are larger at the top due to the higher evaporation rate than at the bottom, and a significant rise is not noticeable. These results were supported in the publication by both systematic investigations of the flow properties of the doughs and by scanning electron microscopy. Similar cohesion problems occur in the production of pasta from these ingredients

Fig. 5.19 Comparison of breads made from different flours, bottom right wheat bread; yeast was used as a leavening agent

[80]. Physicochemically, teff flour remains significantly inferior to wheat. But where does the hunger of Western society for teff come from?

As mentioned several times, the grain is gluten-free, rich in valuable fatty acids (with all the limitations that apply to all plants, see Chap. 2) and has a broad spectrum of minerals. "Free from" but including valuable macro and micronutrients, it must be some kind of superfood. Much to the chagrin of the population in Ethiopia, Eritrea, and other African regions, as teff is the most important staple food there. And it would be in the interest of sustainability, regionality, and ethics if it were eaten where it is grown, rather than being praised as a panacea for their unclear ailments by questionable personalities from film and radio.

Similar effects were already seen during the boom of quinoa from South America: The pseudo-cereal was exported to North America and Europe, and prices rose. Those who did not benefit were the farmers, neither economically nor culturally. The population lacked an important staple food on which they depended. Hunger and malnutrition were the consequences. This is fatal for countries and regions that can barely meet their nutrient needs with great difficulty.

A glance at Table 5.2 immediately shows: The often awarded baking properties of teff flour are overestimated. The supposed superfood cannot compete with gluten, at least in terms of amino acid composition. In the wheat belt of this world, wheat with its gluten remains superior to all other grains. Teff also fails to overcome the shortcoming of not being able to form long-chain, unsaturated fatty acids like all plants (see Chap. 2 and Fig. 2.6) Apart from α-linolenic acid (ALA), there are no omega-3 fatty acids that would justify exploiting Ethiopian farmers. Our ancestors knew this since the Neolithic period when they cultivated wheat without modern analytical methods. Otherwise, they would not have continued wheat cultivation, focusing their breeding efforts on other, more valuable foods to ensure survival.

5.6 Carbohydrates: Structure and Digestion

5.6.1 Complex Carbohydrates

The increased motivation to consume whole grains comes, among other things, from the idea that "slow and complex carbohydrates" cause a slower rise in blood sugar than "simple carbohydrates" such as glucose or sugar, and that this prevents the pancreas from releasing insulin too quickly, resulting in longer-lasting satiety than with white bread. The glycemic index (GI) serves as a measure for this, as discussed in Sect. 4.8.4 and Fig. 4.25. Glucose can be absorbed immediately, the glucose level in the blood (serum) rises rapidly and drops quickly after insulin is released. In the case of complex foods, however, the blood sugar level rises more slowly, is less high, and remains at a higher level for a longer time, which is supposed to be an indirect measure of satiety. These ideas emerged in the 1980s [81], when researchers began to determine glucose levels in response to food [82]. For example, differences in the rise of blood sugar were observed when comparing a minced meat dish once with mashed potatoes ("fast") and another time with bean puree ("slow") [83], leading to conclusions about the interaction with dietary fibers, packaging of starches, and complex carbohydrates during digestion.

5.6.2 Whole Grain Flour during Intestinal Passage

Following the explanations of the molecular aspects of successive digestion (see Fig. 4.36), this assumption sounds plausible. The starch in beans is cooked, and the structure of amylopectin has melted, but it remains packaged in cells and is not immediately accessible to amylases. In potatoes, however, the starch granules are free and burst during cooking. Amylose and amylopectin emerge and cannot resist the rapid access of enzymes.

How does this apply to bread? The variety of breads is immense, so it must be precisely defined which breads are being compared. Comparing typical (German) baked goods of different flour milling degrees with breads made from whole grains, astonishing results regarding the glycemic index emerge [84], which are summarized in Fig. 5.20.

This result is surprising in that the differences between whole grain and milling degree of the whole grain hardly stand out, especially when considering the statistical deviations, which are not included in Fig. 5.17 for the sake of clarity. Only the glycemic index of whole grain rye bread with visible intact grains and sunflower seeds was identified as lower in the publication. Both the whole grain spelt wheat and the rye wheat sourdough bread (not shown for the sake of clarity) have only a medium glycemic index, while conventional pretzels have a comparatively high glycemic index. Remarkable, however, are the very close values of the breads, which hardly differ. If the error bars are taken into account, practically no difference in the course of the measured values can be determined.

This is also evident in other, independent experiments. No differences in terms of the glycemic index can be detected between breads made from whole grain meal, whole grain flour, and white bread within the margin of error, as can be seen in Fig. 5.21. Consuming whole grain baked goods does not necessarily lead to slow carbohydrate digestion. With regard to the glycemic index and the insulin load on the pancreas, it is completely irrelevant whether one eats whole grain or white flour. Only with pumpernickel does there seem to be a slight advantage.

From a physical point of view, however, this result is less surprising, as starch remains starch. As soon as the starch is free and gelatinized, and thus also in the ground and crushed grain, the speed of digestion is determined only by enzyme activity. Since the starch is located in the flour body of the grain, the amount to be digested is the same, regardless of whether whole grain or white flour is present. Only whole grains do not allow the enzymes to access the starch. Then the glycemic index is somewhat lower: The starch enclosed therein remains indigestible.

The assumption of "slow, complex carbohydrates" thus turns out to be not universally valid. The carbohydrates in wholemeal flour and sifted white flour are

5.6 Carbohydrates: Structure and Digestion

Fig. 5.20 Changes in the glycemic index (**a**) and insulin response (**b**) when eating various whole grain breads. Blue: reference glucose, brown: pretzel, red: spelt bread, green: whole grain rye bread, finely ground, yellow: whole grain spelt bread with whole grains

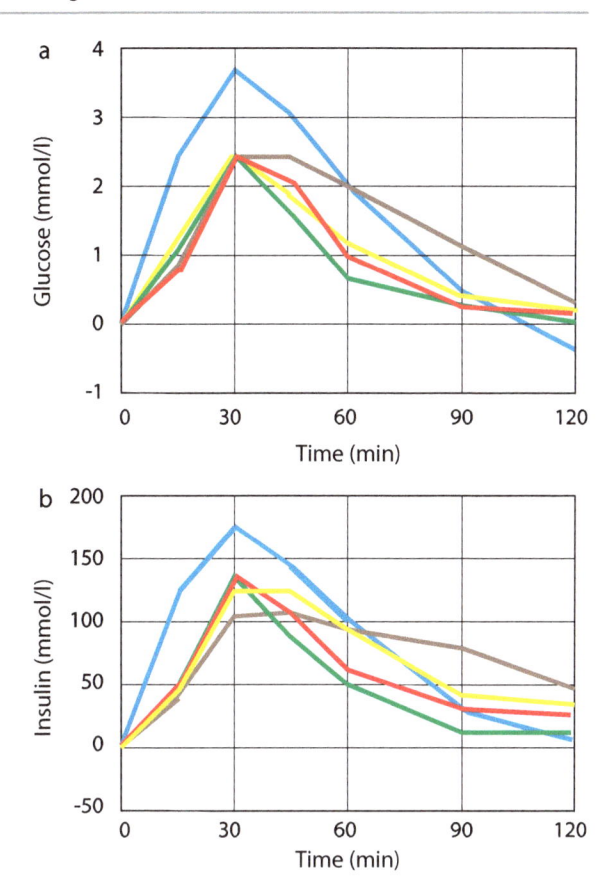

equally simple or—depending on the point of view—equally complex. What is important for digestion is merely whether amylopectin is present in a swollen and melted form, making it accessible to amylases easily.

5.6.3 Dietary Fiber

This does not mean that whole grain bread should be avoided for this reason. Its fiber content is indeed valuable and the other side of the coin of FODMAPs and the ATI. For people who tolerate whole grains and do not experience any discomfort from them, dietary fiber and enzymatically indigestible components contribute to the training of the microbiome and the peristalsis of the intestine. Typical fermentation products from dietary fiber, such as butyrate, appear to be anti-inflammatory [85], provided that the experiments carried out under extreme laboratory conditions in this study can be transferred to real-life situations at all. Initial work in human microbiology, however, suggests this [86].

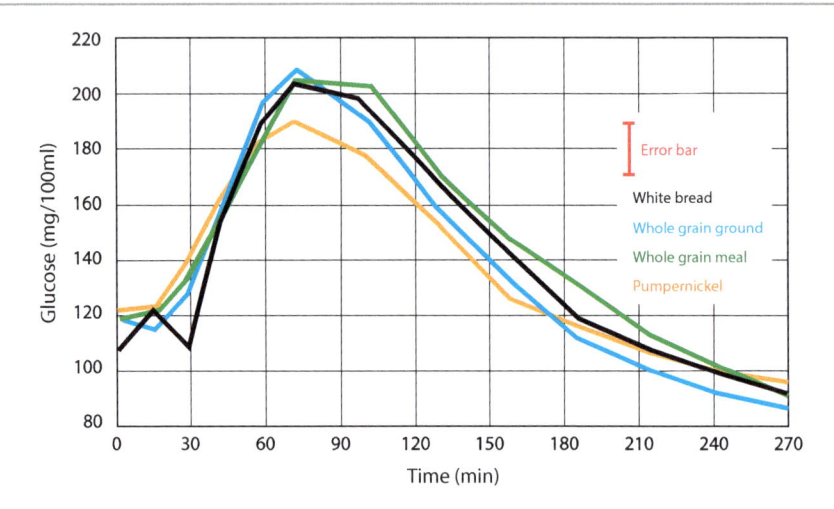

Fig. 5.21 Glucose in the blood after consuming white bread (black), bread made from finely ground whole grain wheat (blue), wheat whole grain meal bread (green), and pumpernickel (yellow). The red bar is the typical statistical measurement error, which affects all curves. Since whole grain flour, whole grain meal, and even white bread are within the margin of error, there is no statistical relevance for systematic differences in the glycemic indices

Secondary plant compounds, fatty acids from the germ, minerals, and other micronutrients, etc., may all be somehow "good" and "positive" with unclear facts. However, they are only incorporated into the overall diet to a small extent by consuming the whole grain. As a universal panacea, as recommended by some nutrition experts, whole grains are not really suitable. The side effects, however, can be clearly described at the molecular level: At least ATI and FODMAPs contribute to discomfort in some people with the corresponding genetic predisposition. The motto "healthy is what tastes good" still carries a germ of truth. Especially when it comes to natural, artisanal products and home-cooked meals.

5.7 Nitrate and Nitrite

5.7.1 Nitrate, Nitrite, and Stoked Fear

Nitrite and Nitrate are other keywords with high potential for irritation, which, for example, fuel fears of cured meat or nitrogen-contaminated vegetables (see Fig. 4.4). These nitrogen compounds are associated with cancer risks. However, there is no conclusive evidence for this, even with intensive searches in scientific databases. Many popular statements are based on assumptions and conjectures. Nevertheless, the connection between nitrate and cancer development has become entrenched in people's minds, leading to regular warnings about cured meat and sausage products or even supposedly healthy vegetables such as beetroot, radish, lamb's lettuce,

lettuce, or spinach. As a result, the terms nitrate, nitrite, and especially nitrosamine, which can form under certain conditions, trigger fears without the chemical relationships being recognized. However, there are many positive aspects of these nitrogen compounds. In cases of cardiovascular problems and especially angina pectoris, "nitroglycerin lingual spray" is given for emergencies, which consists of glycerol trinitrate (better known as nitroglycerin). On a glycerol molecule, which we know from fats, the triacylglycerols, there are three nitrogen dioxides, NO_2. When this is split off, nitrite (NO_2^-) is formed, whose vasodilating effect has long been known and is still used in medicine today [87, 88]. The formation of nitrates and nitrites is an everyday occurrence in metabolism. They are intermediate products in cell metabolism and are therefore not toxic or even carcinogenic.

Moreover, nitrate (NO_3^-) is abundant in nature and is the most important nitrogen source for plants. Without nitrate, they cannot grow, let alone build their amino acids , as nitrogen (N) is present in every amino acid. Therefore, nitrate is used as a fertilizer in agricultural production, which, with intensive use, can lead to high nitrate concentrations in groundwater and thus also in drinking water. This is not a question of artificial fertilizers, but—as is well known—also of the classic and most original form of fertilization using nitrogen-rich agricultural waste products such as manure and animal excrement, which have been carried onto fields for thousands of years.

Many plants also store nitrates in their leaves and roots as a reserve for their metabolism. In this way, "natural" nitrate enters humans through food and adds to the "artificial" nitrate from sausage, ham, and cheese. There is no difference between natural and artificial here. The relevant molecules and corresponding molecular processes are identical, and so far, none of them are harmful. The largest nitrate source for vegetarians is vegetables and plant-based food, some of which, such as leafy vegetables, radish, radishes, and beetroot, can store a lot of nitrate due to their metabolism. However, in previous studies, nitrates ingested through food have been completely inconspicuous. Even with the consumption of meat products and animal products containing curing salt, no connection with cancer development has been observed or proven so far [89]. This should not be surprising, as nitrate and nitrite are involved in every cell metabolism. Even with the consumption of cured products, no negative effects can be seen in non-vegetarians [90].

The hypothesis of the carcinogenic activity of nitrogen compounds is based on the possibility that nitrite can form nitrosamines in an acidic environment— potentially also in the stomach with its low pH values. Some representatives of this group of substances are classified as highly carcinogenic, not because of direct molecular interactions, but because of the action of nitrosamines over longer periods. For the reaction of nitrite compounds to nitrosamines in an acidic environment, as the name suggests, amino acids, amines, are necessary, which are present in the food itself. Therefore, the hypothesis seems to find logical support: curing salts and nitrate from meat and vegetables, plus the cleavage of proteins ingested simultaneously with the food in the stomach via acid-tolerant enzymes, pepsin, and some free amino acids react with nitrite to form the dreaded nitrosamines, especially in processed cured meat, raw sausages or ham, because cooking and fermentation already cause some hydrolysis of proteins during the processing process.

5.7.2 Nitrate, Nitrite—A Search for Traces

In the search for evidence for this hypothesis, however, completely different mechanisms are found, and the obvious does not always occur. For ten years, the suspicion has been increasingly confirmed that the bioactive advantages of nitrate dominate its disadvantages [91]. And even studies involving nitrate critics show the physiological importance of even the "artificial" nitrate [92].

In our body, nitrate comes from two main sources: from the diet (mainly green leafy vegetables and cured products) and from the oxidation of NO-synthase-derived NO, as shown in a very simplified way in Fig. 5.22. The nitrate ingested through food is rapidly actively absorbed by the salivary glands and excreted again via saliva. In the mouth, the nitrate is reduced to the more reactive nitrite anion (NO_2^-) by bacteria present there. In blood and tissue, nitrite is then further metabolized to a series of bioactive nitrogen oxides, including NO, as well as sulfur-containing S-nitrosothiols (thionitrites) and other nitrated compounds. The reduction of nitrite to NO is catalyzed by several proteins and enzymes, including deoxygenated hemoglobin/myoglobin and xanthine oxidoreductase (XOR). Therefore, nucleotides are also actively involved in this process, as in purine metabolism (Fig. 4.10).

A whole range of positive effects on human metabolism results from these bioactive substances (Fig. 5.22). Nitrates and their derivatives have different effects and therefore show a wide range. As already mentioned, nitric oxide NO, which is endogenously produced from the amino acid L-arginine and oxygen by NO synthases, is a signaling molecule involved in cardiovascular and metabolic regulation. This includes lowering blood pressure, improved endothelial function, increased exercise performance, and containment of metabolic syndrome. These processes thus counteract diabetes. Furthermore, it is shown how the accumulation of fat in the liver fat cells slows down under the influence of S-nitrosothiols, and inflammation values improve.

Various reaction products of nitrites increase the number of brown fat cells [93]. This refers to a class of fat depots characterized by high energy consumption and heat production for thermoregulation. Brown fat essentially serves as an internal heater. In this context, it has been shown that brown fat also improves insulin production [94]. At the same time, the concentration of free triacylglycerols in the blood decreases. Positive effects can also be observed in the function of the pancreas: Insulin production is stimulated by better blood circulation of the so-called Langerhans islet cells. Even in the muscle cells of the skeletal muscles, nitrates have a positive effect. Glucose (and thus energy) can be absorbed more quickly, which directly links to muscle performance and is therefore of interest in sports medicine.

5.7 Nitrate and Nitrite

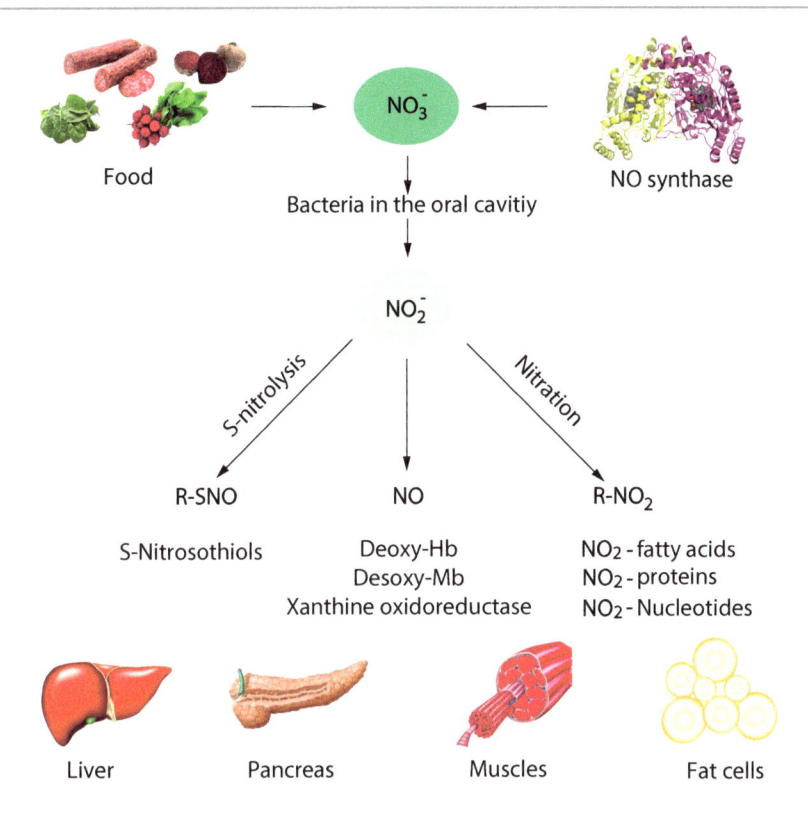

Fig. 5.22 The intake of nitrate and nitrite leads to a whole range of bioactive compounds that have a positive effect on liver function, the pancreas, muscles, and fat cells [94]

From these results, molecular medicine even conceives active substances that specifically produce this stimulant from the various NO compounds (see Fig. 5.22, bottom row). This is obvious, and the advantage is evident, as they are modeled on the physiological processes in the cells [95].

5.7.3 Nitrosamines

What role do carcinogenic nitrosamines play? They are hardly detected in relevant concentrations in studies. Especially not in studies that examined normally healthy and balanced eaters. The mixed diet practiced for millennia—raw, cooked, and fermented—provides sufficient vitamin C and E as well as trace elements such as selenium, which together with secondary plant compounds strongly inhibit nitrosamine formation. Species that do not practice this form of nutrition, including monkeys on which similar experiments were conducted, do not have this protective effect. In certain cases, cancer diseases may then arise, which can possibly be traced back to nitrosamines. *In-vitro* experiments, on the other hand, show a carcinogenic effect of nitrosamines at concentrations irrelevant for daily consumption. Furthermore, one essential point seems to be increasingly confirmed: Nitrate and Nitrite rarely occur in isolation; no one ingests pure nitrate or nitrite. Each food contributes countless other components, especially in a complete meal. Laboratory experiments in animal models therefore suffer from a lack of comparability. Together with a whole range of food-typical ingredients, the formation of nitrosamines is apparently suppressed *in vivo* [96], as summarized in Fig. 5.23.

The results of the studies conducted so far can therefore be summarized under quite positive aspects. A real danger does not arise from the Nitrate in the amounts usually consumed. Here too, the rule applies: Those who consume excessive amounts of cured food and nothing else are guaranteed to have a higher risk. Provided that they do not have other problems due to unbalanced nutrition beforehand. According to the current state of research, it is irrelevant from which source the nitrate originates. Even the much-maligned curing salts used in the food industry, such as potassium nitrate or calcium nitrate, which were used for preservation in higher doses, undergo identical metabolic processes. It is therefore by no means the case that Homo sapiens only ingested nitrate and nitrite since the industrialisation of food. Since the Neolithic period, humans have consumed fertilized plants, i.e., since people practiced agriculture and animal husbandry. Later, in the Middle Ages, natural preservation methods using saltpeter were added [97]. These nitrates obtained by saltpeter boilers allowed for systematic preservation. The historical saltpeter was, by the way, far unhealthier than all industrial nitrates, as it contained lead and barium components in its mixture. In contrast, the nitrates that

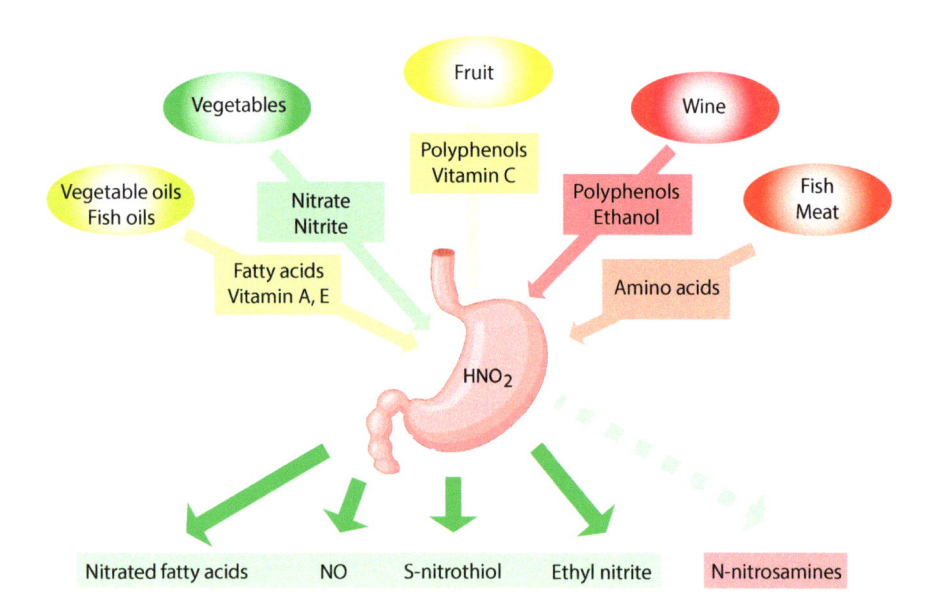

Fig. 5.23 In addition to vitamin C, a whole range of other food ingredients counteract strong nitrosamine formation

are still permitted today and which we develop irrational fears of are downright high-quality preservatives. Our diet is molecular, and our physiology reacts with functional biomolecules and biochemical processes that are controlled exclusively by molecular interactions.

Even if Nitrate and Nitrite did not turn out to be as harmful as supposed, this fact is of course not a carte blanche for over-fertilizing fields and arable land. It is certainly not a carte blanche for the thoughtless spreading of manure and feces from unreasonable and unsustainable mass animal husbandry.

5.8 Raw Food Diet: Physical-Chemical Consequences

5.8.1 Only Raw Food is Healthy? Stories of the *Homo Non Sapiens*

Raw foodists believe that valuable ingredients in food are already destroyed when heated above 42 °C. The food would thus already be cooked dead. Therefore, almost every food is consumed raw, although the benefits of this diet are not convincing from a scientific point of view. The raw food form is thus committed to the upper part of the culinary triangle (see Fig. 2.29), ignoring the other basic pillars of human nutrition. Cultural techniques that have emerged from fire are rejected, while fermentation techniques are allowed. Fermented foods are therefore considered raw, even if this does not correspond to the facts. The raw food concept is thus based on the strict exclusion of the temperature range above 42 °C. Where this strict temperature limit actually comes from is unclear from a scientific point of view. As already mentioned in Chap. 1, robust set of teeth and endurance-oriented chewing muscles are necessary for the good utilization of raw food in order to extract as many macro and micronutrients as possible from raw vegetables, roots, and berries already in the mouth (see Fig. 1.18), because the bioavailability is low in raw food for physical-chemical reasons alone. Even in such simple systems as (raw) milk and its whey protein, it became apparent how much stronger the biological availability of amino acids through the enzymatic digestion of β-lactoglobulin increases with heating (including pasteurization). It always depends on the packaging, the structure of proteins or carbohydrates, the cell structure, and the physical environment as to how exactly digestion takes place. For only the respective food matrix and the composition of the food pulp during intestinal passage define the biological availability in an exact manner.

All foods consist of macro- and micronutrients (Fig. 5.24), which can also be found in corresponding tables; however, as already explained, this says nothing about their biological availability. These tables do not indicate whether these nutrients are also physiologically usable. As mentioned in this chapter, cooking increases the biological and physiological availability of proteins and thus essential amino acids. Many of the micronutrients anchored in plant cells and cell walls can only unfold their physiological effects through heating and grinding.

Fig. 5.24 The components of all foods. The presence of the ingredients says nothing about their thermal properties or their biological availability

Proteins

Vitamins I   Vitamins II

Fat   Secondary Plant substances Polyphenols   **Water**

Flavors   Trace elements
Flavors
Minerals

Polysaccharides

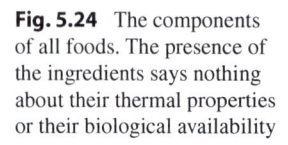

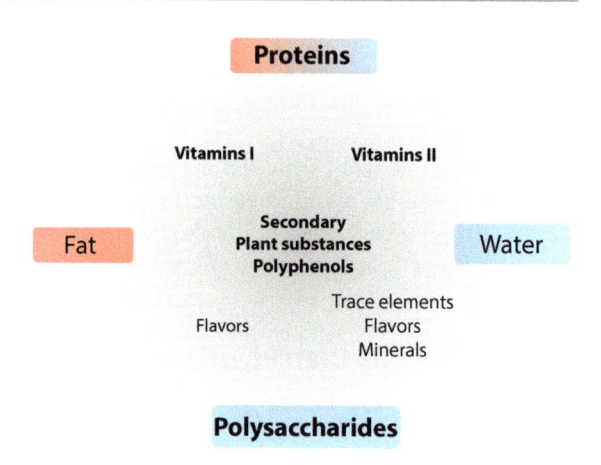

Well-known examples are the carotenoids in carrots or tomatoes. This immediately shows that the hypothetical 42 °C limit has no general validity. Daily experience in the kitchen and at the stove teaches this as well. The cooking temperature of many proteins present in fish muscle is very low at 35 °C, while chicken meat requires between 60 and 78 °C to show significant denaturation effects. Vegetables must be heated to at least 78 °C; below 78 °C, vegetables show neither structural nor textural changes. The cells of the vegetables remain firm and the vegetables crisp, even if they are kept at this temperature for an extended period. Apparently, there is a clear structure-temperature relationship to these phenomena, which can only be recognized at the molecular level. This is precisely the incidental task of macronutrients, proteins, and carbohydrates, as well as indigestible dietary fibers.

5.8.2 Structure and Mouthfeel—Macronutrients and Dietary Fiber

All natural foods consist of water, fat, proteins, and polysaccharides such as starch, as well as other carbohydrates. Since fat and water do not mix, a variety of emulsifiers such as lecithin (phospholipids) are added. The ratio of these molecular components roughly defines the structure and properties of all foods. Protein-rich foods such as eggs, meat, and fish, for example, contain a high proportion of muscle protein, connective tissue protein, and depending on the species, fat and about 80% water. Polysaccharides are bound in membrane proteins as glycoproteins. The consistency of a piece of meat, a fish fillet, or a crustacean is primarily defined by the interaction of these ingredients, but especially by the structural proteins of the muscles. This consistency defines the characteristic mouthfeel and bite when eating carpaccio, sushi, or raw seafood. Vegetables have a much lower protein content. Instead, cell materials (insoluble and soluble dietary fibers) such as cellulose, hemicellulose or pectin dominate in fruits, vegetables, or herbs, as

5.8 Raw Food Diet: Physical-Chemical Consequences

well as carrageenan, alginate, or agarose in algae or other marine plants. The high water content of often over 90%, i.e., the taut tension of the plant cells, also determines the mouthfeel experienced with crunchy vegetables. Vegetarians appreciate crunchy vegetables, raw food enthusiasts who do not reject animal foods appreciate wafer-thin carpaccios with a melting texture, sushi lovers the surprisingly crunchy texture of raw fish, which is slaughtered using the Ikejime method, which has a significant impact on the texture [98]. All of this has exclusively molecular causes. This will be discussed in more detail in Chap. 6.

When heating both protein-rich and cellulosic foods, the texture and thus the mouthfeel change. This change from "raw" to "cooked," usually triggered by temperature changes or the action of acids, alters the texture through molecular rearrangements in these structural molecules: proteins denature, cell material softens. Since the foods have different compositions, as can be seen in fish, meat, vegetables, or fruit, the "raw" state must be defined in a way adapted to the molecular properties. This is clearly visible in Fig. 5.25. There, the denaturation ranges of the structural proteins in animal products and the proteins and polysaccharides of cell material in vegetables are shown for typical food groups.

The definition of "raw" therefore requires not only a cultural-scientific perspective (Fig. 2.31), but also a clear, molecular definition from a natural-scientific perspective, which cannot be captured by a hypothetically postulated temperature limit.

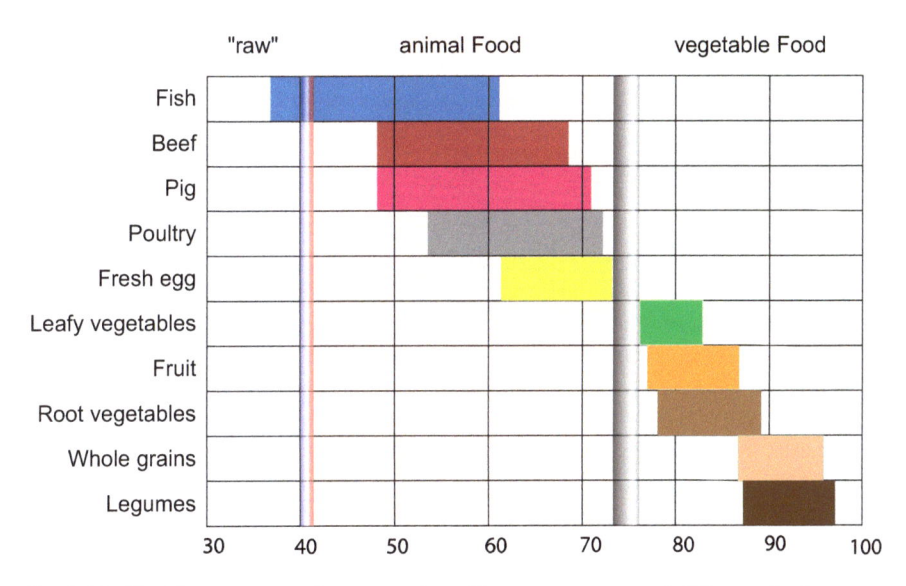

Fig. 5.25 Significant texture changes and cooking temperatures of different food groups. Animal and plant foods (LM) are clearly separated. The upper limit for animal foods is formed by ovalbumins in the egg at 71 °C and the structural protein actin in muscle cells. Cell materials in plants only begin to change significantly from about 80 °C. At the upper limit of the usual raw definition at 41 °C, some muscle proteins of fish denature

5.8.3 Micro-nutrients

Foods that are heated are considered "dead foods." For the raw food movement, the changes in micro-nutrients are therefore one of the most important criteria. This general statement is not scientifically valid, especially with regard to trace elements and minerals. A separate consideration is necessary, as not all vitamins change under the influence of heat. Ascorbic acid (vitamin C) is particularly critical.

More important, however, is the provision of nutrients through heating and thus the reduction of energy expenditure in digestive processes (see Sect. 1.16.1). All these evolutionary advantages are less present in a purely raw food diet. Part of this can be compensated for by molecular cooking processes such as fermenting and germinating at low temperatures [99]. This indirectly also applies to many micro-nutrients and trace elements. Calcium is functionally often bound in cell polymers such as pectin or alginates as ionic bonds. The calcium ions are only freely available and biologically active when they have been sufficiently heated, higher than ionic binding. The same applies to magnesium and other divalent and higher-valent ions that fulfill biological functions due to their charge. When consuming raw food, these components are only available to a limited extent. The same also applies to sulfur, which is found, for example, in the amino acid cysteine.

Similarly, this applies to most trace elements except sodium, potassium, iodine, chloride, and fluoride, which as monovalent ions have no binding properties to structural proteins or polysaccharides, but merely adjust the ionic strength with different ion radii and are necessary for the transport of charges through membranes. Trace elements such as chromium (Cr), cobalt (Co), iron (Fe), copper (Cu), manganese (Mn), molybdenum (Mo), selenium (Se), zinc (Zn), or silicon (Si) occur in the structure formation and complexation of metal-binding proteins. These metal atoms are often trapped in the centers of proteins and must first be exposed to become biologically available.

5.8.4 Temperature and Nutrients

The temperature limit is justified exclusively by the destruction of nutrients, mostly vitamins and enzymes. Various molecule types are lumped together with a blanket temperature limit, which is inadmissible from a physicochemical point of view. With enzymes, this argument is of no scientific significance, as enzymes are nothing more than proteins with special folds that may also carry metal atoms. Most of these enzymes are denatured by the low pH value in gastric juice, as the isoelectric point is exceeded. Therefore, enzymes contribute only marginally to the macronutrients in foods, and there is no measurable gain for nutrition, especially since most enzymes in plants have tasks that humans do not face.

The situation is more complicated with vitamins, as summarized in Table 5.3. Vitamin C is shown to be a very sensitive vitamin, while the loss or oxidation of all others is limited. The statement that all foods are destoyed when heated above 42 °C is not tenable in terms of the facts.

5.8 Raw Food Diet: Physical-Chemical Consequences

Tab. 5.3 Sensitivity of vitamins to oxygen (oxidation), light, and temperature. The only vitamin that is damaged under all influences is vitamin C

Micronutrient	Solubility	Loss due to exposure to		
		Oxygen	Light	Temperature
Vitamin A	Fat	Partial	Partial	No
Vitamin D	Fat	No	No	No
Vitamin E	Fat	Yes	Yes	No
Vitamin K	Fat	No	Yes	No
Thiamine	Water	No	No	>100 °C
Riboflavin	Water	No	Partial	No
Niacin	Water	No	No	No
Biotin	Weak in water	No	No	No
Pantothenic acid	Yes	No	No	Yes
Folic acid	Yes	No	No	>80 °C
Vitamin B_6	Yes	No	Yes	No
Vitamin B_{12}	Yes	No	Yes	No
Vitamin C	Yes	Yes	Yes	Yes

The degradation of vitamin C at higher temperatures, however, strongly depends on the pH value, i.e., the acidity of the food. The degradation of vitamin C at low pH values, i.e., in the strongly acidic range, slows down significantly, as shown in Fig. 5.26. In the alkaline range as well, although this is less relevant for culinary purposes and more for food technology processes.

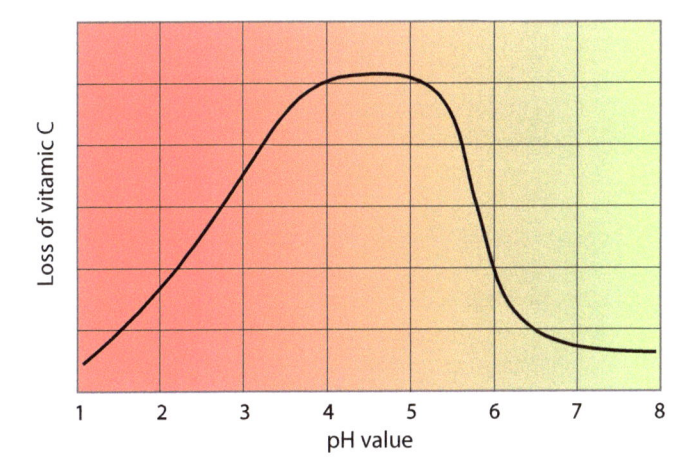

Fig. 5.26 The degradation of vitamin C (in test solutions) depending on acidity. In the red area, vitamin loss is highest

In the range of pH values between 3 and 5 (Fig. 5.26), the degradation is highest. Vitamin C breaks down very quickly into less effective compounds. At lower pH values below 3, the degradation is significantly delayed [100], even during cooking, as is known from sauerkraut, for example. Cooked sauerkraut still contains significantly more vitamin C than cooked (unfermented) cabbage. This fact becomes relevant for all lactic acid-fermented products. During fermentation with lactic acid bacteria, pH values between 5 and 4 are reached after about four weeks of fermentation. Vitamin C also breaks down much more slowly in fruits with higher acid content. This leads to exciting insights for "pseudo-raw food": marinating at low pH values hardly changes the vitamin C content and adds new flavors. This is also good news for fermentation. Since the pH value drops significantly during fermentation, the vitamin C content hardly decreases even during longer storage of fermented vegetables.

5.8.5 Interfaces Between Enzymes and Vitamins

When vegetables are processed without heating, the interface between enzymes and vitamins becomes relevant. Many enzymes are located in most cases in the cell membranes directly underneath the cell walls; there they are anchored and wait for their tasks. When cutting, grating, or pureeing, cells are destroyed by knives, and the enzymes are released. This includes, for example, ascorbic acid oxidase in fruits and vegetables, which converts vitamin C, ascorbic acid, into dehydroascorbic acid. Both have a high antioxidant potential, but dehydroascorbic acid is very unstable and quickly reacts to form new chemical compounds, the benefits of which are negligible or no longer present compared to vitamin C. The complexity of the processes is illustrated by the example of broccoli, which has a considerable vitamin C content. However, the bioavailability and stability of vitamin C depend heavily on the history and treatment of the vegetables. Nevertheless, raw processing is not always advantageous. If the ascorbic acid content, the native form of vitamin C, is determined in freshly chopped, raw broccoli florets, it decreases rapidly. Dehydroascorbic acid is quickly formed from ascorbic acid. In blanched broccoli, however, this does not happen. There, the vitamin C remains as ascorbic acid.

The reason for this is the high activity of the enzyme ascorbic acid oxidase, which rapidly converts native ascorbic acid into the less stable dehydroascorbic acid. The enzyme is firmly anchored in the walls of plant cells, while ascorbic acid is located inside the plant cells. Both are thus separated from each other. Only when cutting, crushing, or chopping the broccoli florets—and thus destroying the cell walls—are the enzymes released. At the same time, the cell sap with the native vitamin C pours out from the inside of the cells. When enzymes and vitamin C come close, the ascorbic acid is oxidized. This happens very quickly, and in a short time, the native vitamin C is converted into the less effective dehydroascorbic acid.

5.8.6 Blanching, Enzyme Inactivation, and Vitamin C

To preserve vitamin C in its native form, the enzyme must be inactivated. As always in kitchen practice, this is done by briefly blanching in boiling water. This is already objectionable in the pure doctrine of raw food. However, systematic experiments show that enzymes can be inactivated between 55 and 70 °C. When blanching broccoli florets at 65 °C, most of the enzymes are already inactivated, and the vitamin C is largely preserved in its native form. When blanched at 70 °C, the oxidation of ascorbic acid is largely suppressed. Nevertheless, no structural changes occur during temperature treatment below 78 °C. This temperature is not sufficient to soften the hard cell walls. The raw texture and, as Tab. 5.3 shows, all vitamins, are largely preserved, and the enzymes are inactive (Fig. 5.27). It is therefore obvious to call the temperature range between the arbitrarily assumed raw food limit of 42 °C and the softening temperature of the cell material, which starts at 78 °C depending on the vegetable, "pseudo-raw". In this temperature range, vitamin C is largely preserved as highly active ascorbic acid [101].

The preservation of vitamin C in its native form can be analytically demonstrated. From Fig. 5.28, the change in vitamin C content is evident. In the upper part of the figure, it can be seen that when chopping raw broccoli florets, all ascorbic acid (shown in dark gray) is converted into dehydroascorbic acid (light gray). Only when the florets are heated does the degradation begin to slow down slightly from 50 °C. From 70 °C, the enzymes are largely inactivated. Then the vitamin C content is better preserved.

When the broccoli is blanched at 70 °C, no conversion of ascorbic acid occurs. This example clearly illustrates the misconception that heating would only bring health disadvantages. In many cases, the opposite is true. Blanching offers an advantage here: ascorbic acid, with its high radical scavenging potential, is practically preserved.

Fig. 5.27 Blanching the broccoli at around 60 °C before cutting, chopping, and shredding inactivates enzymes that break down vitamin C into a less stable form. The area between 42 and 70 °C is called "pseudo-raw". The raw texture is preserved, only the enzymes are inactivated (black curve). The gray curve indicates the weaker enzyme activity in the broccoli stems

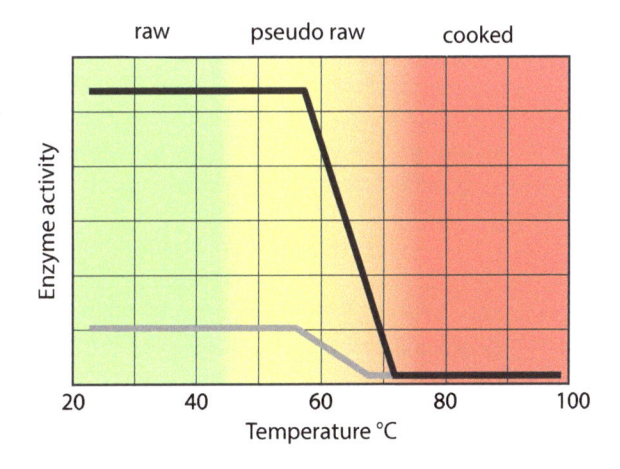

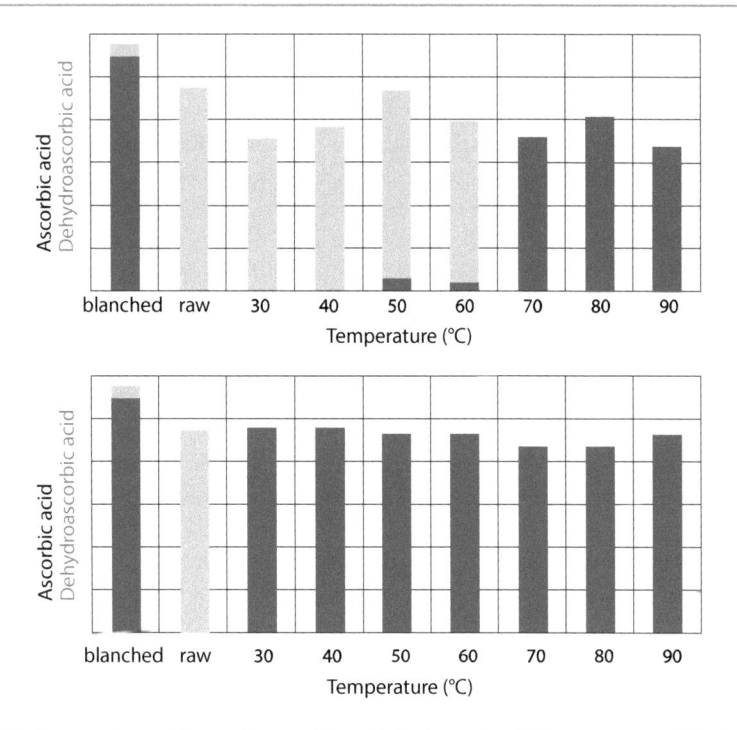

Fig. 5.28 In raw chopped broccoli, ascorbic acid (dark gray) quickly converts to dehydroascorbic acid (light gray) (top). If the broccoli is blanched between 70 and 80 °C, ascorbic acid is preserved even during cooking

In the diverse gourmet cuisine, such aspects are secondary, but it is still important to keep an eye on the "pseudo-raw" temperature range. Moreover, raw broccoli florets are small delicacies. They provide subtle sulfuric aroma notes and contribute a slightly bitter taste to delicious plates with many different components. This unique taste component, texture, and color are reason enough to consciously celebrate broccoli or other vegetables raw or pseudo-raw more often.

Blanching remains prohibited despite the obvious advantages of enzyme deactivation and the resulting benefits for raw food nutrition. However, the microwave offers a method for effective enzyme deactivation. A brief exposure to high power has the advantage of rapid, short-term heating to 60–70°C inside the water-rich fruit and vegetables, so that the enzymes are deactivated, but the structural polymers such as pectin and hemicellulose are not altered. Vegetables treated in this way can then be processed into raw purées without significant vitamin C loss, without enzymatic nutrient loss, and without significant enzymatic browning.

This also helps to avoid enzymatic browning, such as in apples. If a whole, unpeeled apple is placed in the microwave for five to ten seconds at 1000 W, most enzymes are deactivated, but the apple remains raw in terms of its structure. It can then be processed into a raw purée with a clear, unadulterated apple taste without

5.8 Raw Food Diet: Physical-Chemical Consequences

browning and without the addition of lemon juice or ascorbic acid. This is excellent with raw salmon or carpaccio. This method can also be applied to all fruits and vegetables that turn brown and unattractive after cutting and the associated release of enzymes: Jerusalem artichokes, salsify, or artichokes. Polyphenols and other substances oxidize in the presence of enzymes released from cell walls and cell membranes. The resulting enzymatic browning substances absorb light at different wavelengths, causing the cut surfaces to turn brown or gray quickly. Despite the heat pulse, the texture remains crisp and pseudo-raw, and the micronutrients are preserved.

These results also have consequences for the preparation of smoothies. In this case, the application of this method would only be consistent. Modern high-performance blenders, which are specifically designed for high levels of crushing to make as many nutrients as possible biologically available from fruits and vegetables, release particularly large amounts of enzymes, as a high number of cells are damaged. The idea of pureeing fruits and vegetables in the morning to take them to work and then drink them as a "healthy raw food" for lunch is not vitaly correct. During storage, the processes described above take place, enzyme activity is high, and vitamin C is degraded. In such cases, it is indeed advisable to inactivate enzymes beforehand and, instead of the usual raw food smoothie, take a "pseudo-raw" smoothie to work.

5.8.7 Pseudo-Raw Enjoyment—A Matter of Food, Time, and Temperature

These considerations allow for a whole new way of enjoying raw and pseudo-raw foods, adapted to the respective types of meat, fish, and fruits and vegetables.

Let's start with the vegetables. The recently gained insights can be summarized in a consumption diagram (Fig. 5.29). The horizontal time axis is divided into a time range and a temperature range. It becomes apparent that in both parts, a distinction must be made between raw and pseudo-raw, which must be coupled with enzyme activity and temperature.

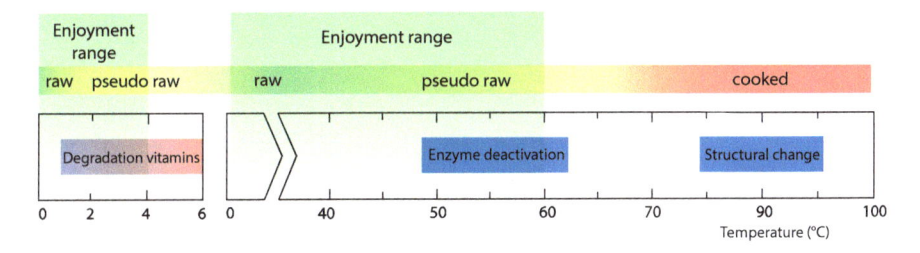

Fig. 5.29 Time and temperature ranges for the enjoyment of vegetables and fruits, taking into account the ascorbic acid content and the change in the structure of plant cells

In the case of animal and thus protein-rich foods, the situation is significantly different. Nowadays, according to Lévi-Strauss, the consumption of raw meat is always preceded by a whole series of cultural acts (see Sect. 2.13). The livestock is slaughtered, wild animals are hunted and shot. After bleeding, the meat is allowed to "mature". During all these post-mortem steps, the meat changes. Rigor mortis is resolved. Through the appropriately long maturation in the cold storage, flavors develop, the water of the meat juices evaporates, the pH value decreases, taste is shaped, aromas form and intensify. The complicated fermentation processes of maturation take place, as well as the enzymes present in the sarcoplasm change the hard muscle structure. The muscle meat becomes tender and tasty and is now suitable for raw consumption, such as very thinly sliced slices. Because of the many processing processes, it is more precisely referred to as pseudo-raw. Raw in the sense of cultural science would only be the warm meat of freshly killed or freshly slaughtered animals. However, with today's chewing muscles, this would hardly be manageable. Therefore, only muscle meat, such as from the loin or back, is suitable for pseudo-raw consumption. Connective tissue-rich meat from the shoulder or leg is hardly suitable for raw or pseudo-raw enjoyment.

The pleasure of pseudo-raw meat (tartare) begins only after a certain maturation time in the cold storage. But even under a 35–40 °C warm pass in the restaurant or under heat lamps to emphasize the taste, the meat does not change its structure. The first protein of the sheer muscle to denature is always myosin. In beef or lamb, the two most important candidates for raw consumption as tartare or carpaccio, myosin begins to change its structure between 48 and 52 °C depending on meat maturation and pH value, while collagen and actin remain completely unchanged. However, the meat already changes its bite in this temperature range. It becomes more gel-like, even more tender, and very pleasant to eat. Since the vitamins in the meat are not very temperature-sensitive, the range up to 52 °C can also be added to the enjoyment area of pseudo-raw. This is summarized in Fig. 5.30.

For the enjoyment of pseudo-raw beef, the factor of time is crucial. Only the meat maturation process results in a good taste and creates a pleasant, tender

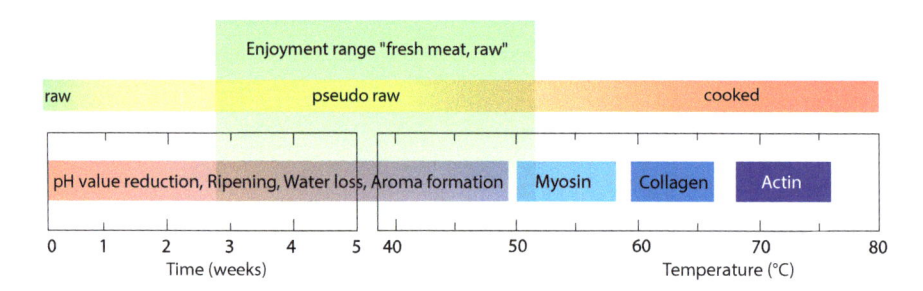

Fig. 5.30 The enjoyment area of pseudo-raw meat. (Over time, maturation takes place, and due to temperature changes, structural changes occur)

texture. Meat can thus be heated up to the denaturation temperature of the muscle protein Myosin (for beef, pork, and lamb, about 48 °C) to be considered pseudo-raw. The slight change in texture without temperature-related water loss, which only begins with the denaturation of Collagen at around 58 °C and the shrinking of the meat, is very beneficial for enjoyment. The change in meat color between 56 and 58 °C is above the pseudo-raw range.

Fish move in water and do not have to bear the entire weight of their body due to buoyancy forces. Although the tiny muscle cells are structured similarly to land animals—the muscle motor proteins Myosin and Actin interlock and thus enable muscle movements—the arrangement of the muscles runs directly from the skeleton to the outer skin. Fish muscles, therefore, have only a small amount of connective tissue (Collagen) between the parallel lamellae. Therefore, fish should be consumed raw immediately after conventional killing and does not need to be hung like beef. Also, the flesh of many fish is rather whitish. The red muscle pigment Myoglobin, which is necessary for oxygen transport during intense muscle exertion, is present in small amounts in many fish. Only in fish that swim long distances, such as tuna, does the musculature turn red.

Due to these anatomical conditions, conventionally caught and slaughtered fish meat does not need to mature. Shortly after conventional killing, off-flavors, fishy aromas, which are often unpleasantly noticeable, develop. One reason for this is the larger amounts of polyunsaturated fatty acids in the fat of marine animals, which easily oxidize and form aroma substances responsible for the typical musty-rancid fish smell. However, this is different when slaughtering according to the Japanese Ikejime method. This directs the maturation processes in entirely different ways. Fish can be matured for up to 15 days, even with an increase in taste and aroma. The texture also does not suffer (see [98]), as discussed in more detail in Chap. 6.

The body temperature of marine animals is significantly lower than that of land animals. Therefore, muscle proteins denature at much lower temperatures. Myosin begins to denature in fish at around 33–35°C, while the denaturation process of Collagen starts at 48–50°C, depending on the fish species. This is used in gastronomy: Many fish, especially salmon or salmon trout, are heated in the sous-vide method (vacuum method) to just above 40–45°C. The color is preserved, the texture becomes gel-like, and there is no water loss. A summary overview of raw and pseudo-raw enjoyment is provided in Fig. 5.31. However, some connoisseurs have fish protein allergies and intolerances, which make the enjoyment of raw or only slightly cooked fish (including seafood) impossible.

At this point, a remark on the culinary triangle is necessary. In the sense of the cultural-scientific definition of "raw," the raw consumption of fish and meat is practically impossible, as a cultural act, namely slaughtering, has always taken place. The closest would be the consumption of warm meat directly after slaughtering and offal by our hunters and gatherers—before the regular use of fire.

Two aspects should not be forgotten: raw is never germ-free, while pseudoraw is. Raw therefore has a higher allergy potential and health risk. When heated, proteins, which are responsible for most allergies, lose their shape and their allergic

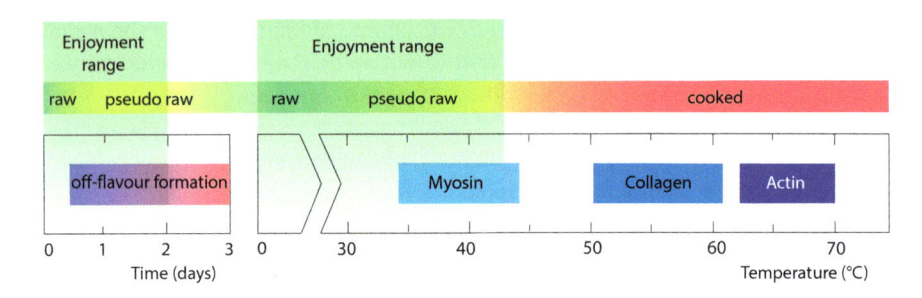

Fig. 5.31 The enjoyment diagram of fish for raw consumption is more complicated. Normally slaughtered fish should not/only briefly "mature" after catching and gutting

effect, as is well known with apples, celery, and nuts, which can be consumed without an allergic reaction after heating. Furthermore, it is also good to become aware from time to time that the real raw food consumption of our ancestors took place naturally without knives and forks, without lightning choppers and high-performance mixers. Anyone who has tried to eat a leek (including porraceous and roots), a cauliflower (of course with leaves and stalk), or an unpeeled celeriac (with peel and leaves) without any kitchen tools in a self-experiment knows what effort and how much time is required. Digestion also causes greater problems for some contemporaries. What is quite simple with the cultivated apple, the cultivated cucumber, or the tomatoes becomes a torment with many vegetables. Neither our dentition nor the chewing muscles or our digestive tract is designed for this primal food today.

Therefore, our current raw food is a large and meaningful part of today's human nutrition, in the sense of the nutrional base raw, cooked, and fermented. Taken by itself and as a complete and sole form of nutrition, it is not particularly recommendable. For purely physical-chemical reasons alone.

5.9 Paleo-Diet

5.9.1 Back to the Stone Age

Is Paleo nutrition a better solution? Since our genes have not changed for 30,000 years, we must eat as we did in the Old Stone Age, so the idea goes. Do we therefore have to throw carbohydrates, milk, cheese, and all other achievements of the Neolithic such as legumes overboard and hunt and gather? A look at popular literature shows which, from a natural science perspective, hardly justifiable ideas are described. The only clear thing is: Every Paleo diet comes with a significant restriction of the food supply and negates the eating biography of Homo sapiens of the last 10,000 years.

What do scientific facts suggest? Let us summarize the development to Homo sapiens*s* from the *Australopithecus*, taking into account its food (as far as secured from archaeological findings) in Table 5.4.

5.9 Paleo-Diet

Table 5.4 Overview of the developmental stages to modern humans, *Homo sapiens*, and their dietary forms and strategies (according to Ströhle [104]). If there are question marks behind certain foods, the archaeological and universally valid evidence is unclear. DQ value = *dietic quality* (see below), P:T ratio: ratio of plant to animal food

Species	Region	Time Period (million years ago)	Nutritional Strategy	Typical Food
Australopithecus	Africa	4.5–2.5	Gathering of plant-based, partly animal-based food	Leaves, fruits, seeds, storage roots (?), insects and invertebrates (?), high crude fiber content, low DQ value
Homo habilis, Homo erectus, Homo ergaster	Africa	2.5–1.5	Gathering of plant and animal food, (carrion?) in combination with hunting	Fruits, seeds, nuts, storage roots (?), carrion (?), mammal meat, use of fire (?), decreasing crude fiber content, increasing DQ value, P:T ratio unclear
Pleistocene *Homo sapiens*	Africa	0.2–0.05	Gathering of plant-based and animal-based food in combination with hunting	Fruits, seeds, nuts, storage roots, mammal meat, use of fire, high DQ value, P:T ratio similar to recent (east-)African hunters and gatherers (about 60–80: 20–40 energy percent)
Pleistoholocene *Homo sapiens*	Africa, Asia, Europe	0.05–0.03	Gathering of plant-based and animal-based food in combination with hunting	Fruits, seeds, nuts, storage roots, mammal meat, use of fire, high DQ value, P:T ratio similar to recent hunters and gatherers in areas with effective temperatures of <13 °C in the old world (variance range 0–90: 0–90 energy percent)
Upper Paleolithic *Homo sapiens*	Worldwide	>0.03–0.008	Gathering of plant and animal food in combination with hunting and fishing	Fruits, seeds, nuts, storage roots, meat from small and large mammals, as well as aquatic resources (freshwater and saltwater fish, shellfish), use of fire, high DQ value. P:T ratio analogous to recent hunters and gatherers worldwide (variance range 0–85: 6–100 energy percent)

In this process, the dietary forms are quantified as far as possible based on archaeological findings and genetic analyses of bones and dental enamel. This allows conclusions to be drawn about the ratio of plant to animal food (P:T ratio) as well as the DQ value, which stands for the English technical term *"dietary quality"*. The DQ value follows a key that is determined by the biological availability of the food components [102]. In Table 5.4, relatively simple keys are used when the proportions A of animal food (meat, insects), B of nutrient-rich plant parts (roots, tubers, fruits), and C of fiber-rich plant parts (leaves) are weighted in the total diet: DQ = C + 2B + 3.5A. Then DQ values from 100 to 350 are possible. In Table 5.4, this shows the systematic shift of food in favor of animal components and an increasing DQ value with the increasing development of the species towards modern humans. The influence of fire and the decreasing proportion of crude fibers also become evident in this better quantified representation. A developing mixed diet, as well as the adaptability of early humans and the associated development of various cultural techniques for food preparation, proved to be a decisive survival advantage [103]. This adaptability allowed modern humans to engage in the niche formation described in Chap. 2 within the given ecological environment and not only to survive with local conditions but even to evolve further.

This niche formation was also the basis for sedentism and cultivation, and above all, the resulting development of new foods, as summarized in Table 5.5.

5.9.2 The Paleo Hypothesis—Adaptation, Maladaptation

The Paleo hypothesis states that humans from the early Paleolithic are not genetically programmed for these foods. Therefore, it would be healthier not to eat these

Table 5.5 The most important achievements in the Neolithic. Foods that people did not know until the late Paleolithic

Food		Period (B.C.)
Grains	Emmer, einkorn	10,000–11,000
	Barley	10,000
	Rice	10,000
	Corn	9000
	Sheep	11,000
Dairy products	Goats/Cattle	10,000
	Legumes	7000–6000
	Oils, Oilseeds	6000–5000
	Salt	6200–5600
Others	Sugar	500
	Wine	5400
	Beer	5000
Alcohol	Distillates	800–1300

foods at all. The causes of many of our civilization diseases are thus already hidden in the cultural achievements of the Neolithic, as there was a maladaptation of the original metabolism.

However, experience teaches the opposite, the lifestyle of the Neolithic was much more advantageous for *Homo sapiens* compared to that of the Paleolithic. Although new disease patterns emerged in the Neolithic, life expectancy increased, food availability was controllable, food safety was better ensured. Trade was made possible, early economic forms crystallized. Reproductive capacity increased, the stock of tribes and clans was secured, the offspring protected (see Chap. 3).

For a more objective assessment, it is therefore useful to consider the nutritional extremes of the Neolithic and compare their nutritional and health data with those of hunters and gatherers. Examples of these extremes were gardeners (mainly plant-based diet), farmers (higher proportion of cereals), and shepherds (predominantly dairy products). Upon sober consideration, it already becomes clear how small the differences must be. Of course, the knowledge of hunters and gatherers was not largely discarded in the eating culture, the cultivation of fields and pastures merely meant more food security, but not yet abundance (as we know it today). Physical work in the fields was necessary and led to the rapid burning of carbohydrates. Thus, data from findings can be compared again. This already shows how shaky the Paleo hypothesis stands. The physical data differ marginally, the physical performance is similarly excellent, even cholesterol levels do not differ [104]. Only in diabetes mellitus prevalence did hunters and gatherers show slightly higher values than planters and farmers. Shepherds showed no signs of disease markers. However, all within the data uncertainty. The Paleo hypothesis would thus be strictly off the table. But again and again, there are newer observational studies that try to prove the opposite. Even in the most recent times, as will be shown in the next section.

5.9.3 For or Against Paleo? What Does Science Say?

In a highly acclaimed meta-analysis, evidence was presented that was supposed to support the many benefits of the Paleo diet [105]. The authors concluded that modern nutritional science easily explains the metabolic benefits of a (moderate) restriction of carbohydrates, the absence of high glycemic index products, a low ratio of n-6/n-3 fatty acids, and a reduction in salt intake in patients with insulin resistance and metabolic syndrome. On the other hand, it is less clear whether avoiding cereal and dairy products is a prerequisite for optimal metabolic control. Although numerous, less representative pathophysiological studies have shown the potential negative effects of cereal and dairy products on health, large epidemiological data suggest that the consumption of whole grains and (fermented) dairy protects against diabetes. Although the summary of the publication is vaguely optimistic in favor of the Paleo diet, a closer look at the data strongly relativizes this. In fact, this meta-analysis also stands on shaky ground when applying the hard mathematical laws of statistics and data interpretation [106]. The summary of the study exaggerates its results so much that the impression quickly arises that the Paleo diet would help people with metabolic syndrome significantly better than control diets without specific Paleo orientation. However, the statistical insignificance of the interpretations can be demonstrated. For example, the average differences for most of the primary outcomes in the meta-analysis by Manheimer et al. [105] are sobering. A decrease in diastolic blood pressure by 2.5 mmHg, a decrease in HDL cholesterol by 0.12 mmol/l, a lowering of fasting blood sugar by 20.16 mmol/l, and a decrease in systolic blood pressure by 3.6 mmHg are determined. These values are neither of clinical nor practical significance in everyday life, as they are subject to the usual fluctuations. In addition, systematic errors in the evaluation can be identified [107]. The results are therefore, like those of many other observational studies, statistically unconvincing. The conclusions were read out because they wanted to read them out, because they were weighted incorrectly, because the statistical relevance is not guaranteed [108, 109].

This meta-analysis thus joins the criticism of Ioannidis [8, 9] of such observational studies already mentioned in earlier chapters [110]. However, this leaves open whether it is actually health-promoting, as claimed in popularly defined Paleo advice, to avoid dairy products as excellent sources of protein, calcium, and phosphorus, fiber- and protein-rich legumes, and cereals and instead eat only nuts, meat, and vegetables. In most Paleo recipe books, Western diets simply adapt to the restrictions of the Paleo diet; for example, desserts made with supposedly "paleo-friendly" alternatives like almond flour and honey instead of sugar and wheat flour. As already shown in Sect. 4.8, replacing sugar with honey is pure deception. Also, most plant proteins are inferior to wheat gluten in many respects, as already shown in this chapter. It is therefore highly unlikely that the philosophy of the Paleo diet will significantly improve public health.

The actual problem is therefore the ambiguity of the statements: A little "good" meat, fish, eggs, plenty of vegetables and fruits, only little sugar and "satiating side dishes" already sound very much like the much-praised (and often misinterpreted) Mediterranean diet. Contemporary Paleo nutrition seems to be nothing fundamentally different from the usual recommendations [111].

5.10 Vegan—Exclusive Exclusion and Missing Links

5.10.1 The Radical Food Elite

A very elitist and at the same time radical form of nutrition is the complete renunciation of animal products. A strictly vegan diet entails a strong restriction of the food supply, so that a deficiency in certain nutrients and micronutrients is not uncommon [112].

The radicality of the way of thinking occasionally brings about absurd, inhumane forms of life philosophies [113], as are well known in religions when it comes to the only true faith [114]. There is no question: Nowadays there are many

5.10 Vegan—Exclusive Exclusion and Missing Links

very good reasons to eat little or no meat: the spiral of Western nutrition, the ever more thoughtless meat consumption, the abundance and the resulting factory farming, which has been spinning faster and faster since the post-war period. Animal products are becoming cheaper and cheaper, which is reflected in the declining meat quality as well as in the constant anonymization of the packaged supermarket meat from unacceptable mass production. The consequences are clear: meat scandals, additives in meat products for better water binding, rapid ripening methods for ham, sausage, and cheese. Processes that increasingly distance us from the artisanal techniques that organizations like "Slow Food" propagate today.

On the other hand, only a few consumers want to pay the real prices that compensate farmers for breeding, keeping, and feeding animals and, moreover, secure their livelihood. Trading corporations shamelessly dictate prices. Scandalously low meat and milk prices are the result; no farmer can live on them. Subsidies are necessary; regulations and ordinances that make life difficult or impossible for small farmers. For many operations, the flight into mass production was the only alternative in order for agriculture to be worthwhile at all.

The history of veganism began between the two world wars. In 1924, Donald Watson gave up meat. The violence against animals during slaughter in a factory made him the first activist for animal rights. In 1944, he introduced the term "vegan," derived from the English *"vegetarian,"* and thus founded the lifestyle that consistently rejects any use of animal products and by-products. Beyond vegetarianism, this also includes eggs, milk and dairy products, honey, and consequently also leather and products containing animal products.

In return, rights are granted to all living beings, and discrimination of living beings solely based on their species affiliation is inadmissible. Moral dignity is not only due to humans but also to animals. Inflicting suffering on living beings is morally and ethically reprehensible. "Veganism represents a manifestation of ethical rationalism, which is alien to any mystical or religious inclination. The advantage of this self-understanding is obvious. It allows consistent vegans to distance themselves from everything sectarian and fashionable. Instead, they believe they represent a generalizable ethical system. According to their premises, the reverse conclusion also applies: Anyone who consumes proteins of animal origin, although they do not have to, acts ethically reprehensible. This would prove: Vegans are more ethical than other people," writes Cantone in the *NZZ* [115].

This attitude is presumably the reason for the strong missionary, fundamentalist-like stance of vegans, and probably also the cause of why some vegans become heavily radicalized, celebrating the death of humans, as is no different in racism [116] or Salafism [117] when "different" and "unbelievers" become victims.

This attitude is also the cause of the contradiction that lies in practiced veganism. The original, heroic approach of vegan nutrition [118] is based on a world without suffering, a peaceful coexistence of animals, including the species *Homo sapiens*. This is in stark contrast to radical veganism when butcher shops are devastated. The contradiction goes even deeper, as a life without suffering is fundamentally impossible. Life itself is characterized by suffering, illness, and death, in plants, animals, and humans. Traditional religions had already recognized this and promised people eternal, paradisiacal life after death. Veganism wants a kind of paradise here and now, which has neither been possible nor existed since the emergence of life on this earth. Neither in the animal kingdom nor in the plant world, just as evolution and human history teach before the appearance of the first hominids.

5.10.2 Health Benefits?

Whether there are health benefits to a vegan diet is highly controversial. The data situation is thin, also because of the shortness of the observation time of the comparatively young trend and the still small number of people living vegan. Small numbers of subjects make the already uncertain evaluations of observational studies many times more difficult. At first glance, the vegan diet seems to be superior [119]. Vegetarians and vegans show on average lower cholesterol levels. In one study, they had only a level of 172 mg/dl compared to 206 mg/dl in meat-eaters and still borderline 190 mg/dl in vegetarians [120]. The relevance is unclear, the arbitrarily set limit of 200 mg/dl is no longer considered mandatory by cardiologists, as long as no other risk factors are present. Vegans are on average slimmer and have a very favorable body mass index, well below the population average, according to a British study [121]. Blood pressure also proves to be more favorable on average in vegans in this publication.

However, laboratory values are only one point; the other side of the coin is observations from clinical practice [122], which suggest an undersupply of certain nutrients. A clinical study concludes that vegetarians have a significantly higher risk of bone fractures compared to omnivores [123]. Although this study mainly concerns vegetarians, it can be applied to vegans. If the complete avoidance of animal products is the reason, the result for vegans would be at least the same or even worse.

In the case of cardiovascular diseases, lower risks can be expected due to significantly better laboratory values. Moreover, the higher intake of secondary plant compounds should have a positive effect on mortality. In a study, 73,308 participants from the five groups of meat-eaters, flexitarians, ovo-lacto vegetarians, pesco-vegetarians, and vegans were examined for their risk of death. The risk for a specific disease in omnivores was set at the relative value of 1, and the risks of the other groups were compared to it [124]. It turned out that vegans had no advantage over omnivorous people in terms of heart death, but rather had an almost 39% higher probability of dying from a heart attack, with differences between men and women being observed, but this cannot be assessed more precisely due to uncertainty and statistical relevance. Vegetarians who still ate fish had the best results with a roughly halved heart attack rate. Even the risk of pure vegetarians was significantly higher than that of pesco-vegetarians. The same study also examined the risk of cancer. There was no significant difference between omnivores and vegans, while in vegetarians who still included dairy products or fish in their diet, cancer mortality was slightly reduced.

These observations are indeed confusing, as according to general assumptions, secondary plant compounds such as anthocyanins, flavonoids, or polyphenols provide strong protection against cancer. At least their antioxidant effect can be demonstrated *in vitro* in laboratory experiments. In vegans, the intake of secondary plant compounds is significantly higher than in meat-eaters and vegetarians. As a result, other factors must play a significant role *in vivo* in *Homo sapiens* that are not found in vegan meal plans [120]. These can only be micro- and macronutrients that are present in animal foods and have defined human nutrition for several hundred thousand years. Anything else would be illogical. Although the publications mentioned in the last section are among the less reliable observational studies and therefore only provide rough indications and must be critically questioned, the observed results can be supported by molecular considerations.

In an early study, the nutritional status regarding micronutrients was examined in 15 Swedish women and 15 Swedish men who followed vegan and meat-eating diets [125]. Although the vegans had a higher intake of vegetables, legumes, and dietary supplements and a lower intake of foods such as cakes, cookies, candies, and chocolate compared to the "omnivores," many serum micronutrient values were below the average requirement, such as riboflavin, vitamin B_{12}, vitamin D, calcium, and selenium. The intake of calcium and selenium remained low even with the use of dietary supplements. There was no significant difference in the prevalence of low iron status between vegans (20%) and omnivores (23%). The authors concluded: The dietary habits of vegans do not meet the average minimum requirements for a range of micronutrients. The major differences are summarized in Table 5.6.

Selenium is usually incorporated into animal foods in place of sulfur in the amino acid cysteine (selenocysteine), while in plants it more often replaces sulfur in methionine (selenomethionine) (Fig. 5.32). Thus, (in this order) beef, veal, and poultry meat, egg yolk, herring, lobster, tuna, redfish, and trout are good sources of selenium, while behind them whole wheat, Brazil nuts, coconuts and coconut flakes, sesame, and porcini mushrooms carry the trace element. This already makes it clear from a chemical point of view why omnivores have a higher selenium supply: On the one hand, there is a higher protein content *per se* in animal foods, and on the other hand, they are more readily available. Since selenium is incorporated into specific amino acids, these proteins must be enzymatically digested to the

Tab. 5.6 Some examples of the different proportions of micronutrients in vegans and omnivores. For selenium and vitamin D, neither of the dietary forms meet the currently valid daily requirements

Micronutrient	Omnivores (mg/day)	Vegans (mg/day)	Recommendation (mg/day)
Vitamin B_{12}	5.0–6.0	0.0–0.1	3
Vitamin D	5.1–7.7	2.0–3.7	20
Zinc	11.0–16.5	7.8–10.0	7–10
Selenium	27–40	10–12	60–70

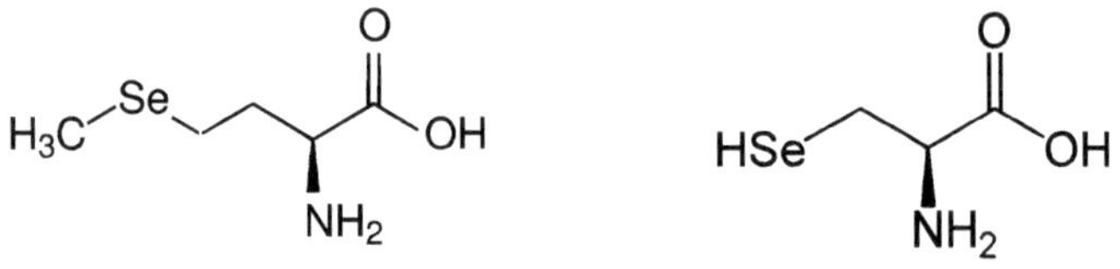

Fig. 5.32 In selenomethionine and selenocysteine, sulfur is replaced by selenium (Se). The bioavailability is therefore determined by the enzymatic cleavage of the proteins

point where small peptides or even free amino acids are present. Only then can further chemical reactions release selenium. In plants, especially raw plants and nuts, the still native proteins are only slightly accessible to the enzymes.

5.10.3 Vitamin D

When talking about vitamin D, the active form cholecalciferol (sometimes also called vitamin D_3) is actually meant. The fact that plants do not contain vitamin D is due to their physiology: it is not necessary in plants, unlike in animals and humans, and can therefore be "saved" in plant physiological pathways. Nature is more than economical; nothing happens that is not necessary. However, the chemistry of the formation of active vitamin D in the human and animal body reveals deep insights into nutrition that are often forgotten, for example, why vitamin D should be supplemented in a vegan diet.

As already mentioned several times, human biochemistry is economically oriented. There are only a few basic molecules from which various functions can be derived. One of these is cholesterol, which, in addition to its function in cell membranes (Sect. 3.1.2), also provides perfect services for bile acids and is thus crucial for fat digestion (Sect. 4.9). Moreover, cholesterol is the basis for vitamin D when it gets the chance to do so in several organs, as schematically shown in Fig. 5.33.

The formation of vitamin D_3 is thus linked to the presence of 7-dehydrocholesterol (provitamin D_3), which is derived directly from cholesterol. This means it can only be part of an animal metabolism. Phytosterols, plant cholesterol, cannot form these precursors of vitamin D. When ultraviolet radiation from sunlight acts on the skin on provitamin D_3, a precursor structurally closer to the active vitamin D is formed, called cholecalciferol. This molecule is converted in the liver to calcifediol (or 25-hydroxy-vitamin-D). Only in the last step, the active vitamin D_3, calcitriol, also known as $1\alpha,25(OH)_2$-cholecalciferol, is formed in the kidney. These processes are common to humans and all mammals.

This synthesis taking place in the organs also shows something else, namely the high quality of liver and kidney as food. These organs, still wrongly referred to as "highly burdened detoxification organs" by some, are in fact high-quality foods. They are the best suppliers of precursors to vitamin D and should therefore be included more frequently in the diet. No other food, neither dairy products nor

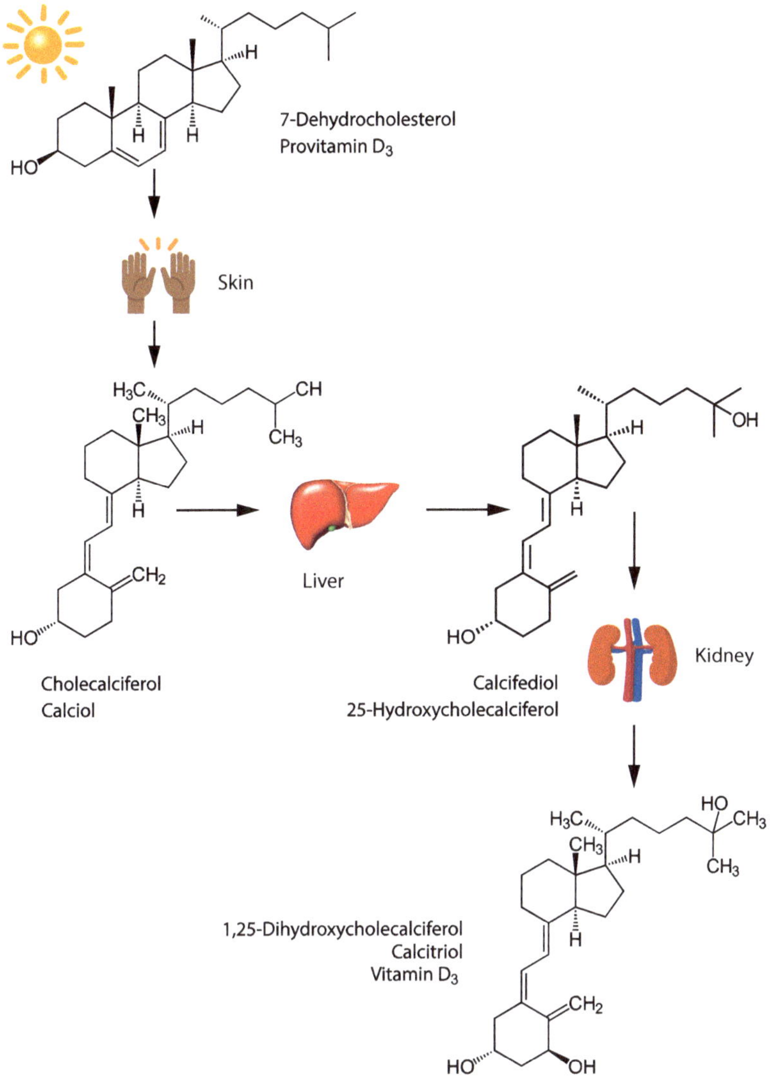

5.10 Vegan—Exclusive Exclusion and Missing Links

Fig. 5.33 From 7-dehydrocholesterol, provitamin D$_3$, cholecalciferol is formed through sunlight exposure on the skin, from which calcifediol or 25-hydroxy-vitamin-D is formed in the liver, and only then the active vitamin D$_3$, calcitriol, in the kidney

egg yolk, can compare to liver and kidney. In fact, fish livers are at the top in this regard, but more on that later.

These points alone show that vegans often have an undersupply of vitamin D (see Table 5.6). Even omnivores and carnivores could do more for their vitamin D supply if they opted for liver and kidney more often instead of neck steak and loin. Unfortunately, offal has fallen out of fashion as a valuable food and is no longer consumed as a matter of course. This was quite different for our ancestors, the hunters and gatherers (Chaps. 1 and 3). Since vitamin D is involved in a whole range of biochemical and physiological processes, a good supply of it was also a basis for human development on the path to *Homo sapiens*.

5.10.4 Essential Fatty Acids: EPA and DHA

In the case of essential long-chain fatty acids, the supply situation for vegans is critical due to the exclusion of animal food. For the reasons already shown in Fig. 2.6, no land plant can incorporate the truly essential omega-3 fatty acids docosahexaenoic acid (DHA) and eicosapentaenoic acid (EPA) into its cell membrane. These fatty acids do not occur in land plant physiology, as they would be rapidly oxidized during the plant's life and impose unnecessary stress on the plant. Such malfunctions, if they occurred spontaneously, were eliminated in time during the evolution of organisms from the sea to land. This also explains why most algae do not contain these essential fatty acids. Many algae do not grow permanently in water, but on the beach or in warmer climates. Only microalgae and nori algae in small quantities manage this balancing act. Some microalgae achieve high efficiency in the synthesis of EPA and DHA through carbon dioxide metabolism and high photosynthesis rates. Therefore, efforts are being made to systematically use the biomass of microalgae for fatty acid production [126].

Indeed, these points are of high relevance for the vegan diet, as contrary to the old assumption, humans cannot convert these two truly essential fatty acids EPA and DHA in sufficient quantities from plant-based α-linolenic acid (ALA), neither in healthy individuals nor with high doses of ALA, as shown in clinical studies [127, 128]. The yield in human physiology remains much too low.

Considering the process required for this [129], it becomes clear why: The C 18:3 fatty acid would have to be elongated in the cells in the endoplasmic reticulum and desaturated at several points, as schematically shown in Fig. 5.34. Many enzymes are necessary for each of these steps, and each step costs relatively much energy in the form of adenosine triphosphate (ATP). This effort is not worthwhile for human physiology, as hominids became omnivores early on. Then the essential fatty acids were supplied through the diet, and physiology no longer had to expend this energy. Other species, such as primates that feed exclusively on plants, cannot do this. Therefore, these organisms must produce EPA and DHA from ALA.

5.10.5 Supplementation Necessary

A vegan diet without supplementation with appropriate preparations is not reasonable for the reasons mentioned so far, especially before and during pregnancy. This would correspond to a permanent malnutrition, because not only vitamin B_{12}, but also bioavailable iron or the long-chain Omega-3 fatty acids EPA and DHA are missing in this case for the supply and development of the fetus [130]. Above all,

5.10 Vegan—Exclusive Exclusion and Missing Links

Fig. 5.34 The long and energy-rich path from ALA to EPA and DHA. Each step of elongation (via elongases) and desaturation (desaturases) costs energy

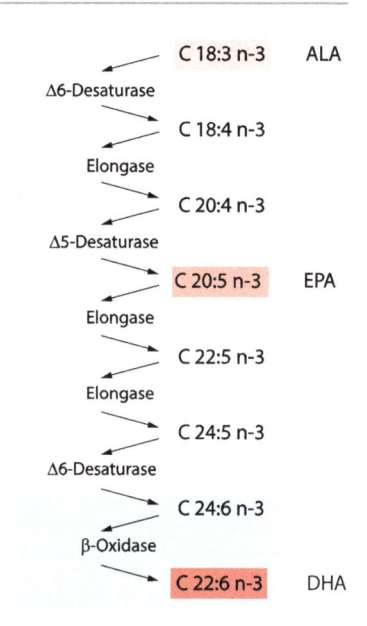

```
                    C 18:3 n-3   ALA
        Δ6-Desaturase
                    C 18:4 n-3
        Elongase
                    C 20:4 n-3
        Δ5-Desaturase
                    C 20:5 n-3   EPA
        Elongase
                    C 22:5 n-3
        Elongase
                    C 24:5 n-3
        Δ6-Desaturase
                    C 24:6 n-3
        β-Oxidase
                    C 22:6 n-3   DHA
```

an adequate development of the brain and the formation of the necessary cognitive abilities are not guaranteed [131]. This would be a lifelong torment for the offspring, as the consequences of times of malnutrition in humanity teach. This also fundamentally contradicts the pain- and torment-free approach, which is at the top of the basic principles of the vegan idea. No question, a well-nourished, individual person who has grown up in opulence can survive the deficiency for a long time. However, compensating for the deficiency in the offspring is not possible for elementary biological reasons. The body physiology is fully focused on self-preservation and survival.

The vegetarian diet does not have these deficits to this extent. Milk, dairy products, and eggs provide enough essential macro- and micronutrients, even if these foods are not consumed daily. This is precisely what distinguishes them from flexitarians. The minimal consumption of high-quality animal products from sustainable production, such as fish and meat, is sufficient to incorporate the necessary nutrients. The propagation of misunderstood "pegan" nutrition (pegan = paleo + vegan) [132] should be inherently rejected. Malnutrition and insufficient nutrient utilization are pre-programmed, similar to raw vegan nutrition [133].

5.10.6 Fermentation and Germination as Systematic Methods in Plant-Based Nutrition

Fermentation is also an essential part of vegan nutrition, as fermentation breaks down antinutritive substances present in raw vegetables and converts them into nutritive substances, such as chlorogenic acid or the previously mentioned FODMAPs. Non-nutritive plant defense substances such as saponins, lectins, and alkaloids are also broken down or significantly reduced. These methods are not only of interest in vegan nutrition but also for connoisseurs who like to try new foods beyond classic products, ferment tofu or other nut and almond preparations themselves to make cheese-like products. In veganism, germination and fermentation are absolutely necessary to at least partially cover the need for macro and micronutrients. As already explained in Sect. 2.10, the number of essential amino acids increases through the formation of enzymes, while essential secondary plant substances are formed that are not yet present in oilseeds, legumes, etc. No matter which diet and personal preference is chosen: fermented foods are part of it.

Our ancestors already benefited from this, but for different reasons. In the past, the positive effects of fermented foods on health were unknown, so people used fermentation primarily for preserving food and improving taste. Today, more is known, and the molecular processes involved in fermentation are increasingly well understood [134]. During fermentation, the microorganisms involved not only synthesize the vitamins already mentioned [135], but proteases also produce bioactive peptides [136], which contribute to taste in many ways and also provide some health benefits.

Some microorganisms can do even more. They produce conjugated linoleic acids (CLA) (Chap. 1), which are attributed to have blood pressure-lowering effects, lactic acid bacteria produce thickeners, so-called exopolysaccharides [137]; these act as dietary fiber, presumably with prebiotic properties [138]. Sphingolipids, which are incorporated into the cell membrane along with phospholipids, can have anticancer and antimicrobial properties. Bioactive peptides show antioxidant, antimicrobial, or antiallergic effects, and even opioid antagonists (Sect. 4.10.8) have blood pressure-lowering properties. Many of these assumptions are not proven in detail and can therefore only be considered as indications. However, the long tradition of fermentation shows that fermented products did not harm the development of *Homo sapiens*. Neither did cooking.

5.10.7 Example Nattō

The fermented soybean popular in Japan, Nattō, is a prime example of nutrient enhancement, as in this example, the cultural techniques of cooking and targeted fermentation are combined to increase the nutritional value of the soybean and make the resulting micro- and macronutrients, including bioactive peptides, available [139]. For the production of Nattō, the soybeans are first cooked. The storage proteins denature, the starch swells the beans, then *Bacillus subtilis* is added as a starter culture, which develops its endospores under aerobic conditions. Traditionally, Nattō was produced over rice straw, on which this bacterium predominantly settles. After inoculation with *Bacillus subtilis*, the fermentation process begins. Storage proteins are broken down, resulting in peptides that change the taste. Thus, a bitter taste and a remarkably weak umami taste for a fermented product are formed. This is mainly due to umami-tasting peptides, while glutamic acid takes one of the few other paths.

Overall, Nattō shows a taste profile less culturally rooted in Western cuisine, even though the preparation is reminiscent of very ripe cheese. The main reason for this is a special protease from the bacterium, the Nattokinase, which cleaves proteins at the amino acid serine. At the same time, an aroma mixture of mushroom-like, earthy, yogurt-like smells and intense roasted notes reminiscent of chocolate and toast is formed, also a very unusual aroma combination that can be attributed solely to the genetics and resulting activity of the bacterium [140]. Typical for the *Bacillus* is also the high production of vitamin K, so this microbiological process serves as a model for the production of this vitamin [141].

However, Nattō forms a kind of mucus (Fig. 5.35), which is perceived as unpleasant at least in the Western world, but again provides very deep insights into the function of fermentation and food. This slimy consistency is produced by a very special biopolymer, the Poly-γ-glutamic acid. This is a long chain molecule, the only building block of which is the glutamic acid. The polymer thus consists exclusively of the flavor enhancer for umami.

The actually restrained Umami taste of Nattō can be detected by the average concentration of free glutamic acid: Among most (Asian) fermented products, it has the lowest value, as can be seen in Table 5.7. This is reflected in the slime,

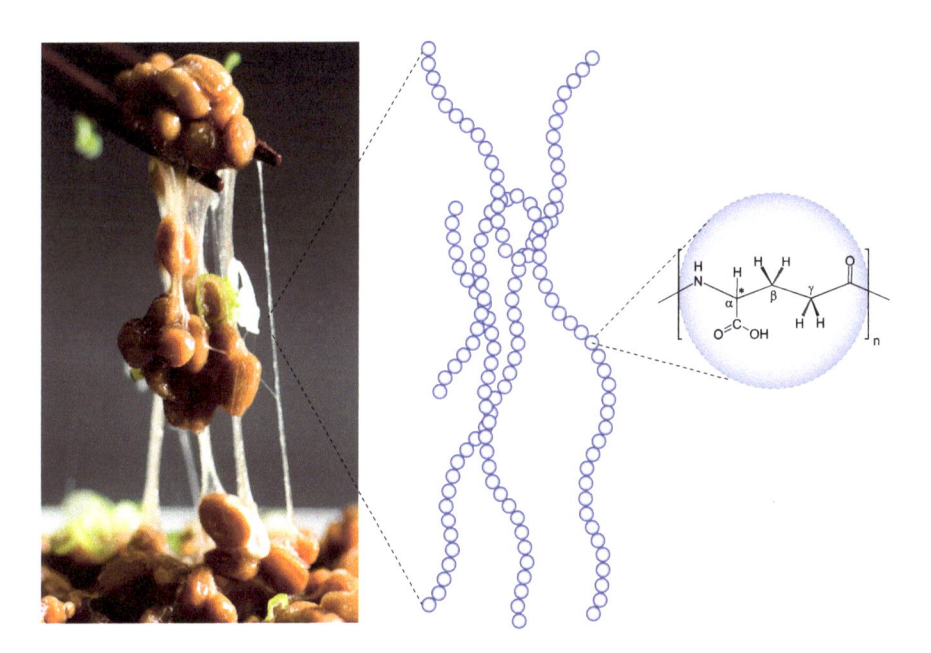

Fig. 5.35 Nattō forms a stringy, viscoelastic biofilm (left), which consists of Poly-γ-glutamic acid. The electrically charged polyelectrolytes form a weakly interlocked network (center). The basic building block (right) consists of γ-linked glutamic acids. (Photo source: Vicky Wasik https://phinemo.com/wp-content/uploads/2018/05/natto.jpg)

Table 5.7 Average proportion of free glutamic acid in various fermented products. Nattō has the lowest proportion, thus lacking the basic trigger of umami taste

Fermented Food	Free Glutamate (%)
Fish sauce	1.383
Korean soy sauce	1.264
Anchovy	1.200
Douchi	1.080
Tempeh	0.985
Chinese soy sauce	0.926
Japanese soy sauce	0.782
Garum	0.623
Miso	0.5–1
Ham	0.340
Sake	0.186
Nattō	0.136

as the extracellular glutamic acid released during fermentation is polymerized to poly-γ-glutamic acid [142]. However, only free glutamic acid triggers the umami taste, while the glutamic acids bound in the long polymer chains can no longer stimulate the receptors. Nattō therefore tastes significantly less savory than other fermented products.

The poly-γ-glutamic acid explains the special, slimy texture: The molecular chains are negatively charged, as the OH group (hydroxyl group) in Fig. 5.35 can deprotonate, releasing the hydrogen as a proton. The resulting negative charges on the oxygens, O^-, along the molecular chain can bind a lot of water, because like glutamic acid, poly-γ-glutamic acid is also highly water-soluble. At the same time, cell water is released from the soybeans during fermentation. This contains salts that shield the charges. The chains are not repelled as strongly, but form a dynamic and viscoelastic network, as shown schematically in the molecular model (middle) in Fig. 5.35.

5.10.8 Free From Animal—Vegan Substitute Products

Surrogate products that resemble meat or animal-based foods are trendy. Burgers, meatballs, sausages, Munich (weisswurst), even shrimps, prawns, and cheese can be purchased animal-free in supermarkets, even in organic quality. For this purpose, proteins from cereals, peas, soy, potatoes, or lupins are used. Many of these raw materials come from industrial by-products, for example, when proteins are produced as waste during starch production, such as with potatoes, or during oil extraction from oilseeds and nuts. In principle, there is nothing wrong with this, on the contrary. Every step away from intensive and mass animal farming is a step in the right direction. Especially when it comes to mass catering. Many thoughtlessly and habitually eaten meat products can be easily replaced by plant-based products. Also, BeyondMeat-, *impossible meat-,* soy-, pea- or lupin-based burgers are certainly not out of place—ultimately, regardless of whether one eats meat-free or enjoys a good piece of pasture-fed beef from time to time.

The deficits of plant-based, technologically produced surrogates are often clearly noticeable. The vegan chorizo with reconstructed fat pieces shows high and unpleasant elasticity, vegan shrimps are neither chewable nor reminiscent of the original, and vegan processed cheese is often low in fat and elastic or leathery. Vegan substitute products and convenience products often have a common denominator: They require a number of additives with E-numbers for texture and flavor adjustment, as well as many clean-label additives that serve taste and aroma control. It is not long ago when analog cheese on pizzas, gratins, or other convenience products was criticized. Instead of Parmesan, the cheese substitute "Gastromix" was used on ready-made pizzas, a product almost perfectly industrially constructed from a physical point of view, consisting of water, fat, starch, plant or whey proteins, emulsifiers, salts, flavor enhancers, and aromas. The balancing of the components allowed, for example, a gentle melt during baking at high temperatures in the oven without fat and water separating, as with cheese.

The fact that this still does not taste good, that texture and flavor are not right, is due to several fundamental, physicochemical problems that are little considered in food technology. Neither taste nor aroma or texture can be easily simulated and reproduced when the basic products, the proteins from which they are imitated, do not fit on a molecular scale. This is particularly true in the vegan area. Pea proteins cannot be compared to the proteins in muscles. Their biological function is completely different. Seafood and fish substitute products made from plant components do not contain long-chain, polyunsaturated fatty acids. Therefore, fish never smell like fish [143]. The fish smell or, in general, the typical smell and thus the aroma of a food, its taste, and its texture are the result of a long history that has to do with plant and animal genetics, living conditions, slaughter and harvest methods, and the time afterwards. Taste and aroma can only be formed from what is present in the living organism: amino acids, fatty acids, sugars, enzymes. Even if external, foreign enzymes are added, such as in meat maturation or fermentation, only what is present can serve as a substrate. That is why fish smells like fish, chicken smells like chicken, and kohlrabi smells like kohlrabi. That is also why a ham from a Spanish Iberico pig tastes significantly different from Black Forest ham, even if all macroscopic process parameters are identical. It depends on the animal, the species, the husbandry, the feeding (Chap. 2). Reconstructing Iberico ham in detail from a leg of Pietrain pig from mass animal husbandry is a currently unsolvable technical challenge. Reconstructing Iberico ham from textured plant proteins requires a high degree of technology. Wouldn't a thick, well-cooked soup made from intact peas without bacon be better? This is also vegan and free from the energy expenditure of the by-products and the reassembly so that a plant-based molded ham is made from peas. This is not really food culture, and even less sustainable.

5.10.9 Industrial Processes for Surrogate Products

Substitute products for animal-based foods require a whole series of complicated steps. With plant-based sausages or plant-based meat, the situation is complex, especially when nutrients come into play. As already demonstrated with gluten, no protein mixture of non-wheat proteins can achieve the nutritive (and physical) properties of gluten without great effort. It is even more difficult to replicate animal protein from plant-based raw materials. This is simply impossible, due to the completely different biological function of animal and plant protein and consequently a whole range of other properties. In Fig. 5.36, two albumins are compared as examples: the ovalbumin of chicken egg and the storage protein 2S-albumin of soybean. The spheres each represent an amino acid, with only the charged amino acids being shown. They also differ significantly in their overall length. While the ovalbumin of chicken egg white has a considerable 386 amino acids, the 2S-albumin of soybean has only 158. Ovalbumin has only one cysteine double bond, but three free cysteine sites suitable for cross-linking. 2S-albumin has eight free cysteines along its chain. These are all suitable for cross-linking, which on the one hand ensures good strength, but on the other hand, due to the short mesh size of the resulting networks, leads to very low elongations, as can occasionally be seen in tofu.

The chicken protein ovalbumin, on the other hand, is significantly longer in its denatured state and binds at fewer sites along its molecular chain. The resulting networks become more open-meshed and thus more stretchable, as is known from egg white gels (not overheated, coagulated protein) in the kitchen.

From this alone, the differences in texture, taste, and aroma arise. Therefore, modern industrial processes are necessary at all levels to address the problem to some extent. As a result, it is technically much easier to produce vegetarian

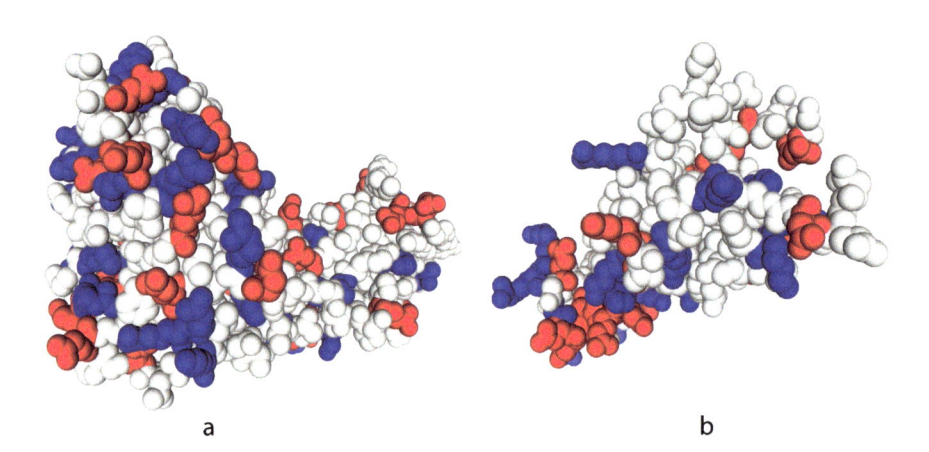

Fig. 5.36 Chicken egg white ovalbumin (**a**) and a soybean storage protein 2S-albumin (**b**) in direct comparison. Only the charged amino acids are color-coded (red = minus, blue = plus)

5.10 Vegan—Exclusive Exclusion and Missing Links

substitute products with egg white than vegan ones without egg white. The cause, as always, lies at the molecular level and the properties predetermined there.

Vegan substitute products must therefore be provided with a mix of starch, proteins, water, fats, emulsifiers, binding and gelling agents, plant fibers, etc., until the texture becomes somewhat consistent. Without techniques, as are possible and common in the food industry, there is little chance. Therefore, many analog products like meat or fish alternatives can hardly be described as artisanal anymore—in contrast to traditional sausages, which are based on a millennia-old cultural technique.

It is therefore not surprising that classic, artisanal scalded sausages (such as Franfurter or Lyoner type sausages), which we will discuss in more detail in Chap. 6, differ significantly from their previous vegan analogues. The multitude of muscle proteins and their diverse molecular properties allow for completely different textures than the limited number of suitable plant proteins. Above all, animal proteins show extraordinary properties during the manufacturing process due to their denaturation temperatures [144], which we are familiar with from cooking. This is because the manufacturing processes for sausages are quite precisely adapted to this [145]. A conventional scalded sausage recipe relies on a balanced interplay between meat (protein), fat (back fat from pork), and water, which is added in the form of ice for cooling during the cutting process, so that nothing happens to the proteins during the frictional heat, which begin to denature at 48 °C. Some muscle proteins therefore only partially lose their shape, bind water, and emulsify the animal fat during the cutting process. These protein structures are thus preserved until the boiling process. The fat, usually lard, also remains solid during the cutting process. This emulsification of cold, solid fats is crucial for the final mouthfeel. Boiling is done at 70 °C, the disulfide bridges are formed, the fats liquefy, but remain at the positions determined during the cutting process in the meat-water matrix. After cooling, the fats recrystallize, and the elastic, crispy structure is established.

The mouthfeel is well-known: boiled sausages are slightly elastic, break open when bitten, and release flavor and aroma. At the same time, the sausage pieces warm up when bitten and chewed, and only then does the still solid part of the pork fat melt, at about 28–32 °C. The aromas dissolved in it are released, fat coats the tongue and palate, and the sensory experience lingers until the fat film on the tongue dissolves and most of the aromas have evaporated.

In many cases, this does not work well with plantbased sausages. Plant protein mixtures are combined under pressure in extruders under thermodynamic non-equilibrium conditions to create an elastic structure. The vegetable oil is already liquid at these temperatures, so it can be emulsified extremely poorly at high temperatures. Therefore, only a small amount is used in the technical processes, as well as less water, since many plant proteins can bind it less well. What remains is a low-fat, overly elastic, rubbery structure in many cases, and this is due to purely physical reasons.

For taste and aroma, assistance is needed due to the non-animal starting products. Of course, taste and aromas must be added so that the product comes close to real meat. Therefore, reconstructed meat aromas of vegan origin must be produced. For this purpose, nature-identical aroma mixtures are used, and the inevitable differences are masked with spices. This inevitably leads to the methods of industrially produced convenience foods, but at least it is supposedly politically correct "fake-food". The transition from the industrial age to the Anthropocene has also taken place on snack plates.

This creates a strong contradiction at the meta-level and in the philosophy, shifting the dividing lines between "good" and "evil". The flavor enhancers of fake products are well-known and mainly trigger the umami taste, which has fascinated us humans since the beginning of our evolution. However, until recently, glutamate, yeast extract, and food industry aromas were considered harmful, but these additives are now hailed as a blessing in vegan nutrition. In the past, we were "poisoned" and "made sick" [146], if not driven out of our minds [147].

However, one question is certainly justified: Wouldn't it be simply better to eat soybeans or even better, native beans, as curry, as stew, as puree? Fermented? To benefit from the nutrients (macro and micronutrients) of the entire beans? This is still vegan and guaranteed whole food. If beans are cooked in wise foresight for storage, they are also somewhat convenient and even more umami-rich, meaning heartier and tastier, after storage time. Even today, the taste of umami does not leave us. In comparison to the whole legume, such "Plant-Based Food", vegan sausage or textured protein lacks what is commonly called the food matrix [148], which makes a significant contribution to food yield [149], even in reconstructed foods [150].

5.10.10 Structuring of Proteins

When texturing proteins, they are combined in extruders under pressure and temperature (*extrusion cooking* [151]). To put it simply, plant proteins are mixed with water in a molding machine (extruder) with a helical thread. This allows proteins and water to be efficiently mixed into a paste. However, the helical thread has a variable number of turns, pitch, and depth. As a result, after mixing, high friction is created at certain process steps, the temperature increases and can be controlled externally at the same time. During transport through the extruder, the protein threads are oriented and then cross-linked, usually via cysteine, due to the sufficiently high temperature. Fibers are formed, which, when viewed with the naked eye, have a meat-like structure [152], as shown in Fig. 5.37 as an example.

These methods can be applied to different proteins and protein mixtures [153], but the results depend, among other things, on the microscopic properties, the primary structure, the number and distribution of the amino acid cysteine, and the charge of the proteins [154]. The goal is to develop textures that come as close as possible to meat fibers. However, this is only possible up to a certain length scale, because within the visible fibers, the ratios are completely different. But it is especially on these length scales from 100 μm to 10 nm that a large part of the sensory experience takes place, both in terms of texture and the release of taste and aroma.

Muscle meat is fibrous because under constant movement, whole fibers grow from individual muscle cells, the sarcomere, which can be longer or shorter

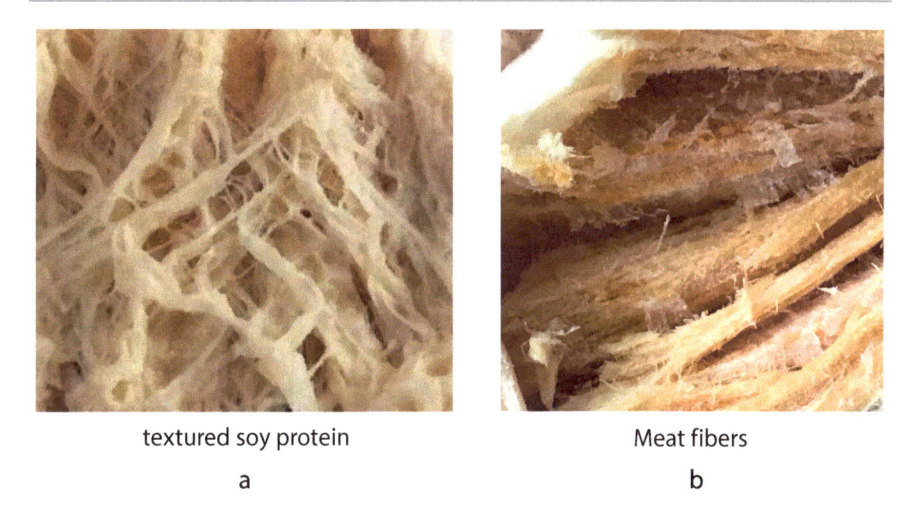

5.10 Vegan—Exclusive Exclusion and Missing Links

Fig. 5.37 On a macroscopic comparison, similarities can be observed between textured plant protein (**a**) and meat (**b**). In this case, the meat is freeze-dried chicken breast. The different structural features between (**a**) and (**b**) are clearly visible even on the macroscopic scale

depending on their biological function. These muscle fibers, in turn, have a very special structure, as they are built from various fibrillar proteins that enable muscle movement. Therefore, muscle meat has a hierarchical structure, as shown in Fig. 5.38.

This naturally grown and unique function-related structure of meat cannot be replicated in all details with plant proteins using technological processes because they have completely different properties. Therefore, for physical-chemical reasons, plant based meat alternatives will never be able to be meat all relevant

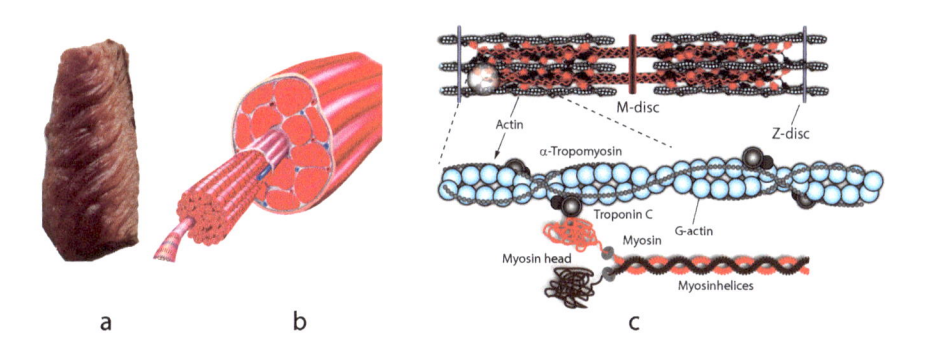

Fig. 5.38 Meat (**a**) is fibrous. Muscles consist of muscle fibers (**b**), which in turn consist of myofibrils (**c**), which are essentially made up of the proteins actin and myosin. In the top right, a muscle cell, sarcomere, is shown. Actin and myosin interlock. The myosin head can grip the actin strands at certain points and move them as a result

time and length scales of meat. Such texturing processes are thus more suitable for certain small-scale applications, such as the imitation of burgers, patties, or Bolognese sauce. In these preparations, the fiber structure is less important, and it is very easy to compensate for the aromatic deficits of plant-based proteins with sauces, spices, or oils.

5.10.11 Leghemoglobin as a Hemoglobin Substitute

Great efforts are being made to make vegan burgers visually similar to meat. In the case of conventional meat burgers, it is, among other things, the red meat juice, sarcoplasm, which provides the appropriate appearance and is a must for omnivores when enjoying it. The meat juice is also a flavor carrier, as many water-soluble flavor components such as sodium ions, calcium ions, free amino acids, peptides, nucleotides, sugars, or the sweetly stimulating adenosine are dissolved in it and round off the taste profile. At the same time, the myoglobin dissolved in the meat juice is responsible for the red color of the muscles and the meat. Myoglobin, which we already know from heme iron, is a specially folded protein that binds the heme group (Fig. 4.3). It is not dissimilar to the hemoglobin of the blood. However, hemoglobin consists of four individual proteins that combine to form a group of four, ensuring sufficient oxygen transport via the blood. The identically structured heme group with the iron ion is responsible for the red color of the muscles and the blood.

The red meat color in vegan burgers is not the problem; it can easily be achieved, for example, with the juice of beetroot and other plant-based colorants. The challenge lies in creating blood-like sensory impressions. Plant-based heme can help with this. The iron ion in the center of the animal heme group binds oxygen and thus contributes to the oxygen supply of the muscles (Fig. 5.39). Some plants also use this mechanism to transport oxygen into cells using leghemoglobin [155].

This mechanism is particularly important for the oxygen balance of plant parts located in the soil, especially in root vegetables [156] and seeds such as legumes [157]. For this purpose, legumes have so-called leghemoglobins (LHb), which, due to their completely different primary structure, are folded entirely differently than myoglobin but carry a heme group, as shown in Fig. 5.39. Another striking feature is that myoglobin and leghemoglobin, despite different folding and primary structure, have identical metal binding sites at the amino acid histidine.

By using leghemoglobin, an attempt is made to come closer to the mouthfeel of minced meat, albeit only to a limited extent [158]. This blood-like flavor results from the combination of salts, heme, and specific hydrolysis products of the heme-carrying proteins. Instead of laboriously extracting leghemoglobins from soybeans and other legumes, they can be produced in larger quantities biotechnologically, for example, through genetically modified microorganisms, specific yeasts [159, 160], which are used for the synthesis of leghemoglobin. The structure of the leghemoglobins is naturally identical, whether they come from the yeast reactor or the soybean. One only needs to make suitable yeasts ferment only these proteins

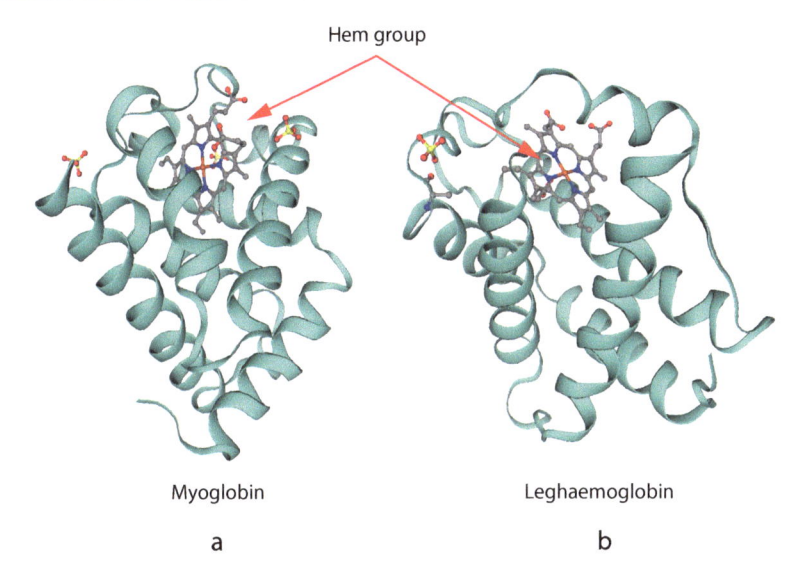

5.10 Vegan—Exclusive Exclusion and Missing Links

Fig. 5.39 An animal myoglobin (**a**) and a plant leghemoglobin (**b**) bind the heme group

and nothing else. This can only be achieved by intervening in the genetics of these microorganisms and introducing the corresponding genes of the soy plant into the DNA of the yeasts. This is better for the environment and does not additionally impair health when consuming plant-based burgers.

5.10.12 The Modified Culinary Triangle of Modern Industrial Culture

Apart from the already mentioned differences in nutritional values of plant and animal proteins, Vleisch lacks the natural food matrix of meat with all its macro and micronutrients. This includes not only the multitude of skeletal muscle proteins, but also the countless components in meat juice, the dissolved proteins and enzymes, the iron, the minerals and the (metallic) trace elements, which are found there as complex and structure-forming agents for enzymes and proteins. The best surrogate products thus reach the limits of the basis of the culinary triangle (see Chap. 2), biotechnological methods even exceed it.

This raises the question of whether such surrogate products with a high number of molecular manipulations can be located in the culinary triangle at all. The required technological steps, such as the extraction of food components, their high purification and treatment, the reconstruction of artificial food matrices, and the addition of flavor components beyond seasoning, suggest a systematic expansion of the Lévi-Strauss approach. To do this, new dimensions and levels must be introduced that take into account processing levels and food technology steps,

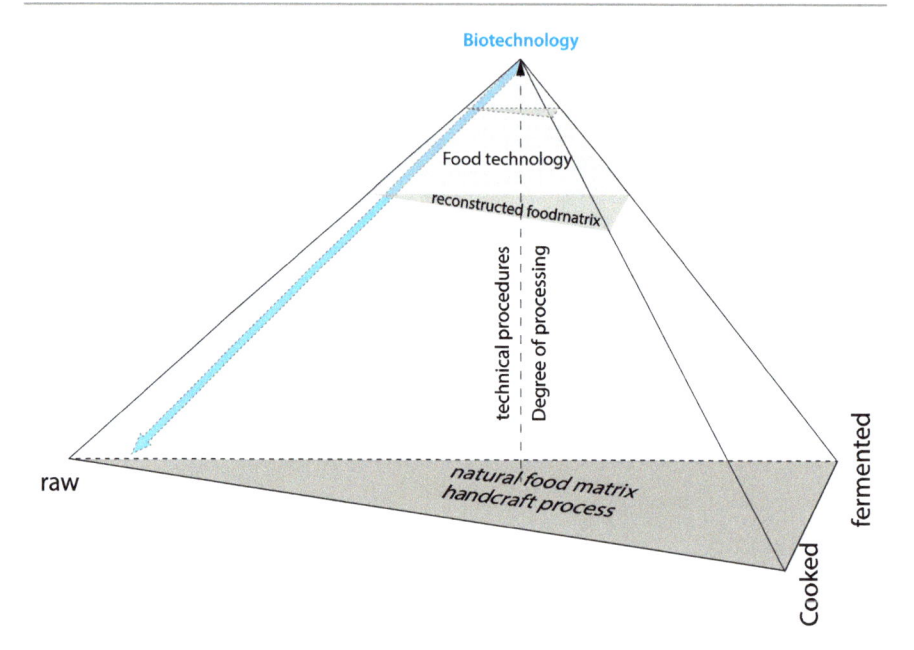

Fig. 5.40 The new technologies require the introduction of a third dimension that captures new technologies. The original triangle lies in the plane, the extension is shown in the "culinary pyramid". The blue arrow indicates the future possibility of new stem cell/3D bioprinting technology to create new, "raw" foods with natural food matrices that can be subjected to artisanal cultural techniques

such as extruding, reshaping, and flavoring. In Fig. 5.40, an extension is proposed. The original culinary triangle lies with the culinary basis of raw, cooked, and fermented on the plane. These cultural techniques use only the natural food matrix to achieve the desired changes. Modern technologies extend the triangle by one dimension. The triangle becomes a pyramid, with the top representing biotechnology and thus also processes that can only be applied in bioreactors, e.g., for *in-vitro*-meat. However, the extension is already necessary for the reconstruction of food, e.g., when new food matrices are created for meat analogues by combining proteins from various plants, thickeners, and gelling agents, etc. These food matrices no longer have any correspondence in nature and are therefore hardly producible on a craft basis.

5.11 Clean Meat—Cultured Meat from the Petri Dish

5.11.1 Meat without Animals

Consideringthe deficits that surrogate products show in flavor and nutrients, meat grown from stem cells and nutrient solutions in bioreactors seems to be a viable alternative. No animal would have to die, and the full nutrient content would be guaranteed. Meat, not only of high quality but also of the highest ethical purity, would be available to all social classes. Society seems to be divided in terms of acceptance [161, 162]. The fundamental decisions and motivations are summarized in the diagram in Fig. 5.41.

Society's attitude is ambivalent. While environmental protection, climate change, concern for animal welfare, and factory farming argue for a decline in meat consumption, supported by the resulting health concerns, a minimum trust in science and biotechnology is necessary for the acceptance of lab-grown meat. Even among population groups that value meat because it has—see evolution—been beneficial to humans since the Paleolithic era, there is a higher acceptance of incorporating lab-grown meat into their diet. For people who, due to urbanization, have come to see meat as a packaged anonymous commodity and have little to no knowledge of meat production or animal husbandry, *in-vitro*-meat would also not be a problem.

On the other hand, there are significant concerns, especially among technology- and science-averse groups. Particularly when health concerns are derived from biotechnological production and doubts about the taste for the (still) relatively high price arise. It certainly seems strange to a broad section of the population when muscles grow in the lab without a recognizable organism. Especially if lab-grown meat were to be produced on a large scale, and the reactors take on the appearance of a chemical plant, acceptance could suffer.

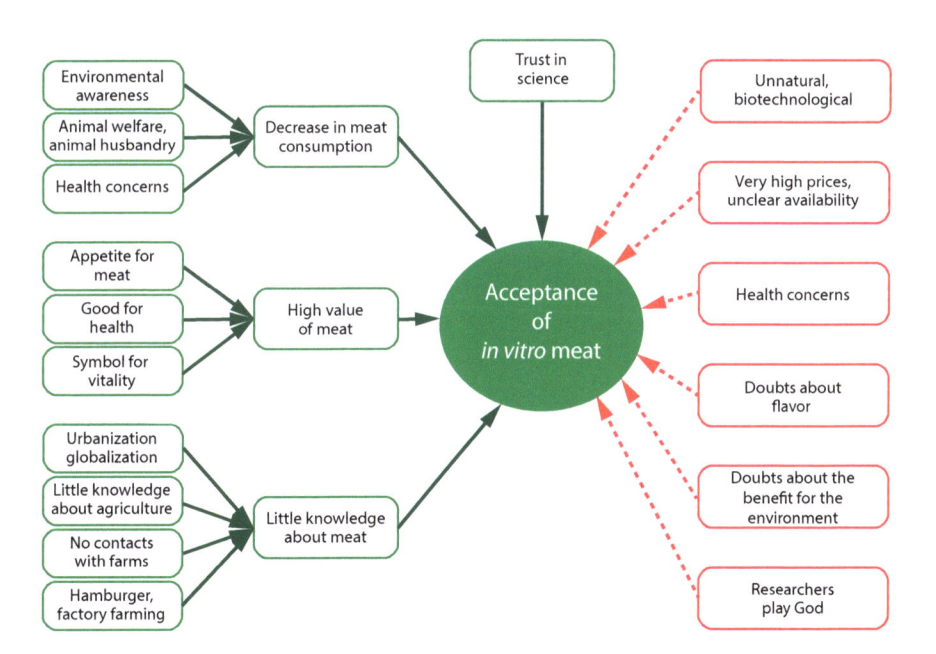

Fig. 5.41 The arguments for and against *in-vitro* meat have different causes

5.11.2 Technical Problems and Solutions

However, this does not work quite so easily, as muscles are very complex in structure (see Fig. 5.38). Countless structural proteins and sarcoplasmic proteins must be synthesized and, above all, brought to the right places. In addition, the muscle must grow under movement, as in animals. However, very specific conditions are necessary for this. Yeasts or even more complex modified microorganisms cannot do this so far. For this, as in the fetus and in living animals, stem cells as starting cells and nutrient solutions are needed. These still come from freshly slaughtered calves. A stain is also found in Clean Meat. However, solutions for this are emerging in the near future.

To get an idea of what such production facilities for on-site meat production might look like, a flow chart is shown in Fig. 5.42 [163]. Stable starter cultures must first be developed in small steps from the stem cells, with which industrial upscaling into large batches is successful. The bioreactors must always be moved, shaken, and stirred. Only then can meat production begin in a sterile environment with oxygen supply through cell growth. Muscle cells form, which after a certain time must be connected and precipitated from the liquid solution. Therefore, transglutaminase and other binding proteins are added. The aggregates of muscle cells sediment, the clear supernatant is removed, and the meat is pressed into a cake. Subsequently, the meat is, for example, chopped and minced until it reaches the end consumer. Further, slightly modified production processes are possible [164], which, however, proceed similarly in essence, as shown in Fig. 5.42.

It is clear that in order to maintain the current meat supply, Clean Meat would have to be produced on an industrial scale in large bioreactors containing more than 20 cubic meters of nutrient solution at many locations. However, it is also clear that the muscle fibers in the current processes are relatively short and are essentially only suitable for burgers, minced meat, or Bolognese sauce. Without transglutaminase or other meat glueing enzymes, as well-known from formed ham or formed meat, the myofibrils cannot be precipitated.

The methods used so far do not allow the incorporation of intramuscular fat. This is not possible with embryonic stem cells in these reactors. Furthermore, contrary to some ideas, naturally, neither shank slices nor flank steaks nor the universally popular loins can be produced biotechnically. A stirred reactor with embryonic stem cells and amino acid solutions does not replace animal physiology, nor does it replace the growth of a cow embryo and simulate the long life of a grazing ox.

5.12 Insects

Insects are indeed an excellent food source for modern humans [165]. Our ancestors already benefited from the higher-quality proteins and numerous micronutrients of insects compared to plants [166] (see Chap. 2). As already mentioned

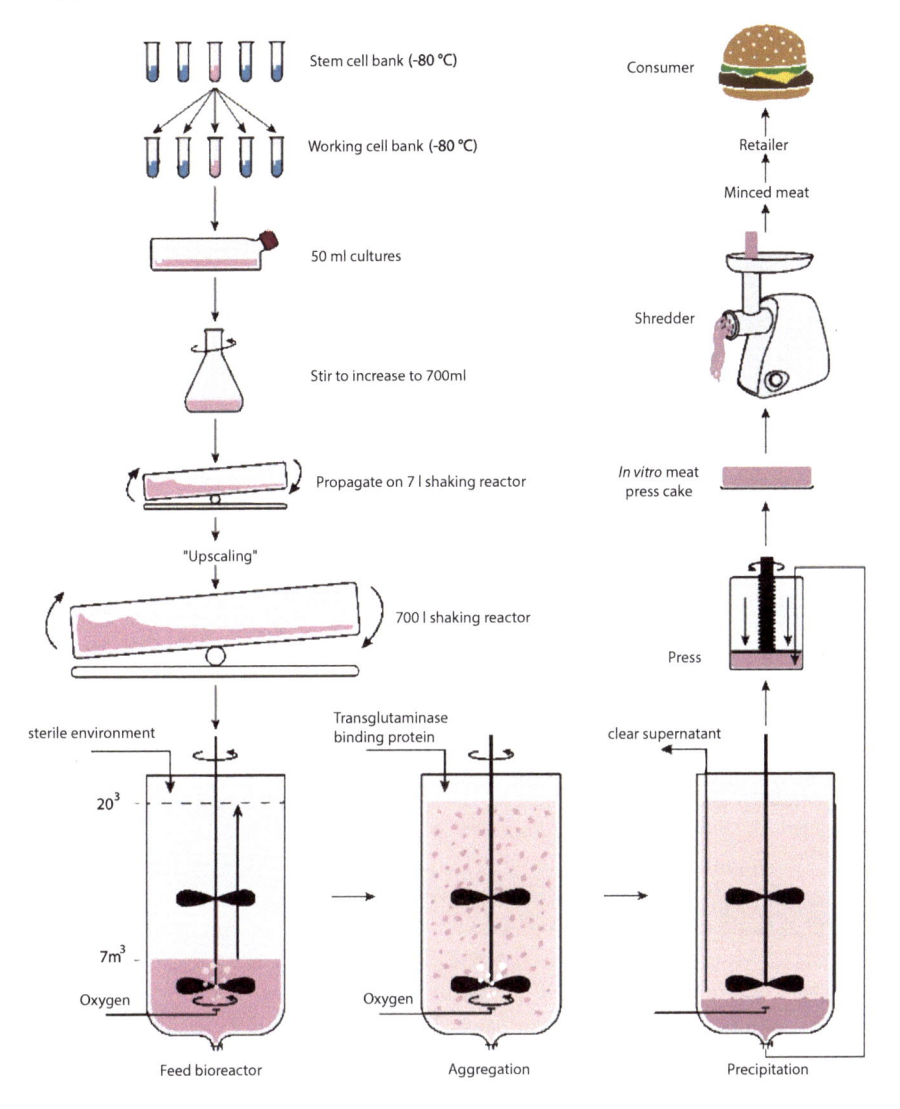

Fig. 5.42 Typical flow chart of a production facility for *in-vitro*-meat. (After van der Weele and Tramper [163])

there, insects provide the essential polyunsaturated fatty acids DHA and EPA, which are not available in plant physiology (Fig. 2.6). Insect proteins are also more equipped with essential amino acids compared to purely plant-based proteins and are more easily accessible to human digestive enzymes due to their structure, making them more biologically available.

The protein content of insects is on par with that of animal products when related to dry mass and edible mass [167], as can be seen in Fig. 5.43. This also

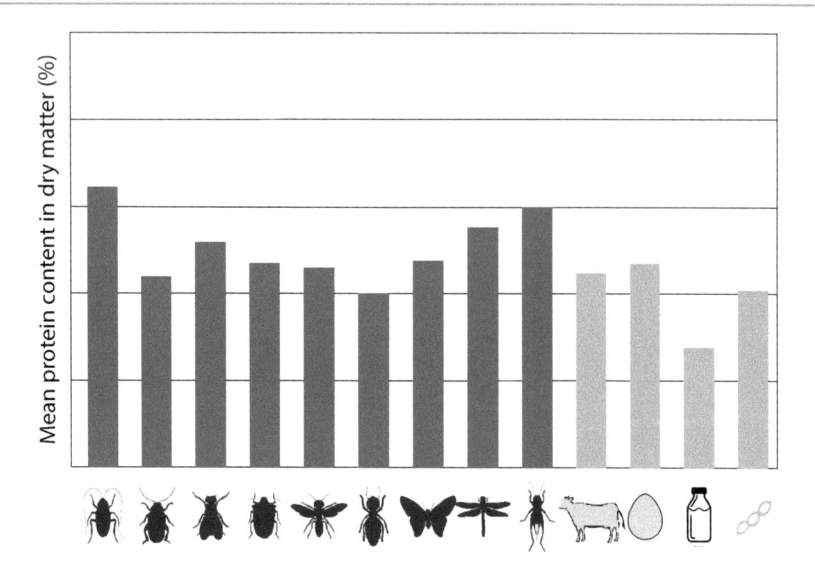

Fig. 5.43 The relative protein content of different insects compared to beef, chicken eggs, milk, and soybeans

underlines the advantage of insect nutrition in evolution. The only problem is obtaining the total mass of protein.

Burgers with a high proportion of insects, such as mealworms (the larvae of flour beetles), are repeatedly offered [168]. At the same time, the sustainability of production is promoted, which concerns both space requirements and water consumption, greenhouse gases, and the CO_2 footprint [169]. Many arguments, therefore, speak in favor of food made from insects.

The only disadvantage in European and Western dietary culture is the disgust for insects, which can be circumvented if proteins from insects are isolated to subsequently "texturize" them [170]; however, this means a higher energy expenditure than, for example, frying mealworms, deep-frying them, or processing them into burgers. Or they are heated, dried, and processed into powder. This protein flour can be used, for example, as a high-quality component in gluten-free baked goods or gluten-free pasta. Not only would the increase in nutrient content compared to non-wheat cereals or pseudocereals be a gain, but also the improved properties during processing and water binding would offer advantages.

In this sense, approaches to modern regional menus in the rapidly developing gastronomy of Denmark can also be seen [171]. Insects, such as ants, are repeatedly added [172], even alive. Often, however, as a paste that shows the unique culinary potential of the contained formic acid. At the latest then, it is worth overcoming feelings of disgust, as always when eating with pleasure. Also, the consumption of insects is a consistent step when talking about Paleo nutrition.

Only one question of bioethics remains to be clarified: Is an insect's life worth less than that of a pasture-raised cow? So what is ethically more justifiable? To kill thousands of mealworms and insects for an insect burger or to slaughter one cow for 1000 burgers?

5.13 Mushroom Proteins—New Research Findings

It is certainly questionable to break down food with great effort and then reconstruct it with further energy expenditure into Plant-Based Food. This may be useful in some cases if sustainable by-products of the industry can actually be used. It would be better to design proteins for food and ferment them using microorganisms. Quorn, the mycoprotein of certain fungal spores, is an example of this [173]. Recently, new systems have been developed that can be a promising start to entirely new ideas. Mycoproteins can be fermented from specific fungal spores [174], which can then be processed with a mixture of thickening and gelling agents such as κ- and ι-carrageenan, locust bean gum, rice starch, konjac gum, citric acid, modified starch, sodium citrate, sweet potato concentrate, and spices and salts after mixing with water to create sensorially acceptable vegan sausages that come closer to a conventional, animal sausage in comparison to conventional analog sausages made from sunflower proteins or soy proteins. A significant advantage is that these sausages fall under craftsmanship again. The sausages for sensory testing were cooked in a standard kitchen appliance, a Thermomix®, and can even be made by oneself, provided that the protein-carbohydrate mixture, the mycelium, is available from the fungi.

Furthermore, oyster mushrooms are generally of great interest: The protein mixture (so far 17 known proteins and enzymes) was obtained from oyster mushroom spores *(Pleurotus sapidus)*. The main proteins are the copper-binding enzymes laccase and laccase2, proteins consisting of 521 and 531 amino acids, respectively [175]. Both proteins have five cysteines, which provide a sufficiently wide-meshed and elastic network. This is mainly because some of the cysteine is located at the ends of the molecular chains. This is also one of the decisive advantages of high-molecular-weight glutenin in bread baking and the associated network formation for the crumb structure. Laccases also cross-link with polysaccharides, such as hemicellulose and the arabinoxylans found in cereals, so that in the future these oyster mushroom proteins will find their functional place in some of the applications discussed in this book, such as gluten-free baked goods, vegetarian and vegan surrogates, and others. The fungi or their spores show biophysical synergies between the cell material β-glucan and proteins. The resulting food products have high stability and excellent textures.

Another advantage is the strong water solubility and function of laccase, which is ensured by a high number of charged amino acids (Fig. 5.44). The positive and negative amino acids are almost evenly distributed along the protein chains,

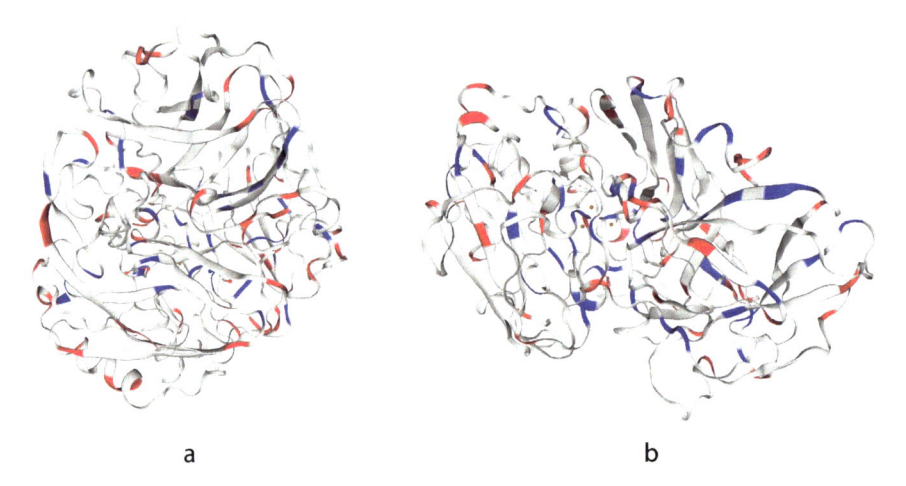

Fig. 5.44 Laccase (**a**) and Laccase2 (**b**) from oyster mushroom spores. The electrically charged areas of the two enzymes are marked in red (minus) and blue (plus)

which arrange themselves in a selected manner with the negatively charged gelling agents κ- and ι-carrageenan.

The resulting loose network structure, together with the bound water, allows for a crispy, sausage-like texture. Once again, it becomes apparent that function and sensory properties are always defined by the physical properties at the molecular level.

5.14 Fast Food, Highly Processed Foods—Curse or Blessing?

Regardless of the perspective from which it is viewed: Alternative products are highly processed products made through industrial processes, and the question repeatedly arises in science: Does the high degree of processing, whether in the convenience food sector, with Fast Food or with reconstituted foods, harm health in any way? So far, nutritional sciences have only provided conjectures from observational studies. A sensational and carefully conducted study on a mouse model was published only recently [176]. In this study, mice were fed a Western diet, i.e., high in sugar, high in fat, and especially too much of it. The animals subsequently developed significant inflammation that spread throughout the body and resembled an infection by dangerous bacteria rather than poor nutrition. An unexpected increase in certain immune cells in the blood was observed, which was interpreted as an indication of the involvement of bone marrow precursor cells in the group of mice that were fed this diet. Therefore, the bone marrow was examined, and it was found that the strong defense activity was genetically activated. Among other things, the genetic material for the multiplication and maturation of immune cells was affected. In the mice, fast food led to a strong defense reaction that was not triggered by infections. In the long term, even immune cells in the DNA are virtually reprogrammed. Certain areas in the DNA sequence can then be read more frequently, and the immune defense is permanently activated (Fig. 5.45). The molecular genetic pathways of these mechanisms can be elucidated in detail. In parallel, blood tests were carried out on humans, and similar mechanisms were observed through the serum.

In Fig. 5.45, the relationships are greatly simplified. With a mixed diet without an excess of fat, sugar, and carbohydrates, no anomalies in inflammation values can be detected. An extraordinary defense reaction is not triggered. The situation is different with excessive consumption of fast food and the associated excess of fat and sugar. The immune system, especially inflammasomes, cytosolic protein complexes in macrophages and neutrophil granulocytes, are apparently stimulated by fast food or the excess of sugar or fat. As a result, precursor cells are formed in the bone marrow, which manifest themselves through a reprogramming or change in the shape of the DNA. This, too, is part of the physical properties, because in order to pack the approximately 1 m long DNA into the chromosomes of the cells, it must be cleverly wound up, as shown schematically in Fig. 5.46.

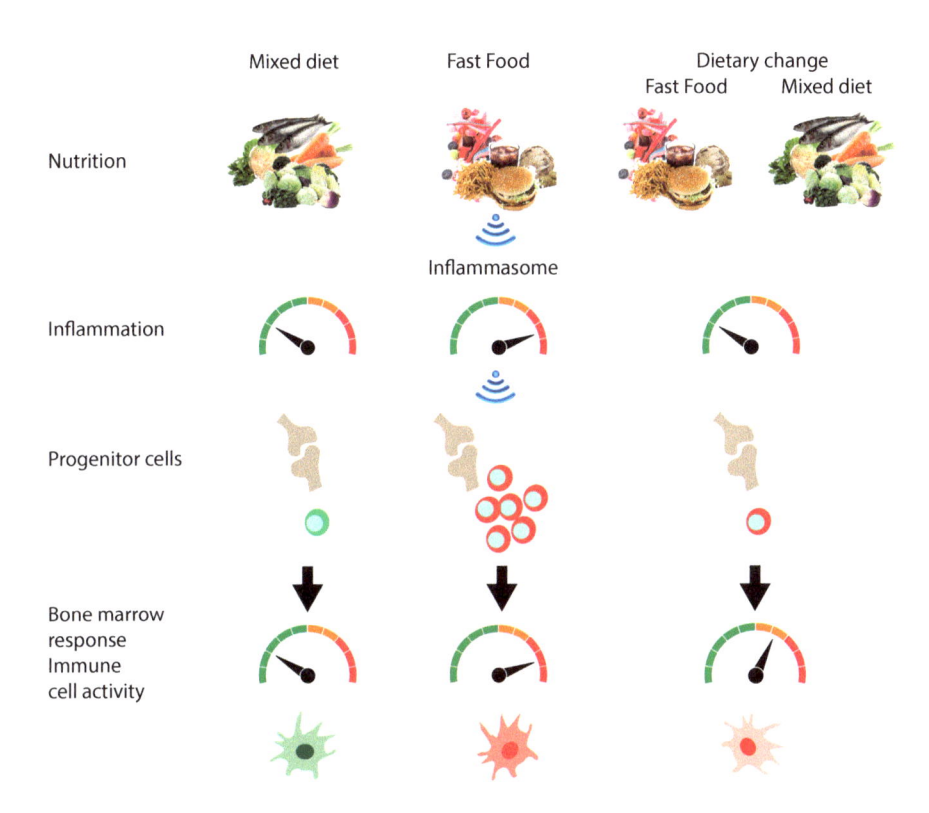

Fig. 5.45 The difference between a mixed diet, as is known, for example, from a Mediterranean cuisine, and Fast Food, rich in sugar, fat, and carbohydrates, can be measured quantitatively

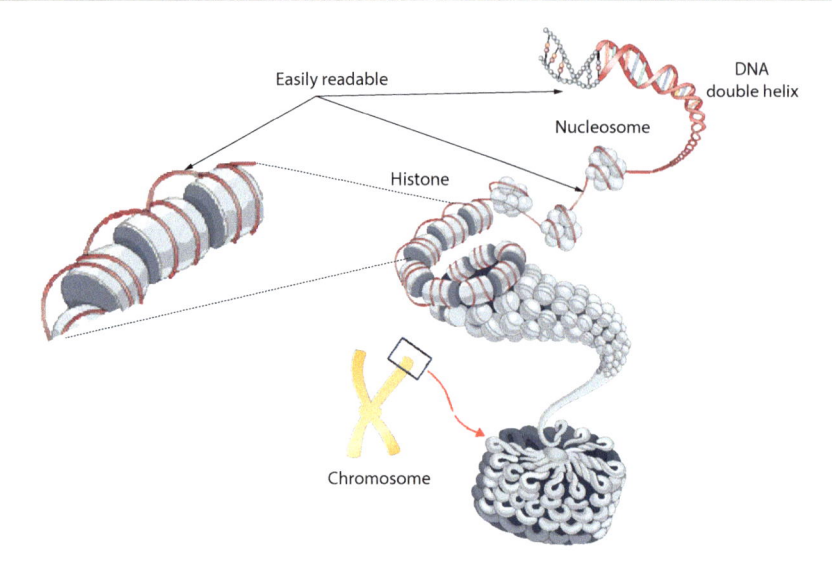

Fig. 5.46 The hierarchical structure of DNA

In the chromosome, there is a detailed hierarchy in structure and length scales, showing how the double helix wraps around histone and nucleosomes. This saves length. In this resting state, however, many of the pieces of DNA are not readable, as they lie tightly on the proteins of the histone. Apparently, these pieces are loosened for reading, so that now parts of the information in the exposed sequences of base pairs can be read, ultimately leading to the reprogramming of the cells.

It is also worth mentioning that even after a change in diet, the inflammation mechanisms persist for a considerable time (right column in Fig. 5.45). This is not surprising, as the changes in a complex macromolecule like DNA as well as the highly complicated interactions are not achievable within a short period of time.

This is indeed remarkable, as for the first time, the effects of "poor nutrition" or even "unhealthy eating" seem to be measurable at the molecular level. Previously, there were only opinions on this, which could hardly be justified or verified. However, it still needs to be checked to what extent the amounts with which the mice and the control group were fed can be transferred to humans and also, over what period fast food and only fast food must be eaten until effects occur. A burger meal with fries once a month definitely does not cause inflammation, but daily consumption accompanied by sweet drinks, sweets, and convenience meals is more likely. Fast food and some convenience foods are therefore only a limited blessing for modern humans. Whether there is a real causal link to an increased mortality rate is not yet known, although there are clear epidemiological indications for it [177].

Our brain also plays a significant role in this, because when we eat well, we are rewarded. The reward center is responsible for this—and, in simplified terms, the biochemistry of neurotransmitters. Schymanski [178] has developed a very

5.14 Fast Food, Highly Processed Foods—Curse or Blessing?

illustrative scheme for this, which seamlessly integrates into the useful triangular relationships repeatedly mentioned here (see Figs. 2.29–2.32, 4.1, 5.1 and 5.40) as shown in Fig. 5.47.

Even with this highly simplified representation, it becomes clear how quickly one can fall into the vicious cycle of constantly needing to eat in a society of abundance with an excess of food. In the supermarket, the colorful candy packaging tempts, the bakery's bread aroma, and the smell of broths at the sausage stand. A little hunger is present, the appetite is great. So, food is bought and eaten. The full expectation and reward storage is right, releasing dopamine, and the reward center reacts positively. The person is happy. Dopamine is broken down from the reward center, glutamate to γ-aminobutyric acid (GABA) is broken down. GABA is an endogenous messenger substance that is formed as a biogenic amine by decarboxylation of glutamic acid and is transferred to the appetite center. If the concentration of GABA there exceeds a threshold, the appetite increases, one becomes hungry and greedy again, and the motivation increases. Activity increases when other incentives are lacking, be it hunting for hunters and gatherers, worthwhile work for planters, farmers, and shepherds, sporting successes, praiseworthy business for merchants, or promising research tasks for scientists. If external incentives or successes are lacking, the way out is to reward oneself with food. The

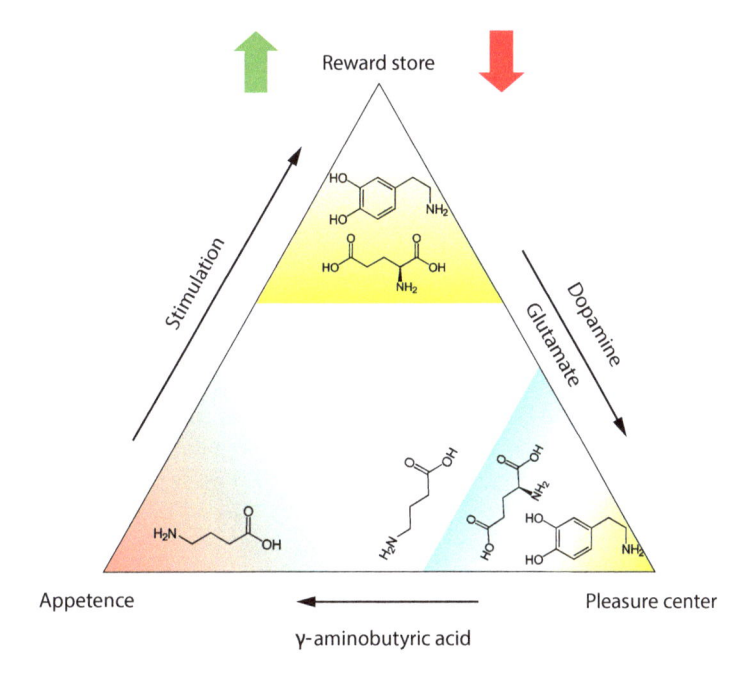

Fig. 5.47 The reward triangle according to Schymanski with the appetence, the drive (left), the storage for neurotransmitters such as glutamate (glutamic acid) and dopamine (top) as well as the pleasure center (reward center), in which glutamic acid is chemically converted to GABA, the γ-aminobutyric acid, and transferred to the appetence center. The glutamate-/dopamine storage is only sufficiently replenished during rest periods (without stimuli)

reach for a snack or a sweet follows. The cycle starts anew. As long as the dopamine-/glutamate storage has been well filled in the meantime, this is not a problem. Satisfaction is achieved through the small snack, the candy, or the little piece of chocolate. If the storage is not sufficiently filled, the snack becomes larger, and at the same time, the intervals between snacks become smaller, especially when other stimuli of this fast-paced time are added, such as likes on social media, favorite music on the radio, or the text message from the beloved life partner. All these rewards draw on the storage, emptying it. Only in rest phases can it be steadily and abundantly filled. If rest phases are lacking, self-rewarding with fast food, convenience bars becomes more likely. It doesn't matter whether it's organic, vegan, labeled with "honey instead of sugar" or even as superfood, all of this ultimately provides completely unnecessary amounts of energy. Vulnerable people are drawn into this cycle. This works particularly well with glucose and glucose-rich foods. However, this only economically rewards those who produce such superfluous products, including those who make products from wheatgrass, oat flakes, sunflower seeds, *Spirulina,* goji berries, and anticarcinogenic turmeric enrichment. One of the simplest, most banal ways to break this vicious cycle would be to enjoy the emerging hunger and recognize it as a luxury. There is an abundance of food in the Western world, and that is precisely the problem.

The "artificial" glutamate in food is not to blame. The neurotransmitter glutamate, and thus the glutamic acid, is produced in the brain according to biochemical laws from glutamine whenever necessary. The origin of the structurally identical functional parts of the glutamate and glutamic acid molecules is irrelevant. Incidentally, a new idea of insight emerges here: everything we think and feel, our moods, are ultimately nothing more than finely tuned, individual biochemistry and thus the result of chemical reactions in the brain. This is also evident in another area.

At the neurological level, there are also remarkable findings [179] that deal with the ways in which we assess food in the first place. Central questions are how the food we swallow and pass on to the gastrointestinal tract is processed in terms of signaling beyond gustatory perception. Much suggests that two separate different systems influence food selection. One system directly links to the nutritional value of foods and is based on metabolic signals that reach the brain. This nutrient sensor system seems to play a crucial role in regulating the neurotransmitter dopamine, determining the value of foods, and selecting foods. In the second system, conscious perceptions such as taste and beliefs about calorie content, cost, and health of foods are also important in the selection.

As tomographic studies have shown, calorie-rich foods activate the striatum in the human cerebrum, and the strength of these reactions is regulated by metabolic signals. Blood sugar levels rise, particularly after consuming carbohydrate-containing beverages and foods—the intensity of the reaction can be correlated with the sight and anticipated sweet taste of the beverage or food. Glucose, as a necessary fuel for cells, is therefore necessarily associated with a metabolic signal, namely the release of dopamine.

5.14 Fast Food, Highly Processed Foods—Curse or Blessing?

Observations in humans suggest that the strength of metabolic signals in the brain is independent of conscious perceptions, for example, how real the food is. The same striatal reactions to the calorie-promising taste, which were so closely linked to changes in plasma glucose levels, were not related to the preferences of the study participants. This is consistent with additional magnetic resonance imaging studies showing that the actual energy density, not the estimated energy density, determines the willingness to purchase food along with the reward response. The situation is similar for high-fat products.

The independence of the neural evaluation systems, the gustatory and the metabolism-related, can be impressively demonstrated in experiments with rats: If glucose or fat is administered directly into the intestine, the dopamine release increases just as strongly as with the oral administration of the corresponding nutrient solution, suggesting a direct communication via physiological signaling pathways between the intestine and neurons in the brain. However, if the vagus nerve is severed in the animals, which is responsible for taste sensations and tactile stimuli on the tongue, only the appetite for fat is inhibited, while the appetite for carbohydrates is not affected and is therefore independent.

The discovery that a non-conditioned, stimulating food intake arises from a signal that is independent of sensory pleasure may seem surprising at first glance. After all, all organisms must supply themselves with energy to survive, even if some brain functions that support conscious food intake no longer work. The brain then relies on signals that transmit information about the properties of the food from the intestine to central circuits in the brain. Nutrition is then regulated independently of consciousness, so that at least the supply of the main fuel for cells, glucose, is possible. Information from the intestine to the brain is therefore crucial for an accurate appreciation and a plausible assessment of the energy of food. These new findings from research are summarized in Fig. 5.48.

These are bad news for the consumption of highly processed foods. In many cases, they are much too energy-dense compared to well-known natural foods. At the same time, they are presented and constructed in such a way that people can hardly resist and escape the reward cycle (Fig. 5.47). The nutrient density drives the neural signals further, much more is consumed than actually corresponds to the energy requirement. The result is overweight due to misinterpreted signal transmission.

In addition, parts of the convenience industry place great emphasis on taste, texture, and mouthfeel, which are controlled by combinations of caloric and non-caloric additives. Beverages or dairy products thus contain energy carriers such as glucose and fructose, calorie-free sweeteners such as sucralose and acesulfameK or stevia extracts (see Sect. 4.8.1). Energy content and taste are completely out of balance. As a result, according to the signaling pathways, more is consumed and drunk than is physiologically necessary, the brain is overstimulated, the constantly released dopamine causes a long-term reprogramming of food selection. The total calorie intake increases, obesity and its consequences become more likely. A stricter restraint on ready-made products is therefore not a mistake.

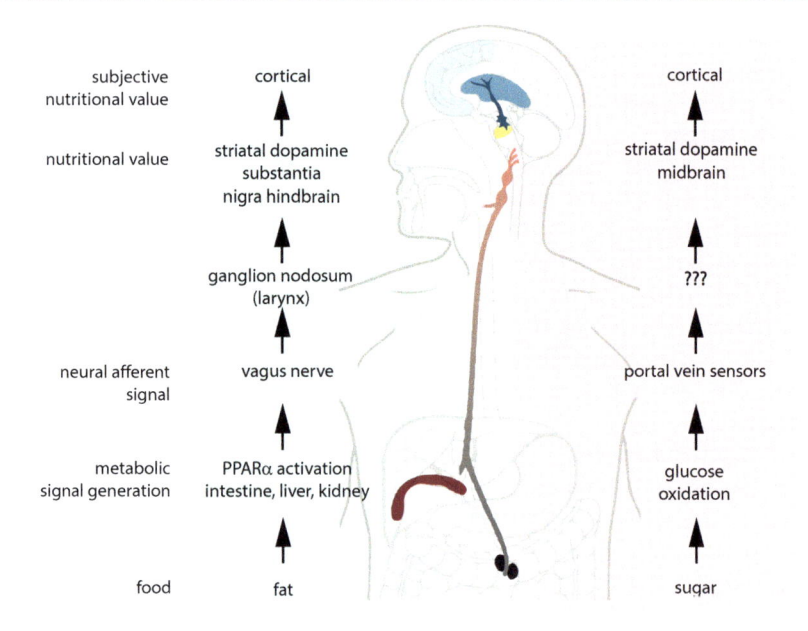

Fig. 5.48 The signaling pathways for fat and sugar run through different systems [177]

5.15 Superfoods

What does science say about superfoods? Apparently indispensable foods with extraordinary nutrient contents that promise health and long life [180]. Chia from South America is preferred over domestic flaxseed [181], berries from distant countries are almost given medicinal status [180], so that native wild blueberries fade in comparison—despite a similarly high content of anthocyanins. Even the simple kale is offered in summer under its Anglo-Saxon name "Kale" in the non-English speaking parts of the world, just because glucosinolates became known as apparently functional ingredients, which are also abundantly present in all cabbages and onion plants.

In these superfoods, it is mainly about those secondary plant substances to which superpowers are apparently attributed. However, this idea is basically not wrong in the light of evolution. The ancestors of hominids already ate wild berries and fruits of the respective season, as long as their sense of taste did not prevent them. If the berries and fruits were too sour, they were unripe. The content of plant defense substances (FODMAPs) was still much too high at this stage, so not particularly healthy, as these in excess act toxic, at least highly antinutritive. If the berries and fruits were bitter, they were not suitable for nutrition. This part of the seasonal diet continued up to hunters and gatherers. In addition to food from hunting, fruits and berries formed a large part of the developing food culture. It was

5.15 Superfoods

not until the Neolithic period that breeding successes increased the supply of these foods. The stock and supply improved significantly.

Essentially, apart from minerals and trace elements, the so-called superfood is mainly about the antioxidant properties of certain substances contained in berries, fruits, and seeds. This also happens systematically in nature, as berries, fruits, tubers, and seeds contain the offspring of the plant and thus the most valuable parts of the plants. From this, the deeper meaning of the standard phrases "many ingredients in the apple are located in and directly below the peel" (polyphenols) as well as "the peel of potatoes should not be eaten because of their toxicity" (solanines, lectins, agglutinins) becomes clear. That are the places where predators attack first. The secondary plant compounds are initially nothing more than the chemical weapons of fruits and vegetables against insects and other herbivores. In the case of berries and fruits, they are also exposed to the sun and thus to the energy-rich ultraviolet (UV) radiation, which quickly causes sunburn in fair-skinned people. Therefore, molecules must be incorporated that absorb harsh UV radiation before the sun-exposed plant parts are damaged.

Since every cell, every life consists of self-organizing molecules, oxidation—chemical electron transfer processes—indeed plays a decisive role. Therefore, practically everything that is present in the cells can be oxidized, lipids, proteins, and of course the DNA in the cell nucleus. Responsible for this are free radicals, which can form under spontaneously occurring and practically unavoidable processes (see Fig. 1.5). These free electrons or molecule parts, in which, for example, an electron is missing, are highly reactive, as they strive for bonding and neutralization. The missing charges must be replaced quickly. These free radicals, therefore, have a high affinity and take charges from neighboring amino acids, phospholipids in the cells, DNA, which can be damaged. Sometimes new free radicals are created, a whole cascade of electron exchange begins—with dire consequences if the oxidation processes

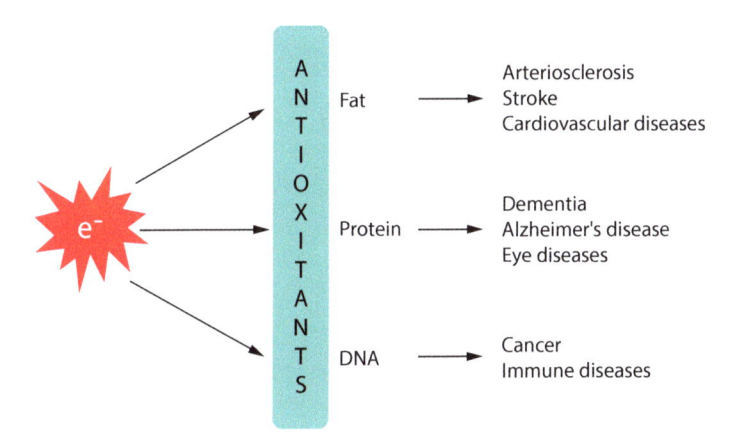

Fig. 5.49 Oxidation processes in different functional systems cause diseases and trigger inflammatory processes

are not stopped (Fig. 5.49). If such charge transfer processes occur too frequently, the repair mechanisms are no longer sufficient. The affected cells lose their function.

Such oxidation processes also play a major role in aging [182]. People are afraid of this and reach for antioxidants. These figuratively stand in the way of free radicals and stop their activity. Typical antioxidants include polyphenols and terpenes, as plants also protect themselves with these molecules from oxidation, energy-rich UV radiation, and sometimes even from predators, as polyphenols have a diverse mode of action—and cause a bitter taste on our tongue. If the chemistry is right, astringency is also triggered, that "dry" and "astringent" mouthfeel that we know very well from red wine, tea, and walnuts.

5.16 Secondary Plant Compounds

5.16.1 Polyphenols

Phenols and Polyphenols are a class of substances in chemistry. Their name is derived from the simplest phenol, the benzene ring, as shown in Fig. 5.50.

The special feature of this typical phenol structure is evident in the behavior of the electrons [183]. The π-electrons, which are indicated in the dumbbells in Fig. 5.50b, will delocalize in the case of several successive conjugated double bonds (in the structural model, Fig. 5.50a) and show their highest quantum mechanical probability of presence above and below in ring-shaped tubes. The intuitive reason for this delocalization is simple: The positions of the double bonds can be arbitrarily interchanged as long as the conjugation is maintained. The benzene ring thus already shows a basic chemical prerequisite for activity as a radical scavenger: It has electrons that are delocalized in a π-orbital system and are therefore, very naively speaking, more readily available than quantum mechanically firmly bound

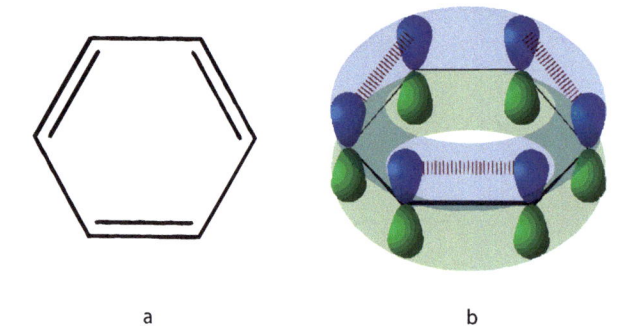

Fig. 5.50 A benzene ring—**a** structural formula, **b** orbital model – already shows the basic principle of the antioxidant effect (even though the depicted compound benzene has carcinogenic effects for other reasons). The delocalized π-electron system (blue and green "rings") plays a major quantum chemical role

5.16 Secondary Plant Compounds

electrons in an atom or molecule. The simplest phenol (carbol, hydroxybenzene), despite its "alcoholic" name (-ol as a suffix), is chemically an acid of benzene because there is a hydroxyl group (OH) on one of the carbon atoms.

Polyphenols, of which there are several thousand different structures in plants, now have several of these benzene rings in their structure and all have a high potential of π-electrons. Therefore, polyphenols are perfect free radical scavengers or antioxidants. In the natural product chemistry of plants [184], all phenolic secondary plant substances can be classified under this umbrella, as indicated in the scheme in Fig. 5.51.

The plant-based polyphenols, which still act as strong antioxidants in food, have fundamentally different tasks in the plant: Due to their richness in electrons, they are primarily used as easily oxidizable metabolites. This allows them to bind metal ions and precisely control metabolic processes in the cell. However, for many of these substances, *in vitro* effects have been demonstrated, and it has been concluded that their consumption helps humans stay healthy [185]. Countless studies on the positive effects of fruits, vegetables, and tea have been published, which cannot be cited here without exceeding the scope. There is much truth in this, which basically does not need to be verified, because even before human evolution, these foods were the basis of the diet of hominids and their ancestors in the immediate environment of gathering. And chemistry is right: The willingness of these phenolic structures to provide their π-electrons helps to stop oxidation processes, even spoilage processes, or at least slow them down. That is precisely

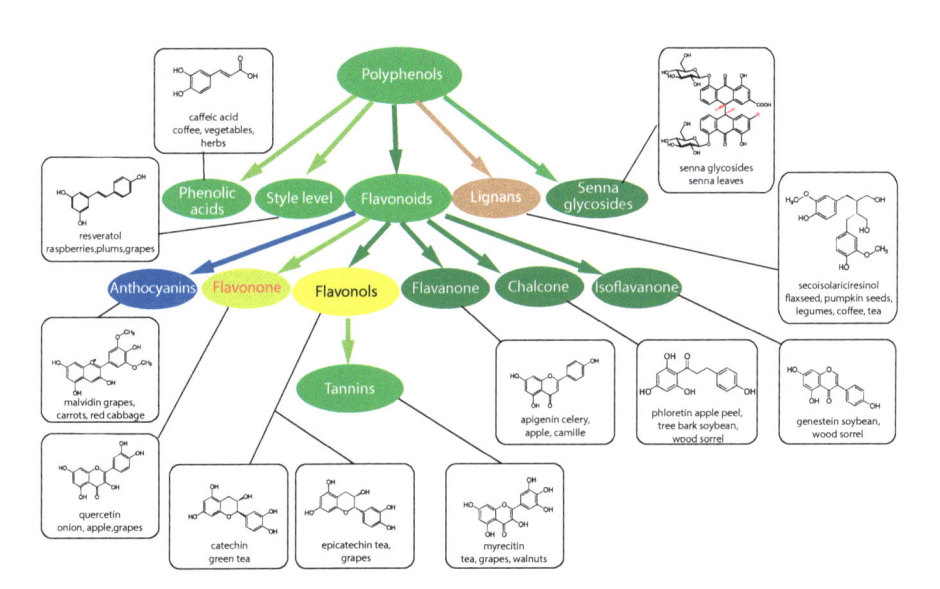

Fig. 5.51 Examples, occurrence, and classification of the various polyphenols. The indicated colors also show the contribution to the coloring of fruits and vegetables

why many antioxidants also serve as natural preservatives. They oxidize first before free radicals oxidize the food molecules—thus preserving the food [186]. Polyphenol-rich rosemary extract (Rosmanol, see Fig. 3.38) even protects against bacterial contamination in raw sausages, ascorbic acid in flour, just to name two examples.

Learning the bitter taste is therefore of great importance, as many of these polyphenols taste bitter and are astringent with a certain chemical structure (see Fig. 5.52). Only the acceptance and assessment of the bitter taste, apart from the alarm signal, makes it possible to eat vegetables and bitter-tasting roots. This taste had to be learned in order to make an assessment between poison and well-being. To this day, this is the case with small children: The bitter taste is the last taste sensation to be learned in later childhood and adolescence.

Bitterness and astringency are fundamentally different perceptions, even though both causes can be traced back to polyphenols. In this case, subtle differences in the molecular structure play a central role [187–190]. Astringency, the "dry tightening in the mouth" when enjoying certain vegetables, tea, red wine, or chocolate, is not a taste stimulus. This sensation is chemically triggered by stimulation of the trigeminal nerve, also by specific phenols, the tannins. The difference between bitter stimuli and astringency is expressed by the number of adjacent OH groups on the phenol rings. While only two OH groups are essential for the bitter taste, at least three OH groups must be present on at least one benzene ring for the astringent trigeminal stimulation, as shown in Fig. 5.52.

Bitter taste and astringency can therefore be precisely distinguished from a molecular perspective. This is also true physiologically, as the stimuli are triggered on completely different receptors and in different physiological systems. In red wines that have been left on the mash for a long time and aged in barriques, this difference between bitter notes and astringency can be most easily experienced. Astringency is also found in many types of vegetables, with gallic acid being the main cause of chemosensory astringency, such as in vegetables like eggplants, cardoons, cucumbers, papayas, purslane, and even in soybeans. Thus, the chemical

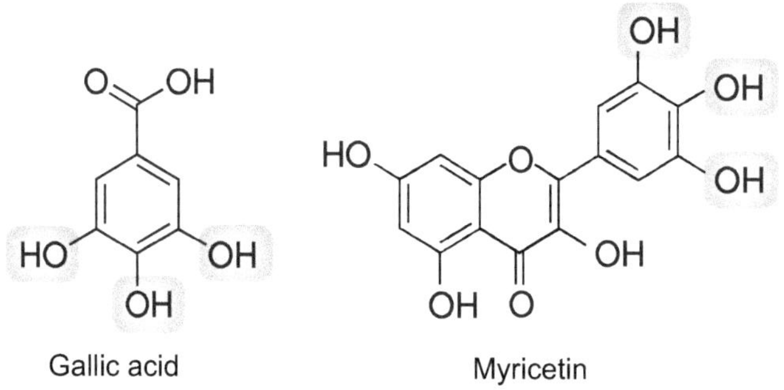

Fig. 5.52 (Poly-)Phenolic compounds with a group consisting of a benzene ring with three OH groups are responsible for the (chemical) astringency. Gallic acid has the strongest astringent effect, and the compound myricetin is shown as another example. They belong to the tannins

sensory experiences of taste and astringency provide targeted information about the ingredients of food.

5.16.2 Carotenoids: Delocalized π-Electron Systems

The universality of this idea of conjugated π-electrons is also demonstrated by carotenoids, whose main function in leafy vegetables is to support photosynthesis. The energy of light must be collected by the plant so that it can be converted into another form of energy that benefits the plant. To do this, electron-rich molecules are provided that are specifically receptive to the wavelengths of sunlight, i.e., visible light, and thus can convert the energy of light waves into chemical energy for plant growth. A whole series of proteins are embedded in the cell membrane, including so-called light-harvesting proteins—known as light-harvesting complex (LHC)—which, due to their special shape, can incorporate a large number of different dye molecules, as shown in Fig. 5.53 [188]. However, it is not very useful if the molecules are not present in high concentrations or are arranged randomly, as the (sometimes weak) light for the plant's metabolism must be effectively collected. Therefore, these electron-rich dyes must be well-aligned and highly concentrated in the leaves and fruits. This alignment is achieved by proteins whose natural design is completely geared towards the integration of dyes.

Fig. 5.53 The antennas of a light-harvesting complex in the membrane in side view (bottom) and top view (top). The membrane is indicated by the two horizontal lines. The protein can capture many dye molecules due to its shape. The planar structures (visible in the top view) symbolize the various chlorophylls, the linear ones the various carotenoids

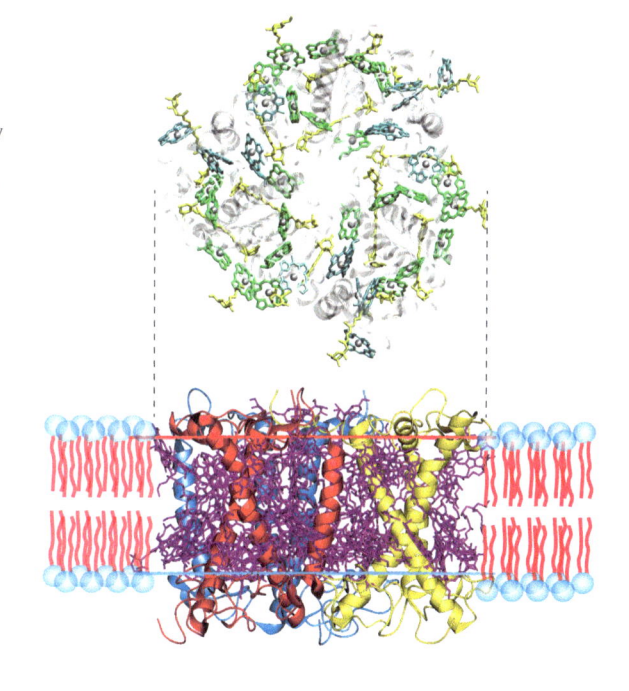

In order to maximize the yield, several individual light-harvesting proteins organize themselves in the membrane into a ring-shaped complex in the cell membrane, so that a significant amplification of the antenna effect can occur. Six to eight of these individual proteins combine to form complexes in the membrane, locally increasing the concentration of dye molecules to a very high density, which would not be possible without the interaction of the light-harvesting proteins in the membrane. The diameter of the light-harvesting complex is 7.3 nm. It is composed of three identical proteins that contain various chlorophylls, lutein, neoxanthin, violaxanthin, antheraxanthin or zeaxanthin, β-carotene and lycopene in high density to collect a broad light wave spectrum with high efficiency. The special arrangement of the proteins, as indicated in Fig. 5.53, allows the color-active molecules to be packed at a very high density.

Natural vegetable colors are therefore determined by the behavior of light-harvesting complexes (LHC II) in the cell membrane. The high concentration of dye molecules in plant leaves is essential for the plant's survival [189]. In this photosynthesis apparatus, the energy of light, which is proportional to the frequency, is collected via the electrons of the dyes and converted into chemical energy in complicated processes, which is required for plant growth.

In order for the light waves or light quanta to be received by these dye antennas at all, the carotenoids, like the chlorophylls, must be electron-rich. As is known from quantum physics, the absorption and emission of light occur through changes in electron states [190]. Light (photons of different energy, or light waves with different wavelengths) excites electrons to states of higher energy; when they fall back, they emit light. If this is absorbed, this energy can be used by the cells to keep plant chemistry and cell physiology running. Therefore, carotenoids (and thus fruits and vegetables) shimmer in typical colors, as different carotenoids absorb different wavelengths/energies (from red to blue). This is achieved by carotenoids and chlorophylls with tiny differences in electron structure, as exemplified in Fig. 5.54.

The tiny structural changes in the carotenoids make the vegetables (as well as flowers) shine in bright colors. Nutritionally, these pigment molecules are highly welcome. The many conjugated double bonds allow the electrons to delocalize along the molecular structure, so they act, very similar to polyphenols, as strong antioxidants. Behind these different systems is the same physical-chemical principle. Colorful fruits and vegetables are therefore the predestined foods for the basic supply of antioxidants. Animal foods do not contain these substance classes at all or only very limited, but have other advantages. It is exclusively the original biological function in the organism of plant and animal that gives exactly these special properties of the respective macro- and micronutrients. These properties must be seen in the entire molecular context. What is important with β-carotene is not only that it reacts under oxygen reaction to two vitamin-A- (retinol-) molecules, but the whole environment. As a densely packed crystal lattice in the chloroplasts, as it is in the raw carrot or generally in the raw food, it is difficult to be

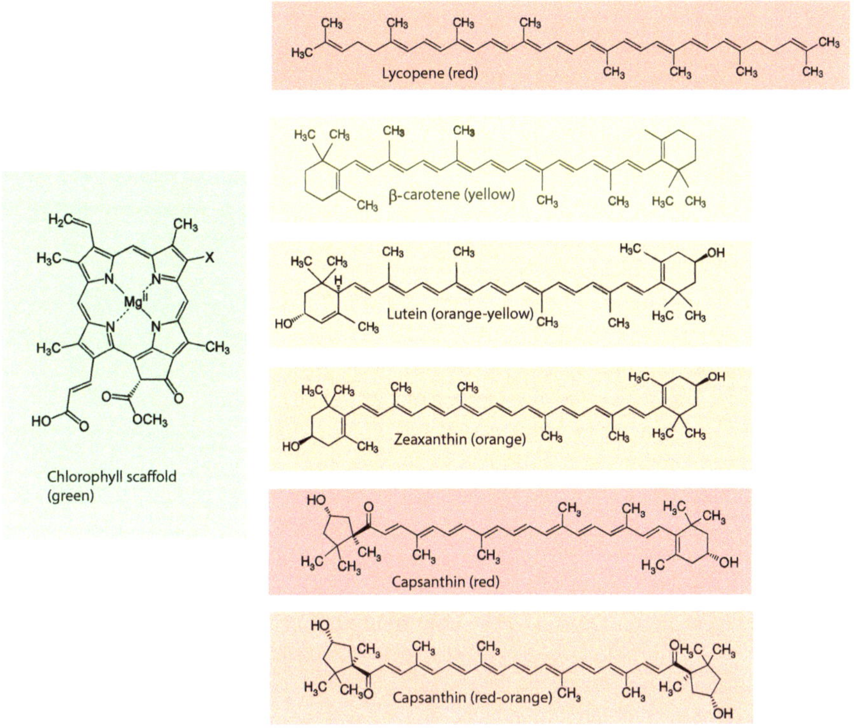

5.16 Secondary Plant Compounds

Fig. 5.54 Shimmering fruit and vegetable colors as a result of electron-rich carotenoids and chlorophylls in the light-harvesting complex and in the chloroplasts, the dye containers of root vegetables and fruits. The many conjugated double bonds allow electrons to delocalize along the molecule

biologically available. Hardly any vitamin A is formed from provitamin A after consuming raw carrots.

But the food matrix also plays a role in physiology: Therefore, it is not surprising that β-carotene taken in isolation as a supplement has little to no effect, sometimes even counterproductive [191]. The interactions are much more complicated, as impressively demonstrated in model experiments [192]. The molecular relationships, and only these, should serve as a guide for a healthy diet.

5.16.3 Why Plants and Seeds Cannot Be Superfoods

The plants and fruits chosen as superfoods remain what they are: fruits and vegetables. Green cabbage does not get better when it is renamed kale. Chia seeds brought from distant countries are by no means superior to local flax seeds.

Even if chia seeds have a slightly higher proportion of α-linolenic acid, this is irrelevant for nutrition in the overall balance and the time average. The conversion to the essential long-chain, polyunsaturated n-3-fatty acids DHA or EPA is not more effective. Also, their storage protein is never as high-quality as animal proteins from egg or whey, solely because of their completely different molecular and biological function. They can only be "super" in a slightly higher proportion of minerals or micronutrients. But this does not matter in a varied and above all enjoyable diet. The natural fluctuations of the values are drowned out by the noise of the other foods consumed.

Food is not medicine. Healthy eating does not cure colds, measles, or cancer, nor does it generally prevent serious illnesses. Healthy eating does not prevent any of this either, although a diverse, varied diet has clear advantages. Otherwise, we would have to ask ourselves how many of the healthy walnuts, aronia berries, or kale leaves do we need to consume daily? Three? Four? Or 20? Before or after eating? These questions are, of course, nonsense.

It is much easier if we consume natural foods from trees, bushes, fields, farms, lakes, seas, and stables. They have helped *Homo sapiens* to achieve their incredibly magnificent development with today's fantastic life expectancies.

The only chance, however, lies in the nutrition-related slowing down of a strong methylation of DNA [193]. This methylation (Fig. 5.55) can have various causes: chronic stress, high-performance sports, hunger, or constant overnutrition—influences of all kinds. Responsible for this are certain enzymes, DNA methyltransferases, which chemically bind methyl groups (CH_3) to the bases. Once methylated, epigenetically reprogrammed mammalian cells can permanently change their function. The DNA becomes more hydrophobic, and the sequences can no longer be read correctly, for example. Due to the strong hydrophobicity, DNAs accumulate and form aggregates in the cell nucleus through hydrophobic coalescence; as a result, the DNA molecules become physically less mobile and less readable. All of this is ultimately based on biophysics, which is based on physical-chemical interactions.

This is indeed a reason to consume as many natural foods as possible and to otherwise lead a sensible lifestyle. The traces in the DNA remain visible even in the brain [194].

However, the individual genetic program encoded in the DNA through our genetic makeup in the form of the sequences of nucleotides of each individual cannot be outsmarted with healthy eating. Unfortunately, "bad genes" weigh far more than nutrition. There is really no remedy for that.

Fig. 5.55 The methylation prevents the correct reading of DNA sequences, while the DNA becomes more hydrophobic the more methyl groups are attached to it

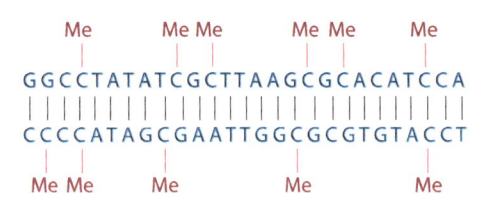

```
         Me         Me Me        Me Me        Me
          |          |  |         |  |         |
      G G C C T A T A T C G C T T A A G C G C A C A T C C A
      | | | | | | | | | | | | | | | | | | | | | | | | | | |
      C C C C A T A G C G A A T T G G C G C G T G T A C C T
          | |        |              |                   |
         Me Me       Me             Me                  Me
```

5.17 The Value of Natural Science in Nutrition

The discrepancy between science, perception, and belief has accompanied humans for a long time. Belief requires unshakable dogmas, no matter how illogical and unscientific they may be. Some absurd theses during the coronavirus pandemic clearly demonstrated this. But since time immemorial, this has been a problem. This becomes apparent with none other than Galileo Galilei. Galileo, one of the founders of modern natural sciences, observations, experiments, and the search for logical and mathematical connections, was eventually ostracized by the then-dominant Catholic Church with an inquisition process. The scientist was not rehabilitated until 1992, and even in 2008, the Church felt compelled to distance itself once again from its own actions over the past 400 years. One might think that was the end, but far from it, as the ever-increasing number of creationism supporters in a high-tech country like the United States shows. Despite all the scientific advances and findings—from experimental evidence of the Big Bang through cosmic background radiation [195] and the detection of the Higgs boson [196, 197] as proof for the standard model of our universe [198], supplemented by the measured gravitational waves, the first visible evidence of a black hole [199] and even the detection [201] of the first molecule after the Big Bang, the helium hydride ion (HeH^+)—to claim that the world is the center of a 6,000-year-old universe must be paired with audacity to "make politics" with such ideas. Evolution and its results are simply denied [202].

What do these remarks have to do with food and nutrition? At first glance, not much, if it were not for similar mechanisms that are currently apparent in the topic of nutrition and health. The theses, "organic is healthier," "vegan saves the world," "vegetables are healthy," "saturated fatty acids and cholesterol are unhealthy," etc., have become dogmas. As mentioned at the beginning of this chapter, these dogmas move in opinion loops and are thus immovable, not correctable. Their scientific core or even the origin of these theses is no longer questioned. Most of these theses are based on similarly weak scientific legs as the holy scriptures of this world.

The reason for this is usually misconceptions that are already detached from scientific findings. These are already close to the opinion loop at the border of the one-sided permeable filter (see Fig. 5.1). If they either become a factual misunderstanding or a "true" explanation from an alternative worldview (alternative facts), the paths diverge. In the case of factual misunderstandings, for example, uninformedness is formed. If a faulty instruction is added, the uninformedness is confirmed and largely immovable, depending on the authority of the instruction (see the scheme in Fig. 5.56).

If the worldview is based on alternative facts, there are three possibilities. The worst of these is the active hindrance of understanding, questioning by authorities or those who are considered as such. The step towards irreversible opinion loops is indispensable. This is also the case when something has been "understood," but without addressing facts, new insights, or a change in the factual situation. The myth of slow carbohydrates, the evil cholesterol in eggs, or the permanently harmful

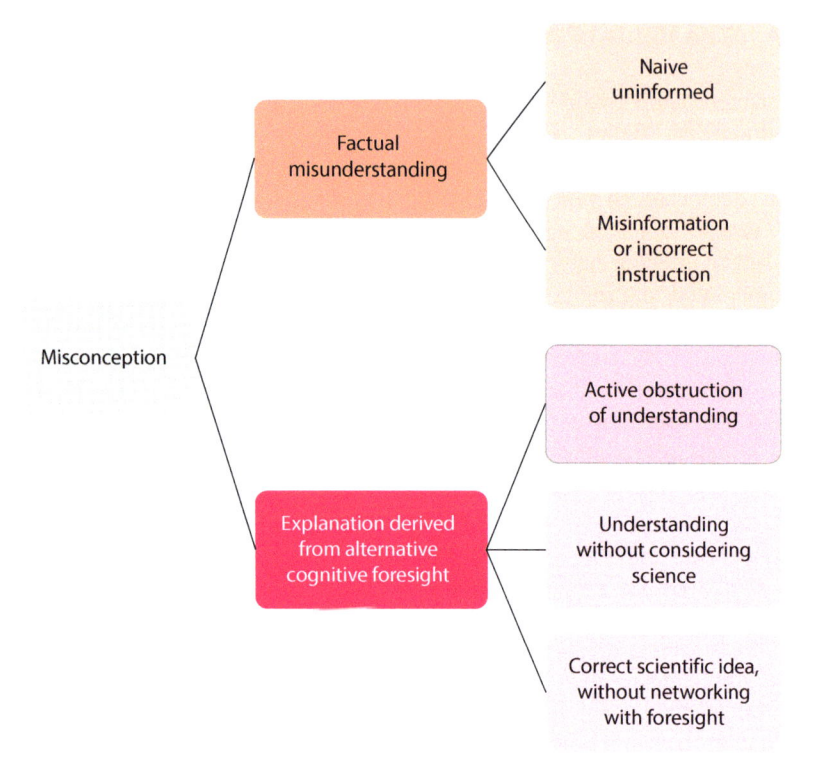

Fig. 5.56 Two basic paths for dubious worldviews result from misconceptions (From Williams [200])

saturated fatty acids are just the most well-known examples. The belief in Superfood appears particularly impressive in the light of this opinion formation. The persistence of belief in it shows absurd traits. Whether organic food is really healthier than conventionally grown food is still unclear and very difficult to prove. Consider the many small farmers whose farms cannot economically afford the certification effort, or the large organic farm that "sprays" its cultivation areas with toxic but approved copper, which enters the plant metabolism. Blanket dogmas are neither solution nor truth, especially in view of the experimental difficulties in verifying these claims. Even the infinite meta-study with infinitely many data would not bring any clarification. If only because of the *per se* built-in errors in observational studies.

5.18 What Does "Healthy" Actually Mean?

At the end of this chapter, the central question arises: what is actually healthy? Can this be clearly defined? Or is it more a matter of opinion? To anticipate: A clear definition in terms of natural sciences and medicine is impossible. The criteria commonly used in general medicine between "sick" and "healthy," for example, after simple fever measurement or a severe, clear diagnosis, can hardly be applied to food. Whether food is considered healthy or unhealthy is a very individual and subjective matter [203]. Food selection is based on preconceived and trained opinions [204], which depend on the information sources from which the study participants were fed, whether they followed diets or not. Social trends [205] and body ideals are increasingly playing a role. The term "healthy" is in no way measurable in terms of analytical methods—even though all food and people consist only of molecules.

Only one point is clear: A one-sided diet is not particularly healthy. Whether one feeds on healthy nuts for years or has grilled food on the plate three times a day. This gradually leads to an undersupply of macro and micronutrients. Then the unhealthy diet becomes obejectable on many levels.

References

1. Schoenfeld, J. D., & Ioannidis, J. P. (2012). Is everything we eat associated with cancer? A systematic cookbook review. *The American Journal of Clinical Nutrition, 97*(1), 127–134.
2. Siehe https://www.eufic.org/article/de/lebensmittelsicherheit-qualitat/sichere-lebensmittelhandhabung/artid/angst-vor-dem-essen/.
3. Schmidbauer, W. (2005). *Lebensgefühl Angst: Jeder hat sie. Keiner will sie. Was wir gegen Angst tun können*. Herde.
4. https://www.sprachaktivierung.de/index.php/grundlagen.html
5. Kast, B. (2021). *The Diet Compass: The 12-Step Guide to Science-Based Nutrition for a Healthier and Longer Life*
6. Wenzel, S. (2018). Der Ernährungskompass. *Ernährung & Medizin, 33*(04), 184–186.
7. https://www.medpagetoday.com/blogs/revolutionandrevelation/75045.
8. Ioannidis, J. P. (2018). The challenge of reforming nutritional epidemiologic research. *JAMA, 320*(10), 969–970.
9. Ioannidis, J. P. (2013). Implausible results in human nutrition research. *British Medical Journal, 14*(4), 401–410.
10. https://www.aerztezeitung.de/panorama/ernaehrung/article/961348/neue-eu-regelung-acrylamid-gefaehrlicher-stickoxide.html.
11. Tareke, E., Rydberg, P., Karlsson, P., Eriksson, S., & Törnqvist, M. (2000). Acrylamide: A cooking carcinogen? *Chemical Research in Toxicology, 13*(6), 517–522.
12. Tareke, E., Rydberg, P., Karlsson, P., Eriksson, S., & Törnqvist, M. (2002). Analysis of acrylamide, a carcinogen formed in heated foodstuffs. *Journal of Agricultural and Food Chemistry, 50*(17), 4998–5006.
13. Mottram, D. S., Wedzicha, B. L., & Dodson, A. T. (2002). Food chemistry: Acrylamide is formed in the Maillard reaction. *Nature, 419*(6906), 448.
14. Krishnakumar, T., & Visvanathan, R. (2014). Acrylamide in food products: A review. *Journal of Food Processing & Technology, 5*(7), 1.
15. EFSA Panel on Contaminants in the Food Chain (CONTAM). (2015). Scientific opinion on acrylamide in food. *EFSA Journal, 13*(6), 4104.
16. Daniali, G., Jinap, S., Hajeb, P., Sanny, M., & Tan, C. P. (2016). Acrylamide formation in vegetable oils and animal fats during heat treatment. *Food chemistry, 212*, 244–249.
17. Granvogl, M., Koehler, P., Latzer, L., & Schieberle, P. (2008). Development of a stable isotope dilution assay for the quantitation of glycidamide and its application to foods and model systems. *Journal of Agricultural and Food Chemistry, 56*(15), 6087–6092.

18. Mustațeă, G., Ppoa, M. E., & Negoiță, M. (2015). A case study on mitigation strategies of acrylamide in bakery products. *Scientific Bulletin. Series F. Biotechnologies, 19*, 348–353.
19. Dybing, E., Farmer, P. B., Andersen, M., Fennell, T. R., Lalljie, S. P. D., Müller, D. J. G., et al. (2005). Human exposure and internal dose assessments of acrylamide in food. *Food and Chemical Toxicology, 43*(3), 365–410.
20. Martins, S. I., Jongen, W. M., & Van Boekel, M. A. (2000). A review of Maillard reaction in food and implications to kinetic modelling. *Trends in Food Science & Technology, 11*(9–10), 364–373.
21. Mestdagh, F., Maertens, J., Cucu, T., Delporte, K., Van Peteghem, C., & De Meulenaer, B. (2008). Impact of additives to lower the formation of acrylamide in a potato model system through pH reduction and other mechanisms. *Food Chemistry, 107*(1), 26–31.
22. Fredriksson, H., Tallving, J., Rosen, J., & Åman, P. (2004). Fermentation reduces free asparagine in dough and acrylamide content in bread. *Cereal Chemistry, 81*(5), 650–653.
23. Arima, K., Sakamoto, T., Araki, C., & Tamura, G. (1972). Production of extracellular l-asparaginases by microorganisms. *Agricultural and Biological Chemistry, 36*(3), 356–361.
24. Hill, J. M., Roberts, J., Loeb, E., Khan, A., MacLellan, A., & Hill, R. W. (1967). l-asparaginase therapy for leukemia and other malignant neoplasms: Remission in human leukemia. *JAMA, 202*(9), 882–888.
25. Xu, F., Oruna-Concha, M. J., & Elmore, J. S. (2016). The use of asparaginase to reduce acrylamide levels in cooked food. *Food Chemistry, 210*, 163–171.
26. Alam, S., Pranaw, K., Tiwari, R., & Khare, S. K. (2019). Recent development in the uses of asparaginase as food enzyme. In S. Alam, K. Pranaw, R. Tiwari, & S. K. Khare (Hrsg.), *Green bio-processes* (S. 55–81). Springer.
27. Cho, I. H., Lee, S., Jun, H. R., Roh, H. J., & Kim, Y. S. (2010). Comparison of volatile Maillard reaction products from tagatose and other reducing sugars with amino acids. *Food Science and Biotechnology, 19*(2), 431–438.
28. Koehler, P. E., Mason, M. E., & Newell, J. A. (1969). Formation of pyrazine compounds in sugar-amino acid model systems. *Journal of Agricultural and Food Chemistry, 17*(2), 393–396.
29. Wong, K. H., Abdul Aziz, S., & Mohamed, S. (2008). Sensory aroma from Maillard reaction of individual and combinations of amino acids with glucose in acidic conditions. *International Journal of Food Science & Technology, 43*(9), 1512–1519.
30. Delgado-Andrade, C. (2014). Maillard reaction products: Some considerations on their health effects. *Clinical Chemistry and Laboratory Medicine, 52*(1), 53–60.
31. Selvam, J. P., Aranganathan, S., Gopalan, R., & Nalini, N. (2009). Chemopreventive efficacy of pronyl-lysine on lipid peroxidation and antioxidant status in rat colon carcinogenesis. *Fundamental & Clinical Pharmacology, 23*(3), 293–302.
32. Michalska, A., Amigo-Benavent, M., Zielinski, H., & del Castillo, M. D. (2008). Effect of bread making on formation of Maillard reaction products contributing to the overall antioxidant activity of rye bread. *Journal of Cereal Science, 48*(1), 123–132.
33. Somoza, V., Lindenmeier, M., Wenzel, E., Frank, O., Erbersdobler, H. F., & Hofmann, T. (2003). Activity-guided identification of a chemopreventive compound in coffee beverage using in vitro and in vivo techniques. *Journal of Agricultural and Food Chemistry, 51*(23), 6861–6869.
34. Wang, H. Y., Qian, H., & Yao, W. R. (2011). Melanoidins produced by the Maillard reaction: Structure and biological activity. *Food Chemistry, 128*(3), 573–584.
35. Yilmaz, Y., & Toledo, R. (2005). Antioxidant activity of water-soluble Maillard reaction products. *Food Chemistry, 93*(2), 273–278.
36. Benjakul, S., Lertittikul, W., & Bauer, F. (2005). Antioxidant activity of Maillard reaction products from a porcine plasma protein–sugar model system. *Food Chemistry, 93*(2), 189–196.

References

37. Gu, F. L., Kim, J. M., Abbas, S., Zhang, X. M., Xia, S. Q., & Chen, Z. X. (2010). Structure and antioxidant activity of high molecular weight Maillard reaction products from casein–glucose. *Food Chemistry, 120*(2), 505–511.
38. Tagliazucchi, D., & Bellesia, A. (2015). The gastro-intestinal tract as the major site of biological action of dietary melanoidins. *Amino Acids, 47*(6), 1077–1089.
39. Raters, M. & Matissek, R. (2012). The big bang acrylamid: 10 Jahre Acrylamid—Rückblick und Status quo, LCI, 184–189. https://www.lci-koeln.de/download/dlr-acrylamid-raters-matissek.
40. Lee, W. T., Weisell, R., Albert, J., Tomé, D., Kurpad, A. V., & Uauy, R. (2016). Research approaches and methods for evaluating the protein quality of human foods proposed by an FAO expert working group in 2014. *The Journal of Nutrition, 146*(5), 929–932.
41. Bossel, H. (1990). Nährstoffbedarf, Nährstoffkreisläufe, Boden. In H. Bossel (Hrsg.), *Umweltwissen* (S. 55–68). Springer.
42. Sharma, S., Singh, R., & Rana, S. (2011). Bioactive peptides: A review. *International Journal of Bioautomation, 15*(4), 223–250.
43. Lafarga, T., & Hayes, M. (2014). Bioactive peptides from meat muscle and by-products: Generation, functionality and application as functional ingredients. *Meat Science, 98*(2), 227–239.
44. Pihlanto-Leppälä, A. (2000). Bioactive peptides derived from bovine whey proteins: Opioid and ace-inhibitory peptides. *Trends in Food Science & Technology, 11*(9–10), 347–356.
45. Bhat, Z. F., Kumar, S., & Bhat, H. F. (2015). Bioactive peptides of animal origin: A review. *Journal of Food Science and Technology, 52*(9), 5377–5392.
46. Maestri, E., Marmiroli, M., & Marmiroli, N. (2016). Bioactive peptides in plant-derived foodstuffs. *Journal of Proteomics, 147*, 140–155.
47. Vilgis, T. A., Lendner, I., & Caviezel, R. (2014). *Ernährung bei Pflegebedürftigkeit und Demenz: Lebensfreude durch Genuss*. Springer.
48. Volm, C. (2010). *Rohköstliches Gesund durchs Leben mit Rohkost und Wildpflanzen*. Ulmer.
49. Harkness, M. L., Harkness, R. D., & Venn, M. F. (1978). Digestion of native collagen in the gut. *Gut, 19*(3), 240–243.
50. Knupp, C., & Squire, J. M. (2005). Molecular packing in network-forming collagens. In J. P. Richard (Hrsg.), *Advances in protein chemistry* (Bd. 70, S. 375–403). Academic Press.
51. Shigemura, Y., Akaba, S., Kawashima, E., Park, E. Y., Nakamura, Y., & Sato, K. (2011). Identification of a novel food-derived collagen peptide, hydroxyprolyl-glycine, in human peripheral blood by pre-column derivatisation with phenyl isothiocyanate. *Food Chemistry, 129*(3), 1019–1024.
52. Asserin, J., Lati, E., Shioya, T., & Prawitt, J. (2015). The effect of oral collagen peptide supplementation on skin moisture and the dermal collagen network: Evidence from an ex vivo model and randomized, placebo-controlled clinical trials. *Journal of Cosmetic Dermatology, 14*(4), 291–301.
53. Raj, U. L., Sharma, G., Dang, S., Gupta, S., & Gabrani, R. (2016). Impact of dietary supplements on skin aging. *Textbook of Aging Skin*, 1–13. Springer-Verlag Berlin Heidelberg 2015M.A. Farage et al. (eds.), Textbook of Aging Skin, DOI https://doi.org/10.1007/978-3-642-27814-3_174-1
54. Roh, E., Kim, J. E., Kwon, J. Y., Park, J. S., Bode, A. M., Dong, Z., & Lee, K. W. (2017). Molecular mechanisms of green tea polyphenols with protective effects against skin photoaging. *Critical Reviews in Food Science and Nutrition, 57*(8), 1631–1637.
55. Felixberger, J. K. (2018). Proteine—Essenzieller Bestandteil unserer Ernährung. In A. Ghadiri, T. A. Vilgis, & T. Bosbach (Hrsg.), *Wissen schmeckt* (S. 169–195). Springer.
56. Richter, C. K., Skulas-Ray, A. C., Champagne, C. M., & Kris-Etherton, P. M. (2015). Plant protein and animal proteins: Do they differentially affect cardiovascular disease risk? *Advances in Nutrition, 6*(6), 712–728.
57. Revel, A., Jarzaguet, M., Peyron, M. A., Papet, I., Hafnaoui, N., Migné, C., et al. (2017). At same leucine intake, a whey/plant protein blend is not as effective as whey to initiate a

transient post prandial muscle anabolic response during a catabolic state in mini pigs. *PLoS One, 12*(10), e0186204.
58. Boland, M. J., Rae, A. N., Vereijken, J. M., Meuwissen, M. P., Fischer, A. R., van Boekel, M. A., et al. (2013). The future supply of animal-derived protein for human consumption. *Trends in Food Science & Technology, 29*(1), 62–73.
59. Li, Y., Xin, R., Zhang, D., & Li, S. (2014). Molecular characterization of α-gliadin genes from common wheat cultivar Zhengmai 004 and their role in quality and celiac disease. *The Crop Journal, 2*(1), 10–21.
60. Bromilow, S., Gethings, L. A., Buckley, M., Bromley, M., Shewry, P. R., Langridge, J. I., & Mills, E. C. (2017). A curated gluten protein sequence database to support development of proteomics methods for determination of gluten in gluten-free foods. *Journal of proteomics, 163*, 67–75.
61. Schwefel, D., Maierhofer, C., Beck, J. G., Seeberger, S., Diederichs, K., Möller, H. M., et al. (2010). Structural basis of multivalent binding to wheat germ agglutinin. *Journal of the American Chemical Society, 132*(25), 8704–8719.
62. https://www.uniprot.org/uniprot/?query=AGGLUTININ&sort=score.
63. Gundry, S. R. (2017). *Böses Gemüse*. Beltz.
64. Schaufler, M., & Drössler, W. A. (2017). *Lektine—Das heimliche Gift*. Riva.
65. https://praxistipps.focus.de/weizenkeimoel-wirkung-und-tipps-zur-anwendung_101131.
66. van Buul, V. J., & Brouns, F. J. (2014). Health effects of wheat lectins: A review. *Journal of Cereal Science, 59*(2), 112–117.
67. Schuppan, D., & Zevallos, V. (2015). Wheat amylase trypsin inhibitors as nutritional activators of innate immunity. *Digestive Diseases, 33*(2), 260-263.
68. Strunz, U. (2015). *Warum macht die Nudel dumm?: Leichter, klüger, besser drauf: No Carbs und das Geheimnis wacher Intelligenz*. Heyne.
69. Perlmutter, D., & Loberg, K. (2014). *Grain brain: The Surprising truth about wheat, carbs, and sugar--your brain's silent killers,*. Little, Brown Spark
70. Davis, W. (2011). *Wheat Belly* Rodale Press.
71. Schuppan, D., & Gisbert-Schuppan, K. (2019). *Wheat syndromes*. Springer.
72. Cuccioloni, M., Mozzicafreddo, M., Ali, I., Bonfili, L., Cecarini, V., Eleuteri, A. M., & Angeletti, M. (2016). Interaction between wheat alpha-amylase/trypsin bi-functional inhibitor and mammalian digestive enzymes: Kinetic, equilibrium and structural characterization of binding. *Food Chemistry, 213*, 571–578.
73. https://www.schaer.com/de-de/p/landbrot.
74. Schober, T. J., Bean, S. R., Boyle, D. L., & Park, S. H. (2008). Improved viscoelastic zein-starch doughs for leavened gluten-free breads: Their rheology and microstructure. *Journal of Cereal Science, 48*(3), 755–767.
75. Fischer, M. H., Yu, N., Gray, G. R., Ralph, J., Anderson, L., & Marlett, J. A. (2004). The gel-forming polysaccharide of psyllium husk (Plantago ovata Forsk). *Carbohydrate Research, 339*(11), 2009–2017.
76. Thakur, V. K., & Thakur, M. K. (2014). Recent trends in hydrogels based on psyllium polysaccharide: A review. *Journal of Cleaner Production, 82*, 1–15.
77. Reese, I., et al. (2018). Nicht-Zöliakie-Gluten-/Weizen-Sensitivität (NCGS)—ein bislang nicht definiertes Krankheitsbild mit fehlenden Diagnosekriterien und unbekannter Häufigkeit. *Aktuelle Ernährungsmedizin, 43*(06), 479–483.
78. Zhang, W., Xu, J., Bennetzen, J. L., & Messing, J. (2016). Teff, an orphan cereal in the chloridoideae, provides insights into the evolution of storage proteins in grasses. *Genome Biology and Evolution, 8*(6), 1712–1721.
79. Hager, A. S., Wolter, A., Czerny, M., Bez, J., Zannini, E., Arendt, E. K., & Czerny, M. (2012). Investigation of product quality, sensory profile and ultrastructure of breads made from a range of commercial gluten-free flours compared to their wheat counterparts. *European Food Research and Technology, 235*(2), 333–344.

80. Zhu, F. (2018). Chemical composition and food uses of teff *(Eragrostis tef)*. *Food Chemistry, 239,* 402–415.
81. Hadji-Gerogoloulus, A., Schmidt, M. I., Margolis, S., & Kowarski, A. A. (1980). Elevated hypoglycemic index and late hyperinsulinism in symptomatic postprandial hypoglycemia. *The Journal of Clinical Endocrinology & Metabolism, 50*(2), 371–376.
82. Tappy, L., Würsch, P., Randin, J. P., Felber, J. P., & Jequier, E. (1986). Metabolic effect of pre-cooked instant preparations of bean and potato in normal and in diabetic subjects. *The American Journal of Clinical Nutrition, 43*(1), 30–36.
83. Leathwood, P., & Pollet, P. (1988). Effects of slow release carbohydrates in the form of bean flakes on the evolution of hunger and satiety in man. *Appetite, 10*(1), 1–11.
84. Goletzke, J., Atkinson, F. S., Ek, K. L., Bell, K., Brand-Miller, J. C., & Buyken, A. E. (2016). Glycaemic and insulin index of four common German breads. *European Journal of Clinical Nutrition, 70*(7), 808.
85. Zimmerman, M. A., Singh, N., Martin, P. M., Thangaraju, M., Ganapathy, V., Waller, J. L., et al. (2012). Butyrate suppresses colonic inflammation through HDAC1-dependent Fas upregulation and Fas-mediated apoptosis of T cells. *American Journal of Physiology-Heart and Circulatory Physiology, 302,* G1405–G1415.
86. Vital, M., Karch, A., & Pieper, D. H. (2017). Colonic butyrate-producing communities in humans: An overview using omics data. *MSystems, 2*(6), e00130-e217.
87. Mathes, P. (2012). Hilfen durch Medikamente. *Ratgeber Herzinfarkt* (S. 167–189). Springer.
88. Gehring, J., & Klein, G. (2015). Mobilität nach Herzinfarkt oder Bypass-Operation. In J. Gehring & G. Klein (Hrsg.), *Leben mit der koronaren Herzkrankheit* (S. 154–158). Urban und Vogel.
89. Schmid, A. (2006). Einfluss von Nitrat und Nitrit aus Fleischerzeugnissen auf die Gesundheit des Menschen. *Ernährungsumschau, 53*(12), 490–495.
90. Martin, H. H. (2008). Vom Saulus zum Paulus? *UGB Forum, 5,* 245.
91. Bedale, W., Sindelar, J. J., & Milkowski, A. L. (2016). Dietary nitrate and nitrite: Benefits, risks, and evolving perceptions. *Meat Science, 120,* 85–92.
92. Habermeyer, M., Roth, A., Guth, S., Diel, P., Engel, K. H., Epe, B., Fürst, P., Heinz, V., Knorr, D., de Kok, T., Kulling, S., Lampen, A., Marko, D., Rechkemmer, G., Rietjens, I., Stadler, R.H., Vieths, S., Vogel, R., Steinberg, P., & Eisenbrand, G. (2015). Nitrate and nitrite in the diet: How to assess their benefit and risk for human health. *Molecular Nutrition & Food Research, 59*(1), 106–128.
93. Li, Y., et al. (2018). Secretin-activated brown fat mediates prandial thermogenesis to induce satiation. *Cell, 175*(6), 1561–1574.
94. Stanford, K. I., Middelbeek, R. J., Townsend, K. L., An, D., Nygaard, E. B., Hitchcox, K. M., et al. (2012). Brown adipose tissue regulates glucose homeostasis and insulin sensitivity. *The Journal of Clinical Investigation, 123*(1), 215–223.
95. Lundberg, J. O., Carlström, M., & Weitzberg, E. (2018). Metabolic effects of dietary nitrate in health and disease. *Cell metabolism, 28*(1), 9–22.
96. Lundberg, J. O., & Weitzberg, E. (2017). Nitric oxide formation from inorganic nitrate. In J. O. Lundberg & E. Weitzberg (Hrsg.), *Nitric oxide* (S. 157–171). Academic Press.
97. Williams, A. R. (1975). The production of saltpetre in the middle ages. *Ambix, 22*(2), 125–133.
98. Vilgis, T. A. (2018). Ikejime versus karashi jukusei (dry aging): Vielfältige molekulare Umami-Phasen. *Journal Culinaire, 27,* 56–84.
99. Vilgis, T. (2013). Fermentation—Molekulares Niedrigtemperaturgaren. *Journal Culinaire, 17,* 38–53.
100. Bognár, A. (2003). Vitaminveränderungen bei der Lebensmittelverarbeitung im Haushalt. *Ernährung im Fokus, 11,* 330–335.
101. Munyaka, A. W., Makule, E. E., Oey, I., Van Loey, A., & Hendrickx, M. (2010). Thermal stability of l-ascorbic acid and ascorbic acid oxidase in broccoli (*Brassica oleracea* var. *italica*). *Journal of Food Science, 75*(4), C336–C340.

102. Burggraf, C., Teuber, R., Brosig, S., & Meier, T. (2018). Review of a priori dietary quality indices in relation to their construction criteria. *Nutrition Reviews, 76*(10), 747–764.
103. Ströhle, A., & Hahn, A. (2011). Diets of modern hunter-gatherers vary substantially in their carbohydrate content depending on ecoenvironments: Results from an ethnographic analysis. *Nutrition Research, 31*(6), 429–435.
104. Ströhle, A., Wolters, M., & Hahn, A. (2009). Die Ernährung des Menschen im evolutionsmedizinischen Kontext. *Wiener klinische Wochenschrift, 121*(5–6), 173–187.
105. Manheimer, E. W., van Zuuren, E. J., Fedorowicz, Z., & Pijl, H. (2015). Paleolithic nutrition for metabolic syndrome: Systematic review and meta-analysis. *The American Journal of Clinical Nutrition, 102*(4), 922–932.
106. Fenton, T. R., & Fenton, C. J. (2016). Paleo diet still lacks evidence. *The American Journal of Clinical Nutrition, 104*(3), 844–844.
107. Bland, J. M., & Altman, D. G. (2015). Best (but oft forgotten) practices: Testing for treatment effects in randomized trials by separate analyses of changes from baseline in each group is a misleading approach. *The American Journal of Clinical Nutrition, 102*(5), 991–994.
108. Ioannidis, J. P. (2005a). Why most published research findings are false. *PLoS Medicine, 2*(8), e124.
109. Ioannidis, J. P. (2005). Contradicted and initially stronger effects in highly cited clinical research. *JAMA, 294*(2), 218–228.
110. Zinkant, K. (2019). „Wir müssen die Forscher befreien" (Interview in der Süddeutschen Zeitung, 3. April 2019), https://www.sueddeutsche.de/wissen/wissenschaft-meta-research-ioannidis-1.4394526?reduced=true.
111. Ströhle, A., Behrendt, I., Behrendt, P., & Hahn, A. (2016). Alternative Ernährungsformen. *Aktuelle Ernährungsmedizin, 41*(02), 120–138.
112. Leitzmann, C., & Behrendt, I. (2015). Vegane Ernährung. *Erfahrungsheilkunde, 64*(02), 76–83.
113. https://www.spiegel.de/panorama/gesellschaft/bulle-toetet-bauer-radikale-veganer-erklaeren-das-tier-zum-helden-a-1015210.html und https://www.morgenpost.de/berlin/article212340909/Nach-Pumpgun-Foto-Koch-Hildmann-von-der-Polizei-vorgeladen.html.
114. https://www.sueddeutsche.de/panorama/frankreich-veganer-tiere-speziesismus-1.4193393.
115. Cantone, D. (2017). Veganer glauben moralisch überlegen zu sein. *Essay, Neue Zürcher Zeitung, 26*(04), 2017.
116. Hund, W. D. (2015). *Rassismus*. Transcript.
117. Bauknecht, B. R. (2015). *Salafismus-Ideologie der Moderne*. Bundeszentrale für Politische Bildung.
118. Potts, A., & Armstrong, P. (2018). VEGAN 27. *Critical Terms for Animal Studies*, 395.
119. Dinu, M., Abbate, R., Gensini, G. F., Casini, A., & Sofi, F. (2017). Vegetarian, vegan diets and multiple health outcomes: A systematic review with meta-analysis of observational studies. *Critical Reviews in Food Science and Nutrition, 57*(17), 3640–3649.
120. Bradbury, K. E., Crowe, F. L., Appleby, P. N., Schmidt, J. A., Travis, R. C., & Key, T. J. (2014). Serum concentrations of cholesterol, apolipoprotein AI and apolipoprotein B in a total of 1694 meat-eaters, fish-eaters, vegetarians and vegans. *European Journal of Clinical Nutrition, 68*(2), 178.
121. Spencer, E. A., Appleby, P. N., Davey, G. K., & Key, T. J. (2003). Diet and body mass index in 38 000 EPIC-Oxford meat-eaters, fish-eaters, vegetarians and vegans. *International Journal of Obesity, 27*(6), 728.
122. https://www.dr-schmiedel.de/macht-vegan-krank/.
123. Appleby, P., Roddam, A., Allen, N., & Key, T. (2007). Comparative fracture risk in vegetarians and nonvegetarians in EPIC-Oxford. *European Journal of Clinical Nutrition, 61*(12), 1400.
124. Orlich, M. J., Singh, P. N., Sabaté, J., Jaceldo-Siegl, K., Fan, J., Knutsen, S., et al. (2013). Vegetarian dietary patterns and mortality in adventist health study 2. *JAMA Internal Medicine, 173*(13), 1230–1238.

125. Larsson, C. L., & Johansson, G. K. (2002). Dietary intake and nutritional status of young vegans and omnivores in Sweden. *The American Journal of Clinical Nutrition, 76*(1), 100–106.
126. Sastre, R. R., & Posten, C. (2010). Die vielfältige Anwendung von Mikroalgen als nachwachsende Rohstoffe. *Chemie Ingenieur Technik, 11*(82), 1925–1939.
127. Greupner, T., Kutzner, L., Nolte, F., Strangmann, A., Kohrs, H., Hahn, A., et al. (2018). Effects of a 12-week high-α-linolenic acid intervention on EPA and DHA concentrations in red blood cells and plasma oxylipin pattern in subjects with a low EPA and DHA status. *Food & Function, 9*(3), 1587–1600.
128. Egert, S., Baxheinrich, A., Lee-Barkey, Y. H., Tschoepe, D., Stehle, P., Stratmann, B., & Wahrburg, U. (2018). Effects of a hypoenergetic diet rich in α-linolenic acid on fatty acid composition of serum phospholipids in overweight and obese patients with metabolic syndrome. *Nutrition, 49*, 74–80.
129. Ferdinandusse, S., Denis, S., Mooijer, P. A., Zhang, Z., Reddy, J. K., Spector, A. A., & Wanders, R. J. (2001). Identification of the peroxisomal β-oxidation enzymes involved in the biosynthesis of docosahexaenoic acid. *Journal of Lipid Research, 42*(12), 1987–1995.
130. Sutter, D. O. (2017). *The impact of vegan diet on health and growth of children and adolescents–Literature review*. Doctoral dissertation, University of Bern.
131. Masana, M. F., Koyanagi, A., Haro, J. M., & Tyrovolas, S. (2017). n-3 Fatty acids, Mediterranean diet and cognitive function in normal aging: A systematic review. *Experimental Gerontology, 91*, 39–50.
132. Zoe, J., & Weyer, F. Pegan: Paleo + Vegan. (2016). Hyman, M. (2015). Why this health expert recommends a Paleo-Vegan diet. https://www.elephantjournal.com/2015/06/why-this-health-expert-recommends-a-paleo-vegan-diet/.
133. Eisenhauer, B. (2019). Vegane Influencerin outet sich: Wer weiß schon, was sich hinter der digitalen Maske verbirgt? *Frankfurter Allgemeine Zeitung*. https://www.faz.net/aktuell/stil/leib-seele/fans-der-veganen-influencerin-rawvana-fuehlen-sich-verraten-16104911.html. Zugegriffen: 30. März 2019.
134. Şanlier, N., Gökcen, B. B., & Sezgin, A. C. (2017). Health benefits of fermented foods. *Critical Reviews in Food Science and Nutrition, 2017*, 1–22.
135. Tarvainen, M., Fabritius, M., & Yang, B. (2019). Determination of vitamin K composition of fermented food. *Food Chemistry, 275*, 515–522.
136. Daliri, E., Oh, D., & Lee, B. (2017). Bioactive peptides. *Foods, 6*(5), 32.
137. Schmid, J. (2018). Recent insights in microbial exopolysaccharide biosynthesis and engineering strategies. *Current Opinion in Biotechnology, 53*, 130–136.
138. Lynch, K. M., Zannini, E., Coffey, A., & Arendt, E. K. (2018). Lactic acid bacteria exopolysaccharides in foods and beverages: Isolation, properties, characterization, and health benefits. *Annual Review of Food Science and Technology, 9*, 155–176.
139. Sanjukta, S., & Rai, A. K. (2016). Production of bioactive peptides during soybean fermentation and their potential health benefits. *Trends in Food Science & Technology, 50*, 1–10.
140. Liu, Y., Song, H., & Luo, H. (2018). Correlation between the key aroma compounds and gDNA copies of *Bacillus* during fermentation and maturation of natto. *Food Research International, 112*, 175–183.
141. Mahdinia, E., Mamouri, S. J., Puri, V. M., Demirci, A., & Berenjian, A. (2019). Modeling of vitamin K (Menaquinoe-7) fermentation by *Bacillus subtilis natto* in biofilm reactors. *Biocatalysis and Agricultural Biotechnology, 17*, 196–202.
142. Hsueh, Y. H., Huang, K. Y., Kunene, S., & Lee, T. Y. (2017). Poly-γ-glutamic acid synthesis, gene regulation, phylogenetic relationships, and role in fermentation. *International Journal of Molecular Sciences, 18*(12), 2644.
143. Hafner, U. (2014). Vische stinken nicht. *Neue Zürcher Zeitung*. https://www.nzz.ch/wissenschaft/bildung/vische-stinken-nicht-1.18327936.
144. Zielbauer, B. I., Franz, J., Viezens, B., & Vilgis, T. A. (2016). Physical aspects of meat cooking: Time dependent thermal protein denaturation and water loss. *Food Biophysics, 11*(1), 34–42.

145. Vilgis, T. A. (2016). Brühwurst Warm- und Kaltfleischverarbeitung. *Journal Culinaire, 22,* 50–70.
146. Grimm, H. U. (2012). *Vom Verzehr wird abgeraten: Wie uns die Industrie mit Gesundheitsnahrung krank macht*. KnaureBook.
147. Grimm, H. U. (2003). *Die Ernährungslüge: Wie uns die Lebensmittelindustrie um den Verstand bringt*. Droemer HC.
148. Capuano, E., Oliviero, T., Fogliano, V., & Pellegrini, N. (2018). Role of the food matrix and digestion on calculation of the actual energy content of food. *Nutrition reviews, 76*(4), 274–289.
149. Palzer, S. (2009). Food structures for nutrition, health and wellness. *Trends in Food Science & Technology, 20*(5), 194–200.
150. Turgeon, S. L., & Rioux, L. E. (2011). Food matrix impact on macronutrients nutritional properties. *Food Hydrocolloids, 25*(8), 1915–1924.
151. González, R. J., Drago, S. R., Torres, R. L., & De Greef, D. M. (2016). 12 Extrusion Cooking of. *Engineering Aspects of Cereal and Cereal-Based Products,* 269.
152. Lin, S., Huff, H. E., & Hsieh, F. (2002). Extrusion process parameters, sensory characteristics, and structural properties of a high moisture soy protein meat analog. *Journal of Food Science, 67*(3), 1066–1072.
153. Palanisamy, M., Franke, K., Berger, R. G., Heinz, V., & Töpfl, S. (2019). High moisture extrusion of lupin protein: Influence of extrusion parameters on extruder responses and product properties. *Journal of the Science of Food and Agriculture, 99*(5), 2175–2185.
154. Osen, R., Toelstede, S., Eisner, P., & Schweiggert-Weisz, U. (2015). Effect of high moisture extrusion cooking on protein–protein interactions of pea (Pisum sativum L.) protein isolates. *International Journal of Food Science & Technology, 50*(6), 1390–1396.
155. Wilson, D. O., & Reisenauer, H. M. (1963). Determination of leghemoglobin in legume nodules. *Analytical Biochemistry, 6*(1), 27–30.
156. Appleby, C. A. (1984). Leghemoglobin and *Rhizobium* respiration. *Annual Review of Plant Physiology, 35*(1), 443–478.
157. Hyldig-Nielsen, J. J., Jensen, E. Ø., Paludan, K., Wiborg, O., Garrett, R., Jørgensen, P., & Marcker, K. A. (1982). The primary structures of two leghemoglobin genes from soybean. *Nucleic Acids Research, 10*(2), 689–701.
158. Fu, Y., Bak, K. H., Liu, J., De Gobba, C., Tøstesen, M., Hansen, E. T., et al. (2019). Protein hydrolysates of porcine hemoglobin and blood: Peptide characteristics in relation to taste attributes and formation of volatile compounds. *Food Research International., 121,* 28–38.
159. Robinson, C. (2018). The impossible burger: Boon or risk to health and environment? *GMO Science,* https://gmoscience.org/2018/05/16/impossible-burger-boon-risk-health-environment/
160. Dance, A. (2017). Engineering the animal out of animal products. *Nature Biotechnology, 35*(8), 704–707.
161. Hinzmann, M. (2018). *Die Wahrnehmung von In-Vitro-Fleisch in Deutschland*. TU-Berlin.
162. Siegrist, M., Sütterlin, B., & Hartmann, C. (2018). Perceived naturalness and evoked disgust influence acceptance of cultured meat. *Meat Science, 139,* 213–219.
163. van der Weele, C., & Tramper, J. (2014). Cultured meat: Every village its own factory? *Trends in biotechnology, 32*(6), 294–296.
164. Arshad, M. S., Javed, M., Sohaib, M., Saeed, F., Imran, A., & Amjad, Z. (2017). Tissue engineering approaches to develop cultured meat from cells: A mini review. *Cogent Food & Agriculture, 3*(1), 1320814.
165. Shockley, M., & Dossey, A. T. (2014). Insects for human consumption. In M. Shockley & A. T. Dossey (Hrsg.), *Mass production of beneficial organisms* (S. 617–652). Academic Press.
166. Rumpold, B. A., & Schlüter, O. K. (2013). Potential and challenges of insects as an innovative source for food and feed production. *Innovative Food Science & Emerging Technologies, 17,* 1–11.

References

167. Churchward-Venne, T. A., Pinckaers, P. J., van Loon, J. J., & van Loon, L. J. (2017). Consideration of insects as a source of dietary protein for human consumption. *Nutrition Reviews, 75*(12), 1035–1045.
168. Fiebelkorn, F. (2017). Insekten als Nahrungsmittel der Zukunft: Entomophagie. *Biologie in unserer Zeit, 47*(2), 104–110.
169. Holst, K. (2019). Von Entomophobie zu Entomophagie. *Hamburger Journal für Kulturanthropologie (HJK), 8,* 85–98.
170. Smetana, S., Pernutz, C., Toepfl, S., Heinz, V., & Van Campenhout, L. (2019). High-moisture extrusion with insect and soy protein concentrates: Cutting properties of meat analogues under insect content and barrel temperature variations. *Journal of Insects as Food and Feed, 5*(1), 29–34.
171. Dicke, M., & van Huis, A. (2015). Six-legged protein. *Oxygen, 26,* 68–71.
172. Tresidder, R. (2015). Eating ants: Understanding the terroir restaurant as a form of destination tourism. *Journal of Tourism and Cultural Change, 13*(4), 344–360.
173. McIlveen, H., Abraham, C., & Armstrong, G. (1999). Meat avoidance and the role of replacers. *Nutrition & Food Science, 99*(1), 29–36.
174. Stephan, A., Ahlborn, J., Zajul, M., & Zorn, H. (2018). Edible mushroom mycelia of Pleurotus sapidus as novel protein sources in a vegan boiled sausage analog system: Functionality and sensory tests in comparison to commercial proteins and meat sausages. *European Food Research and Technology, 244*(5), 913–924.
175. Linke, D., Bouws, H., Peters, T., Nimtz, M., Berger, R. G., & Zorn, H. (2005). Laccases of *Pleurotus sapidus:* Characterization and cloning. *Journal of Agricultural and Food Chemistry, 53*(24), 9498–9505.
176. Christ, A. K., et al. (2018). Western diet triggers NLRP3-dependent innate immune reprogramming. *Cell, 172*(1–2), 162–175.
177. Schnabel, L., Kesse-Guyot, E., Allès, B., Touvier, M., Srour, B., Hercberg, S., et al. (2019). Association between ultraprocessed food consumption and risk of mortality among middle-aged adults in France. *JAMA internal Medicine.* https://doi.org/10.1001/jamainternmed.2018.7289.
178. Schymanski, I. (2015). *Im Teufelskreis der Lust: Raus aus der Belohnungsfalle!* Schattauer.
179. Small, D. M., & DiFeliceantonio, A. G. (2019). Processed foods and food reward. *Science, 363*(6425), 346–347.
180. Hancock, R. D., McDougall, G. J., & Stewart, D. (2007). Berry fruit as 'superfood': Hope or hype. *Biologist, 54*(2), 73–79.
181. Cassiday, L. (2017). Chia: Superfood or superfat. *Inform, 28*(1), 6–13.
182. Berlett, B. S., & Stadtman, E. R. (1997). Protein oxidation in aging, disease, and oxidative stress. *Journal of Biological Chemistry, 272*(33), 20313–20316.
183. Federle, S., Hergesell, S., & Schubert, S. (2017). Aromaten. In S. Federle, S. Hergesell, & S. Schubert (Hrsg.), *Die Stoffklassen der organischen Chemie* (S. 43–65). Springer Spektrum.
184. Quideau, S., Deffieux, D., Douat-Casassus, C., & Pouységu, L. (2011). Pflanzliche Polyphenole: Chemische Eigenschaften, biologische Aktivität und Synthese. *Angewandte Chemie, 123*(3), 610–646.
185. Pandey, K. B., & Rizvi, S. I. (2009). Plant polyphenols as dietary antioxidants in human health and disease. *Oxidative Medicine and Cellular longevity, 2*(5), 270–278.
186. Papuc, C., Goran, G. V., Predescu, C. N., Nicorescu, V., & Stefan, G. (2017). Plant polyphenols as antioxidant and antibacterial agents for shelf-life extension of meat and meat products: Classification, structures, sources, and action mechanisms. *Comprehensive Reviews in Food Science and Food Safety, 16*(6), 1243–1268.
187. Schöbel, N., Radtke, D., Kyereme, J., Wollmann, N., Cichy, A., Obst, K., et al. (2014). Astringency is a trigeminal sensation that involves the activation of G protein-coupled signaling by phenolic compounds. *Chemical Senses, 39*(6), 471–487.

188. Kühlbrandt, W., Wang, D. N., & Fujiyoshi, Y. (1994). Atomic model of plant light-harvesting complex by electron crystallography. *Nature, 367*(6464), 614–621.
189. Liu, Z., Yan, H., Wang, K., Kuang, T., Zhang, J., Gui, L., An, X., & Chang, W. (2004). Crystal structure of spinach major light-harvesting complex at 2.72 Å resolution. *Nature, 428*(6980), 287–292.
190. Haken, H., & Wolf, H. C. (2013). *Atom-und Quantenphysik: Einführung in die experimentellen und theoretischen Grundlagen*. Springer.
191. Goralczyk, R. (2009). ß-Carotene and lung cancer in smokers: Review of hypotheses and status of research. *Nutrition and Cancer, 61*(6), 767–774.
192. Vrolijk, M. F., Opperhuizen, A., Jansen, E. H., Godschalk, R. W., van Schooten, F. J., Bast, A., & Haenen, G. R. (2015). The shifting perception on antioxidants: The case of vitamin E and β-carotene. *Redox Biology, 4,* 272–278.
193. Hahne, D. (2012). Epigenetik und Ernährung. Folgenreiche Fehlprogrammierung. *Deutsches Ärzteblatt, 109*(40), A-1986/B-1614/C-1586.
194. Bittner, N., Jockwitz, C., Mühleisen, T. W., Hoffstaedter, F., Eickhoff, S. B., Moebus, S., et al. (2019). Combining lifestyle risks to disentangle brain structure and functional connectivity differences in older adults. *Nature Communications, 10.* https://doi.org/10.1038/s41467-019-08500-x.
195. Fabris, J. C., Piattella, O. F., Rodrigues, D. C., Velten, H. E., & Zimdahl, W. (2016). The cosmic microwave background. *Astrophysics and Space Science Proceedings, 45,* 369.
196. Patrignani, C., Weinberg, V., et al. (2016). Review of particle physics. *Chinese Physics B, 40,* 100001.
197. Anastasiou, C., Duhr, C., et al. (2016). High precision determination of the gluon fusion Higgs boson cross-section at the LHC. *Journal of High Energy Physics, 2016*(5), 58.
198. Cheng, T. P., & Li, L. F. (1984). *Gauge theory of elementary particle physics*. Clarendon Press.
199. https://eventhorizontelescope.org/.
200. Güsten, R., Wiesemeyer, H., Neufeld, D., Menten, K. M., Graf, U. U., Jacobs, K., et al. (2019). Astrophysical detection of the helium hydride ion HeH$^+$. *Nature, 568*(7752), 357.
201. Graf, D. (Hrsg.). (2010). *Evolutionstheorie-Akzeptanz und Vermittlung im europäischen Vergleich*. Springer.
202. Williams, J. D. (2010). Evolution und Kreationismus im Schulunterricht aus Sicht Großbritanniens. Ist Evolution eine Sache der Akzeptanz oder des Glaubens?. In *Evolutionstheorie-Akzeptanz und Vermittlung im europäischen Vergleich* (S. 99–118). Springer.
203. Bucher, T., Müller, B., & Siegrist, M. (2015). What is healthy food? Objective nutrient profile scores and subjective lay evaluations in comparison. *Appetite, 95,* 408–414.
204. Laska, M. N., Hearst, M. O., Lust, K., Lytle, L. A., & Story, M. (2015). How we eat what we eat: Identifying meal routines and practices most strongly associated with healthy and unhealthy dietary factors among young adults. *Public Health Nutrition, 18*(12), 2135–2145.
205. Cairns, K., & Johnston, J. (2015). Choosing health: Embodied neoliberalism, postfeminism, and the "do-diet". *Theory and Society, 44*(2), 153–175.

Pleasure and Nutrition 6

Abstract

The role of enjoyment in nutrition plays a larger role than previously assumed. In this context, attitude and handling of food, as well as their appreciation, are of central importance for enjoyment. These parameters are ultimately determined by the composition of the food, as they determine its enjoyment potential, but also its nutritional value.

6.1 Hygiene and Pleasure

Compared to our ancestors, we live in a sterile world. Strict hygiene regulations make life difficult for producers and consumers from time to time. There is no question that a high food safety has highest priority, especially for mass-produced products from industrial production. It is fatal when entire batches of cheese regularly contain Listeria [1], packaged chickens from factory farming are contaminated with Campylobacter [2] and large quantities of goods have to be recalled and destroyed. These reports keep coming up and the statements stick in the brain. The conclusion that food cannot be trusted is subjectively quickly drawn. It is therefore not surprising that foods such as raw milk and raw milk cheese come under the scrutiny of regulatory authorities and their prohibition is sought. Raw sausage products are also increasingly viewed critically, and wild fermentations are only possible under strict conditions or at home under one's own responsibility. However, these measures have strong consequences for pleasure. The difference in taste and aroma, texture and flavor is clearly noticeable. This is objectively understandable, as the microorganisms present in raw milk, for example, emit enzymes that provide a much broader aroma spectrum and a much better and deeper taste, as can be experienced in raw milk butter and raw milk curd [3] as well as raw milk cheese [4]. The number of bioactive peptides also differs significantly in favor of raw milk and the products made from it [5]. As a result, raw milk cheeses have a higher biofunctionality than comparable products made from pasteurized, largely germ-free milk.

Of course, raw milk has a higher germ count (see Chap. 4), but as long as the germs remain below a critical concentration, they are generally not dangerous for *Homo sapiens*. On the contrary: There is clear evidence from clinical studies that children who drink raw milk at a very early stage after weaning or even grow up on farms have a significantly higher resistance to the development of allergies or respiratory diseases compared to more sheltered children [6–8].

This "farm effect" can protect children from developing allergic reactions. In the context of these studies, it can be shown that children with and without allergies differ significantly in their microbiome, in the totality of microorganisms that inhabit every human body and co-determine its immune defense. The microbiome is influenced by living conditions and diet (see Sect. 4.6.1 and 5.6.3). Responsible for this are apparently also certain components of the air we breathe, as they occur on farms [9]. Children are trained through these allergens and develop better defense forces. Apparently, "dirt" makes children immune or more immune, as does administering other typical foods with high allergy potential, such as peanuts, to infants [10].

Now, due to the natural germ content, drinking raw milk cannot be generally recommended. Therefore, in a new study, elaborate methods were developed together with a dairy to remove the germs but maintain the raw milk character. Among other things, the proportion of long-chain, polyunsaturated omega-3 fatty acids EPA and DHA seems to play a crucial role again, and it is clear that some of it is oxidized in thermal processes (see Fig. 1.5) and is therefore no longer present. An important aspect of these ideas is thus to keep the milk as natural as possible and only remove the pathogens, such as from the udder, milking equipment, or contamination from the stable floor, but otherwise leave the milk as raw as possible—and thus also preserve all exosomes and messenger or messenger RNAs (mRNA, see Sect. 4.10.9).

These results and ideas are in diametrical opposition to reports that raw milk is a highly explosive cocktail from which all kinds of dangers emanate [11], which were partly addressed in Sect. 4.10.9. However, there are also opposing ideas about the signaling effects of mRNAs [12], so these controversial questions are not at all resolved, not in observational studies [13] and certainly not the individual molecular processes necessary for a fundamental understanding.

Unfortunately, unclear statements on this topic are also found in popular literature [14] and promote the static persistence in opinion loops (see Fig. 5.1). This has nothing to do with "more enjoyment." And even less so with a balanced and varied diet. For some, this is another step towards the compulsion to eat exclusively healthily. But this is precisely what makes people sick.

6.2 A Dilemma of Food Production

A large part of the abundance Western world lives in a phase of abundance. People are completely oversupplied with food thanks to high productivity, globalization of trade, transport, and cooling possibilities. Stockpiling with modern refrigerators is also not a problem for private households. Many people are inclined to buy more than necessary, a first step towards food waste. The innate urge of *Homo sapiens* for food supply is stronger than the rational consideration of what would be sufficient for the next few days. Basically, this is not new, because with the development of preservation methods, but also home preservation, the canning of fruits, jams, pickling of vegetables, people have always preserved more than was necessary. This made it possible to use the harvest abundance of good years for crop failures in bad years.

This has not been necessary since the economic miracle after World War II. Supermarkets offer so many food items that not all of them can be consumed. Food waste has its first cause in overproduction and in the production of sometimes questionable food items. The impression occasionally arises that some foods are no longer intended for consumption, but merely serve the food industry for annual financial statements. This not only affects the food industry but also many companies in the organic sector, which devote themselves to production that is devoid of food culture and is defined by superfood, body optimization, and hardly redeemable health promises. Organic fitness bars, organic smoothie powders, and other highly processed products do not differ significantly from comparable conventional products. Food analogs only work with the use of industrial technologies and are as far removed from craftsmanship as conventional convenience products. Vegan substitute products (often positively connotated) and convenience products (often negatively judged) differ only slightly in the manufacturing process— except for the origin of the ingredients, which plays an increasingly smaller role as the degree of processing increases. Even with foods labeled "organic," the processing steps are so high (see Fig. 5.37) that they are hardly distinguishable from convenience products.

One reason for this is the standardization of (industrial) food. Rolls come from large bakeries, sausages from industrially operating large companies, and ready-made pizzas must not only taste good but always be the same. Today as yesterday, tomorrow as today, etc. This does not work naturally; every craft business that uses natural raw materials can tell a tale of the fluctuating properties of natural products. But every craftsman can react individually to this. The dough from bad wheat years can be adjusted with water, salt, and flour until the dough properties are correct again. A machine, an industrial robot, cannot do this yet. The raw material must be identical for the process to work reliably. Recipes must therefore be standardized down to every detail, up to the reproducibility of the flavor [15].

If convenience foods are to become cheaper, high-quality raw materials must be saved. Instead of more expensive meat protein, cheaper protein, such as milk protein, is used whenever possible and permitted. Although the required protein content is correct, it is mixed from different protein sources. In this way, a surrogate product is created at a low price. This happens step by step. The situation is similar in the organic/vegan sector. Pea protein is added, some soy protein, and for a satisfactory texture formation, lupine protein, which can technologically replace egg proteins, milk proteins such as whey protein, and casein in vegan convenience food up to a certain extent. Although these proteins come from an organically produced product, the extracted and highly purified molecules no longer show this. As mentioned earlier, it is a pity for every pea and bean that is no longer eaten as a whole with all its ingredients but whose valuable parts are degraded into raw materials for questionable products. This no longer corresponds to a wholesome, natural diet.

Only a multitude of these high-tech processes allows for the abundance and the resulting oversupply of food. Often, these are edible materials that have little to do with their original in terms of nutritional value, flavor, and texture, apart from the appearance. These problems are homemade. There is hope that these technologies will prove useful for the challenge of feeding the constantly growing human population.

This oversupply overwhelms many people, the production processes are not very transparent, additive declarations contain chemical terms that sound dangerous but are not. Only specialists understand their deep molecular function on nanoscales, the tasks to be accomplished there; but not even all food technologists anymore, as the sometimes contradictory list of hydrocolloids, emulsifiers, plant fibers, etc. suggests. The boundless product variety resulting from these technological possibilities is another problem: Many people nowadays no longer know what they want to eat. The choice between several, almost identical products makes the decision difficult to impossible. Escaping to organic supermarkets hardly makes it any better. Anonymous products with various claims (free from, *clean label,* without additives), including pre-packaged meat with origin traceability, merely soothe the conscience, even if the prices there are still so low that one basically has to apologize to the organic producer. The price no longer reflects the technological and agricultural effort.

6.3 Detoxifying Food and a Few Contradictions

It is not surprising that food has gained an increasingly bad reputation in recent decades, leading to the belief that humans need to detoxify from it. It is therefore no wonder that many suggestions for detox therapies come up, where the line between science and belief is blurred. Detox is also known as cleansing or detoxification [16, 17], and these methods actually prove to be questionable and unworthy of enjoyment.

Therefore, let us once again recall the basic taste directions [18], which are triggered by sugar, salt, acids, glutamate, and bitter substances (Fig. 6.1) and which have guided the *Homo sapiens* well through the millennia.

6.3 Detoxifying Food and a Few Contradictions

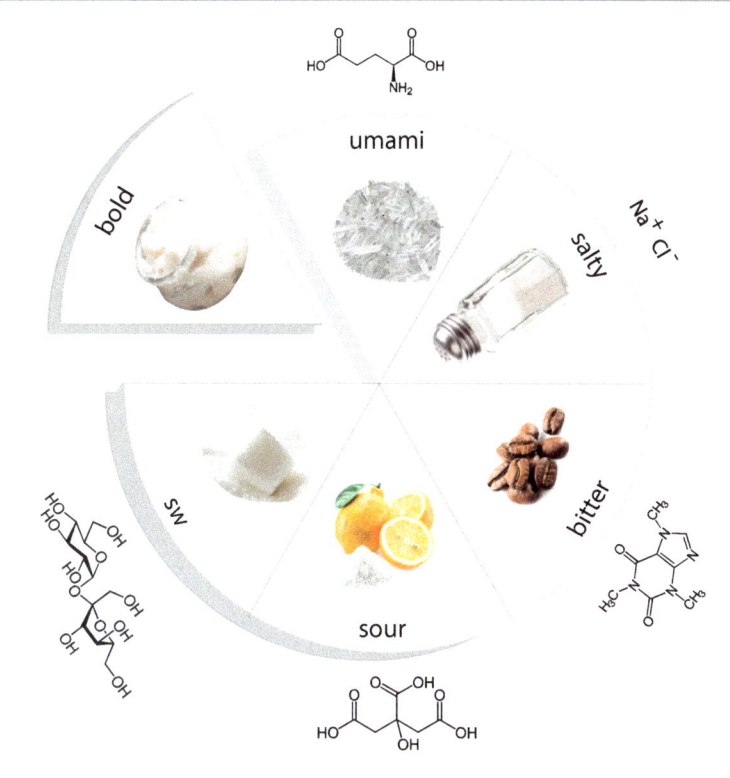

Fig. 6.1 The basic ingredients responsible for the five basic taste directions as well as fats have become supposed toxins: sucrose (sweet), glutamic acid (umami), dissociated sodium chloride (salty), caffeine (bitter), and citric acid (sour)

The consequences of the thesis that fat is harmful, but protein is good, are clearly visible: pig breeds with a high fat percentage were banned, lean pigs bred. The meat was declared healthier due to its lower fat content. As a result, demand increased, overproduction got out of hand. Meat production had to be faster, animals were fattened with protein-rich grains, and in many cases, feed additives were used to help. The problem of the so-called "Western Diet" [17] took its course.

The assumption that red meat causes cancer led to a rapid increase in demand for white meat. The consequences are mass-produced chickens and broiler chickens, completely overbred turkeys and wild turkeys. The extremely lean breast meat became popular, while the rest of the meat from the animals was less in demand. Feeding and breeding made the breasts grow, the biomechanics of the animals became unbalanced, and the poultry could neither stand nor walk. However, the meat is cheap and available to everyone. Evolution teaches us even more that we do not need meat every day to survive. On the contrary, long and irregular phases of hunger and scarcity made the *Homo sapiens* what it is: As an omnivore, it could adapt to the respective current nutritional situation. This was crucial in developing appropriate cultural techniques beyond the control of fire to make macro and micronutrients of the respective food biologically available. Only this ability is the distinguishing feature from other creatures that did not and could not take this step.

From this point of view, any exclusion of food and foodstuffs that have proven themselves since the Neolithic period is a step back in evolution. Throughout the history of nutrition, new foods have always been added, as long as they served humans and satisfied them. There is therefore not much reason to question lard, which has been the basis of nutrition alongside vegetable oils for thousands of years, and to ban the delicious lard bread with salt from the menu. A major disadvantage of abundance is that we begin to ban real food. Nevertheless, the search for new foods is essential, but we should be careful not to preserve what is physiologically proven. The ability to eat everything allowed the *Homo sapiens* a broader spectrum of food intake than many other species. Only this ability made it possible to experience emotions, and thus pleasure in eating.

How contradictory and irrational our handling of "toxic foods" is can be best seen in the example of alcohol. We know about the danger of the neurotoxin alcohol and are aware of the number of alcohol-related deaths. This is in stark contrast to the purely theoretical, added glutamate deaths, acrylamide deaths, and E-number deaths. Based on these, entire research programs are developed, legal limits are established, and bans are demanded. However, there is hardly any call to limit or even ban alcohol. If the ethanol molecule were discovered today, alcoholic beverages would not receive regulatory approval as food. It is therefore contradictory when authors criticize the industrial kitchen [19] while simultaneously publishing books on strong alcoholic beverages [20] – a problem of measuring with different standards. Despite this knowledge, we still enjoy alcohol, condemn fat, salt, sugar, and glutamate. While evolution enlightens us about the function of salt, sugar, glutamate, and fat (Sect. 1.7), it also shows for some people the possibility of extremely moderate enjoyment of alcohol (Sect. 1.8) – long before the Neolithic Revolution and the accompanying selective lactose tolerance.

6.4 Understanding of Research is Dwindling

6.4.1 Recognizing Connections

Among the major problems are the systematic inaccuracies of associative observational studies. They leave much room for interpretations and speculations. Depending on one's perception and viewpoint, interpretations can be derived that fit one's own belief system. This results in inconclusive and, unfortunately, often pseudoscientific opinions that, although well-intentioned, cause many misunderstandings and untruths. Refuting these is often difficult, as on the one hand, many observational studies leave a lot of room, and on the other hand, beliefs weigh more heavily than facts.

The many ambiguities, the weaknesses of the data, escalating statistical errors, and the resulting contradictions and ambiguities of observational studies in nutritional sciences and medicine create, in most cases, little basic understanding for the general public, but rather fuel doubts. It would be simpler and more comprehensible for everyone to take a look at the eating biography of Homo sapiens. The conclusion is quite simple: Our ancestors, since the first hominids, have done little wrong in their diet and meal plan. Because as we now know, the "poison alarmism" is true in very few cases or only with strong restrictions. Most of these discussions concern ready-made products, convenience food, and industrially produced foods.

In fact, it is easy to almost stop poisoning discussions and simply leave these products on the corresponding shelves in the supermarket and instead go to the weekly market and buy directly from the producers. Many people buy these products, however, because they do not care what they eat (probably the majority), because the products are cheap (also the majority), or because they are available in the supermarket.

But it also follows that not only "the industry" that produces analog food, overly sweetened drinks, sausages made from the criticized separator meat [21] and water-binding hydrocolloids is ultimately the sole culprit, but also "the consumers" themselves are to blame. As long as these products end up in shopping carts as a matter of course, profit can be made from them; as long as capital is generated, there is no reason for manufacturers not to produce these products.

6.4.2 Why Mechanically Separated (or Deboned) Meat is Good in Essence

At this point, it is worth considering the much-maligned mechanically separated meat. It is the meat that everyone who makes their own broths and stocks is already familiar with. Mechanically separated meat is the tissue that binds muscle meat to the bones, simply put, it consists of proteins that define the transition between muscles and bones (Fig. 6.2). It is structured slightly differently than muscle, which is evident from its function alone. Mechanically separated meat is richer in connective tissue, which, as gelatin or its bioactive peptides, is supposed to prevent skin aging (see Sect. 5.4.4). But it is still meat, it consists of protein, it comes from the slaughtered animal, and its use is nose-to-tail in the best sense, and from this perspective, there is nothing to criticize.

Fig. 6.2 The mechanically separated meat is located at the transition between muscles and the collagen of tendons, which are firmly attached to the bone (marked in red). It thus consists of muscle meat attachments and connective tissue

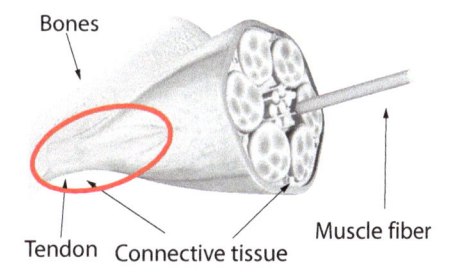

In its uncooked state, it must be mechanically removed using suitable machine processes due to the high tensile strength of the collagen. This is not disgusting, but meat that can also be used as an ingredient in recipes for sausages, provided it is declared according to food law. It contains, apart from the small amounts of the amino acid tryptophan (see collagen, Sect. 5.4.3), a significant content of essential amino acids [22] and can be incorporated into boiled sausages [23] without loss of quality. It would be downright reprehensible, in the sense of our Paleolithic ancestors, not to eat this protein content. The only difference from that time can be found in the industrial techniques used today to completely remove the meat from the bone without prior cooking.

If you want to produce inexpensive boiled sausages for the discount sector from pure mechanically separated meat, this works less well due to the lack of muscle content. Then, additional proteins, hydrocolloids, binders, starch, or transglutaminases (see "lab-grown meat", Sect. 5.11.1) must be used to help.

6.5 Tradition and Enjoyment: Meat and Sausage from Home Slaughtering

6.5.1 What We Know About Meat and Sausage

Good meat is more expensive because it does not come from factory farming, has not been transported across the country, and the slaughter animal died largely stress-free, in the best case even during a home slaughter in a small butcher's shop. Sometimes, farmers are even allowed to slaughter themselves and process the animal on site together with a butcher. Not too long ago, pigs were allowed to be kept in small businesses without strict hygiene measures and were slaughtered and processed directly on the farm. Afterwards, strong, intensely umami sausage soups were made from meat cuts, skins, bones, head meat, snouts, and tails, blood sausages (black pudding) were produced, while the meat for the boiled sausages was still processed while warm from slaughter. This traditional cultural technique of warm meat processing provided qualities that are hardly available today, especially not in industrially operating slaughterhouses. The taste of warm meat preparations is equipped with intense umami potential, the texture is indescribably firm, to the bite, the fat is completely emulsified in small droplets, which leads to an excellent mouthfeel. Above all, the warm meat processing can do without any additives such as phosphates. For all these positive sensory aspects, there are solid physical-chemical arguments that can be easily derived and understood from the relationships discussed so far.

Handcrafted boiled sausages contain a high proportion of muscle meat in addition to fat and water. Fat mainly serves the mouthfeel and as a flavor carrier. The water is added for biophysical reasons in the form of ice and serves for cooling, while the muscle meat serves for binding. Each muscle fiber consists of myofibrils (see Sect. 5.10.10, Fig. 5.38). The essential features for the following are simplified again in Fig. 6.3.

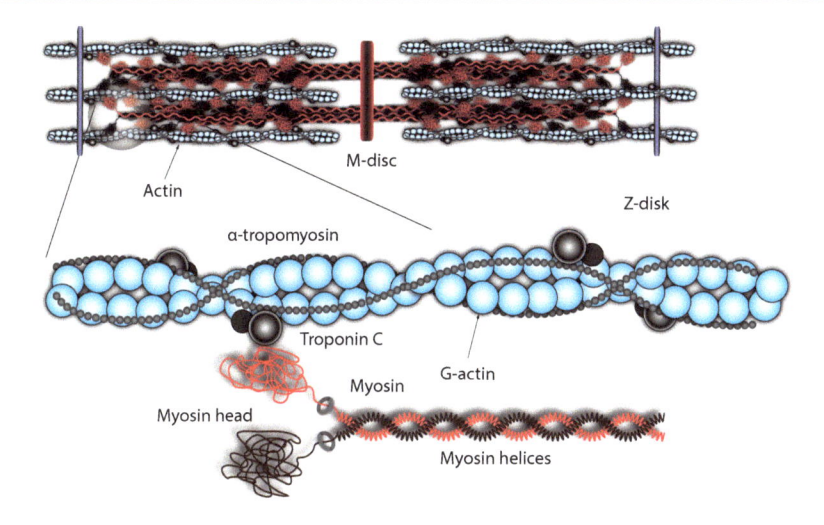

Fig. 6.3 Highly simplified representation of the interplay between actin and myosin in the myofibril of a muscle cell (sarcomere)

Fibrillar actin (blue twisted bead strands in Fig. 6.3) consists of helically arranged globular F-actin proteins (light gray beads, surrounded by α-tropomyosin are. ATP causes the release of binding sites to which the myosin head can bind and cause muscle movement. Myosin itself consists essentially of two helices, which in turn are wound into helices with a larger pitch length. They end in the myosin heads. The various proteins denature, or "cook," at different temperatures in the range of 48 °C and 75 °C.

In fact, myosin is the main molecular actor in the difference between warm and cold meat processing. To understand this more precisely, the postmortem processes in meat must first be considered, as they are simplified in Fig. 6.4.

During slaughter, blood supply is stopped and thus oxygen supply. Nevertheless, adenosine triphosphate (ATP) persists for a short time before it decomposes according to the purine metabolism (see Sect. 4.4.3) and is enzymatically converted into flavor enhancers. Therefore, up to three or four hours after slaughter, myosin and actin are still in their physiological state, they are still separate. Gradually, however, oxygen-poor fermentation sets in, the pH value begins to decrease, and ATP is degraded, causing myosin and actin to irreversibly bind at the heads. The rigor mortis forms, and the muscle cells becomes firm.

The cleavage of glycogen (animal starch, carbohydrate energy storage of the muscle) into glucose and subsequently into lactic acid lowers the pH value of the meat to about pH 5. This slightly changes the shape of the proteins, but the actin-myosin complex remains intact. These exact conditions define the physical difference between warm meat processing and the widely used cold meat processing.

If the meat is processed while still warm from slaughter, all the physico-chemical properties of the intact muscle before rigor mortis can be utilized.

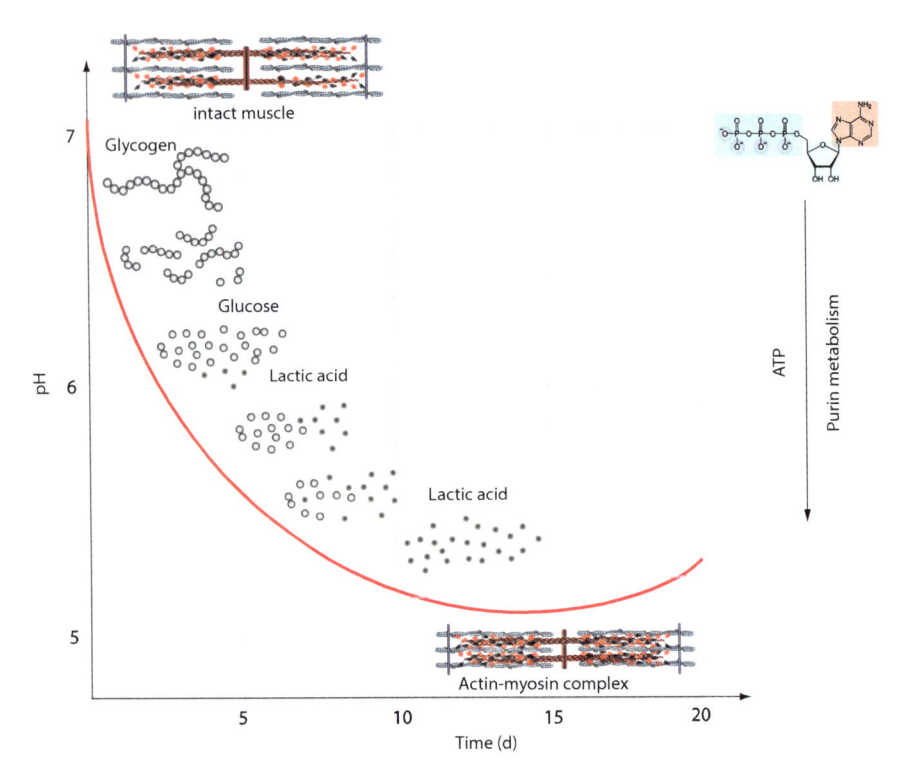

Fig. 6.4 The postmortem processes in meat after slaughter are mainly manifested macroscopically in the temporal course of the pH value and the anaerobic lactic acid fermentation inside the animal body

Biophysically, this term means that the actin-myosin muscle motor is still functional, so the actin and myosin fibrils are not yet rigidly connected, as in the case of developed rigor mortis. Myosin and actin can therefore be separated. The muscle proteins myosin and actin are used largely separately in warm meat. Therefore, myosin can take on its emulsifying tasks in the cooked sausages. In cold meat processing, both are firmly connected in each myofibril to form an actin-myosin complex. This is the main reason for the necessary use of cutter processing aids.

For the scalded sausages, the meat must be minced together with fat under the addition of ice to form a sausage meat (emulsion). The cutter speeds (and thus the shear rates) primarily determine the resulting particle size of the muscle meat and fat, as well as their size distribution. Fine boiled sausages require high shear rates and longer times for good homogenization. At the low temperatures maintained by the ice cooling (<10–12 °C), fat remains almost solid, meat is minced, muscle proteins are released and can contribute to the emulsifying effect. The sausage meat must stably bind both the added water (as finely crushed ice) and the fat [24]. This stability must be ensured during boiling. If this structure of the emulsion is maintained under solidification during the temperature increase, the sausage texture is perfect [25].

6.5.2 Warm Meat Processing: Fundamental Advantages on a Molecular Scale

The adenosine triphosphate (ATP) is hardly degraded up to six hours after slaughter, so it is abundantly available. The proteins are virtually still in their native state, the solid actin-myosin-complex has not yet formed. Myosin and actin can be easily separated. The pH value of the meat is still at pH 7 and when minced with salt, sufficient myosin is dissolved, which, in addition to the emulsifiers (phospholipids, lecithin) released from the cell membranes, has an excellent emulsifying effect. The myosin heads show a distinct hydrophobicity, so they are more inclined towards fat, while the helical parts of the myosin are more hydrophilic and therefore protrude into the water-rich parts of the sausage meat. Each myosin is thus an emulsifier. The heads of the free myosin largely attach to the fat surfaces of the still solid fat particles during mincing. Furthermore, myosin has a high number of bindable cysteine (Fig. 6.5), which is capable of forming disulfide bridges.

Since the myosin in warm meat is not bound to actin, there is enough free myosin available after chopping to coat the fat particles like a larger emulsifier, as shown schematically in Fig. 6.6. At the same time, globular proteins of the sarcoplasm are released by cutting many muscle cells. Undamaged muscle fragments and, depending on the chopping speed, more or less small solid meat particles are also present in the sausage meat after chopping. Collagen is partially transported from the muscle fibers into the sausage meat, as well as fibrillar actin, since there are no bound actin-myosin complexes in warm meat. Water from the melted ice is distributed in the sausage meat.

During scalding (usually in several temperature stages from 50 °C to a final 75–80 °C), the fat begins to melt first. As a result, the hydrophobic amino acids of the myosin head can dive further into the fat as soon as the heads begin to denature. During the denaturation of the myosin head, the cysteine sites are exposed (colored yellow in Fig. 6.5). Since the concentration of free myosin is high in warm meat, the fat droplets are densely covered with myosin, so that the slowly

Fig. 6.5 Representation of the structure of a myosin head. The amino acid cysteine is located at the yellow marked spots. This cysteine is available for the formation of disulfide bridges after the unfolding of the head

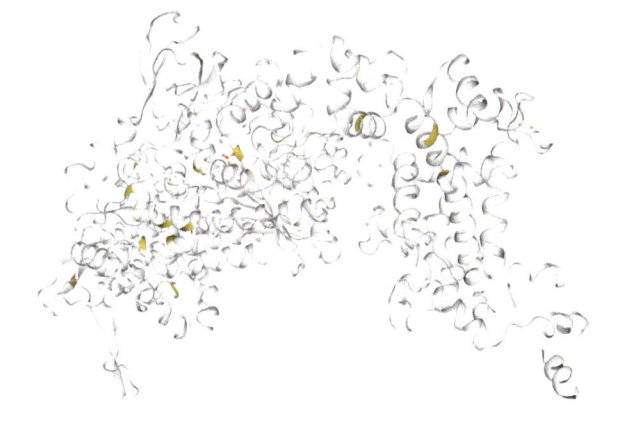

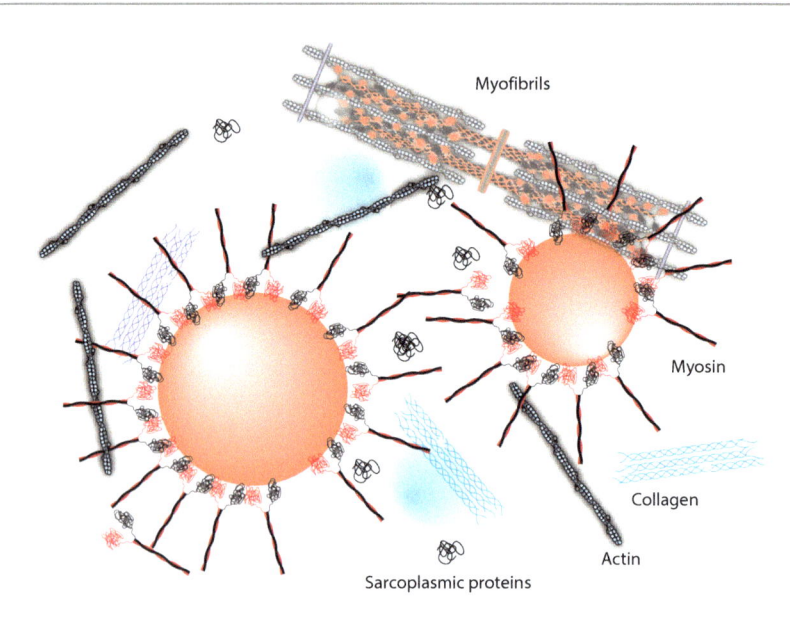

Fig. 6.6 Schematic model of a warm-processed sausage meat for a cooked sausage. Due to the immediate processing after slaughtering and the strong cooling with ice during chopping, most proteins remain largely native

unfolding myosin heads of different chains can come close to each other at the interface of the fat droplets. As the temperature continues to rise, disulfide bridges form quickly (see Sect. 3.4.6, Fig. 3.26). Myosin thus takes on the main task of stabilizing the complex emulsion, because as indicated in Fig. 6.7 by the black circles, the myosin heads network at the surfaces of the fat droplets and "fix" the emulsion. The rest is pure structural cosmetics: at higher cooking temperatures, sarcoplasmic proteins denature, intertwine and network with denaturing myosin helices, collagen denatures to gelatin and binds water.

After cooling, the firmly bound fat partially crystallizes again, and the structure is fixed through a diverse network formation between myosin-myosin and the sarcoplasmic proteins. The fresh boiled sausage therefore shows its characteristic elastic-crunchy bite.

Furthermore, in the warm meat processing, the glycogen is still present. This large, branched starch molecule ensures a high viscosity of the sausage meat. The individual components can only diffuse very slowly, as the typical diffusion coefficients are inversely proportional to the viscosity and thus small. The meat's inherent hydrocolloid glycogen contributes to a much greater extent than gelatin to both water binding and the high viscosity of the aqueous (continuous) phase, and fat droplets diffuse considerably more slowly. The formation of free, unstructured fat droplets and their coagulation into large, non-emulsified fat droplets are excluded.

In addition, the pH value in slaughter-warm meat is close to pH 7, the myosin is far from its isoelectric point and therefore has a higher total charge. The

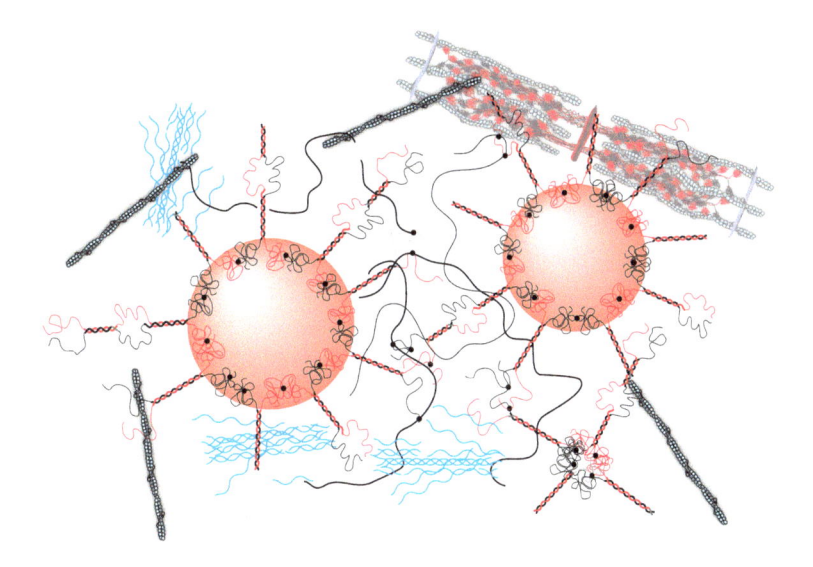

6.5 Tradition and Enjoyment: Meat and Sausage from Home Slaughtering

Fig. 6.7 Schematic representation of the scalded sausage structure after cooking. The fat is solid again, network points (black circles) are formed. Connective tissue (blue helices) denatures, as do the helices of myosin

individual myosin proteins repel each other in the aqueous phase via their helix parts: They arrange and organize themselves at approximately equal distances around the fat droplets as an effective emulsifier. As temperatures continue to rise in the next boiling stage, the myosin heads, which are located at the interface to the fat droplets, begin to network. As a result, myosin heads form a firmly networked and thus fixed emulsifier structure, and fat can no longer escape from the droplets even at higher temperatures, with hardly any free, non-emulsified fats forming.

The taste of scalded sausages produced using the warm meat method is also less sour, as the pH value of the warm meat sausage meat is at pH 7. Short glucose chains and glucose already enzymatically formed in the living animal from the glycogen give a slightly sweet taste, and the aroma of the spices is perceived more intensely.

6.5.3 Cold Meat Processing: Systematic Physical Deficits in Microstructure

If scalded sausages are produced using the cold meat processing, as is common today, they differ significantly in texture and taste for purely physicochemical reasons. The rigor mortis has already occurred, myosin and actin are present in a solid, insoluble complex. Therefore, there is hardly any free myosin available as a multifunctional emulsifier. In addition, the pH value drops sharply with storage, which changes the taste of the scalded sausages, making them sensorially much more acidic. Part of the texture differences compared to warm meat sausages can also be attributed to the low pH value when myosin and various sarcoplasmic proteins are near their isoelectric point. They are already denatured, are already in a non-native conformation, and therefore show a significantly different networking behavior. The water-binding properties also change, not least because of the no longer present glycogen, the animal starch, which is why it is obvious to occasionally add physically modified (e.g., pre-swollen) potato or corn starch or other binders to the sausage meat, which take over the role of glycogen in thickening, i.e., viscosity adjustment, of the sausage meat.

As already noted, the actin-myosin complex is fully developed in cold meat. However, since myosin plays a central role in structure formation in the emulsion during hot meat processing, it is necessary to separate the actin-myosin complex. The obvious idea is therefore the use of diphosphate (Fig. 6.8), which is a tetravalent ion and can mimic the electrostatic interactions of ATP in the solidified muscle.

However, not completely, the reason for this is the different molecular shapes: The tetravalent ion diphosphate corresponds only to the ionic part of the adenosine triphosphate (ATP) in terms of charge strength (see Sect. 4.4.3, Fig. 4.10). The non-ionic part of ATP, i.e., the hydrophilic ribose and the hydrophobic adenine, cannot be simulated with it. Consequently, diphosphate has a different local (quantum chemical) interaction with the amino acids in the myosin head. The diffusion of the molecule also does not correspond to that of ATP. In living muscle (and therefore also in warm meat), however, this detailed chemical structure of this group plays an essential role in the interaction with the corresponding, precisely fitting amino acids in the myosin heads. Therefore, less free, emulsifiable myosin is available for the sausage matrix. Furthermore, the pH value of cold meat is significantly lower. The myosin heads denature more easily, and the first micelles form. The structure of the cold meat batter is thus less homogeneous, less sticky, and characterized by poorer water binding. Fat is less well integrated into the batter matrix. However, the emulsifying ability is supported by the actin-myosin complexes released during chopping and the associated increase in viscosity, which occurs in warm meat through glycogen; but not its water-binding capacity. Therefore, cold meat products can be added with some starch (or maltodextrin, and thus hydrolyzed starch) to achieve higher water binding in cold meat. The micro- and nanostructure of cooked sausages made from cold meat are therefore noticeably and tastefully different from those of warm meat sausages, as indicated in Fig. 6.9.

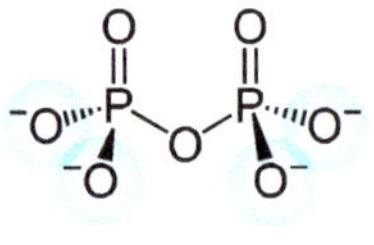

Fig. 6.8 The structure of the tetravalent diphosphate ion can only very limitedly simulate ATP, therefore the effect of phosphates on the dissolution of the actin-myosin-complex is rather low

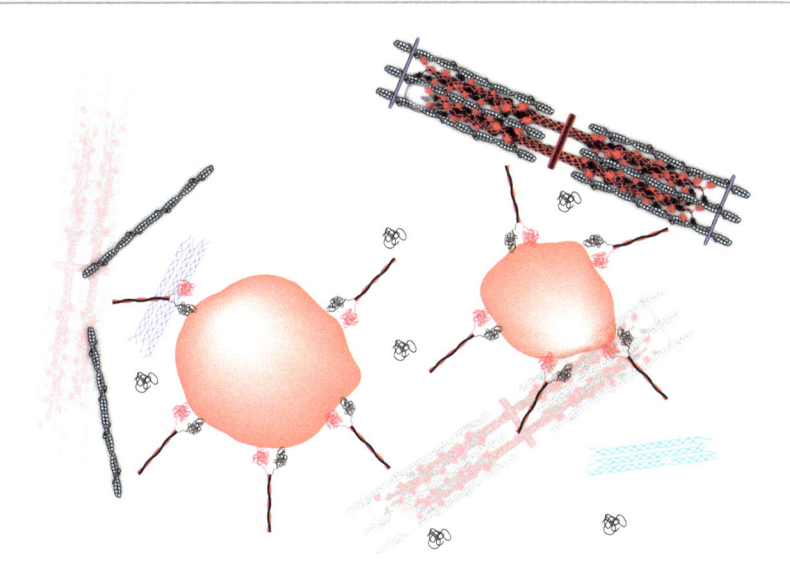

6.5 Tradition and Enjoyment: Meat and Sausage from Home Slaughtering   399

Fig. 6.9 Highly schematic and not to scale representation of some elements of the cold meat batter. The myofibrils are more numerous due to the fixed actin-myosin complexes. Less free myosin is available for emulsifying the fat

Under the temperature increase of cooking, the liquidized fat cannot be completely emulsified by the small amount of myosin. Thus, the droplets become smaller despite the increase in surface area, and excess fat is squeezed out of the droplets, creating free fat, as shown in Fig. 6.10.

Despite the reduced effect of sufficiently detaching myosin from the actin-myosin complex, diphosphate, due to its fourfold negative charge, has a positive effect on cold meat. It dissolves and binds with the doubly positively charged calcium ions, which are responsible for the cold shortening of muscles [26, 27]. The reduced swelling capacity of muscle proteins due to cold shortening is partially reversed by the use of diphosphate.

This is another example of how well-developed, traditional cultural techniques are lost due to industrial criteria. Taste and sensory experience suffer under price pressure, hygiene regulations, and the decline of craftsmanship. A partial return to smaller businesses and warm meat processing would thus be a significant step towards greater enjoyment, away from the largely standardized mass taste. Not to mention the associated sustainability in animal husbandry, slaughter, and local processing.

At this point, it becomes particularly clear how different cooked sausage substitutes are from cooked sausages made of meat and animal fat. If the differences between warm and cold meat processing are already so large on the scale of micro- and nanometers, a structuring and texturing of plant proteins or egg white can never be replicated so analogously. The proteins in primary, secondary, tertiary, and quaternary structure are much too different for that.

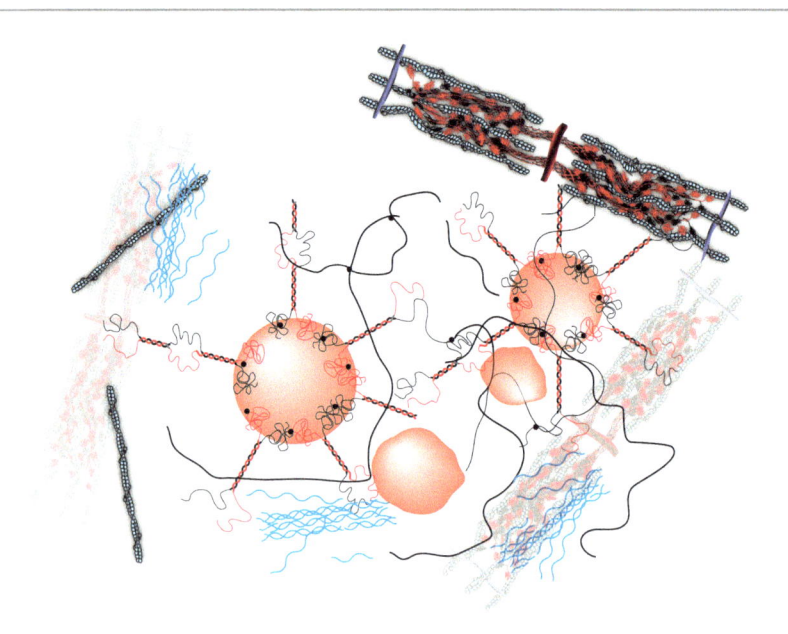

Fig. 6.10 Schematic structure of a cold meat processed cooked sausage. Free, non-emulsified fat is formed. Not all emulsifying myosin heads can network with others on the fat surface (non-emulsified fat droplets). The structure is less stable

6.6 Home Cooking is Worth Its Weight in Gold

6.6.1 Control It, Do It Yourself

Then there is the other faction of people who stand in the kitchen and prepare food themselves. These pleasure-seekers would not think of sugar, salt, and fat as harmful to health. Sugar, salt, and fat are welcome flavor enhancers, and the tongue retains control over the added amount through taste. Any discussion about the harmfulness of salt, sugar, or fat becomes irrelevant, because those who cook seasonally and regionally themselves maintain control over everything in their food: not only the use of fat, salt, and sugar, but also product quality, product origin, and their lifestyle. If one is not sure whether the goods are fresh enough, have been stored poorly, or the farmer is known for not skimping on pesticides, one has a choice: The next stand is just a few meters away. However, it may be that the celery root then costs a few cents more than at the discount store. There is no such control in the supermarket. Trade dictates prices and conditions to many producers, and the lowest bid usually wins, as consumers want to shop cheaply and trade operates for maximum profit.

Cooking for oneself essentially makes every guide, all nutritionists, and all health books (including this one) unnecessary. In essence, that's all it takes. The most beautiful reward is, moreover, one's own creativity, which also succeeds in breaking the reward cycle (see Fig. 5.47) in a very simple way, instead of continuing to eat and snack while getting less and less satisfaction and reward.

6.6.2 Forgotten Vegetables, Secondary Vegetable Cuts, Root-to-Leaf

A visit to the market is not more expensive compared to shopping at the supermarket, on the contrary. Especially if you eat everything that radishes, celery, kohlrabi, etc. have to offer. Just like with meat, you can discover "secondary cuts" with vegetables, which provide whole meals for a few days. It is quite simple: Have you ever tried soup green puree? A soup green at the market costs little. It contains a quarter to half a celery with greens, some parsley, a carrot, a leek – you can already smell the vegetable soup just by looking at the bundle. Most of the time, the soup vegetables are B-grade, Secondary vegetable cuts, so to speak, which were damaged during harvesting, but that doesn't matter, it still tastes good.

Indeed, with soup greens and a little culinary competence, a wonderful puree can be made that is suitable for many occasions of vegetarian main courses or starters. It is not even time-consuming: Clean the vegetables for a soup with a brush, cut them into small pieces, add them to the broth with one to two teaspoons of salt and about a third of the weight of the vegetables in good sweet cream butter, and for structure, add three to five tablespoons of sunflower seeds (alternatively macadamia nuts) to the ever-present Thermomix®, cook it there for 45 minutes and puree it very finely. The result is a creamy, satisfying puree that will accompany the following menus for the next few days with great adaptability, without effort, at a low price, with a lot of taste and high satiation potential. It is also low carb, has one of the best fatty acid distributions that nature offers with the butter used, and an extremely intense vegetable taste. The provitamin A of the carrot is largely biologically available due to the long cooking process, as are the polyphenols of the vegetables. The puree goes well with the rest of the boiled beef, the liver or the chickpeas, as well as with a poached egg, the lukewarm goat cheese or a dollop of full-fat yogurt. And if you need the puree to be more umami the next day, add some glutamate, some soy sauce or some oyster sauce to taste when reheating, then it will accompany the trout from the local stream as well as the briefly grilled smoked tofu. Cold, the butter is now partially crystalline, it even spreads on canapés and beats many vegetarian spreads from the organic supermarket in many respects. This way, you have eaten the soup vegetable puree for seven days in a wide culinary context without feeling bored without missing any pre-prepared pizza or the delivery service.

The extra vegan version also works with extra virgin olive oil if the puree is stored in the refrigerator. The oil is then crystalline, excellently emulsified in the vegetable components, and the basis for homemade vegetarian and vegan vegetable spreads. Learning effects are guaranteed, and the foundation for the next step of "do all yourself" is laid.

6.6.3 Pure Pleasure

If the butcher of confidance happens to have a wonderfully aged, densely marbled Wagyu beef, which he only gets once or twice a year, it can become quite expensive – provided the origin, quality, and fat composition are right. Then the money is very well invested: in perfect flavor and best quality. Above all, the meat has a lot of learning potential.

When you run your finger over the fat marbling, the intramuscular fat of genuine Wagyu or Kobe beef will quickly melt under the slight temperature increase of the fingertip. This is an unmistakable sign of a high proportion of unsaturated fatty acids, as normal beef tallow only melts above the temperature at the extremities of the human body, which is a few degrees below body temperature. Thus, the intramuscular fat of many fat-rich cattle breeds (including Angus) melts at similar temperatures to goose or pork lard, indicating a high proportion of unsaturated fatty acids in the fat.

For those who find these empirical observations insufficient, they can turn their attention to the scientific journal *Science* [28]. There, the proportion of different fatty acid types in mortality was statistically estimated using epidemiological methods based on observational studies. Of course, the results are fraught with large errors, and they do not say anything about the individual, but they at least reflect trends, as shown in Fig. 6.11.

It clearly shows the strong influence of *trans* fatty acids (see Fig. 3.12). Even with saturated fatty acids, there is a slight increase, which, however, is practically offset by the positive contribution of monounsaturated fatty acids, as can be seen from the arrows in Fig. 6.11. This is not surprising when we recall the discussion about cell membranes in Sect. 2.4 in connection with Figs. 2.8 to 2.10. Saturated and monounsaturated fatty acids in phospholipids provide the basic flexibility of the cell membrane and thus its function.

This is good news insofar as all natural fat sources, whether animal or plant, have a balanced balance between saturated and monounsaturated fatty acids, as can be found, for example, in olive oil. The very positive influence of polyunsaturated fatty acids is obvious and also results from the physicochemical discussions in Chaps. 2 and 3.

In such model calculations, it is of course difficult to separate the difference between omega-3 and omega-6 fatty acids, which is why no objective statement can be made about this. Furthermore, it should be pointed out once again that there is no difference for physiology in the origin of the fatty acids: whether animal or plant-based makes no difference (see Sect. 1.3.4, Fig. 1.6).

In the case of Wagyū fat, the average fatty acid composition was analyzed, and the broad spectrum of saturated and unsaturated fatty acids was impressively confirmed [30], for example, where the percentage of monounsaturated fatty acids

6.6 Home Cooking is Worth Its Weight in Gold

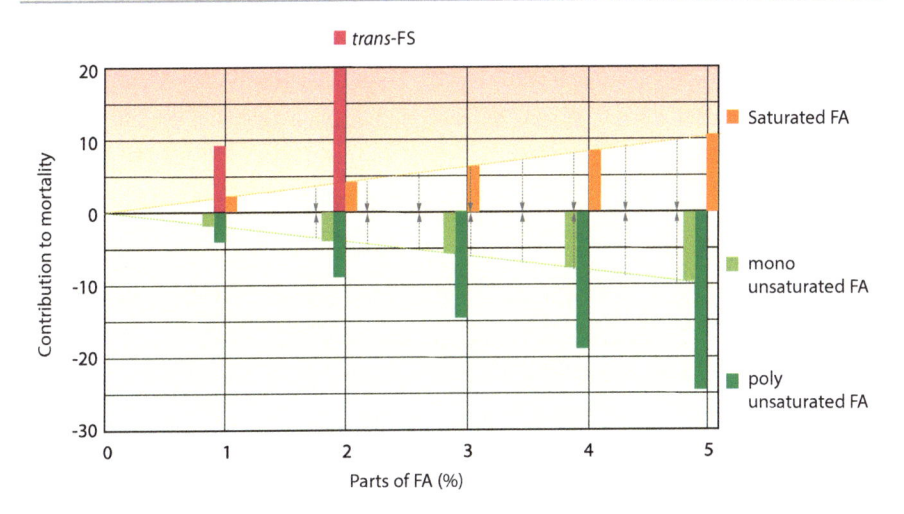

Fig. 6.11 The calculated contribution of certain fatty acid classes to mortality (taking into account age, diet, gender, smoker/non-smoker, etc.). It is noteworthy that the weakly negative influence of saturated fatty acids(FS) is practically offset by the weakly positive effect of mono-unsaturated fatty acids over the entire range (representation based on data from Wang [29])

(C 14:0, C 16:0, C 18:0) is on average higher than the proportion of saturated fatty acids (C 14:1, C 16:1, C 18:1) (Table 6.1).

This should by no means imply that a fatty beef steak should be on the plate every day. However, this aspect also shows that it is by no means "unhealthy" to eat some of it from time to time. Of course, this meat has its exceptionally high price, but the animals have been raised in more than appropriate conditions, neither overfed nor slaughtered under stress. This pleasure is therefore rare and is thus part of the appropriate diet of the *Homo sapiens*.

6.6.4 Intramuscular Fat: Fancy Flavour Enhancer

Without fat, food usually tastes bland – well known from the lean meat of fat-free pigs, breeding poultry, or low-fat cheese. Lean, low-fat muscle meat remains somewhat boring regardless of the cooking method and hardly forms specific flavors. The difference to marbled meat is enormous. The aromatic differences can already be recognized during orthonasal smelling, especially when the meat, e.g., in beef, has been aged for a long time: Oxidation of fatty acids forms distinct green, grassy, slightly cheesy, and above all mushroom-like notes that are missing in any lean piece. The difference becomes even more striking after cooking: While meat with a high degree of marbling melts between the tongue and palate, lean pieces, depending on the muscle fiber length, need to be chewed and processed more. Marbled meat releases a multitude of dissolved flavors while chewing. As already noted, it also facilitates chewing and swallowing. Fat acts as a lubricant in

Table 6.1 The average composition of fatty acids in Wagyū cattle. The essential long-chain polyunsaturated fatty acids EPA and DHA are present in significant proportions

Fatty Acid	Name	Proportion (%)
C 14:0	Myristic Acid	4.1
C 14:1	Myristoleic Acid	1.3
C 16:0	Palmitic Acid	29.8
C 16:1	Palmitoleic Acid	5.1
C 18:0	Stearic Acid	9.2
C 18:1	Oleic Acid	41.1
C 18:2	Linoleic acid	1.1
C 18:2 9c 11t	CLA	
C 18:3 (n–3)	ALA	0.3
C 20:3 (n–6)	Eicosatrienoic acid	0.1
C 20:4 (n–6)	AA	4.0
C 20:5 (n–3)	EPA	0.2
C 22:4 (n–6)	Adrenic acid	0.5
C 22:5 (n–3)	DPA	0.7
C 22.6 (n–3)	DHA	0.5

the mouth [31]. The friction between the tongue and palate is reduced, making the "oral processing" enjoyable [32].

In animals with a high proportion of intramuscular fat (IMF), fine-fibered muscle bundles are formed, and a long-fibered meat structure is already prevented during growth in favor of tenderness. Furthermore, the intramuscular fat surrounds the muscle fibers. Meat juices can escape less during cooking, as the meat juices can hardly penetrate the fat barriers around the fibers. Water losses are therefore lower, and juiciness is preserved. Tenderness is thus ensured by the interplay between intramuscular fat and meat juices. Even more, the roasting flavors that develop during frying or grilling also dissolve to a large extent in the fat. They remain in the meat and end up in the mouth when enjoyed.

The intramuscular fat is located in the muscle bundles and is the last fat depot to be formed during the growth and movement of the animals. The IMF is crucial for sensory perception and meat enjoyment, as it determines a large part of the texture and aroma release. In Japan, the percentage content of the IMF is measured according to Japanese standards in twelve marbling grades. From these, five quality levels are derived in the USA. Lean German beef is consistently below average according to these standards in terms of texture, taste, and aroma (Fig. 6.12).

The IMF of a well-marbled piece of beef consists of a mixture of triacylglycerols (triglycerides), saturated, monounsaturated, and polyunsaturated fatty acids of different lengths, as well as omega-3 fatty acids, whose exact composition depends on the breed and feeding. Since the IMF must serve as a rapidly available energy store in the muscle, its proportion of shorter, monounsaturated, and polyunsaturated fatty acids is significantly higher compared to tallow: It therefore melts

6.6 Home Cooking is Worth Its Weight in Gold

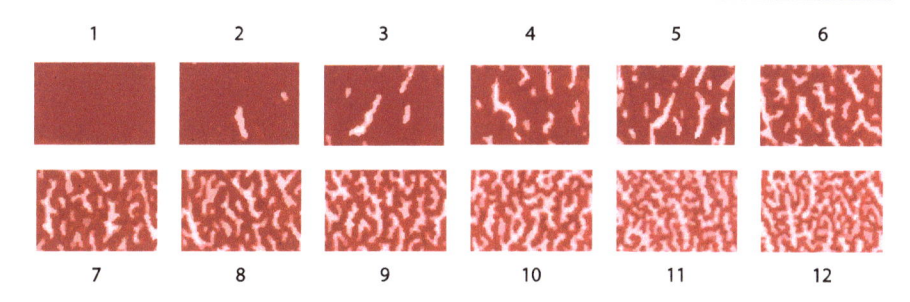

Fig. 6.12 The twelve marbling grades used in Japan for cattle. Red: muscle, white: intramuscular fat

at significantly lower temperatures. That is why the fat of Wagyu tartare already melts on the tongue.

Not every cattle breed is *per se* suitable for the storage of a high IMF content. Naturally, Wagyū is at the top, but also Angus, Murray Grey, or Shorthorns are characterized by higher IMF contents. Typical meat breeds, such as Charolais, Holstein, or Simmental, are significantly behind. In general, an advantage of female animals can also be observed. Heifers always have a significantly higher proportion of IMF than bulls. Heifers are also noticeably finer-fibered than bulls. Steers lie in between for IMF incorporation due to hormonal changes.

A very decisive influence is the high fat and IMF content on meat maturation and the associated aroma formation. Unsaturated fatty acids play a particularly important role here. The double bonds are chemically less stable, which is why during maturation, some of them are more likely to break. Small fragments split off, forming fragrant aroma compounds and contributing significantly to the typical aroma of matured meat. Fatty meat, especially when more unsaturated fatty acids are stored, as in Wagyū or other breeds with high IMF content, has a much more intense, specific *dry aged* "beef scent" than lean meat. The fat also indirectly increases the mouthfeel, kokumi, which was already mentioned during ripening and fermenting as a constant companion of the umami taste. Long-chain unsaturated fatty acids are therefore much more than just healthy; they create enjoyment in a completely different way.

From the long-chain fatty acids that split off during meat maturation, so-called oxylipins are formed, which, similar to braising and fermentation techniques, cause a significant increase in the kokumi effect. The oxidation of fatty acids not only forms welcome and characteristic aromas [33], but also contributes to moutfilling and mouthfeel [34, 35] (Fig. 6.13).

Such oxylipins also contribute to the extraordinary mouthfeel of sardines and anchovies or anchovies preserved in oil. The small fish are briefly fried and then stored in high-quality oils, even for many years in the case of vintage sardines. Over time, during the storage years, these fatty acid derivatives responsible for kokumi are formed through oxidation processes. The mouthfeel increases significantly, as does the taste intensity.

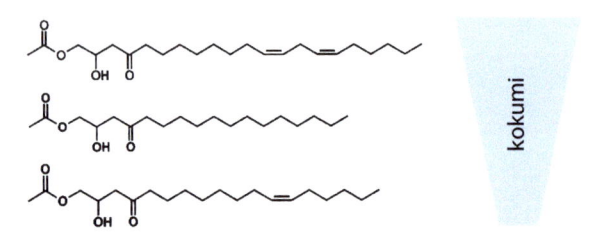

Fig. 6.13 Examples of oxylipins, fatty acid derivatives responsible for the kokumi impression. The perceived kokumi intensity is structure-dependent. The intensity (according to objective sensory tests) increases from bottom to top

Fat has another very special and, apart from the friction-reducing texture influences, largely unknown and unconsidered sensory side effect: It acts as an "oral aroma storage." Fat coats parts of the tongue and palate in a thin film (*oral coating*) and traps dissolved aromas for a certain time, which season the next bite, be it meat, bread, or vegetables. In many ways, fat acts as a sustainable and molecular "flavor enhancer" and pleasure enhancer. And all this with higher satiation, even with very small portions of meat.

6.7 Nose-to-Tail, Taken Seriously

However, attention should not be paid exclusively to the noble parts of the animals. It is also urgently necessary, in the interest of sustainability, to take care of all edible parts of the animal. As was the case until recently. There is much more edible than is generally offered. Regional differences also become apparent, for example, when considering the sour tripe, i.e., the parts of the rumen of cattle and sheep, which are widespread in southern Germany but are hardly eaten north of the Main. In France (*Tripes*); in Italy (*Trippa*), in Poland (*Flaki*), in Turkey (*şkembe çorbası*) and in other countries, they are still available as a delicacy. The recipes vary depending on the region and range from simple country cuisine to gourmet versions created by star chefs like Paul Bocuse [36]. Highly decorated chefs thus bestow the highest culinary honors on simple tripe. There is no reason not to eat them.

In addition to the common edible offal such as heart, liver, and kidney, whose virtues have already been praised, there is also spleen. A precious piece that our ancestors feasted on. It is only eaten in Bavaria in the form of milt sausage, and in parts of Languedoc in France, it is the main ingredient of a sausage, the Mèlsa, whose consistency is similar to that of coarse liverwurst. The name Mèlsa means milt in the Occitan regional language.

Even collagen-rich parts, such as head meat (*tête de veau*) and ox muzzle, were (and are) part of the diet. The famous ox muzzle salad, the aspics, the pickled, strongly acidified meat are forgotten in many regions. No one thought that collagen and gelatin were biologically inferior proteins with limiting amino acids (see

Sect. 5.4.2). The contradiction is obvious anyway: Nowadays, people prefer to drink tinctures with bioactive peptides from connective tissue for skin tightening. In these dishes, they would get them, with a high degree of satiation and enjoyment value, free of charge.

The tongues of animals are wonderful, precious, largely forgotten, and high-quality cuts of meat. The tongue of a fully grown cow can satisfy a family for several days. Mind you, during these days, no other cuts of meat would be necessary for meals, i.e., schnitzel or other supposedly premium cuts would not need to be purchased. Likewise, long-braised oxtails, shanks, or shoulder pieces taste significantly more intense than briefly fried or grilled loins. Our ancestors already knew this, as the resulting umami taste was a crucial driving force of evolution and the ever-evolving cooking culture. It is no wonder that dishes like *pieds et paquets,* long-braised lamb feet wrapped in lamb tripe, can still be found on the menus of bistros and Michelin-starred restaurants in Mediterranean countries [37]. The deep umami taste, the acidity of the tomatoes, the delicate, collagen-dominated consistency of the sauce, the tender tripe, and the meat of the feet are culinary experiences that can open one's eyes to a somewhat old new diet of the future. A step back from industrial to artisanal would be a good start.

In Austria, star and toque chefs around Vienna celebrate delicious culinary creations called *Wiener Bruckfleisch*; this consists of all parts of the animal that must be processed first during slaughter [38]. This includes everything that has high taste and nutritional value: offal, savory meringues and macarons made from blood, bone marrow, and even the main artery, the aorta of the cow, long-braised and thinly sliced, it visually and texturally resembles fried squid rings that adorn the plate.

Then delicacies such as sweetbread (thymus gland) and the long-forgotten testicles would also be a topic. Both are inevitable consequences of the Neolithic period and the resulting milk and cheese economy, whether they come from a bull, sheep, or goat buck. The parts of male animals are not popular on plates today, but they provide excellent meat. Eating the testicles is also part of the Nose-to-Tail concept. Entire cooking cultures have even formed around testicle cuisine [39].

Even the bravura *Andouilles* and *Andouillettes,* the sausages made from the stomachs and intestines of pigs and calves, which are part of the dining culture throughout France, originating from the Lyonnais region and the area around Troyes, are delicacies that are unparalleled. Closely related is the Palatinate stuffed pig's stomach. The pig's stomach essentially serves only as a container for the meat farce and vegetables, but when you taste it, another step into the edible parts of animals is revealed [40]. This also includes the brain. Alongside the liver, it is a primary source of truly essential fatty acids that humans cannot produce sufficiently on their own (see Sect. 5.10.4, Fig. 5.34). Eating the brains of hunted and killed animals was good for our ancestors. A highly recommended and instructive overview of all edible parts can be found on the website of a specialized butcher company [41], which is strongly recommended for intensive study before kitchen practice.

6.8 Game (Venison), Organic Meat from the Forest

6.8.1 Meat Doesn't Get More Natural Than This

For thousands of years, the forest has provided some of the best and most natural foods and is still the number 1 pantry for hunters and gatherers for mushrooms, berries, herbs, and of course game meat. Meat of wild animals is characterized by high nutritional values, convinces with its natural taste, and is guaranteed to be "organic". The quality is very high, as long as the game was shot from a stand and not chased through the woods in senseless driven hunts, exposing it to stress. The meat quality is noticeably worse due to stress hormones, the resulting higher enzyme activity, and faster pH value reduction.

Meat of shot game, unlike that of slaughtered animals, is not always completely bled out (as is also found in suffocated pigeons and wild poultry). Animals can only bleed out as long as the heart is beating. There is still more blood in the muscles, the flavor impression is slightly metallic to "liverish", which is part of the typical game taste. In addition, oxygen is still bound in the blood via the haem iron (see Sect. 4.3.4), which causes oxidation depending on the remaining blood in the muscles. Game meat, unlike beef, does not need to be aged for a long time. Early hunters and gatherers already took advantage of these benefits. The meat could be consumed almost immediately and largely preserved over the smoke of the fire. Storage options were hardly available anyway. Therefore, experienced hunters today make sure to place the shot in such a way that there is still some heart activity and that the veins are opened as quickly as possible to allow for extensive bleeding. With a targeted lung shot, the heart and brain remain active for a short time, and the animal bleeds out extensively.

An important difference, however, lies in the muscle structure. Wild animals are both escape and endurance animals, while domesticated animals hardly need to have escape reflexes. These different movement patterns require a fundamentally different structure of muscle fibers, as can be understood, for example, in domestic and wild pigs.

6.8.2 Red and White Muscle Fibers

This different release of movement energy from the muscles is regulated by the structure of muscle fibers and myofibrils. Red muscle fibers(also called Type-I fibers or *slow-twitch fibers* (*ST fibers*)) are designed for endurance and holding work, such as in grazing or fattening cattle, and respond slowly to movement. Typical time scales for contraction are about a slow 80 milliseconds, but they fatigue less quickly. Red fibers, the name is not coincidental, have more blood vessels (Fig. 6.14), are much more heavily perfused, as they must constantly be supplied with oxygen bound to the iron ion of the heme group via hemoglobin (four interlocked proteins, each with one heme group). Myoglobin, a single protein (monomer) and

6.8 Game (Venison), Organic Meat from the Forest

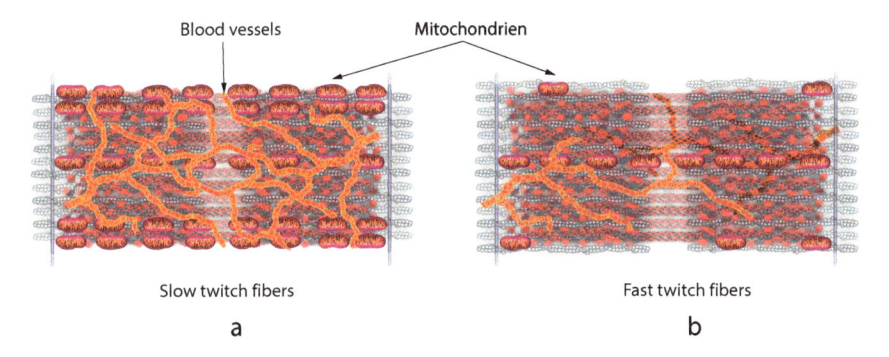

Fig. 6.14 Schematic representation of red and white muscle fibers. Red fibers (**a**) contain far more mitochondria than white (**b**), the fibers are thinner and much more heavily perfused by blood vessels

very similar to a single haemoglobin, is also more abundant in red muscle fibers. It also stores oxygen, which can be rapidly released in the muscles. Simplified, myoglobin absorbs oxygen from the hemoglobin delivered in the blood and stores it in the muscle. Red muscle fibers thus cover their energy needs from an aerobic metabolism. Therefore, there is much less glycogen in the red fibers, and the ability to form lactate (lactic acid formation) is low. The activity of the enzyme LDH (lactate dehydrogenase) is low in these red fibers. This enduring and long-lasting muscle activity through the red fibers requires a high energy turnover, which is why there is a high number of mitochondria in the ST fibers (Fig. 6.14), ensuring a sufficient production of the muscle fuel adenosine triphosphate(ATP) in constant concentrations.

White fibers are designed for rapid, fast, and strong contraction, and are therefore responsible for the *fast twitch*. On the other hand, they fatigue quickly and are therefore also referred to in the technical literature as *fast-fatigue* fibers (*FF-fibers*). Typical time scales of contraction are 30 ms. The rapid contraction and stretching ability are guaranteed by a particularly rapid breakdown of ATP via the corresponding enzyme ATPase at the myosin heads. Myosin and actin bind and detach extremely quickly. White fibers are less heavily perfused and have significantly fewer mitochondria (Fig. 6.14). The much higher ATP demand in white muscle fibers is mainly met by anaerobic Glycolysis obtained: Glucose is enzymatically obtained from the starch-like polymerized energy storage Glycogen, which is synthesized to ATP via corresponding enzymes, such as the already mentioned LDH. White fibers therefore have high glycogen reserves, as well as creatine and creatine phosphate, which together with the enzyme creatine kinase can rapidly regenerate ATP in white muscles from the degradation product ADP. However, white fibers are only equipped with small amounts of the oxygen storage enzyme myoglobin. Mitochondria are significantly less present in white muscle fibers due to the completely different metabolism. The different muscle types and their characteristics are summarized for clarity in Table 6.2.

Table 6.2 Overview of the main characteristics of red and white muscle fibers. The taste-relevant properties are highlighted in red. The texture is determined by the properties highlighted in blue

	Red fibers *(slow twitch)*	White fibers *(fast twitch)*
Function	Postural work, endurance	Rapid movement
Time scale of muscle motor	80 ms	30 ms
Metabolism	Oxidative	Glycolytic
Mitochondria	Many	Few
Myoglobin content	High	Low
Glycogen content	Low	High
ATP	Low	Very high
Creatine content	Low	High
Diameter	Small	Larger
Collagen content	High	Low

Thus, it becomes clear that (grazing) cattle, with their weight and associated endurance, have a high proportion of red muscle fibers. The same applies to wild animals, whose lives require high endurance. In barn animals and domestic pigs, which require only minimal exercise, more intermediate fibers are formed, which roughly lie between the two extremes. However, wild animals also need a proportion of fast white fibers for escape reflexes in case of danger. Therefore, wild animals are well equipped with both red and white fibers [42, 43]. The exact distribution of fibers depends heavily on the animal species, for example, the proportion of white fibers in springboks is higher than in wild boars [44].

In these details of the differences between muscle fibers, however, many enjoyment parameters are also hidden, as can be seen from Table 6.2 and the findings from Chap. 4. The taste-relevant parameters are colored red in the table. More myoglobin means that metallic "blood-flavor" through the heme iron, which, for example, must be added to vegan burgers and other "fake" products via leghaemoglobin; the glycogen content describes the potential for lactic acid formation and decides between sour and sweet, depending on the time between the shot and consumption. Meat pieces with *fast-twitch* fibers often taste slightly sweeter. Most important, however, is the ATP content, as this defines the umami potential of the meat. This is significantly higher in white *(fast-twitch)* fibers than in red fibers, as the flavor enhancer IMP (inosine monophosphate) is formed from ATP. If the purine metabolism continues, IMP is converted to adenosine, which is then known for its enhancement of sweet taste [45]. Creatine, on the other hand, is known for its bitter potential in hot broths, soups, and sauces [46], while it tastes slightly sweet at cooler temperatures [47].

However, texture and possible preparation techniques can also be deduced from Table 6.2: The different fiber thicknesses between white and red fibers provide information about the texture, especially bite force and tenderness. The thinner red fibers require less force to bite through [48] than the thicker *slow-twitch* fibers. On the other hand, the many thin fibers have a higher overall surface area, so the collagen content is higher. This is clearly noticeable in endurance animals such as wild boar, deer, and roe deer; the meat is tender and requires very short cooking times. Even the pieces suitable for braising cook faster than similar pieces from cattle or sheep. The high collagen content due to the high proportion of red muscle fibers in wild hares allows for long braising, which creates small taste and texture wonders, such as *lièvre à la royale* (roughly "royal wild hare ragout"), a true traditional of French game based cuisine. The deep umami taste becomes perfect: The high enzyme activity causes pronounced protein hydrolysis and thus a high amount of free glutamic acid, the high concentration of ATP from the white fibers significantly boosts the umami taste.

Of course, everything is eaten in game as well: The giblets (heart, liver, and kidneys) traditionally go to the hunter who killed the animal. If these rarities are ever offered fresh, it is essential to seize the opportunity. Apart from the excellent and delicate taste and the enchanting aromas, the offal of game also offers everything the *Homo sapiens* needs: essential long-chain, polyunsaturated fatty acids, precursor stages of vitamin D_3, rapidly biologically available amino acids, bioactive peptides, high amounts of creatine for muscle metabolism, and in high quantities, as only the offal of animals can provide naturally.

There are many reasons to include game in the diet of the *Homo sapiens* of the future, especially since the maintenance of forests requires a certain amount of hunting.

6.9 The Salt Issue

6.9.1 Salt in the Kitchen

"Ilsebill put on more salt salt." With this remarkable sentence, the novel *The Flounder* by Günter Grass [49] starts, which was first published in German in 1977. A year earlier, Paul Bocuse's *La cuisine du marché* was published in France, which was released in German translation around the same time as *The Flounder*. In 1972, L. K. Dahl published a summary article on blood pressure and salt consumption [50] in the journal *The American Journal of Clinical Nutrition*, which the author had begun in the 1960s [51, 52]. The discussion about salt began, and salt gradually became a poison responsible for cardiovascular diseases over the last decades. Did Ilsebill make a big mistake? After all, the German Society for Nutrition (DGE) recommends 6 g of salt per day, and the World Health Organization (WHO) recommends 5 g per day. Adding more salt would be difficult.

Since salt enhances the taste of many foods, many pleasure products end up on a blacklist of dangerous food products. These include everyday products such as chips, French fries, burgers, and most convenience products, as well as natural products like bread, cheese, sausages, sauerkraut, or seasoning sauces, regardless of whether they are "organic" or not. In all these products, salt performs essential functional tasks that have been forgotten due to a lack of knowledge about processes and manufacturing methods. No cheese ripening, no safe fermentation, and no water binding can take place without salt and the resulting interactions with proteins. Therefore, in addition to "hidden fats," there is also talk of "hidden salt" in foods. Salt is much more than a flavoring ingredient; it functions at the molecular level, as this brief list already shows.

How much salt is actually consumed through food has been published through the evaluation of worldwide data [53]. A reliable measure for this can be the concentration of sodium ions excreted in urine, which was also used in the sensational epidemiological study [54]. It quickly becomes apparent how high salt consumption really is. In all countries, in all cultures, it is significantly higher than the recommended 5-6 g/day, as indicated in Fig. 6.15. This clearly illustrates how reality

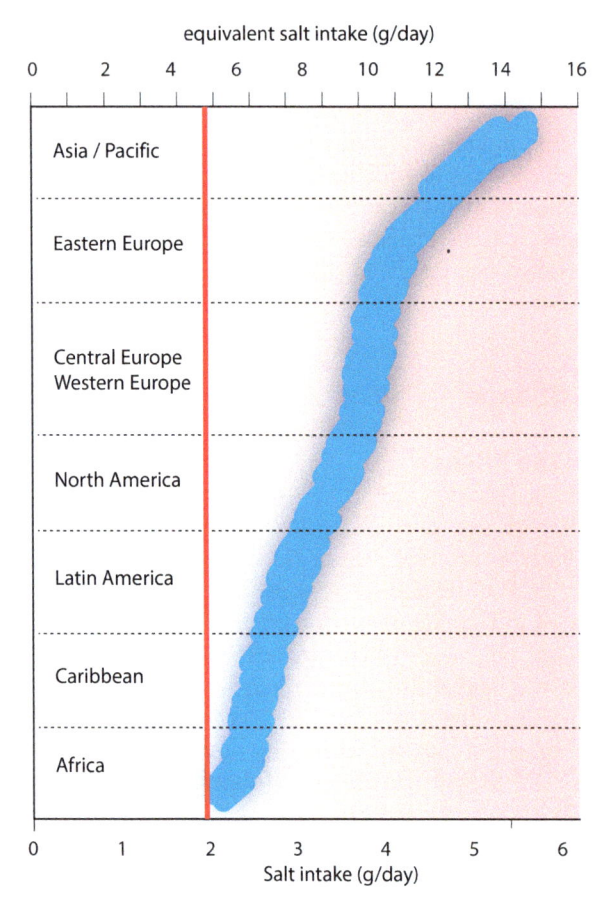

Fig. 6.15 Salt consumption in different regions and cultures is everywhere higher than the WHO proposed limit of a maximum of 5 g table salt per day. The equivalent table salt consumption is higher when all sodium-containing foods (on average) are converted to pure table salt (upper scale)

6.9 The Salt Issue

deviates from the guidelines. Not even in countries and regions where traditionally little table salt is used are the average figures lower than the World Health Organization (WHO) guideline. This is only to a small extent a consequence of increasing industrialization, as elaborated in the very carefully conducted studies [53, 54], but largely due to cultural history. In many parts of Asia, fermented products that are highly salted have been consumed for more than 1000 years. The Japanese fermented Umeboshi pears, which are currently finding increased culinary use in Germany, contain over 10% salt and far exceed traditional sauerkraut with its fermentation-related 1.5-2%. Eastern Europe bordering Asia also uses more salt than its western neighbors. Central and Western Europe (including Germany) are in the lower middle range in terms of per capita salt consumption compared to many Asian regions. Nevertheless, mortality is not necessarily correlated with this salt content, as life expectancy in many Asian countries is comparable and higher than in Western Europe. Reason enough to take a closer look at the effect of salt on our health.

On the other hand, many studies have been supporting for years that a permanently high, excessive salt consumption raises blood pressure and thus increases the risk of stroke and cardiovascular diseases [55]. Complete abstinence or too strong a salt reduction also appears to be not particularly beneficial. This is expressed through a J-shaped curve of salt consumption and blood pressure [56], as schematically shown in Fig. 6.16. The question of which range is really appropriate for salt consumption cannot be precisely defined. Least of all, what "excessive" means for individuals.

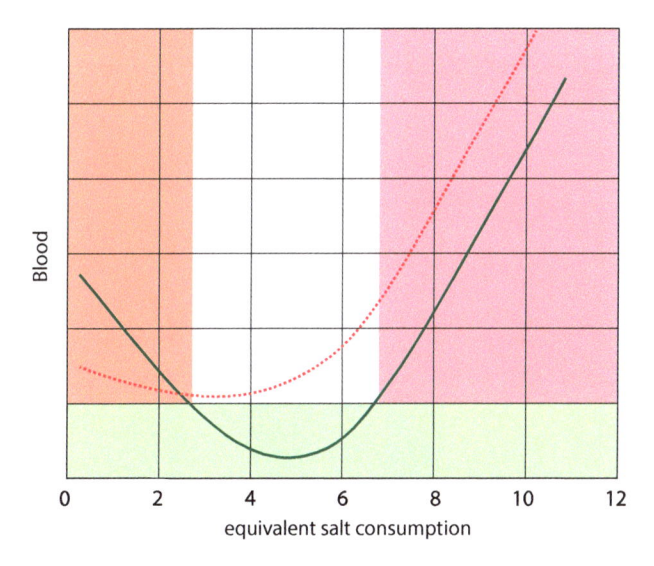

Fig. 6.16 The J-curve of salt consumption for normal healthy people (green curve) and for people with high blood pressure (dashed red curve)

It turns out that the average blood pressure in humans is in the green range between 4 g and about 6-7 g sodium chloride per day. Hypertensive patients are unlikely to achieve these ideal blood pressure values solely through variations in salt consumption. Even an extremely low salt intake of less than 5 g per day is not conducive to average life expectancy, as an increase occurs again, although it is somewhat weaker in hypertensive patients.

6.9.2 Salt and Osmosis

Now it is generally not easy to determine how much sodium is actually absorbed. Sodium is an integral part of intracellular metabolism in all cells, both animal and plant. The monovalent and simultaneously small sodium ion therefore physically serves as a charge transporter and charge switch on all scales below cell size. Sodium is thus present in every food in the physiologically required concentrations. Even simple school experiments on osmotic pressure make the blood pressure idea plausible: When table salt, NaCl, dissociates in water into its ions Na^+ and Cl^-, they generate an osmotic pressure, as shown in Fig. 6.17.

When sodium chloride crystals (top left in Fig. 6.17) are added to one side of the tubes separated by a semipermeable membrane, the crystals dissolve into positive sodium and negative chloride ions. The ion concentration increases in the left tube. According to the laws of thermodynamics, a concentration balance is sought, and the osmotic pressure increases. In this process, water molecules can penetrate the membrane and migrate from one side to the other. The ions cannot do this because they carry a bound hydration shell with them, making these structures much too large for passage through the membrane. Water therefore passes through the membrane to reduce the salt concentration on the other side by dilution. In doing so, an osmotic pressure builds up, which can be measured directly as a difference in the height of the water columns.

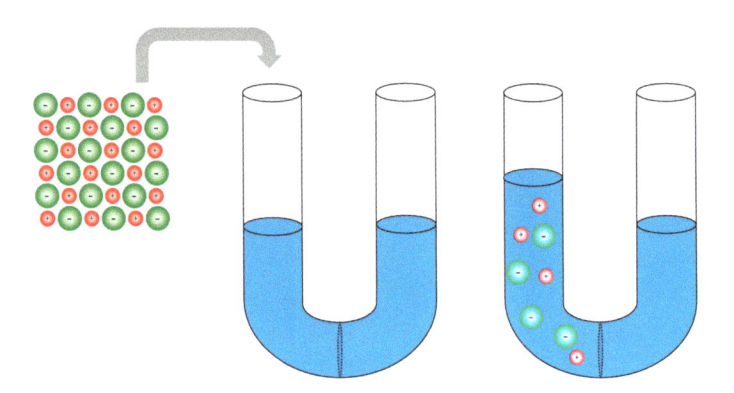

Fig. 6.17 Osmotic pressure in two connected tubes separated by a semi-permeable membrane, only water can pass through this membrane

6.9.3 Salt and Humans

Of course, salt does not behave that simple in blood vessels. It is by no means the case that a high salt consumption increases the concentration of sodium and chloride ions in the blood. This cannot be the case for biophysical reasons, as the salt ions would bind water from the blood serum and could even interact with the electrically charged amino acids of the proteins in the blood. As a result, the dissolved proteins would change their shape, as would the flow properties of the blood. Therefore, the concentration of ions in the blood is practically constant, with only tiny fluctuations, and adapted to the required physiology. Excess, unneeded sodium and chloride ions must therefore be excreted as effectively as possible through the kidneys. As already noted, the concentrations of these ions in the urine are a reliable measure and thus a clearly defined medical criterion for salt consumption.

Therefore, it is not surprising that only a small increase of 2.8 mm of mercury scale pressure is found for 1 g more salt per day, and this only in people who already consumed more than 5 g of salt per day. Only a very high salt consumption led to a greater risk of stroke, which was mainly the case in China, where the average consumption is 14 g/day and more. However, when life expectancy is correlated with salt consumption, this statement is significantly relativized [57], as shown in Fig. 6.18.

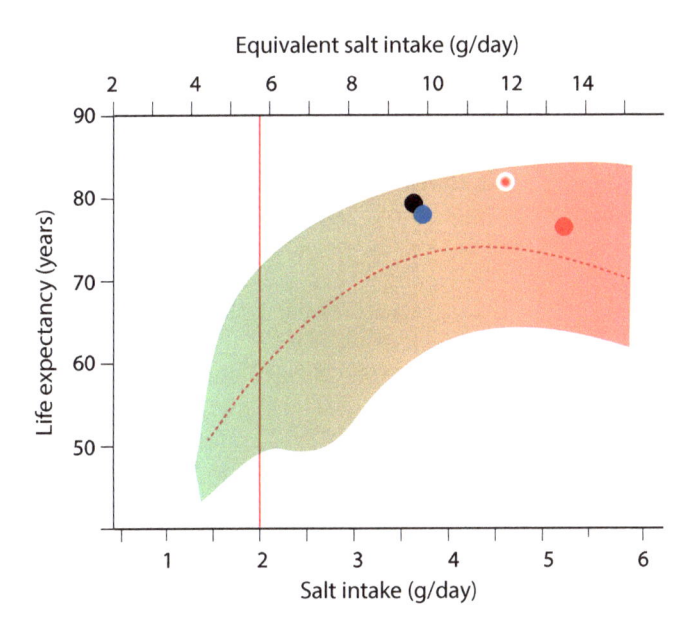

Fig. 6.18 Country- and culture-specific salt intake and life expectancy (based on Messerli et al. [57], life expectancy updated to 2018). The hatched area contains all regions examined in Fig. 6.15 and the associated study, dashed: statistical average. Specifically mentioned are: Germany (black), USA (blue), Japan (red-white), China (red). The WHO limit: vertical red line

As it turns out, higher salt consumption on average up to an equivalent of over 10 g/day is not correlated with a lower life expectancy. In Japan, despite a mean equivalent salt intake of over 11 g/day, life expectancy is the highest among the examples shown, while life expectancy in China is lower. The USA and Germany have a life expectancy with a moderate salt consumption from this perspective above that in China, but significantly below the average life expectancy in Japan. Whether these results can ultimately be attributed exclusively to salt consumption is completely unclear.

However, what turns out to be an obvious mistake is to consider only sodium ions and to disregard potassium ions, as the cited works also examined the correlation of the two monovalent positive sodium and potassium ions with cardiovascular diseases. It was found that, regardless of salt intake, potassium reduces all risks for heart attack, stroke, and overall mortality. Even in patients with high salt intake, potassium can reduce the risk. Potassium is mainly found in fruits, vegetables, and nuts. Focusing solely on a single parameter is basically insufficient in nutritional issues and does not allow for conclusive statements. Consequences cannot be derived from this. Strict regulations for salt reduction would therefore not be necessary. However, mixtures in which part of the sodium chloride is replaced by potassium chloride are a step in the right direction for better potassium supply.

6.9.4 Salt and Interactions at the Atomic and Molecular Level

These are indeed good news, because as already mentioned several times, salt is not only a flavoring ingredient, but also performs physical tasks in all food preparations where proteins are important: bread, sausage, cheese, tofu, substitute products. Proteins consist of amino acids, which can be electrically positive and negative. It is precisely these electrical charges that interact with sodium and chloride ions: opposite charges attract, like charges repel; of course, also when these charges are located in proteins on the respective amino acids. Thus, positively charged sodium ions gather around the negatively charged amino acids, while the negatively charged chloride ions rather seek the positively charged amino acids. In the case of wheat protein in bread, this is even of technical importance when the charges of the amino acids determine the shape of glutenin and thus the processability of the doughs. Salt plays a crucial role in this, as can be seen from Fig. 6.19.

As just mentioned: The positive sodium ions preferentially accumulate around the negatively charged amino acids (glutamic acid, aspartic acid) while the negative chloride ions preferentially accumulate around the positively charged amino acids (arginine, histidine, and lysine). The charges of the proteins are therefore neutralized. The stretching properties of doughs improve with salt, as do the flow properties of sausage meat or process properties of textured protein preparations for vegan products.

Sodium chloride sodium chloride also has a whole range of effects on the water content of food. Water itself is polar, it has a dipole and is slightly negatively

6.9 The Salt Issue

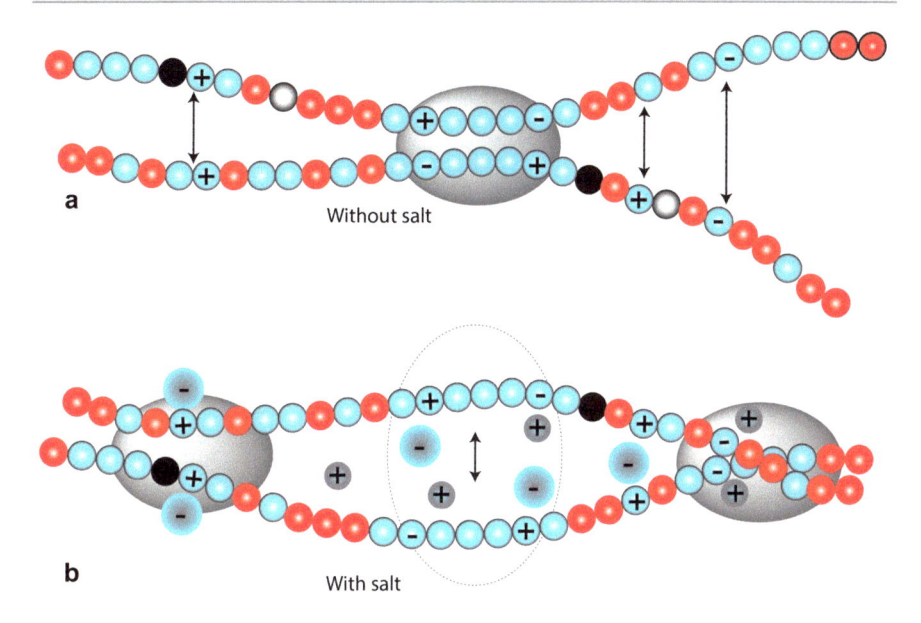

Fig. 6.19 Salt has a decisive influence on the structure of proteins. Without salt, negative and positive charges attract, they form complexes or repel each other strongly (**a**). With salt, charges on proteins (e.g., gluten) are shielded (**b**). The structure of the protein network changes

charged at the oxygen atom, slightly positive at the two hydrogens. Therefore, ions can bind water on very small time scales according to their charge and orient it according to polarity, as indicated in Fig. 6.20.

These hydration shells, marked with blue shading in Fig. 6.20, are relatively strong for sodium and chloride and can only be dissolved with slightly higher energy. The water is therefore "bound". Table salt in food binds water, prolongs shelf life, and prevents too rapid drying or too rapid staling of bread and other baked goods. Salt thus serves far more than just taste. A demand for consistent salt reduction, without thinking more deeply about the physical consequences, would be counterproductive for dough, baking, and bread properties.

The physics of the hydration shell shown in Fig. 6.20 is ultimately also the reason why potassium, compared to sodium, has a weak but measurably different effect on blood pressure as well as other physiological properties despite having the same charge (+1). The hydration shell is weaker despite the larger ion radius [58], as shown in Fig. 6.21.

Thus, the different effects of sodium and potassium on simple physical properties can also be traced back to this. Many views on nutrition and health can easily be stripped of their esoteric foundation, as the next section shows.

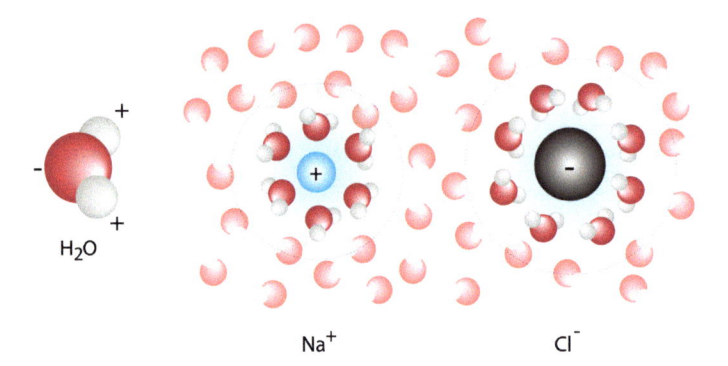

Fig. 6.20 The polarity of water (left) and the different charges of sodium and chloride ions cause the formation of strong hydration shells around the ions (right). The charge in the center orients the water molecules according to polarity on average. The water molecules in the immediate vicinity are highlighted

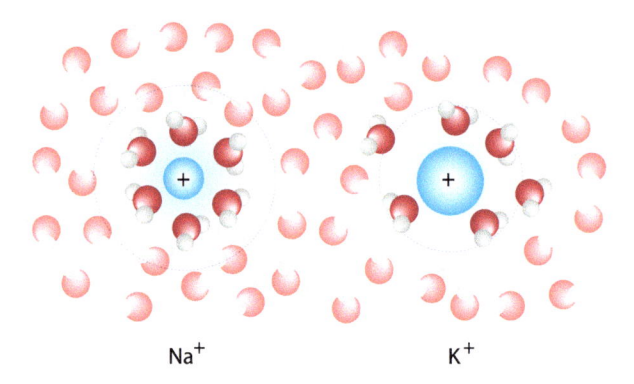

Fig. 6.21 Sodium ions (Na$^+$) and potassium ions (K$^+$) differ significantly in their diameter and ability to bind water in hydration shells despite having the same charge

6.9.5 Salt is Not Equal to Salt

Many myths surround salt, culminating in the belief that refined sodium chloride (NaCl) is "dead salt," while sea salts, *fleur de sel*, or unrefined rock salts are healthier [59]. The occasionally praised Himalayan salt [60] is even attributed with healing properties, storing light quanta from primordial times and all sorts of other pseudoscientific nonsense.

All these salts consist of more than 97% sodium chloride. The rest are embedded minerals, which appear as vacancies and other crystal defects [61]. Nutritionally, these do not carry significant weight when considering the comparatively high mineral content of all foods. With the usual small amounts of salt used for seasoning, their mineral content is far below that of the foods being salted.

Claims that salts can store light energy, photons, or even cosmic life force [62] cannot be true for physical reasons, as a brief look into a textbook on solid-state physics [63] shows.

6.9.6 Salt is Not a Poison

In this type of discussion, one tends to forget the elementary facts: The kidneys of healthy people are capable of excreting up to 20 g of sodium and chloride ions per day [64]. Both sodium and chloride ions are necessary for cell metabolism and cell function, and have been since the existence of the first biological cells on this Earth. Therefore, there is a very specific filter and regulation system for all cells that keeps the water and ion balance (across all minerals) constant over a wide range, independent of the ingested and offered salt and fluid quantity. The kidneys are able to compensate for a water and/or electrolyte deficiency relatively quickly, as well as to correct a water or salt surplus quickly, in both directions, with too few and too many ions. In the lower direction, this only works to a limited extent, as storage is quickly depleted, hence the increase in risks with too low sodium intake.

Contrary to many views, NaCl is chemically neither toxic nor aggressive. The ions circulate in all physiological body fluids with almost constant concentration. Through thermodynamic equilibria, Na^+ and Cl^- – among many other tasks – also serve to maintain cell pressure; through osmosis, they ensure the correct volume ratios of cells and a balance between intracellular and extracellular water. According to Fig. 6.16, the daily intake of sodium chloride should not fall below the limit of 2.5 g/day in order to maintain vital body functions. The physiological limits are high in the upward direction. Only from a salt intake of 0.5 g/kg body weight per day does it become really critical. At a body weight of 70 kg, this would correspond to an extremely unpalatable 35 g/day. According to Fig. 6.15 and 6.19, salt intake in Germany is about 6-8 g of table salt per day, still below the mentioned dangers.

6.10 Lot Makes You Full, Complex Makes You Satisfied!

6.10.1 More is Not Always Better

Sometimes the impression arises that having more food on plates makes one feel fuller. Some plates are loaded to the brim at buffets, and during promotions like *All you can eat*, people stand in long queues reminiscent of food distribution points after the war. However, people were not obese at that time. Hunger cannot be the driving force in today's times; people are generally full, even if the last meal was "already" four hours ago. The feeling of satiation after food intake seems disturbed, signals no longer work (see Sect. 5.14, Fig. 5.45). Ultimately, the behavior is hardly reprehensible; it is a primal instinct of *Homo sapiens* to consume available food.

Eating behavior and food preferences are interesting. In many cases, the familiar and beloved is chosen, dishes that one already likes to eat, and plenty of them, which contradicts the much-praised diversity. As a result, most bites selected are very similar. Like with a plate of soup, when each spoonful tastes the same; like with a burger, when each bite differs only marginally from the previous one. The layers (of the fast-food burger) are designed precisely this way. However, a strong variation in sensory properties is lost. Aromatics and taste hardly change in many common preparations from spoonful to spoonful. Eating is done mechanically, without reflecting on what exactly is being eaten. The repetitions tire the reward center. The release of dopamine, glutamate, and other biomarkers [65] occurs more hesitantly (see Chap. 5, Fig. 5.47). The meal was not an experience. Soon after, one allegedly feels hungry again. One approach to breaking this eating loop would be more components, more courses, but significantly smaller portions. More variety in components creates more reward, more attention to the food, and, incidentally, a significantly better supply of macro- and micronutrients.

6.10.2 Variety and Combinatorics – Complexity on the Plates

The possibility of achieving higher satiation with less energy input would be a completely different type of plate design than is the case with classic presentations [66]. It offers the connoisseur a multitude of possibilities that can be mathematically formulated. For example, if there are only two elements on the plate, there is only the possibility of tasting each of these components individually or both together. The eater thus has exactly three options.

If there are three distinguishable components – a, b, c – on the plate, the theoretical number is already at seven possibilities: Each element by itself results in three two-component combinations – ab, ac, and bc – as well as all three components abc together, making a total of $n = 7$ possibilities for combining plate components on a spoon or fork.

For a number of n elements, a subset of $1 \leq k \leq n$ elements can consequently always be tasted together. The respective number is determined by the binomial coefficient $\binom{n}{k}$ given. The mathematical definition is provided by

$$\binom{n}{k} = \frac{n!}{(n-k)!k!} = \frac{1 \cdot 2 \cdot 3 \cdot \ldots \cdot n}{(1 \cdot 2 \cdot 3 \cdot \ldots \cdot (n-k))(1 \cdot 2 \cdot 3 \cdot \ldots \cdot k)}$$

given. These binomial coefficients are, for example, also known from the lottery game "6 out of 49", where there are $\binom{49}{6}$ possibilities of distinguishable number combinations. The exclamation mark behind a natural number $n!$ is the factorial of this number, i.e., $n! = 1 \cdot 2 \cdot 3 \cdot \ldots \cdot n$, the product of all natural numbers from 1 to n.

These various subsets arranged on spoons reflect the diversity of the plate with their different combinations. The spoons are sensorily distinguishable and offer the brain many opportunities to respond to the reward. Consequently, the eater has an increasing number of possibilities to multiply their enjoyment with the number of plate elements. This complexity can therefore be represented by the sum of all different possibilities of binomial coefficients

$$N = \binom{n}{1} + \binom{n}{2} + \binom{n}{3} + \ldots \binom{n}{n} = \sum_{k=1}^{1} \binom{n}{k} = 2^n - 1$$

With modern cooking techniques in haute cuisine, it is therefore no problem to arrange $n = 10$–15 or more elements on plates. This means between $N = 1023$ to $N = 32{,}767$ theoretical possibilities to taste the plates, for some a true overload.

Such large numbers suggest the definition of a new logarithmic quantity, the "entropic complexity" of the plate K

$$K = \ln N \approx n \ln 2$$

as is customary in physics (and corresponds to information entropy). The natural logarithm (ln) was chosen as the base of Euler's number (e = 2.718.281.828.459.045.235…), as is common in statistical thermodynamics and information theory.

However, even with just a few components that can be specifically selected, there is more variety, as the simple example in Fig. 6.22 shows.

The entire dish [67] is shown in the center of the illustration; it consists of broccoli florets, mushrooms, sauce, and a crispy tomato biscuit. The portioning and texture of the elements allow for the tasting of different combinations per spoon, which are grouped around the plate. Each of these spoons presents a different taste and aroma profile, and the textures also vary. Within a single plate, there is a high variation of flavors. This is precisely what is beneficial for our receptors, as they "measure" not constant stimuli, but their temporal changes. There is more to process in the brain, the reward center receives various stimuli, and a satisfying satisfaction sets in. A habituation to a consistently identical taste with each spoon, as with a plate of soup, is therefore less likely.

An important aspect is the distinguishability of the components. They must therefore be distinguishable in this definition in that they have a different taste, a different texture, or a different aroma, but can still be picked up with the eating utensil, a fork or a spoon. Powdery components, for example, consist of a large number of individual granular particles, but they all have a uniform taste, texture, and aroma. The portioning of the individual components and the respective total number thus define complexity and repeatability of the spoons. Small-scale arrangements are therefore an additional complexity factor that was examined more closely in a subsequent study. Intuitively, however, it is already clear that two or three embedded caviar pearls in a plate of high complexity have a much deeper

Fig. 6.22 The complete culinary combinatorics of a simple plate with four components: broccoli, mushrooms, sauce, and a crispy tomato biscuit. Each of the 15 possible spoons has its own culinary, aromatic, and textural play. The complexity would be $K = \ln 15 \approx 2.7$

thought-out sensory function than a generous portion of caviar on a classic plate with noble fish of low complexity.

6.10.3 Excitement and Variety

In a modern, pleasure-oriented kitchen, as is common in gastronomy, it is desirable to create plates with greater complexity. The available cooking techniques from classic cuisine, molecular cuisine, or avant-garde cuisine offer all possibilities for this. Even components of the same genus, such as a vegetable, which are served raw, cooked, or fermented according to the cornerstones of the original culinary triangle, increase the entropic complexity with a total number of n components by the factor $(n + 2)$. If additional techniques of avant-garde cuisine, such as creams, gels of different textures, or foams, are used, the complexity can be increased without using another ingredient. The different projections through costs from different parts of the dish always result in other aspects. Connoisseurs thus have the choice between their own compositions, which they bring to their mouth with each fork or spoon. However, this example is still very simply structured. The three possible combinations of this arrangement are repeatable. If one of the combinations is found to be particularly exciting, it can easily be put together and eaten again. Therefore, this dish can ultimately still be reduced to the classic forms of homogeneous structures.

A very complex dish by the German pastry chef Christian Hümbs, which makes use of practically all elements of the culinary triangle, is listed in Fig. 6.23 [68]. None of the exemplarily shown spoons can be repeated due to the arrangement. Each spoon is thus final in enjoyment, in terms of texture, temperature, and aggregate state, as well as the resulting different aroma, taste, and texture images. Therefore, the question for the eaters is which path is the "right" one. However,

6.10 Lot Makes You Full, Complex Makes You Satisfied!

Fig. 6.23 A few exemplary examples of non-repeatable spoons of a highly complex dish (shown in the center) made from strawberries (in gel form), asparagus, and parsley cream, served on a light cream. Other elements include rosemary and nut powder

there is no answer to this, as personal preferences and cultural influences play a significant role. Enjoyable eating can thus become very exciting. The reward center is occupied, and the number of components in a multi-course menu ensures the complete nutrient requirement with high enjoyment value.

That this is also physiologically advantageous was examined in a first study [69]. For this purpose, voluntary subjects were presented with two well-known dishes, once in the classic arrangement and once in an avant-garde and thus more complex presentation, as illustrated in Fig. 6.24: once a vegetarian dish and once a classic meat dish with chicken.

For the simple version of the vegetarian dish, all the vegetables were cooked with the broth in a pressure cooker. The cooked vegetables were then pureed in a blender together with the broth and refined with cream, butter, and seasoned with pepper and salt. The plate was garnished with parsley.

The complex version of the vegetable meal consisted of four elements. For the mashed potatoes, the potatoes were diced and cooked in water, then mixed with parts of the broth and a small part of the cream to make a puree and seasoned with pepper and salt. The onions were cut into rings and slowly fried with the butter in a small pan. The peas were simply steamed and seasoned. Carrots and celery were cut into julienne and blanched in the remaining vegetable broth. The remaining cream was mixed with a little salt and pepper and stirred with the parsley to make a "dip." A serving ring was used to lay a layer of mashed potatoes. The onions fried in butter were arranged on top. The julienne vegetables were arranged around it. The parsley cream was also arranged on the plate. A small jug filled with vegetable broth added a liquid component. All plates of the respective dish had the identical energy input as well as the same amount of all seasoning ingredients.

Fig. 6.24 Vegetarian (pea soup): "simple" (**a**), "complex" (**b**). Consisting of 100 g potatoes, 100 g carrots, 100 g green peas, 100 g celery root, 50 g onions, 50 g crème fraîche, 20 g butter, 200 ml vegetable broth, 1 tablespoon chopped parsley, salt, pepper. Chicken: simple (**c**) and complex (**d**). The basic ingredients are two equally sized chicken legs, oil for frying, 100 g tomatoes (diced), 50 g onions (diced), 10 g garlic (diced), 40 g couscous, 5 g mint leaves, 5 g coriander leaves, 20 g lemon fruits (diced), 10 g lemon zest, 80 mL olive oil, spices

After eating the two dishes, psychological tests were conducted with the subjects, which were supposed to evoke different reactions. Although the results suffered from statistical reproducibility due to the small number of subjects, statistically weak correlations were found in measurable parameters such as eating time, serum glucose levels, and salivary cortisol levels. Cortisol serves as a marker for insulin production. The most striking results are summarized in Fig. 6.25.

In the process, it was found that eating complexly prepared dishes takes significantly longer, which naturally increases enjoyment, as complexly arranged plates force slow and conscious eating. A rapid "gobbling" as with simple plates leads to faster eating. Since the first satiety signals only set in gradually, satiety can already be noted during eating with complex plates, while with simple plates of the same energy, satiety only becomes noticeable after eating. The somewhat slower increase in blood sugar levels with the complex presentation method can also be seen. The longer eating time can also contribute to this. The cortisol level does not allow for a clear assessment in this study, so studies with a higher number

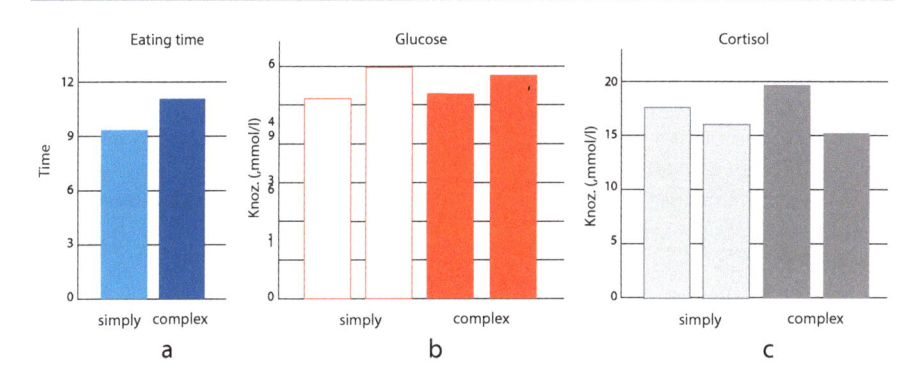

Fig. 6.25 The differences between a simple and complex meal (**a**) can be seen in the eating time, glucose levels (**b**), and cortisol levels (**c**)

of subjects would be necessary to confirm the other results as well. Nevertheless, trends can be clearly seen.

6.11 Hunger—A Western Luxury

6.11.1 The Forgotten Hunger

The abundance and the industrialization of food in Western nutrition, coupled with the power of advertising and marketing, the desire for body optimization, the compulsion for the one true healthy diet, no longer allow a fundamental feeling of our ancestors: hunger. Even after a lavish, conscientiously optimized breakfast (in the light of evolution, the most nonsensical meal for adults anyway), snacks are taken, as well as a short time after lunch, sometimes because the stomach growls, or out of exaggerated fear of normal healthy people of hypoglycemia [70]. The recommendation "better five small meals than three large ones" that was expressed until recently promoted this not very general view. However, there have been counterexamples for a long time. In many Mediterranean countries, breakfast is not a big deal, especially in Spain, where, in addition, dinner is hardly taken before 10 pm. People here practically eat only twice a day, but in the evening or/and at noon at least three courses, appetizer, main course, dessert, with rather small, by no means too lavish portions. In between, there is nothing, an extensive breakfast is dispensed with. Fruit is integrated into the dessert, vegetables (raw, cooked, fermented) into the appetizer and main course. Mountains of carbohydrate-rich side dishes are not necessary, a little carbohydrates come in France via the baguette, in Italy via the *primi piatti,* the pasta courses. If hunger pinches between 10 and 11 am, there is no power bar, but an espresso. At the latest then, one is happy about the empty stomach and the prospect of a three-course lunch menu away from sandwiches, smoothies & Co.

Between the main meals, "fasting" takes place, there are no snacks, no bars, no nibbles. The digestive system is still busy. What was normal there some time ago got the name intermittent fasting.

The eating breaks and thus the daily intermittent fasting, as was inevitably the case with early humans, are much better understood today. People back then had to cope with constant food shortages, which did not only affect macroscopically visible events but reached down to the most elementary units of life. The physiological term for this is autophagy, the sophisticated molecular recycling system on nano- and micro-scales in cells.

6.11.2 Autophagy

The autophagy is a mechanism that analyzes, recycles, and returns molecular materials in the cell for reuse, and, if that is not possible, safely disposes of them as "waste"; the mechanism is very roughly illustrated in Fig. 6.26. This mechanism has been and is crucial for the life of plants, animals, and humans, and thus essential for all cells. No (essential) amino acid, no chemical molecule or molecule group that is still important for the human organism or can be used for proteins or messenger substances is wasted. Autophagy is triggered, among other things, by spermidine [71], a biogenic polyamine that acts as a signaling substance but is also physiologically involved in various processes. The molecule got its name because of its high concentration in sperm, where it was discovered.

Waste within the cell, represented as protein in Fig. 6.26, is slowly enveloped by forming phagophores. These phagophores consist of a lipid bilayer, as is known from cell membranes or, for example, exosomes (see Sect. 4.10.9, Fig. 4.56). They gradually close, enveloping the molecules and molecular residues to be processed, and form closed autophagosomes. Enzyme containers in the form of lysosomes, which are also surrounded by a lipid bilayer, are also present in the cell. Thus, autophagosomes and lysosomes can connect via physical mechanisms in the lipid bilayers, and the enzymes penetrate the autophagosomes. There, they begin the systematic processing of the molecules present and convert them chemically. Proteases, for example, cleave proteins into amino acids, from which new proteins can be synthesized. The comprehensive process of autophagy is therefore an extensive (bio-) chemical program and enables the cell to survive even in lean times through recycling and waste disposal. By the way, these processes are also set in motion during the constant exchange of consumed (oxidized) cell material, such as phospholipids or membrane proteins, as well as in the elimination of odorants in olfactory cells.

Indeed, the benefits of intermittent fasting and longer breaks between meals can be demonstrated and supported by hard measurement data [72, 73]. For the generation of these data, mice are systematically fed twice a day, at 11 a.m. and 7 p.m., with an isocaloric diet in animal experiments, and various parameters in the organs are determined using analytical methods. The most important effects are graphically summarized in Fig. 6.27.

6.11 Hunger—A Western Luxury

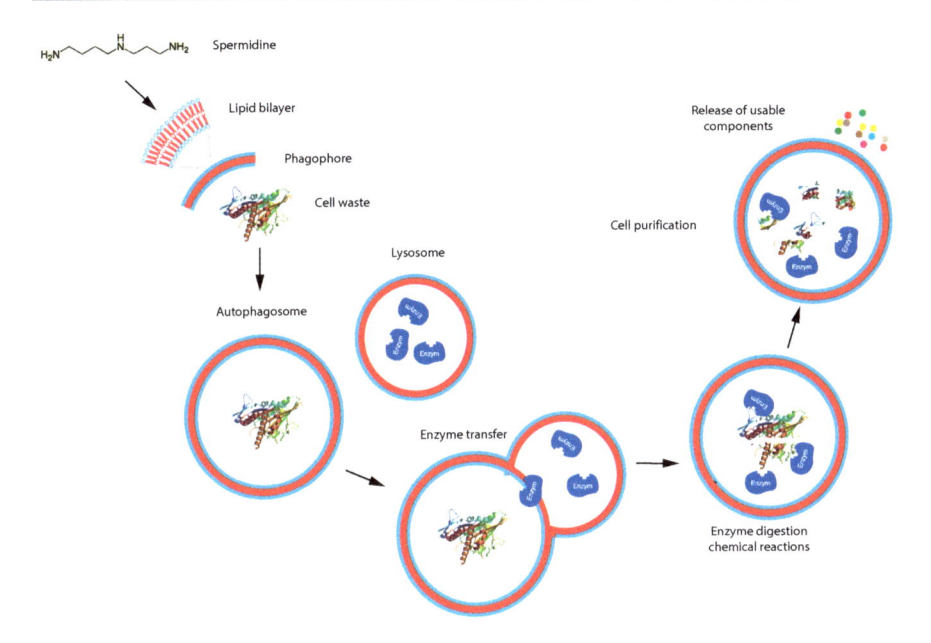

Fig. 6.26 Simple representation of the important steps of autophagy, triggered in part by spermidine: from "encapsulating" the residual material to its enzymatic conversion into physiologically usable material, which is then released to the cell for further processing

Control is primarily exerted through specific proteins and neurons (proopiomelanocortin) in the brain. For questions about nutrition, the processes in the liver are initially of importance. It is shown that more mitochondria, the powerhouses of all cells, are formed. At the same time, the breakdown of (excess) fatty acids increases through an increase in β-oxidation, fat synthesis decreases, and fat in the liver is reduced. In the fat cells, more brown fat cells are formed, which are responsible for regulating body temperature. At the same time, sensitivity to leptin, which regulates satiety and hunger, increases. As a result, the tendency towards obesity decreases. In the muscle, muscle mass increases, with a significant increase in *fast-twitch* fibers (white muscle fibers, see Chap. 5), whose glycolytic metabolism stores more glucose in glycogen. This glucose is therefore no longer free, and blood sugar levels decrease. The data also show positive effects in diabetic patients.

A very special role is played by fat digestion, lipophagy [74]. These fat-specific digestive processes (in the liver) follow a similar course as autophagy, but target fats (triacylglycerols or triglycerides) more specifically. These are also trapped in autophagosomes. Lipase-rich lysosomes, i.e., organelles richly equipped with fat-digesting lipases, connect in a similar way and specifically digest fat. Excess fatty acids are broken down in the liver through β-oxidation. In this context, it is remarkable that this process takes place purely physically and functions through

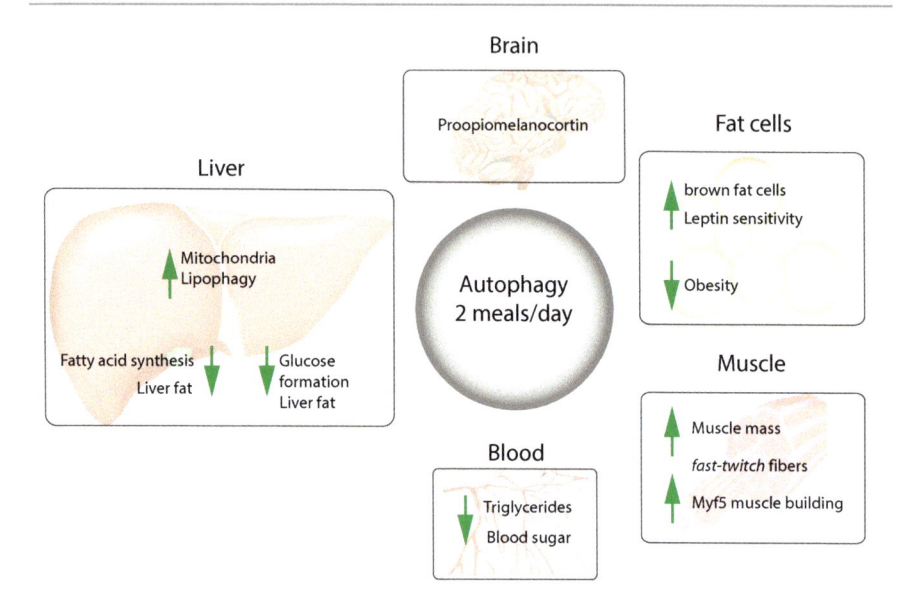

Fig. 6.27 Autophagy has positive effects on muscle formation, blood values, and fat metabolism

nothing more than the interaction of lipid bilayers. The basis of these processes is therefore nothing more than the physics of fat droplets and colloids [75].

The biogenic polyamine spermidine, which accelerates autophagy, is formed during digestion in the large intestine via the microbiome and apparently has an effect on cell aging, slowing down rapid cell degradation and thus possibly slowing down aging processes [76]. The process also seems to have a positive effect on memory performance in old age [71], and spermidine also has a heart-protective effect [77].

Since the process of autophagy is universal and necessary in every cell, whether animal or plant, for good biological reasons, spermidine is also present in our food. It is therefore logical to investigate the effects of orally administered spermidine, and clear evidence of positive effects has been found in many systematically conducted animal experiments [78]. As it turned out, the spermidine concentration in the microbiome of mice can be increased by oral administration of spermidine. Therefore, it makes sense to pay attention to the spermidine concentrations in food. In Fig. 6.28, the most important culinary items are shown in increasing spermidine concentration.

Immediately, the usual healthy foods can be recognized again: green vegetables such as broccoli, apples and pears, nuts, umami suppliers like peas and mushrooms, as well as fermented and matured products like ripe cheese. Soybeans also provide good amounts of spermidine, which is further increased by their fermentation into nattō. Sprouted wheat is also rich in spermidine.

Fig. 6.28 The concentration of the biogenic polyamine spermidine in natural foods. The concentration increases from left to right

However, for the cell cleaning processes in humans to run smoothly, a constant supply must not enter the cells. This is where intermittent fasting comes in. Sufficient breaks between main meals are essential for this. Constant snacking and function-free eating is precisely the wrong approach for this autophagy process. The cells are overloaded by the constant supply of food and, figuratively speaking, can no longer keep up with cell cleaning. Ultimately, this also means that the process taking place during intermittent fasting and food breaks should not be interrupted by the intake of spermidine-rich foods, as all the foods listed in Fig. 6.28 also provide other components that leave their molecular traces in the cells before the cell clearance is completed. Therefore, the integration of these components into meals is important. Nuts for an aperitif, sprouts or soybeans in the starter or main course, cheese after the main course, and fruit finally in the dessert. Then, at the same time, diversity is ensured, and hunger is satisfied until the next meal, many hours later.

6.12 What We Will Eat in the Future

6.12.1 We Eat What We Used to Eat

What we will or must eat in the future is the most frequently asked question these days. The answer is currently being sought in many novel foods, reconstructed substitute products, or alternative proteins. It may be that these ideas are occasionally successful. However, a step back is also a step into the future. Many people today have forgotten how to eat many excellent foods. Dishes like tripe, sour lung, *rognon blanc*, sweetbread, Berlin liver, and kidneys are no longer eaten. As already discussed in the topic "Nose-to-Tail" (Section 6.7), these must be reintegrated into future, especially regional, nutrition if we are to talk about ethics in slaughtering and nutrition. These meat parts, despised for decades and declared as slaughter waste, are nutritious and have a higher biological value than some plant-based foods.

In this context, it is also important to rediscover long-forgotten delights in plants. Roots that make their starch biologically available after cooking. Slowly, tubers like chervil roots, turnips of all kinds, legumes, which enrich daily nutrition with pleasure, are finding their way back into the domestic market. The same applies to many vegetables: Root-to-Leaf. If everything edible were actually eaten, it would quickly become apparent that there is much more food available than the vegetable department of the standard supermarket suggests. All these approaches are feasible for everyone in every household at any time. These would be small but effective steps, also in the sense of eating and cooking culture.

These steps are, of course, associated with "home cooking" No standard production could adopt these small measures in the convenience sector. But they would be far more sustainable and the first step in a chain of measures to solve the nutrition problem before discussing new food sources where ethical questions are unclear, such as insects.

6.12.2 Insects, But Not Only

As already known from the considerations in Chaps. 1 and 2, insects (and mollusks) were an essential step on the way to *Homo sapiens*, also because of the essential long-chain fatty acids. Insects also have high-quality proteins with all essential amino acids, in a mix and biological availability superior to plant proteins. In particular, the larvae of meal beetles (mealworms) have moved into the center of food and protein production. They are easy to breed, even on nutrient substrates in the tightest of spaces, without the population suffering. Their water consumption is low, and their CO_2 footprint is acceptable. Above all, everyone could breed their own mealworm larvae at home and process them by cooking, frying, deep-frying, or pureeing. This is no longer envisaged in most Western cultures. However, there was once a standard dish of cockchafer soup in Germany—especially in years of cockchafer infestation. Before they destroy the harvest, they are eaten—like locusts in other parts of the world. In a figurative sense, this would be no different than eating wild boars, whose numbers must be limited by hunting before the population becomes too high.

The idea of breeding larvae for personal consumption, despite all its advantages, evokes disgust and aversion. Thus, insect powder is more likely to be purchased when it is anonymously packaged, just like meat, which is sold in supermarkets and at counters completely detached from the animal and the slaughtering process. Alternatively, one might opt for protein powder, the industrially obtained isolate from insects, the differences to lupine protein isolate are visually marginal. Thus, only industrial production remains. The use of insects as an important source of human food and animal feed brings with it two major challenges: first, how to transform insects into safe, socially acceptable food; and second, how to produce enough insects at a reasonable price while still being sustainable [79]. The criteria for food safety in the required mass insect breeding must also be clarified and ensured [80].

The second challenge is more of an ethical nature. The question of whether insects can feel pain [81] and thus have the right to dignity in the sense of animal ethics is not raised in current discussions. However, it would be worthwhile to consider this, as it is difficult to imagine how many mealworm larvae would have to be killed to meet the global demand for beef burgers. In light of the recent increase in insect deaths [82] and the newly discovered love for these useful creatures [83], the question arises whether an insect's life is worth less than that of a grazing cow. Is it not a problem to kill 2,000 mealworm larvae, crickets, or moths for the protein patty of an insect burger or to slaughter one cow for 2,000 burgers? Do insects have less dignity than land animals and fish? Does no one speak of "mass worm farming" when mealworm larvae are bred in cramped spaces in their own excrement? It is indeed unclear whether it will be possible to cover the current meat consumption with insect proteins. The question of whether insects can feel pain remains unclear to this day [84, 85] and this question is not purely philosophical [86].

However, approaches to breed insects for use as feed in large-scale aquaculture or even for pigs in mass farming are questionable [87]. This would indeed be ethically wrong, as the current undignified cycle of increasing demand, mass animal farming, and resource consumption would neither be changed nor broken.

6.12.3 New Foods to Discover: Duckweeds

Water lentils, also called duckweed, are rapidly growing plants on water surfaces, such as lakes. Duckweed are undemanding, grow everywhere under almost all climatic conditions, and can therefore be found on all continents. There are many different species, as can be seen in Fig. 6.29. The reproduction and growth rates are extremely high, ensuring an adequate harvest. All that is needed is clean water for cultivation, as duckweed can also accumulate pollutants. The plants do not suffer from this, but they could then no longer be consumed.

For the prerequisites of targeted and controlled cultivation, some criteria are important. Duckweeds prefer standing water, large flow velocities at the surface hinder growth. The water must contain sufficient minerals. Especially with low iron content, the rich green coloration hardly forms, as shown in Fig. 6.29. Duckweeds thrives at air temperatures from—15 °C to 33 °C. It is acid-tolerant and grows even at pH values up to 4. Due to the high coverage of the water surface, further use of the water is hardly possible, as only little light penetrates below the surface.

For some time now, duckweeds have also been discussed as a possible food source for humans [88, 89] and elaborate studies are intensively researching applications and acceptance [90]. In fact, both the macro and micronutrients of duckweeds are interesting. The fatty acids comprise a very wide spectrum [91] of saturated and unsaturated fatty acids, reflecting the wide temperature range of the living conditions of duckweeds. Saturated fatty acids are also very short, such as C 10, with C 16:0 dominating in all duckweeds species. Since duckweeds grow and thrive well at temperatures around 30 °C, some of the saturated fatty acids are very long. Fatty acids with up to 28 carbon atoms, C 28:0 (for example, a component of beeswax), occur. In the case of monounsaturated fatty acids, oleic acid C

Fig. 6.29 Different duckweeds species on a water surface. (*Source:* Wikipedia, Christian Fischer)

18:1 is present at only 2–3%, while α-linolenic acid (ALA) is highly prominent at 35–42%. Linoleic acid C 18:2 is also very common at 19–27%.

For amino acids, glutamic acid and aspartic acid are equally represented at about 9–11%. This shows the extraordinarily high umami potential of duckweeds when considering fermentations or other forms of protein hydrolysis. Of the essential amino acids, leucine and lysine are at the top with 6–8%, followed by valine and phenylalanine, which account for 5%.

In terms of micronutrients, a balanced mineral content is worth highlighting, which can naturally be controlled in breeding tanks through the mineral content of the water. Calcium and phosphorus are on par at 15-20%, depending on the variety, and roughly correspond to the ratios of both minerals in milk, while potassium is at the top with an average of 67%. Due to the high photosynthesis rate of duckweed, carotenoids are present in significant amounts: both *trans* and *cis*-β-carotene (20-30%) as well as zeaxanthin can be found in all species. The most common carotenoid in most duckweed species is lutein, which makes up 40-80% of the carotenoid content in the light-harvesting complex. Vitamin E, α-tocopherol, is also abundantly incorporated into the hydrophobic parts of the cell membrane (lipid bilayer).

Despite these many positive properties, the overall low fat and protein content must be taken into account. Complete meals made from duckweed are therefore not possible. Duckweed can only be considered as a supplement [92], with its advantage certainly lying in its rapid growth and the resulting multiple harvests in all seasons.

6.12.4 Spirulina: Hype or Opportunity?

Spirulina, a multicellular and filamentous cyanobacterium, is also touted as a new superfood with true miracle powers attributed to it [93]. It is said to, for example, activate the immune system, help with allergies (allergic rhinitis) [94] or increase performance [95]. Upon closer examination, these studies are again unfounded. As usual, despite double-blind and placebo tests, the statistical fluctuations are higher than the measured effects, and on the other hand, a study with nine subjects cannot yield a significant result [96, 97].

Spirulina has been considered a food for quite some time [98], as it contains significant macro- and micronutrients [99]. Its protein content and the proportion of essential amino acids are relatively high [100]. The protein isolates and extracts are of great importance for food technology. The proteins have a high water-binding capacity, bind oil due to high hydrophobic protein sequences (via the essential amino acids and in some proteins proline-leucine-isoleucine sequences), and are therefore highly surface-active and can stabilize foams and emulsions [101]. One reason for this is the high number of different proteins contained in the biomass of *Spirulina* .

For economic reasons, it is therefore not surprising that *Spirulina* has already received considerable attention in the health sector, the food industry, and aquaculture [102]. It grows in water, can be easily harvested and processed. It has a very high content of macro- and micronutrients, essential amino acids, proteins, lipids, vitamins, minerals, and antioxidants [103]. This includes a whole range of carotenoids. *Spirulina* is considered a dietary supplement to combat malnutrition in developing countries. In addition, *Spirulina* meets all food safety requirements, so a wide range of applications are being tested [104]. For example, due to the pronounced algae aroma of Spirulina and its derived products, novel forms of vegan sushi, stuffed ravioli, or dried products reminiscent of "Beef Jerky" are being considered [105], to test them for acceptance in large-scale European-wide studies. The development of meat-free alternative products is just beginning, but this is a viable direction based on current knowledge [106].

6.13 *Ikejime*—Gentle, Sustainable, and Umami-Promoting Cultural Technique

6.13.1 Taste-Driven Cultural Technique

Entirely at the beginning of this book, in Chap. 1, the central question was asked, what motives drove the early hominids to take the risk of hunting with modest equipment, instead of staying with roots and largely risk-free gathered food despite fire. Apparently, it was the umami taste that initiated human development. Without a doubt, this protein taste was the godfather of most food refinement techniques. To this day, this can be seen in the fish-rich Japanese cuisine, which lives in the broadest sense from its simplicity [108], and its preparation techniques.

The quality and thus also the taste and texture of killed animals depend very much on the method of killing and the treatment of the animal bodies after death. This is particularly evident in fish. According to popular belief, fish spoil quickly, become rancid, and acquire an unpleasant "fishy" smell. Especially with

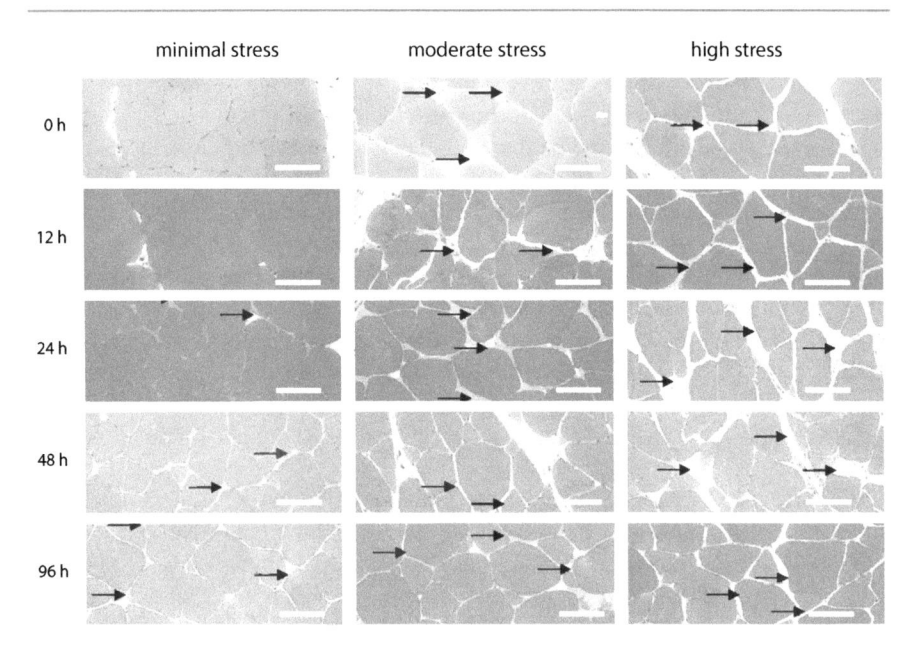

Fig. 6.30 The influence of stress on the change in muscle structure, as seen in light microscopy. From top to bottom: maturation times of the fillets (from immediately after slaughter to 96 hours later. From left to right: varying degrees of stress load. The white bar at the bottom right corresponds to 100 μm)

conventional fishing methods, such as trawling, fish have no chance of a stress-free death, let alone when they suffocate gasping for air or are deliberately suffocated with carbon dioxide. The meat suffers from this stress load with increasing time after killing [109], as shown in Fig. 6.30.

In the experiments, a strong stress-dependent decrease in pH value and a stress-dependent detachment of the myofilaments were observed, as indicated by the arrows in Fig. 6.30. The detachment of the myofilaments from each other in the muscle cells naturally increases with maturation time but is much more pronounced with increasing stress factors. The reason for this is not solely due to pH-value-related local changes in protein interactions (Fig. 6.4), but primarily to an increased activity of the enzyme cathepsin B, which cleaves peptide bonds responsible for myofibril attachment. This causes textural changes and a "softening" of the muscle structure after rigor mortis, which is not desired in fish, but very much so in (beef) meat and its maturation, as indicated in Fig. 6.31.

Due to the rapid pH value reduction with increased stress factors before slaughtering, the enzyme is activated more quickly, the myobrils are also cut at the actin, and the cohesion of the muscle fibers becomes weaker. The fish fillets become softer and lose their typical "bite", the fish meat becomes soft and loses its elasticity. Increased stress factors therefore lead to rapid degradation of myofibrillar cohesion.

6.13 Ikejime—Gentle, Sustainable, and Umami-Promoting Cultural Technique

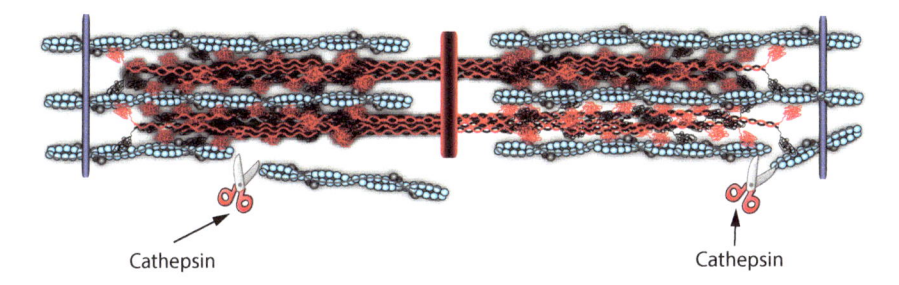

Fig. 6.31 A simple model for the action of the muscle's own enzyme cathepsin B, which is shown here for simplification as a "molecular scissors"

For the results shown in Fig. 6.30, the salmon were filleted before the onset of rigor mortis to recognize the differences between *pre-rigor* and *post-rigor* filleting. With conventional catch and killing methods, another point is added: The fish cannot bleed out at the beating heart, the muscle twitching is strongly pronounced and lasts due to the existing nerve connections to the spinal cord even after death. What this means is discussed in detail in Sects. 4.3 and 4.4. The adenosine triphosphate (ATP) is rapidly degraded, the purine metabolism is initiated, and the umami taste is greatly reduced. Conventional fish catch and mass killings thus result in loss of texture and taste. The fish can only be eaten fresh or is immediately shock-frozen on the ships.

This is exactly where the Japanese slaughtering method Ikejime comes in: The fish are killed individually and by hand, which is referred to as the "most humane" slaughter. The freshly caught fish are skillfully pierced with a thorn in the brain, causing immediate brain death, but the heart continues to beat. Then the gills and the carotid arteries are opened immediately, and the fish begins to bleed out. Immediately afterwards, the fish is cut at the tail and a long thin wire is inserted through the spinal cord canal. The connections of the nerves to the muscles are separated. All post-mortem muscle activities such as twitching stop immediately. The physiological processes also stop immediately. For the final bleeding out, the fish is placed in ice water. The fish muscles remain in this fixed state for a longer time, defining that texture and taste appreciated by lovers of Japanese cuisine when they enjoy fish in excellent restaurants. It is noteworthy that this state of the muscles, fixed by the Ikejime method, hardly changes even after several days of storage, while fish fillets from conventional sources are hardly storable. The flavor even becomes more intense with storage up to 15 days. No off-flavors develop, and above all, the texture of the fish remains crisp; they do not become soft and fibrous.

6.13.2 The Molecular Aspects of *Ikejime* Slaughtering

Thus, the advantages of the Ikejime slaughter method are evident. The killing process is not a mass slaughter; each of the fish delivered in small water containers is slaughtered individually. The fish do not suffocate in the air, minimizing stress. The targeted stab causes immediate brain death, and the separation of the spinal cord nerves from the fillet muscles immediately stops any mechanical movement of the muscles . ATP is not consumed by uncontrolled twitching but is even further produced through the ongoing metabolism. As a result, more inosine monophosphate (IMP) is formed (see Fig. 4.10). By storing at cool temperatures, IMP is preserved for a long time, while the concentration of free glutamic acid increases.

The ATP concentration increases, strong acidification through an accumulation of protons during purine metabolism is avoided. The onset of rigor mortis is significantly delayed due to the still high ATP concentration. The slowed breakdown of ATP into AMP results in a high concentration of IMP. The IMP concentration also remains high during storage compared to conventionally slaughtered fish. Taste and texture are excellent even after several days, which can never be achieved in conventionally killed fish for chemical reasons.

For the storage of fish, regardless of whether they were slaughtered using the Ikejime method or by conventional (preferably stress-free) means, the storage temperature remains another parameter that has not yet been addressed [109]. The preservation of IMP is naturally better at low temperatures, as enzyme activity is weaker. In Fig. 6.32, the relative (normalized) concentrations of IMP (inosine monophosphate) and hypoxanthine (Hx) (see Sect. 4.4.3) are shown for salmon fillets at storage temperatures of 4 °C and 0 °C. The differences are striking: compared to storage at 4 °C, at 0 °C the IMP concentration remains comparatively high even after seven to eight days, while at 4 °C the IMP concentration decreases significantly within the first four days.

The determination of the IMP and Hx content, which is shown in Fig. 6.32, was carried out using nuclear magnetic resonance (NMR; *nuclear magnetic resonance*) methods, a very complex procedure, but one with very high accuracy. A significant (but not further discussed in the work) result is the determination of free amino acids in the muscles. These are of great importance for taste, especially the glutamic acid, which, along with IMP (and other nucleotides), is responsible for the umami taste. Free amino acids are present in every living muscle, as proteins must be constantly renewed and repaired at the site of the cells (for which certain enzymes in the sarcoplasm are responsible), including glutamic acid. However, during storage, more glutamic acid is released through the already mentioned enzymatic cleaving processes. The progression is shown, within the limits of measurement, in Fig. 6.33.

Immediately after slaughter, the value of glutamic acid (Glu) is approximately 7 mg per 100 g of fish muscle, doubling after seven days of storage, while the concentration decreases again after further storage. Since both data, the concentration of IMP as well as that of glutamic acid, were obtained from the same method (and

6.13 Ikejime—Gentle, Sustainable, and Umami-Promoting Cultural Technique

Fig. 6.32 The course of the concentration of IMP (solid lines) and Hx (dotted lines) as a function of storage duration (in days) and as a function of storage temperature (dark blue 0 °C, light blue 4 °C)

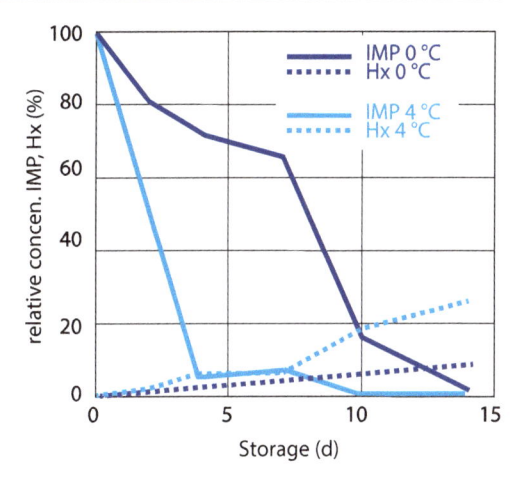

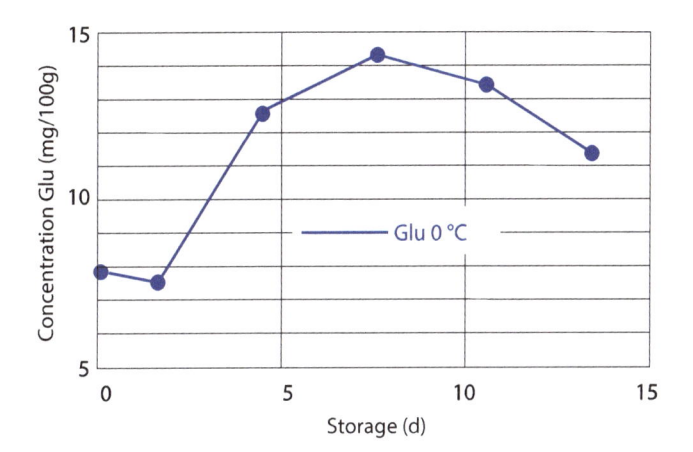

Fig. 6.33 The concentration of free glutamic acid in salmon muscle during storage at 0 °C. (Data from [109])

simultaneously), the concentration ratio is calculated and plotted over the storage period (Fig. 6.34). This results in a measure of the umami taste of the salmon fillets as a function of the storage duration.

Considering the maturation conditions of fish fillets in this light, it quickly becomes clear what makes the taste appeal of the Ikejime slaughter method. Fish slaughtered using the Ikejime method not only impress with their texture after longer storage, but also with a distinctly pronounced umami taste, as can be seen from the synergy curve in Fig. 4.11. Thus, it is clear: Fish slaughtered using the Ikejime method never leaves the intense umami taste range for an extended period, as shown in Fig. 6.35.

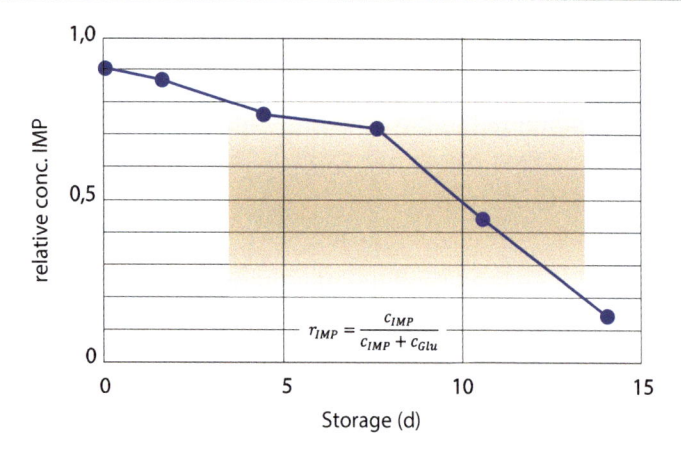

Fig. 6.34 The concentration ratio of IMP in the salmon fillets as a function of storage duration. It decreases continuously. However, the umami taste is highest in the brown hatched area when the IMP concentration ratio, according to Fig. 4.11, is between 20% and 80%

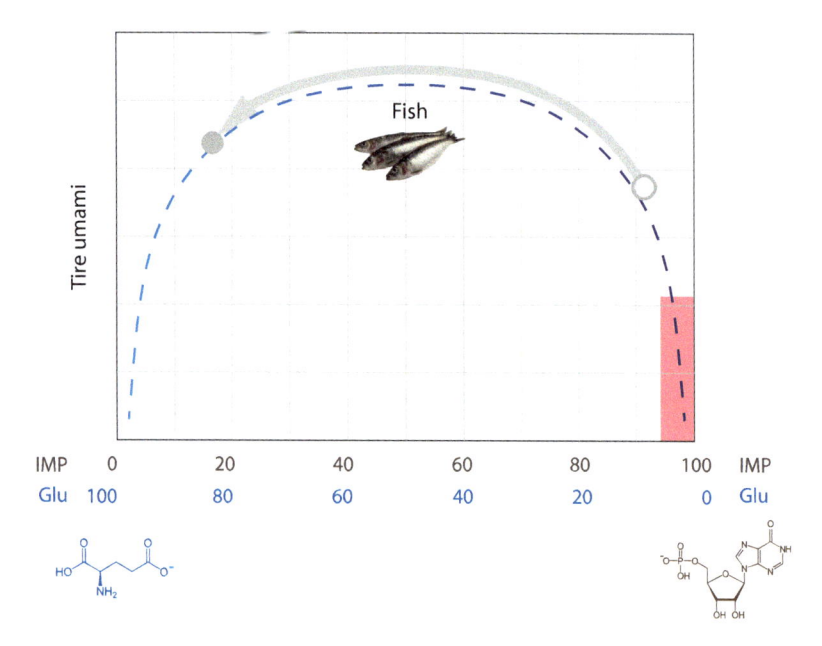

$$r_{IMP} = \frac{c_{IMP}}{c_{IMP} + c_{Glu}}$$

Fig. 6.35 The maturation process of Atlantic salmon slaughtered using the Ikejime method: The umami taste remains high, as the ratio of glutamic acid and IMP shows. The open circle shows the umami value immediately after slaughter. The closed circle after 15 days of maturation. Conventionally treated fish remain in the red area for only a short time

With this, the secret of the Japanese cultural technique is revealed and shows once again: The development of certain cultural techniques is always driven by taste and quality. Another, non-physical learning effect from this Ikejime technique is fundamental: We *Homo sapiens* must treat our food, especially animals, carefully, attentively, and with dignity. All the more so when we kill them to eat them. But this serves the taste.

References

1. https://www.produktwarnung.eu/2019/04/12/oeffentliche-warnung-listerien-riesiger-rueckruf-von-vielen-franzoesischen-kaese/13521.
2. https://www.aerzteblatt.de/nachrichten/100295/In-jedem-zweiten-Haehnchen-im-Handel-befinden-sich-Campylobacter.
3. Rombaut, R., Camp, J. V., & Dewettinck, K. (2006). Phospho-and sphingolipid distribution during processing of milk, butter and whey. *International Journal of Food Science & Technology, 41*(4), 435–443.
4. Beuvier, E., Berthaud, K., Cegarra, S., Dasen, A., Pochet, S., Buchin, S., & Duboz, G. (1997). Ripening and quality of Swiss-type cheese made from raw, pasteurized or microfiltered milk. *International Dairy Journal, 7*(5), 311–323.
5. Pisanu, S., Pagnozzi, D., Pes, M., Pirisi, A., Roggio, T., Uzzau, S., & Addis, M. F. (2015). Differences in the peptide profile of raw and pasteurised ovine milk cheese and implications for its bioactive potential. *International Dairy Journal, 42,* 26–33.
6. Braun-Fahrländer, C., & Von Mutius, E. (2011). Can farm milk consumption prevent allergic diseases? *Clinical and Experimental Allergy, 41*(1), 29–35.
7. Loss, G., et al. (2015). Consumption of unprocessed cow's milk protects infants from common respiratory infections. *Journal of Allergy and Clinical Immunology, 135*(1), 56–62.
8. Wyss, A. B., et al. (2018). Raw milk consumption and other early-life farm exposures and adult pulmonary function in the Agricultural Lung Health Study. *Thorax, 73*(3), 279–282.
9. Stebbins, N., von Mutius, E., & Sasisekharan, R. (2019). *Analytics on farm dust extract for development of novel strategies to prevent asthma and allergic disease. The science and regulations of naturally derived complex drugs* (pp. 79–90). Springer.
10. Fisher, H. R., Keet, C. A., Lack, G., & du Toit, G. (2019). Preventing peanut allergy: Where are we now? *The Journal of Allergy and Clinical Immunology Practice, 7*(2), 367–373.
11. https://www.welt.de/politik/deutschland/plus190797629/Ernaehrungswissenschaftler-Milch-ist-ein-hochbrisanter-Cocktail.html.
12. Sozańska, B. (2019). Raw cow's milk and its protective effect on allergies and asthma. *Nutrients, 11*(2), 469.
13. Swanson, K., Kutzler, M., & Bionaz, M. (2019). Cow milk does not affect adiposity in growing piglets as a model for children. *Journal of dairy science*. https://doi.org/10.3168/jds.2018-15201.
14. Bas Kast. *Der Ernährungskompass: Das Fazit aller wissenschaftlichen Studien zum Thema Ernährung*. C. Bertelsmann.
15. Berk, Z. (2018). *Food process engineering and technology*. Academic.
16. Jentschura, P., & Lohkämper, J. (2003). *Gesundheit durch Entschlackung*. Jentschura.
17. Neu: Carrera-Bastos, P., Fontes-Villalba, M., O'Keefe, J. H., Lindeberg, S., & Cordain, L. (2011). The western diet and lifestyle and diseases of civilization. *Research Reports Clinical Cardiology, 2*(1), 15–35.
18. Chaudhari, N., & Roper, S. D. (2010). The cell biology of taste. *The Journal of cell biology, 190*(3), 285–296.
19. Zipprick, J. (2012a). *In Teufels Küche: Ein Restaurantkritiker packt aus*. Bastei Lübbe.
20. Zipprick, J. (2012). *Die Welt des Cognac*. Neuer Umschau Verlag.

21. Stiebing, A. (2002). Separatorenfleisch im Kreuzfeuer der Kritik. *Fleischwirtschaft, 82*(2), 8.
22. Negrão, C. C., Mizubuti, I. Y., Morita, M. C., Colli, C., Ida, E. I., & Shimokomaki, M. (2005). Biological evaluation of mechanically deboned chicken meat protein quality. *Food Chemistry, 90*(4), 579–583.
23. Pereira, A. G. T., Ramos, E. M., Teixeira, J. T., Cardoso, G. P., Ramos, A. D. L. S., & Fontes, P. R. (2011). Effects of the addition of mechanically deboned poultry meat and collagen fibers on quality characteristics of frankfurter-type sausages. *Meat Science, 89*(4), 519–525.
24. Acton, J. C., Ziegler, G. R., Burge, D. L., Jr., & Froning, G. W. (1983). Functionality of muscle constituents in the processing of comminuted meat products. *CRC Critical Reviews in Food Science and Nutrition, 18*(2), 99–121.
25. Jones, K. W., & Mandigo, R. W. (1982). Effects of chopping temperature on the microstructure of meat emulsions. *Journal of Food Science, 47*(6), 1930–1935.
26. Vilgis, T. A. (2011). *Das Molekül-Menü*. Hirzel.
27. Vilgis, T. A. (2015). Soft matter food physics – The physics of food and cooking. *Reports on Progress in Physics, 78*(12), 124602.
28. Ludwig, D. S., Willett, W. C., Volek, J. S., & Neuhouser, M. L. (2018). Dietary fat: From foe to friend? *Science, 362*(6416), 764–770.
29. Wang, D. D., Li, Y., Chiuve, S. E., Stampfer, M. J., Manson, J. E., Rimm, E. B., Willet, W. C., & Hu, F. B. (2016). Association of specific dietary fats with total and cause-specific mortality. *JAMA Internal Medicine, 176*(8), 1134–1145.
30. O'Fallon, J., Busboom, J., & Gaskins, C. *Fatty acids and wagyu beef*. https://www.researchgate.net/profile/Jan_Busboom/publication/265932588_Fatty_Acids_and_Wagyu_Beef/links/54caafa60cf2c70ce5237b4c.pdf.
31. Pradal, C., & Stokes, J. R. (2016). Oral tribology: Bridging the gap between physical measurements and sensory experience. *Current Opinion in Food Science, 9*, 34–41.
32. Vilgis, T. A. (2015). Lebensmittelkonsistenzen und Genusssteigerung. *Ernährung bei Pflegebedürftigkeit und Demenz* (pp. 137–150). Springer.
33. Vierich, T., & Vilgis, T. (2017). *Aroma Gemüse: Der Weg zum perfekten Geschmack*. Stiftung Warentest.
34. Degenhardt, A. G., & Hofmann, T. (2010). Bitter-tasting and kokumi-enhancing molecules in thermally processed avocado (*Persea americana* Mill.). *Journal of Agricultural and Food Chemistry, 58*(24), 12906–12915.
35. Feng, T., Zhang, Z., Zhuang, H., Zhou, J., & Xu, Z. (2016). Effect of peptides on new taste sensation: Kokumi-review. *Mini-Reviews in Organic Chemistry, 13*(4), 255–261.
36. Rao, H., Monin, P., & Durand, R. (2004). Crossing enemy lines: Culinary categories as constraints in French gastronomy. *Organizational Ecology and Strategy Conference, Washington University in St Louis, April* (pp. 23–24).
37. Zubillaga, M. (2008). Pourquoi j'aime les plats canailles. *La pensee de midi, 4,* 216–219.
38. https://alacarte.at/essen/rindfleischkueche-20-20153/.
39. Vié, B. (2011). *Testicles: Balls in cooking and cultur*. Prospect Books.
40. Louis-Sylvestre, J., Krempf, M., & Lecerf, J. M. (2010). Les charcuteries. *Cahiers de nutrition et de diététique, 45*(6), 327–337.
41. https://www.cds-hackner.de/productoverview.aspx?site=cdsschlachtnebenprodukteuebersicht.
42. Sales, J., & Kotrba, R. (2013). Meat from wild boar (*Sus scrofa* L.): A review. *Meat science, 94*(2), 187–201.
43. Bogucka, J., Kapelanski, W., Elminowska-Wenda, G., Walasik, K., & Lewandowska, K. L. (2008). Comparison of microstructural traits of Musculus longissimus lumborum in wild boars, domestic pigs and wild boar/domestic pig hybrids. *Archives Animal Breeding, 51*(4), 359–365.

44. North, M. K., & Hoffman, L. C. (2015). The muscle fibre characteristics of springbok (*Antidorcas marsupialis*) longissimus thoracis et lumborum and biceps femoris muscle. *Poster presented at the 61st international congress of meat science and technology.*
45. Dando, R., Dvoryanchikov, G., Pereira, E., Chaudhari, N., & Roper, S. D. (2012). Adenosine enhances sweet taste through A2B receptors in the taste bud. *Journal of Neuroscience, 32*(1), 322–330.
46. Schlichtherle-Cerny, H., & Grosch, W. (1998). Evaluation of taste compounds of stewed beef juice. *Zeitschrift für Lebensmitteluntersuchung und Lebensmittelforschung A, 207*(5), 369–376.
47. Panić, J., Vraneš, M., Tot, A., Ostojić, S., & Gadžurić, S. (2019). The organisation of water around creatine and creatinine molecules. *The Journal of Chemical Thermodynamics, 128,* 103–109.
48. Bocuse, P. (1977). *La cuisine du marché*. Flammarion.
49. Grass, G. (1977). *The Flounder, Vintage Classics*
50. Dahl, L. K. (1972). Salt and Hypertension. *The American Journal of Clinical Nutrition, 25*(2), 231–244.
51. Dahl, L. K. (1961). Possible role of chronic excess salt consumption in the pathogenesis of essential hypertension. *The American Journal of Cardiology, 8*(4), 571–575.
52. Dahl, L. K., Heine, M., & Tassinari, L. (1962). Effects of chronic excess salt ingestion: Evidence that genetic factors play an important role in susceptibility to experimental hypertension. *Journal of Experimental Medicine, 115*(6), 1173–1190.
53. Powles, J., Fahimi, S., Micha, R., Khatibzadeh, S., Shi, P., Ezzati, M., Engell, R. E., Lim, S. S., Danaei, G., & Mozzafarian, D. (2013). Global, regional and national sodium intakes in 1990 and 2010: A systematic analysis of 24 h urinary sodium excretion and dietary surveys worldwide. *BMJ open, 3*(12), e003733.
54. Mente, A., et al. (2018). Urinary sodium excretion, blood pressure, cardiovascular disease, and mortality: A community-level prospective epidemiological cohort study. *The Lancet, 392*(10146), 496–506.
55. Graudal, N. (2016). A radical sodium reduction policy is not supported by randomized controlled trials or observational studies: Grading the evidence. *American Journal of Hypertension, 29*(5), 543–548.
56. Stanhewicz, A. E., & Larry Kenney, W. (2015). Determinants of water and sodium intake and output. *Nutrition Reviews, 73*(2), 73–82.
57. Messerli, F. H., Hofstetter, L., & Bangalore, S. (2018). Salt and heart disease: A second round of „bad science"? *The Lancet, 392*(10146), 456–458.
58. Carrillo-Tripp, M., Saint-Martin, H., & Ortega-Blake, I. (2003). A comparative study of the hydration of Na^+ and K^+ with refined polarizable model potentials. *The Journal of Chemical Physics, 118*(15), 7062–7073.
59. https://www.volkskrankheit.net/news/ernahrung/raffiniertes-salz.
60. Hendel, B., & Ferreira, P. (2001). *Wasser & Salz: Urquell des Lebens; über die heilenden Kräfte der Natur*. Ina-Verlag.
61. Mendelson, S. (1961). Dislocation etch pit formation in sodium chloride. *Journal of Applied Physics, 32*(8), 1579–1583.
62. http://www.naturecke.li/himalayasalz.html.
63. Hellwege, K. H. (2013). *Einführung in die Festkörperphysik* (Bd. 34). Springer.
64. https://www.ugb.de/exklusiv/fragen-service/was-ist-himalaya-salz-wie-ist-es-zu-bewerten/?kristallsalz.
65. De Graaf, C., Blom, W. A., Smeets, P. A., Stafleu, A., & Hendriks, H. F. (2004). Biomarkers of satiation and satiety. *The American Journal of Clinical Nutrition, 79*(6), 946–961.
66. Vilgis, T. (2013). Komplexität auf dem Teller – ein naturwissenschaftlicher Blick auf das kulinarischen Dreieck von Lévi-Strauss. *Journal Culinaire, 16,* 109–122.
67. Caviezel, R., & Vilgis, T. (2012). *Foodpairing: Harmonie und Kontrast*. Fona.

68. Hümbs, C. (2012). http://www.sternefresser.de/restaurantkritiken/2012/aromenmenu-huembs-la-mer-sylt/.
69. Schacht, A., Łuczak, A., Pinkpank, T., Vilgis, T. A., & Sommer, W. (2016). The valence of food in pictures and on the plate: Impacts on brain and body. *International Journal of Gastronomy and Food Science, 5,* 33–40.
70. Pfeiffer, E. F. (1990). Unterzuckerung und Stoffwechselentgleisung. *Das Ulmer Diabetiker ABC* (pp. 57–61). Springer.
71. Sigrist, S. J., Carmona-Gutierrez, D., Gupta, V. K., Bhukel, A., Mertel, S., Eisenberg, T., & Madeo, F. (2014). Spermidine-triggered autophagy ameliorates memory during aging. *Autophagy, 10*(1), 178–179.
72. Mattson, M. P., Longo, V. D., & Harvie, M. (2017). Impact of intermittent fasting on health and disease processes. *Ageing Research Reviews, 39,* 46–58.
73. Martinez-Lopez, N., Tarabra, E., Toledo, M., Garcia-Macia, M., Sahu, S., Coletto, L., Bastia-Gonzales, A., Barzilai, N., Pessin, J. E., Schwarz, G. J., & Kersten, S. (2017). System-wide benefits of intermeal fasting by autophagy. *Cell Metabolism, 26*(6), 856–871.
74. Singh, R., & Cuervo, A. M. (2012). Lipophagy: Connecting autophagy and lipid metabolism. *International Journal of Cell Biology, 282041,* 12. https://doi.org/10.1155/2012/282041.
75. Thiam, A. R., Farese, R. V., Jr., & Walther, T. C. (2013). The biophysics and cell biology of lipid droplets. *Nature Reviews Molecular Cell Biology, 14*(12), 775.
76. Madeo, F., Tavernarakis, N., & Kroemer, G. (2010). Can autophagy promote longevity? *Nature Cell Biology, 12*(9), 842.
77. Tong, D., & Hill, J. A. (2017). Spermidine promotes cardioprotective autophagy. *Circulation Research, 120*(8), 1229–1231.
78. Madeo, F., Eisenberg, T., Pietrocola, F., & Kroemer, G. (2018). Spermidine in health and disease. *Science, 359*(6374), eaan2788.
79. Gjerris, M., Gamborg, C., & Röcklinsberg, H. (2016). Ethical aspects of insect production for food and feed. *Journal of Insects as Food and Feed, 2*(2), 101–110.
80. EFSA Scientific Committee. (2015). Risk profile related to production and consumption of insects as food and feed. *EFSA Journal, 13*(10), 4257.
81. Eisemann, C. H., Jorgensen, W. K., Merritt, D. J., Rice, M. J., Cribb, B. W., Webb, P. D., & Zalucki, M. P. (1984). Do insects feel pain? – A biological view. *Cellular and Molecular Life Sciences, 40*(2), 164–167.
82. Landwirtschaft, Bundesinformationszentrum. „Insektensterben in Deutschland." *Bundesanstalt für Landwirtschaft und Ernährung (BLE).* https://www.landwirtschaft.de/diskussion-und-dialog/umwelt/insektensterben-in-deutschland/.
83. Settele, J. (2019). Insektenrückgang, Insektenschwund, Insektensterben? *Biologie in unserer Zeit, 49*(4), 231–231.
84. Adamo, S. A. (2016). Do insects feel pain? A question at the intersection of animal behaviour, philosophy and robotics. *Animal Behaviour, 118,* 75–79.
85. Tye, M. (2016). Are insects sentient? *Animal Sentience: An Interdisciplinary Journal on Animal Feeling, 1*(9), 5.
86. Fischer, B. (2016). What if Klein & Barron are right about insect sentience? *Animal Sentience: An Interdisciplinary Journal on Animal Feeling, 1*(9), 8.
87. Veldkamp, T., & Bosch, G. (2015). Insects: A protein-rich feed ingredient in pig and poultry diets. *Animal Frontiers, 5*(2), 45–50.
88. Appenroth, K. J., Sree, K. S., Böhm, V., Hammann, S., Vetter, W., Leiterer, M., & Jahreis, G. (2017). Nutritional value of duckweeds (Lemnaceae) as human food. *Food Chemistry, 217,* 266–273.
89. Sree, K. S., Dahse, H. M., Chandran, J. N., Schneider, B., Jahreis, G., & Appenroth, K. J. (2019). Duckweed for human nutrition: No cytotoxic and no anti-proliferative effects on human cell lines. *Plant Foods for Human Nutrition, 74*(2), 223–224.

90. de Beukelaar, M. F., Zeinstra, G. G., Mes, J. J., & Fischer, A. R. (2019). Duckweed as human food. The influence of meal context and information on duckweed acceptability of Dutch consumers. *Food Quality and Preference, 71,* 76–86.
91. Appenroth, K. J., Sree, K. S., Bog, M., Ecker, J., Seeliger, C., Böhm, V., Lorkowsky, S., Sommer, K., Vetter, W., Tolzin-Banasch, K., Kirmse, R., Leiterer, M., Dawcynzki, C., Liebisch, G., & Jahreis, G. (2018). Nutritional value of the duckweed species of the genus *Wolffia* (Lemnaceae) as human food. *Frontiers in chemistry, 6.* https://doi.org/10.3389/fchem.2018.00483.
92. Yaskolka Meir, A., Tsaban, G., Zelicha, H., Rinott, E., Kaplan, A., Youngster, I., Rudich, A., Shelef, I., Tirosh, A., Pupkin, D. B. E., Sarusi, B., Blüher, M., Stümvoll, M., Thiery, J., Ceglarek, U., Stampfer, M. J., & Shai, I. (2019). A green-mediterranean diet, supplemented with mankai duckweed, preserves iron-homeostasis in humans and is efficient in reversal of anemia in rats. *The Journal of Nutrition.* https://doi.org/10.1093/jn/nxy321.
93. https://www.zentrum-der-gesundheit.de/spirulina-immunsystem-ia.html.
94. Mao, T. K., Water, J. V. D., & Gershwin, M. E. (2005). Effects of a *Spirulina*-based dietary supplement on cytokine production from allergic rhinitis patients. *Journal of Medicinal Food, 8*(1), 27–30.
95. Kalafati, M., Jamurtas, A. Z., Nikolaidis, M. G., Paschalis, V., Theodorou, A. A., Sakellariou, G. K., Koutedakis, Y., & Kouretas, D. (2010). Ergogenic and antioxidant effects of spirulina supplementation in humans. *Medicine and Science in Sports and Exercise, 42*(1), 142–151.
96. Goodman, S. N., Fanelli, D., & Ioannidis, J. P. (2016). What does research reproducibility mean? *Science Translational Medicine, 8*(341), 341ps12.
97. Benjamin, D. J., et al. (2018). Redefine statistical significance. *Nature Human. Behaviour, 2*(1), 6.
98. Ciferri, O. (1983). *Spirulina,* the edible microorganism. *Microbiological Reviews, 47*(4), 551.
99. Sánchez, M., Bernal-Castillo, J., Rozo, C., & Rodríguez, I. (2003). *Spirulina (Arthrospira):* An edible microorganism: A review. *Universitas Scientiarum, 8*(1), 7–24.
100. Lupatini, A. L., Colla, L. M., Canan, C., & Colla, E. (2017). Potential application of microalga *Spirulina platensis* as a protein source. *Journal of the Science of Food and Agriculture, 97*(3), 724–732.
101. Bashir, S., Sharif, M. K., Butt, M. S., & Shahid, M. (2016). Functional properties and amino acid profile of *Spirulina platensis* protein isolates. *Biological Sciences-PJSIR, 59*(1), 12–19.
102. Soni, R. A., Sudhakar, K., & Rana, R. S. (2017). *Spirulina* – From growth to nutritional product: A review. *Trends in Food Science & Technology, 69,* 157–171.
103. Wu, H. L., Wang, G. H., Xiang, W. Z., Li, T., & He, H. (2016). Stability and antioxidant activity of food-grade phycocyanin isolated from *Spirulina platensis. International Journal of Food Properties, 19*(10), 2349–2362.
104. Mobin, S., & Alam, F. (2017). Some promising microalgal species for commercial applications: A review. *Energy Procedia, 110,* 510–517.
105. Grahl, S., Strack, M., Weinrich, R., & Mörlein, D. (2018). Consumer-oriented product development: The conceptualization of novel food products based on spirulina (*Arthrospira platensis*) and resulting consumer expectations. *Journal of Food Quality.* https://doi.org/10.1155/2018/1919482.
106. Grahl, S., Palanisamy, M., Strack, M., Meier-Dinkel, L., Toepfl, S., & Mörlein, D. (2018). Towards more sustainable meat alternatives: How technical parameters affect the sensory properties of extrusion products derived from soy and algae. *Journal of Cleaner Production, 198,* 962–971.
107. Härtig, M. (2013). *Einfachheit: Kulturreflexion der Kaiseki-Küche Kyōtos.* Doctoral dissertation, Universität Witten.

108. Bahuaud, D., Mørkøre, T., Østbye, T. K., Veiseth-Kent, E., Thomassen, M. S., & Ofstad, R. (2010). Muscle structure responses and lysosomal cathepsins B and L in farmed Atlantic salmon (*Salmo salar* L.) pre-and post-rigor fillets exposed to short and long-term crowding stress. *Food Chemistry, 118*(3), 602–615.
109. Shumilina, E., Ciampa, A., Capozzi, F., Rustad, T., & Dikiy, A. (2015). NMR approach for monitoring post-mortem changes in Atlantic salmon fillets stored at 0 and 4°C. *Food Chemistry, 184,* 12–22.

Conclusion—Or: What Remains?

Abstract

Incorporating scientific aspects from the physics and chemistry of soft matter contributes to a better understanding. It is not enough to read the summaries of the studies. The results must be put into context with the study design, as they are often dependent on it. Studies should never be considered in isolation, but always in connection with existing results. As long as they are not verified with independent methods, they remain open-ended discussion material.

7.1 Reading and Better Understanding Critical Studies

The uncertainty associated with many nutritional studies was already revealed in Fig. 4.2. Depending on the study design, food can be "healthy" or "unhealthy." Classifying such results seems difficult. In such cases, an evidence-based look at new approaches in nutritional research helps. A quick look at the current scientific literature reveals how shaky some of the derived opinions and claims are. The data on meat is completely unclear, regardless of whether the meat is processed or not [1]. Some authors recommend, for example, maintaining the consumption of meat and sausage if it has already been part of one's personal and long-term eating history. For this assessment outside the mainstream, the new methods and criteria of evidence-based research are used, which have only recently been introduced into nutritional research[2] (Fig. 7.1).

It becomes very clear how evidence and relevance of different methods must be classified. It turns out that neither *in-vitro* research combined with *in-vivo* experiments, animal experiments, and computer simulations *(in silico)* nor observational studies have sufficient evidence to make any strict dietary recommendations. When these are combined with randomized studies, the evidence increases—but only if there are as few restrictions as possible in the study group. Inconsistencies

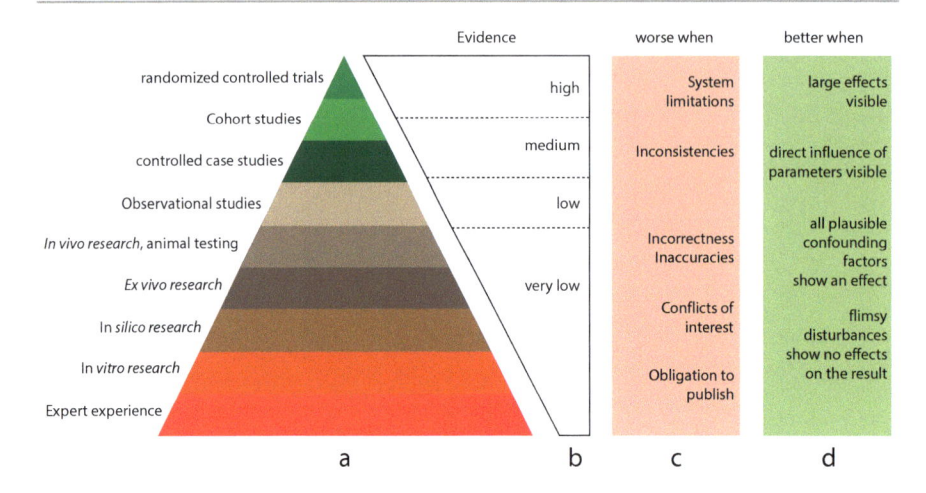

Fig. 7.1 The conventional hierarchy in evidence-based nutritional research (**a**) with its evidence ratings (**b**). Hard criteria (**c**) for decision-making for recommendations have been gradually supplemented over time (**d**)

in methods, incorrect evaluations, and false data collection, as well as inadequate statistics and, one of the biggest problems, inaccuracies in survey studies, also have a negative impact. The selection of study participants is another problem.

High The publication pressure on researchers proves to be very negative. Time and again, studies are published under time pressure and to raise research funds that are basically insufficient or "unfinished." How strongly publication pressure, publication compulsion, and conflicts of interest or even a desire for profiling often present in science can affect the question of whether meat is responsible for diseases or not [3].

In fact, "good" final dietary recommendations can only be made if, ideally, all methods of the pyramid in Fig. 7.1 point in the same direction without contradiction. The criteria for evaluating randomized meta-studies and even more so observational studies thus go far beyond the interpretation of the raw data.

Furthermore, significance, error bars, and data collection methods must be carefully analyzed—combined with the detailed considerations of statistical analyses by Benjamin [3] and Ioannidis, whose working group repeatedly points out non-credible results in nutrition research using mathematical analyses [4]. Overall, this leaves little in the way of solid results from many studies, and they should be taken less seriously. Even when correlations are clear, after verified evidence and clear significance, causality remains unanswered.

Causal relationships can hardly be derived from randomized meta-studies, which is not only due to the still high error bars, but also to the fundamental methods. Causalities can only be established directly through molecular relationships. This is precisely why investigations *in vitro, in vivo,* and *ex vivo* are necessary. With the developments in bioinformatics and the methods of Artificial Intelligence (AI) and *machine learning* [5], computer simulations [6] *(in silico)* are increasingly becoming more important possibilities. These concern laboratory experiments on a molecular scale simulated on the computer, but also for data collection and evaluation, randomized meta-studies can be classified much better [7]. Molecular methods thus provide an increasingly important basis for understanding many relationships at the molecular level, but taken alone, they do not provide evidence for humans, as shown in the classical pyramid in the evidence-based illustration. Therefore, for a far-reaching interpretation, a feedback loop between molecular methods and randomized, controlled studies is always necessary. Only molecular biology, biophysics, and chemistry provide direct evidence, e.g., for the effect of molecules with living cells and biosystems, as is indispensable in pharmacy to test the effect of a specific molecule precisely in all facets and under all molecular conditions.

However, examining and analyzing the effect of an entire food and its effect on a few isolated human cells under laboratory conditions is simply impossible. For nutritional recommendations for complex foods or even meals, conclusions from molecular investigations *in vitro, in silico, in vivo,* and *ex vivo* provide good indications of causalities but must be closely linked to evidence-based methods. A good and instructive example of this is the new work on the effect of egg consumption on cardiovascular diseases published in *Ernährungs Umschau* [26, 27]. On this topic, *Google Scholar* finds 205,000 publications under the search terms "egg intake risk." If you limit the search to the year 2021, you will find over 200 papers dealing with this topic and concluding that there is predominantly no relationship between egg consumption and diseases. The term "predominantly" causes confusion, as all observational studies after many years of egg, fat, and cardiovascular problems and strokes do not find predominantly nothing, but absolutely nothing, as summarized in the most recently published randomized cohort study (PURE) [28]. Now the question arises whether it is not finally enough to raise research funds, research spirit, and time for this topic, as new results or paradigm shifts are not necessarily to be expected. There are far more urgent questions in the field of nutrition.

In the meantime, molecular research is progressing rapidly. In another recent study, the egg is described as a nutrient-rich food, whose proteins and lipids exhibit excellent functional properties and biological activities [29]. In recent years, egg yolk lipids such as egg yolk fats, phospholipids, and fatty acids have been separated and studied, which have anti-inflammatory or antioxidant effects or provide protection for the cardiovascular system and improve memory, resulting in the regulation of cell function and physiological homeostatic balance.

7.2 It's Not Just About the Amino Acid Balance

The prevailing opinion in nutritional science is clear: We need sufficient macro- and micronutrients from whole foods with the entire food matrix, as they contain micronutrients, polyphenols, secondary plant compounds, antioxidants, fiber, etc. [8]. This is (for healthy individuals) always preferable to dietary supplements.

From this, it has been derived that all essential amino acids and all essential fatty acids must be supplied in sufficient quantities, regardless of their source. There is no disputing this, as there are many reasons for it, as has become clear throughout this discussion. However, humans and their molecular metabolism are highly non-equilibrium systems, and there is more to it than previously thought.

However, the actual need goes far beyond this. Current research increasingly shows that it is not just about the amino acids themselves, but about bioactive peptides that have far-reaching functions, as already hinted at in Fig. 4.54 and can be read in detail in the scientific literature [9]. These protein fragments, which are formed during the fermentation of food, raw sausage ripening, meat and cheese maturation, but also form during enzymatic digestive processes or fermentation via the microbiome during gastrointestinal passage. Bioactive peptides are absorbed, activate and trigger receptors, and have a variety of functions, as schematically indicated in Fig. 7.2.

Thus, it is clear that proteins are much more than just pure amino acids deliverers and mere energy suppliers. Specific fragments, whose amino acid sequence is determined by the primary structure of the proteins and which are formed during digestion, have an impact on various levels of physiological well-being and functions. It turns out that peptides containing the aromatic amino acids phenylalanine, tryptophan, and tyrosine, as well as the branched amino acids valine, leucine, and isoleucine, have high bioactivity. Such peptides have a high Fischer quotient [10], which is used to define nutritional status.

What this means for human nutrition can be illustrated by the current example of the boom in surrogate products. Visually and in terms of taste, meat, fish, and cheese substitutes must resemble the original. As shown in earlier chapters, this is not an insurmountable problem. There is nothing wrong with plant-based products; on the contrary, they are an important step in the future to curb the appalling mass animal farming for meat production. This is especially true when it comes to

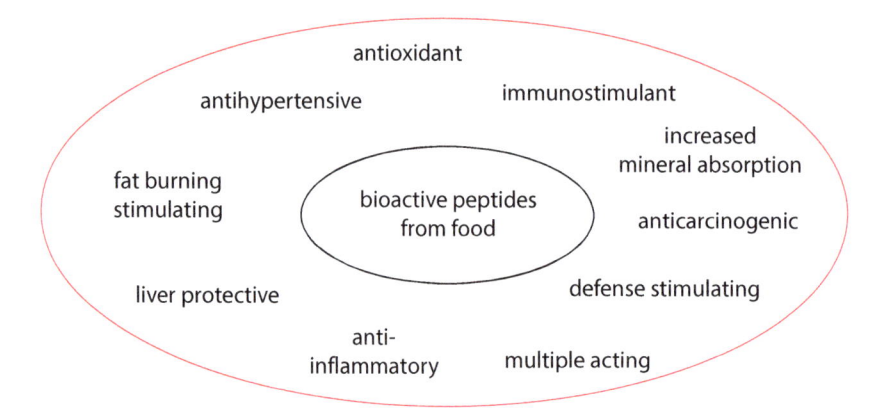

Fig. 7.2 Protein fragments from food perform a whole range of positive physiological functions

7.2 It's Not Just About the Amino Acid Balance

mass consumption, which can be observed daily at burger stands, fast-food restaurants, and discount stores.

Carelessly and heedlessly consumed meat products can be easily replaced by plant-based products, probably even without many eaters realizing that it is a substitute product. This is especially true when the inherent taste of the patty is completely masked by heavily seasoned sauces and creams anyway. Then, Beyond-Meat, soy, pea, or lupine burgers are certainly not out of place. This is logical and ultimately independent of whether one prefers meat-free or consumes meat. In terms of sociology, this can at least describe the societal [11], ethical [12] and political [13] challenges of the food of the future.

Whether these aspects are the ultimate wisdom from the perspective of molecular nutrition, however, remains to be seen. From a molecular point of view, only to a certain extent. The well-known problems still exist, which can only be solved by supplementation, such as long-chain essential fatty acids, which are found only in animal products or microalgae, vitamin B_{12}, which is found in animal products, as well as precursor compounds of vitamin D, which, as cholesterol derivatives, are exclusively present in the animal organism, and so on and so forth. But there is another aspect that has not yet been fully understood in nutritional science, namely the role of bioactive peptides.

Although the amino acid spectrum can be well represented with most plant proteins, e.g. soy and peas, when they are mixed as in various plant-based burgers, as shown in Tab. 7.1.

According to the simplified approach of nutritional science, as shown in Tab. 7.1, it is not a problem to replace animal foods with plant-based ones without

Tab. 7.1 Average amino acid composition of soy and pea protein as well as beef, each based on dry matter

(Semi-)essential Amino acid	Soy protein (mg/100 g) In dry matter	Pea protein (mg/100 g) In dry matter	Beef (mg/100 g) In dry matter
Arginine	2360	3910	3630
Histidine	830	2800	2380
Isoleucine	1780	3800	3630
Leucine	2840	6580	5670
Lysine	1900	6320	6090
Methionine	580	6210	1760
Phenylalanine	1970	1620	1880
Threonine	1490	2670	3080
Tryptophan	450	570	770
Tyrosine	1250	4040	2380
Valine	1760	7510	3980

loss of nutrients, particularly amino acids. However, it is not that simple, as it is now known how important so-called bioactive peptides [14, 15] are for the functioning of physiology [16].

7.3　Bioactive Peptides: Small but Essential Features of Protein Origin

The spectrum of physiological effects attributable to the consumption of bioactive peptides and proteins supplied through the diet is diverse. They modulate the immune system, regulate blood pressure, and have anti-inflammatory and anti-cancer effects. The consumption of foods with a high potential for bioactive peptides is therefore of particular importance for health and well-being. Bioactive peptides help optimize health and prevent infections and even diseases directly or indirectly. Foods provide these bioactive peptides in a very special way, whether from protein-rich plants like soybeans [17], milk or dairy products [18], or meat [19].

The peptides are formed from food proteins through enzymatic cleavage, such as during maturation (meat, cheese), fermentation (yogurt, miso, soy sauce, raw sausages, ham). But above all, they are formed *in vivo* during human digestion when proteins from foods are cleaved by the proteases of the pancreas. And it is here that a whole new aspect of diverse nutrition suddenly reveals itself. Bioactive peptides are well-defined fragments of very specific proteins that carry between three and 20 amino acids and whose sequence follows very specific patterns. These patterns can, for example, stimulate specific functional proteins on cells, trigger molecular switching processes through their charge, or block switching processes through their highly water-insoluble sequences. This shows how important their exact structure is, and above all, where exactly they come from. The longer the bioactive peptides, the more information from the primary structure of the original protein is contained in the respective peptide and its bioactivity. Peptides consisting of only two or three amino acids can come from many proteins. For peptides of seven amino acids or more, the probability that this peptide can only come from a specific protein type increases significantly, and for peptides of 20 amino acids in a very specific pattern, the probability is already practically 100%, as the biological function of a protein is precisely determined by its amino acid sequence.

It is therefore instructive to examine the most important bioactive peptides from pea, soy, and muscle protein and compare them in light of Tab. 7.1 with regard to their functional patterns. Even when considering only the most fundamental properties of amino acids—hydrophilic, hydrophobic, and their electrical charge—completely different modes of action can be identified.

A vivid example is a peptide consisting of nine amino acids in the sequence Isoleucine-Valine-Glycine-Arginine-Proline-Arginine-Histidine-Glutamine-Glycine (see peptide in Fig. 7.3c), which is formed during the digestion of the muscle protein actin, but also during meat and raw sausage ripening. This peptide is triple positively charged via the amino acids Arginine and Histamine

7.3 Bioactive Peptides: Small but Essential Features of Protein ... 451

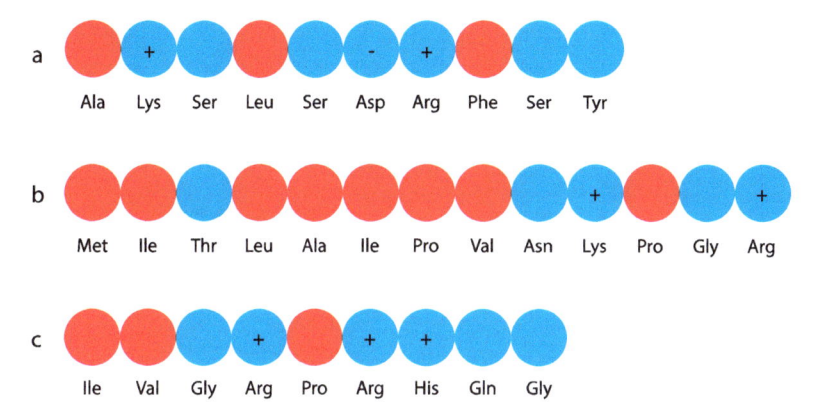

Fig. 7.3 Typical bioactive peptides from different foods show completely different patterns in amino acids (AA). Red: hydrophobic AA, blue: hydrophilic AA (polar; with ± the respective electrical charge): **a** from pea protein, **b** from soy protein, **c** from muscle protein

and is highly water-soluble due to the additional polar amino acids Glycine and Glutamine. This pattern is found exclusively in the muscle protein actin, i.e., in the muscles of animals and insects. There is not a single plant source for this bioactive peptide. The only alternative sources would be certain yeasts, such as *Candida intermedia,* found in winemaking, *Candida mogii,* a xylitol-producing fungus in some miso fermentations, *Mucor ambiguus,* a mold, or the yeast *Hanseniaspora uvarum,* known to winemakers under the synonym *Kloeckera apiculata* and found on berry skins. These yeasts produce very small, nutritionally irrelevant amounts of the muscle protein actin. Also, some pathogens, such as *Cryptococcus neoformans,* which almost always occurs in patients with massive immune deficiency. This shows how exclusive this protein is obtained only from animal sources.

The peptide Alanine-Lysine-Serine-Leucine-Serine-Aspartic acid-Arginine-Phenylalanine-Serine-Tyrosine originates exclusively from pea and chickpea storage protein [20]. This sequence (the peptide in Fig. 7.3a) is contained in the Legumin A of the pea and the similarly structured Legumin of the chickpea. Both proteins serve the legumes as energy and amino acid storage for germination. This water-soluble peptide is overall simply positively charged, as the neighboring amino acids Arginine (+) and Aspartic acid (−) neutralize each other, but this does not mean that both charges cannot interact strongly in very specific receptor pockets.

Furthermore, the bioactive peptide with the sequence Methionine-Isoleucine-Threonine-Leucine-Alanine-Isoleucine-Proline-Valine-Asparagine-Lysine-Proline-Glycine-Arginine (the peptide in Fig. 7.3b) is found exclusively in soybean protein, not even in the proteins of other legumes such as peas or lupins. It is characterized by a stronger hydrophobicity due to the block of five water-insoluble amino acids in the middle of the peptide. This peptide has two positive charges on one side due to Lysine and Arginine, is therefore also active at hydrophilic-hydrophobic interfaces, and can, in addition, be activated only by completely differently structured receptors compared to the other two examples.

These peptides strongly reflect the original functions of the proteins. It is clear that muscle proteins like actin have completely different physiological tasks than storage proteins like Legumin in peas and chickpeas. Bioactive peptides of a certain length are thus clearly assignable to specific foods and food classes.

These three exemplary examples alone show two important facts: It is not enough to just consider the amino acids and conclude that all essential amino acids are, for example, present in surrogate products. In the long run, this would be a fatal misconception, because the sequences of longer bioactive peptides are of at least equal importance. Ultimately, excluding certain foods is counter-evolutionary. Avoiding animal products thus has more far-reaching consequences than initially assumed. Any self-imposed and long-term restriction of certain foods is questionable from the perspective of the molecular aspects of nutrition.

Recently, it has become increasingly clear through modern methods of protein and peptide analysis (proteomics) that longer peptides also have high functionality [21]. Peptides, that is, which have more than just three, four, or five amino acids, but consist of ten to 20 amino acids. These are formed, for example, during the ripening of meat from very specific proteins of the muscle [21], from fish products [22, 23], in fermented dairy products and cheese [15], and generally in protein-rich, animal products [24–29].

This is precisely the crux of the matter, as these long peptides, whose sequences originate from animal proteins, cannot be replaced by plant sequences like individual amino acids—solely due to the different function of the protein. Short bioactive peptides can be, but the more amino acids in a specific sequence the bioactive peptides have, the more unique they become. This may also be another molecularly justified reason why a supplement-free vegetarian diet, in which eggs and dairy products are consumed, is more advantageous than a purely vegan one—despite supplementation.

7.4 Should We Eat Everything?

The conclusion is relatively simple: eat everything and think more rationally about food, drink, and enjoyment, rather than believing. In this book, it has been shown how the answers to controversial and often irrationally debated nutritional questions can, in many cases, be traced back to clear molecular relationships. That, and only that, is the basis for all further discussions and conclusions. In many places, it is reassuring that physics and chemistry play the main role, far ahead of ecotrophology and physiology, before all ideological statements. Our food consists of a wide variety of molecules that can only be recognized and processed through further functional molecules, such as enzymes or bioreactors like microorganisms. That's all there is to it. What happens to our food after swallowing is well-defined and very precisely determined. The second insight from this is more than banal: We *Homo sapiens* are essentially nothing more than highly complex biomachines.

Everything that happens within us is based on fundamental interactions between cell-relevant molecules, ingenious self-organization through competing interactions, and the interplay of entropy and energy, as is common and understandable in the physics of soft matter, even if it sometimes becomes very complex. The second aspect is biochemistry, which is also based on fundamental chemical and quantum chemical principles. All of this is reassuring because we need to worry much less about food than it currently appears.

There are only a handful of elementary "rules":

1. Prepare food yourself, enjoy it.
2. Obtain food directly from producers.
3. Never forget the fundament of human nutrition: "raw, cooked, and fermented."
4. Pay attention to variety and do not restrict food choices for the wrong reasons in the long term.
5. Eat rarely or not at all highly processed foods that are excessively advertised.

There are many good reasons for the last point, which can, of course, be explained on a molecular level.

In plain language, this means: The millions of years old menu of *Homo sapiens* was not the worst, quite the contrary. Intuitively, people knew what was good for them and still is today. A little of everything, not too much of anything! This age-old wisdom still applies today. Despite many new insights on the one hand and pseudoscientific inaccuracies on the other, this remains the only true nutritional formula, no matter what grows, thrives, or is developed by food technologists on this earth in the future.

The ideology-free view and the molecular perspective on food and the human interior confirm this in an impressive, albeit banal way. This is exactly how Fig. 4.2 should be read: There are neither good nor bad foods, especially not those that come from nature. Let us eat a varied diet of everything, without looking through ideologically biased lenses, as long as we feel healthy. Then we will largely remain on the safe side, as the last 4.7 million years have shown in various ways.

References

1. Johnston, B. C., Zeraatkar, D., Han, M. A., Vernooij, R. W., Valli, C., El Dib, R., Marshall, C., Stover, P. J., Fairwheather-Taitt, S., Wójcik, G., Bhatia, F., de Souza, R., Brotons, C., Meerpohl, J. J., Patel, C. J., Djulbegovic, B., Alonso-Coello, P., Bala, M. M., & Guyatt, G. H. (2019). Unprocessed red meat and processed meat consumption: Dietary guideline recommendations from the nutritional recommendations (NutriRECS) consortium. *Annals of Internal Medicine*. Unprocessed red meat and processed meat consumption: dietary guideline recommendations from the Nutritional Recommendations (NutriRECS) Consortium. *Annals of internal medicine, 171*(10), 756-764. https://doi.org/10.7326/m19-1621.
2. Djulbegovic, B., & Guyatt, G. H. (2017). Progress in evidence-based medicine: A quarter century on. *The Lancet, 390*(10092), 415–423.
3. Benjamin, D. J., et al. (2018). Redefine statistical significance. *Nature Human Behaviour, 2*(1), 6. https://doi.org/10.1038/s41562-017-0189-z.

4. Ioannidis, J. P. (2013). Implausible results in human nutrition research. *British Medical Journal BMJ, 347*, f6698. https://doi.org/10.1136/bmj.f6698.
5. Zhao, Y., Singh, G., & Naumova, E. (2019). Joint association of multiple dietary components on cardiovascular disease risk: A machine learning approach (OR06-02-19). *Current Developments in Nutrition, 3*(Supplement_1), nzz039-OR06. https://doi.org/10.1093/cdn/nzz039.OR06-02-19.
6. Yu, Z., Chen, Y., Zhao, W., Li, J., Liu, J., & Chen, F. (2018). Identification and molecular docking study of novel angiotensin-converting enzyme inhibitory peptides from *Salmo salar* using in silico methods. *Journal of the Science of Food and Agriculture, 98*(10), 3907–3914.
7. Mooney, S. J., & Pejaver, V. (2018). Big data in public health: Terminology, machine learning, and privacy. *Annual Review of Public Health, 39*, 95–112.
8. Varzakas, T., Kandylis, P., Dimitrellou, D., Salamoura, C., Zakynthinos, G., & Proestos, C. (2018). Innovative and fortified food: Probiotics, prebiotics, gmos, and superfood. In M. E. Ali & N. N. Ahmad Nizar (Eds.), *Preparation and processing of religious and cultural foods* (pp. 67–129). Woodhead.
9. Aluko, R. (2012). *Bioactive peptides. Functional Foods and Nutraceuticals* (pp. 37–61). Springer.
10. Odagima, C. H., Kimura, Y., Adachi, S., Matsuno, R., & Yokogoshi, H. (1995). Effects of a peptide mixture with a high Fischer's ratio on serum and cerebral cortex amino acid levels and on cerebral cortex monoamine levels, compared with an amino acid mixture with a high Fischer's ratio. *Bioscience, Biotechnology, and Biochemistry, 59*(4), 731–734.
11. Prahl, H. W., & Setzwein, M. (2013). *Soziologie der Ernährung*. Springer.
12. Ploeger, A., Hirschfelder, G., & Schönberger, G. (Eds.). (2011). *Die Zukunft auf dem Tisch: Analysen, Trends und Perspektiven der Ernährung von morgen*. Springer.
13. Lemke, H. (2014). *Politik des Essens: Wovon die Welt von morgen lebt*. transcript.
14. Sharma, S., Singh, R., & Rana, S. (2011). Bioactive peptides: A review. *International Journal Bioautomation, 15*(4), 223–250.
15. Ryan, J. T., Ross, R. P., Bolton, D., Fitzgerald, G. F., & Stanton, C. (2011). Bioactive peptides from muscle sources: Meat and fish. *Nutrients, 3*(9), 765–791.
16. Rutherfurd-Markwick, K. J. (2012). Food proteins as a source of bioactive peptides with diverse functions. *British Journal of Nutrition, 108*(S2), S149–S157.
17. Singh, B. P., Vij, S., & Hati, S. (2014). Functional significance of bioactive peptides derived from soybean. *Peptides, 54*, 171–179.
18. Marcone, S., Belton, O., & Fitzgerald, D. J. (2017). Milk-derived bioactive peptides and their health promoting effects: A potential role in atherosclerosis. *British Journal of Clinical Pharmacology, 83*(1), 152–162.
19. Bhat, Z. F., Kumar, S., & Bhat, H. F. (2015). Bioactive peptides of animal origin: A review. *Journal of Food Science and Technology, 52*(9), 5377–5392.
20. Liao, W., Fan, H., Liu, P., & Wu, J. (2019). Identification of angiotensin converting enzyme 2 (ACE2) up-regulating peptides from pea protein hydrolysate. *Journal of Functional Foods, 60*, 103395.
21. Segura-Campos, M., Chel-Guerrero, L., Betancur-Ancona, D., & Hernandez-Escalante, V. M. (2011). Bioavailability of bioactive peptides. *Food Reviews International, 27*(3), 213–226.
22. Fu, Y., Young, J. F., & Therkildsen, M. (2017). Bioactive peptides in beef: Endogenous generation through postmortem aging. *Meat Science, 123*, 134–142.
23. Atef, M., & Ojagh, S. M. (2017). Health benefits and food applications of bioactive compounds from fish byproducts: A review. *Journal of functional foods, 35*, 673–681.
24. López-Expósito, I., Miralles, B., Amigo, L., & Hernández-Ledesma, B. (2017). Health effects of cheese components with a focus on bioactive peptides. In J. Frias, C. Martinez-Villaluenga, & E. Penas (Eds.), *Fermented foods in health and disease prevention* (pp. 239–273). Academic.

25. Rubin, R. (2020). Backlash over meat dietary recommendations raises questions about corporate ties to nutrition scientists. *JAMA, 323*(5), 401–404. https://doi.org/10.1001/jama.2019.21441.
26. Maretzke, F., Lorkowski, S., & Egert, S. (2020a). Egg intake and cardiometabolic diseases: An update. Part 1. *Ernährungs Umschau, 67*(1), 11–17.
27. Maretzke, F., Lorkowski, S., & Egert, S. (2020b). Egg intake and cardiometabolic diseases: An update. Part 2. *Ernährungs Umschau, 67*(2), 26–31.
28. Dehghan, M., et al. (2020). Association of egg intake with blood lipids, cardiovascular disease, and mortality in 177,000 people in 50 countries. *The American Journal of Clinical Nutrition, 111*, 795–803.
29. Xiao, N., Zhao, Y., Yao, Y., Wu, N., Xu, M., Du, H., & Tu, Y. (2020). Biological activities of egg yolk lipids: A review. *Journal of Agricultural and Food Chemistry, 68*(7), 1948–1957.